T0325592

Principles of Stem Cell Biology and Cancer

Principles of Stem Cell Biology and Cancer: Future Applications and Therapeutics

Edited by

Tarik Regad

The John van Geest Cancer Research Centre, Nottingham Trent University

Thomas J. Sayers

Leidos Biomedical Research Inc., and the Cancer and Inflammation Program, Frederick National Laboratory for Cancer Research

Robert C. Rees

The John van Geest Cancer Research Centre, Nottingham Trent University

WILEY Blackwell

Registered office: John Wiley & Sons, Ltd, The Atrium, Southern Gate, Chichester, West Sussex, PO19 8SQ, UK

Editorial offices: 9600 Garsington Road, Oxford, OX4 2DQ, UK
The Atrium, Southern Gate, Chichester, West Sussex, PO19 8SQ, UK
111 River Street, Hoboken, NJ 07030-5774, USA

For details of our global editorial offices, for customer services and for information about how to apply for permission to reuse the copyright material in this book please see our website at www.wiley.com/wiley-blackwell.

Library of Congress Cataloging-in-Publication Data

Principles of stem cell biology and cancer : future applications and therapeutics / edited by Tarik Regad, Thomas Sayers, Robert Rees.
p. ; cm.
Includes bibliographical references and index.
ISBN 978-1-118-67062-0 (cloth)
I. Regad, Tarik, editor. II. Sayers, Thomas, editor. III. Rees, Robert C., editor.
[DNLM: 1. Neoplastic Stem Cells. 2. Cell- and Tissue-Based Therapy–methods. 3. Neoplasms–therapy. 4. Stem Cells. QZ 202]
RC269.7
616.99402774–dc23

2014049379

A catalogue record for this book is available from the British Library.

Wiley also publishes its books in a variety of electronic formats. Some content that appears in print may not be available in electronic books.

Cover image: Immunofluorescence images of the PC3 prostate cancer cell line stained with E-Cadherin and N-Cadherin. The images originated from research at The John van Geest Cancer Research Centre, Nottingham Trent University, UK.

Typeset in 10.5/12.5pt TimesTenLTStd by Laserwords Private Limited, Chennai, India
Printed and bound in Malaysia by Vivar Printing Sdn Bhd

1 2015

Contents

List of Contributors

Marcio Alvarez-Silva
Laboratory of Stem Cell and Bioengineering, Department of Cell Biology, Embryology and Genetics, Federal University of Santa Catarina, Florianópolis, Brazil

Hideo Baba
Department of Gastroenterological Surgery, Graduate School of Life Sciences, Kumamoto University, Kumamoto, Japan

Jennifer M. Bailey
Department of Surgery, Johns Hopkins University School of Medicine, Baltimore, MD, USA; and The McKusick-Nathans Institute of Genetic Medicine, Johns Hopkins University School of Medicine, Baltimore, MD, USA; and Department of Internal Medicine, Division of Gastroenterology, Hepatology and Nutrition, The University of Texas Health Science Center at Houston, Houston, TX, USA

David Bakhshinyan
McMaster Stem Cell and Cancer Research Institute, McMaster University, Hamilton, ON, Canada; and Department of Biochemistry and Biomedical Sciences, Faculty of Health Sciences, McMaster University, Hamilton, ON, Canada

Graham Ball
The John van Geest Cancer Research Centre, Nottingham Trent University, Nottingham, UK

Maria J. Barrero
CNIO-LILLY Epigenetics Laboratory, Spanish National Cancer Research Center (CNIO), Madrid, Spain

Toru Beppu
Department of Multidisciplinary Treatment for Gastroenterological Cancer, Kumamoto University Hospital, Kumamoto, Japan

Magdalena E. Buczek
The John van Geest Cancer Research Centre, Nottingham Trent University, Nottingham, UK

Akira Chikamoto
Department of Gastroenterological Surgery, Graduate School of Life Sciences, Kumamoto University, Kumamoto, Japan

Shawn G. Clouthier
Comprehensive Cancer Center, Department of Internal Medicine, University of Michigan, Ann Arbor, MI, USA

Jignesh Dalal
Cardiac Transplant Research Laboratory, Section of Bone Marrow Transplantation, The Children's Mercy Hospital, Kansas City, MO, USA

Shoukat Dedhar
Department of Integrative Oncology, British Columbia Cancer Research Centre, Vancouver, BC, Canada; and Department of Biochemistry and Molecular Biology, University of British Columbia, Vancouver, BC, Canada

Loic P. Deleyrolle
McKnight Brain Institute, Department of Neurosurgery, University of Florida, Gainesville, FL, USA

Jos Domen
Cardiac Transplant Research Laboratory, Section of Cardiac Surgery, The Children's Mercy Hospital, Kansas City, MO, USA

Jerome C. Edwards
The John van Geest Cancer Research Centre, Nottingham Trent University, Nottingham, UK

Wafik S. El-Deiry
Laboratory of Translational Oncology and Experimental Cancer Therapeutics, Department of Medical Oncology and Molecular Therapeutics Program, Fox Chase Cancer Center, Philadelphia, PA, USA

Niklas Finnberg
Laboratory of Translational Oncology and Experimental Cancer Therapeutics, Department of Medical Oncology and Molecular Therapeutics Program, Fox Chase Cancer Center, Philadelphia, PA, USA

Neha Garg
McMaster Stem Cell and Cancer Research Institute, McMaster University, Hamilton, ON, Canada; and Department of Surgery, Faculty of Health Sciences, McMaster University, Hamilton, ON, Canada

James Hackland
Centre for Stem Cell Biology, Department of Biomedical Sciences, University of Sheffield, Sheffield, UK

Daisuke Hashimoto
Department of Gastroenterological Surgery, Graduate School of Life Sciences, Kumamoto University, Kumamoto, Japan

Hiromitsu Hayashi
Department of Gastroenterological Surgery, Graduate School of Life Sciences, Kumamoto University, Kumamoto, Japan

Audrey M. Hendley
Department of Surgery, Johns Hopkins University School of Medicine, Baltimore, MD, USA; and The McKusick-Nathans Institute of Genetic Medicine, Johns Hopkins University School of Medicine, Baltimore, MD, USA; and Department of Internal Medicine, Division of Gastroenterology, Hepatology and Nutrition, The University of Texas Health Science Center at Houston, Houston, TX, USA

Katsunori Imai
Department of Gastroenterological Surgery, Graduate School of Life Sciences, Kumamoto University, Kumamoto, Japan

Takatoshi Ishiko
Department of Gastroenterological Surgery, Graduate School of Life Sciences, Kumamoto University, Kumamoto, Japan

Takatsugu Ishimoto
Department of Gastroenterological Surgery, Graduate School of Life Sciences, Kumamoto University, Kumamoto, Japan

Katherine S. Koch
Hepatocyte Growth Control and Stem Cell Laboratory, Department of Pharmacology, School of Medicine, University of California at San Diego, La Jolla, CA, USA

Hasan Korkaya
Comprehensive Cancer Center, Department of Internal Medicine, University of Michigan, Ann Arbor, MI, USA; and Department of Biochemistry and Molecular Biology, GRU Cancer Center, Georgia Regents University, Augusta, GA, USA

Hideyuki Kuroki
Department of Gastroenterological Surgery, Graduate School of Life Sciences, Kumamoto University, Kumamoto, Japan

Caterina A.M. La Porta
Department of Biosciences, University of Milan, Milan, Italy

Hyam L. Leffert
Hepatocyte Growth Control and Stem Cell Laboratory, Department of Pharmacology, School of Medicine, University of California at San Diego, La Jolla, CA, USA

Bigang Liu
Department of Molecular Carcinogenesis, The University of Texas MD Anderson Cancer Center, Science Park, Smithville, TX, USA

Fayaz Malik
Comprehensive Cancer Center, Department of Internal Medicine, University of Michigan, Ann Arbor, MI, USA

Kiran Mall
The John van Geest Cancer Research Centre, Nottingham Trent University, Nottingham, UK

Paul C. McDonald
Department of Integrative Oncology, British Columbia Cancer Research Centre, Vancouver, BC, Canada

Nicole McFarlane
McMaster Stem Cell and Cancer Research Institute, McMaster University, Hamilton, ON, Canada; and Department of Surgery, Faculty of Health Sciences, McMaster University, Hamilton, ON, Canada

Kosuke Mima
Department of Gastroenterological Surgery, Graduate School of Life Sciences, Kumamoto University, Kumamoto, Japan

Harry Moore
Centre for Stem Cell Biology, Department of Biomedical Sciences, University of Sheffield, Sheffield, UK

Shigeki Nakagawa
Department of Gastroenterological Surgery, Graduate School of Life Sciences, Kumamoto University, Kumamoto, Japan

Hidetoshi Nitta
Department of Gastroenterological Surgery, Graduate School of Life Sciences, Kumamoto University, Kumamoto, Japan

Hirohisa Okabe
Department of Gastroenterological Surgery, Graduate School of Life Sciences, Kumamoto University, Kumamoto, Japan

Varun V. Prabhu
Laboratory of Translational Oncology and Experimental Cancer Therapeutics, Department of Medical Oncology and Molecular Therapeutics Program, Fox Chase Cancer Center, Philadelphia, PA, USA

David Preskey
Centre for Stem Cell Biology, Department of Biomedical Sciences, University of Sheffield, Sheffield, UK

Maleeha A. Qazi
McMaster Stem Cell and Cancer Research Institute, McMaster University, Hamilton, ON, Canada; and Department of Biochemistry and Biomedical Sciences, Faculty of Health Sciences, McMaster University, Hamilton, ON, Canada

Robert C. Rees
The John van Geest Cancer Research Centre, Nottingham Trent University, Nottingham, UK

Tarik Regad
The John van Geest Cancer Research Centre, Nottingham Trent University, Nottingham, UK

Brent A. Reynolds
McKnight Brain Institute, Department of Neurosurgery, University of Florida, Gainesville, FL, USA

Thomas J. Sayers
Leidos Biomedical Research Inc., and the Cancer and Inflammation Program, Frederick National Laboratory for Cancer Research, Frederick, MD, USA

Sheila K. Singh
McMaster Stem Cell and Cancer Research Institute, McMaster University, Hamilton, ON, Canada; and Department of Biochemistry and Biomedical Sciences, Faculty of Health Sciences, McMaster University, Hamilton, ON, Canada; and Department of Surgery, Faculty of Health Sciences, McMaster University, Hamilton, ON, Canada

Dean G. Tang
Department of Molecular Carcinogenesis, The University of Texas MD Anderson Cancer Center, Science Park, Smithville, TX, USA

Christian Unger
Centre for Stem Cell Biology, Department of Biomedical Sciences, University of Sheffield, Sheffield, UK

Chitra Venugopal
McMaster Stem Cell and Cancer Research Institute, McMaster University, Hamilton, ON, Canada; and Department of Surgery, Faculty of Health Sciences, McMaster University, Hamilton, ON, Canada

Max S. Wicha
Comprehensive Cancer Center, Department of Internal Medicine, University of Michigan, Ann Arbor, MI, USA

Stefano Zapperi
CNR-IENI, Milan, Italy

Preface

Stem cells are a population of cells capable of differentiating into diverse specialized cell types, or of undergoing self-renewal to produce more stem cells. There are two types of stem cells: embryonic stem cells, isolated from the blastocyst, and adult stem cells, found in different tissues of the body. These cells are essential in generating different cell lineages and thus maintaining the structural and functional integrity of tissues and organs. The decision whether to self-renew or to differentiate is tightly regulated and requires a strict control of cell division and cell-cycle exit. This level of control involves key molecules implicated in cell-cycle regulation, as well as several critical growth factors and cytokines. The balance between self-renewal and differentiation can be the target of oncogenic events, leading to cell transformation and the emergence of 'cancer stem cells', which are thought to be subpopulations of cancer cells responsible for tumour progression, development of metastases, tumour dormancy, cancer relapse and resistance to chemotherapy.

In recent years, the stem cell field has become a subject of extensive research, with many groups focusing on isolating and identifying cancer stem cell populations. This effort relies on identifying molecules expressed preferentially by cancer stem cells, with the aim of developing cancer therapies targeting these specific molecules in this cancer population without affecting the pool of normal healthy stem cells. Although some progress has been made, developing efficient therapies targeting cancer stem cells remains one of the important challenges facing the growing stem cell research community.

This book will provide a detailed introduction to stem cell biology. Part I focuses on the characterization of stem cells, the progress made towards their identification and their future therapeutic applications. Part II focuses on cancer stem cells and their role in cancer development, progression and chemoresistance, and presents an overview of recent progress in therapies targeting cancer stem cells. We believe that this book will be unique in providing compiled information about the link between stem cell biology and cancer.

The contributing authors are renowned experts in the field and will provide a timely book of high quality, outlining the current progress in and exciting future possibilities for stem cell research.

Tarik Regad
Thomas J. Sayers
Robert C. Rees

Part I

Stem Cells

1

Isolation and Characterization of Human Embryonic Stem Cells and Future Applications in Tissue Engineering Therapies

Christian Unger, James Hackland, David Preskey and Harry Moore
Centre for Stem Cell Biology, Department of Biomedical Sciences, University of Sheffield, Sheffield, UK

1.1 Derivation of human embryonic stem cells from the ICM

1.1.1 Early development of the ICM: the cells of origin for hESCs

The mammalian zygote (fertilized ovum) is defined as being totipotent, as it is capable of developing into a new offspring and the placenta required for full gestation. The zygote initially undergoes cleavage-stage cell division, forming cells (early blastomeres) that remain totipotent. With further development to the preimplantation blastocyst stage, a primary cell differentiation results in outside trophectoderm cells (TE) and an inside aggregate of inner cell mass (ICM) cells. The TE forms placental tissue and membranes, while the

Principles of Stem Cell Biology and Cancer: Future Applications and Therapeutics, First Edition.
Edited by Tarik Regad, Thomas J. Sayers and Robert C. Rees.

ICM forms the foetus and extra-embryonic membranes. Therefore, ICM cells are defined as being pluripotent, forming all cells of the developing offspring other than the complete placenta (unless genetically manipulated). Embryonic stem cells (ESCs) are derived *in vitro* from ICM cells, which adapt to specific conducive conditions that enable indefinite cell proliferation (self-renewal) without further differentiation and thereby confer a pluripotent capacity. This *in vitro* pluripotent state is due principally to the induction and maintenance of expression of key 'gate-keeper' genes, including Oct4, Nanog and Sox2, which then regulate one another (Silva & Smith, 2008). The capacity for self-renewal is sustained by high telomerase activity, which protects chromosome telomeres from degradation during mitosis (Blasco, 2007).

Mammalian ESCs were first derived in the mouse (mESC) (Evans and Kaufman, 1981; Martin, 1981). When mESCs are integrated into an embryo and returned to a recipient, they can contribute to all cell lineages, including germ cells. Their utility soon became invaluable for many transgenic procedures. Successful derivation of human (hESC) lines was reported by Thomson *et al.* (1998), who essentially followed the same procedure as used for the mouse. ICMs isolated from preimplantation human blastocysts were plated on to mitotically inactivated mouse embryonic feeders in culture medium with basic fibroblast growth factor (bFGF) and foetal calf serum (FCS). This culture medium was also supplemented with leukaemia inhibitory factor (LIF), a cytokine necessary to maintain mESCs (Smith *et al.*, 1988), although (as is now known) not necessary for standard hESC derivation. Human ESCs display (or lose on differentiation) plasma membrane expression of stage-specific embryonic antigens (SSEAs) that correlate with the preimplantation morphological development of human embryos (Henderson *et al.*, 2002) and form teratomas (benign tumours) in immune-deficient mice that can contain cell phenotypes from the three major cell lineages (endoderm, mesoderm and ectoderm), as well as trophoblast. The differentiation of trophoblast cells indicates that hESCs are not entirely equivalent to mESCs, as usually defined, but align with slightly later LIF-independent mouse epiblast pluripotent stem cells, which have the propensity to differentiate to trophoblast *in vitro* (Brons *et al.*, 2007).

1.1.2 Derivation of hESCs

Success in the derivation of hESCs depends in part on the quality of the human embryos used (usually blastocysts from days 5 to 8), although cell lines have been generated from morphologically poor embryos. Numerous hESC lines have been derived (Figure 1.1) from normal, aneuploid and mutant embryos from patients undergoing treatment for assisted conception (IVF, ICSI) or preimplantation genetic diagnosis (PGD) who consent to donate them for stem cell research. Some of these cell lines have been extensively characterized and compared, enabling international standards to be established (Adewumi *et al.*, 2007).

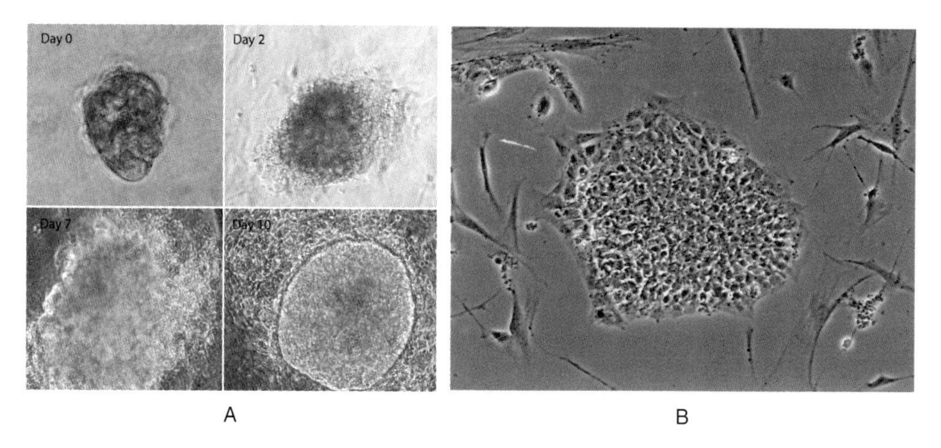

Figure 1.1 (A) Outgrowth of hESCs over 10 days of culture from ICM. In this instance, a clearly defined colony was observed by 10 days, which was mechanically passaged. (B) hESC line Shef1 plated on ECM.

1.1.2.1 Evolution to a more efficient and better-defined derivation method: drivers and technologies Over the last 15 years, continuous improvements have been made in the process of deriving and maintaining hESC lines. The emphasis initially was on improving efficiency and consistency in the stem cell laboratory. But as hESC lines have become readily available for research in many countries, the focus has changed to devising methods for deriving clinical-grade cell lines that comply with health care regulatory authorities (e.g. Federal Drug Administration, FDA; European Medicines Agency, EMA), which can be used as starting materials for potential cell-therapy trials. Xeno-free methods (free of nonhuman animal components) are preferable as they minimize the risk of cross-species contamination with adventitious agents. An important early improvement was the replacement of FCS with a serum extract (knockout serum replacement, KOSR) to reduce hESC differentiation. This modification also minimized batch variation (inherent in FCS) between culture media, and allowed consistency in the proliferation of the cells after passaging (transfer of cells to a new culture vessel). Subsequently, more defined culture media (xeno-free) have been devised, which, in combination with a variety of extracellular matrix (ECM) compositions, facilitate the proliferation and passage of pluripotent hESCs in the absence of feeder cells (mouse or human), which otherwise remain an ill-defined and inconsistent component of the cell culture. Manipulation of the embryo has also changed over time. Initially, the ICM was isolated according to mouse protocols using enzymatic (protease) removal of the zona pellucida (ECM surrounding blastocyst) and immunosurgical lysis of TE with antitrophoblast antibody to prevent TE culture outgrowth from inhibiting early ESC proliferation. However, xeno-free methods using laser-assisted removal of the zona and plating of the intact blastocyst or the ICM on to a defined matrix (e.g. laminin 521) with a defined culture medium is the method of choice, leading to successful feeder/xeno-free cell line production in ~20–40% of attempts with good-quality human embryos (Hasegawa *et al.*, 2010). With further improvements to the cell adhesion matrix and cell medium, the efficiency of hESC line derivation is likely to increase further, although the quality of the embryo used to develop ICM cells remains a crucial factor.

Another important consideration is the genetic character and stability of the hESC line. Generally, most hESC outgrowths and initial cell lines derived from unselected embryos (i.e. not PGD selected) are determined to be karyotypically normal within the precision of the chromosomal analysis. However, hESCs acquire genetic mutations in culture, which may endow them with a selective cell culture advantage, so that mutated cells predominate (Baker *et al.*, 2007). Since derivation and ESC passage represent key stress events for ESC cultures, minimization of selective pressure on cells at these stages may help to maintain their normal karyotype. For example, the proliferation of cells by mechanical division of hESC colonies into smaller aggregates may be preferable to enzymatic disaggregation to single cells, which will initiate apoptotic stress pathways unless inhibited from doing so by a chemical inhibitor (i.e. ROCK inhibitor).

1.1.3 Regulation of embryo research and hESC derivation

The destruction of the preimplantation human embryo in order to derive hESC lines has prompted fierce ethical debate in many countries, especially on religious grounds, which to some extent remains unresolved and irresolvable. The result is the implementation of policies of ethical oversight, regulation and permission for hESC research, which vary from country to country, and even within a country (the United States). In the United Kingdom, early introduction of laws related to human embryo research and the formation of a regulatory body (Human Fertilisation of Embryology Authority, HFEA) provided a framework (and important public confidence) for continuation of hESC research. Clinical-grade hESCs must meet compliance with conditions set by the EMA and overseen in the United Kingdom by the Human Tissue Authority. In the United States, the FDA and National Institutes of Health (NIH) undertake this responsibility. Since the development of cell therapies using pluripotent stem cells is novel, it remains to be determined exactly how regulatory authorities will implement conditions of compliance.

The induction of pluripotency in mouse and human somatic cells in 2006–07 using retroviral vectors to introduce four genes to reprogramme the genome (*Oct4*, *Sox2*, *Klf4*, and *c-Myc*) and enable the derivation of induced pluripotent stem cells (iPSCs) (Takahashi *et al.*, 2007) radically changed the landscape of human pluripotent stem cell (hPSC) research (Yamanaka, 2012). This technology not only provides a potential route for the creation of patient-specific stem cell lines for use in cell therapies but also makes pluripotent cell lines available to many more laboratories, with seemingly fewer ethical bottlenecks. However, hESCs remain the current gold standard as their cellular reprogramming events are those that are normally evoked in the early embryo, rather than artificially induced, and they are therefore less likely to be subject to aberrant epigenetic effects on their gene function. Moreover, ethical issues related to obtaining informed consent from donors to use tissue samples to derive iPSCs still persist. Progress in the use of hESCs (or iPSCs) for therapy will depend on whether robust protocols for their expansion and differentiation to a precise and economic manufacturing level can be devised, and a key aspect in meeting this objective is the implementation of reliable and accurate assays of cell type and quality.

1.2 Basic characterization of hESCs

Immediately following their derivation, hESCs are identified fundamentally on the basis of their indefinite capacity for self-renewal, their ability to form derivatives of all three embryonic germ layers and, usually, their ability to maintain a euploid karyotype over extended periods in culture. However, not every derivation procedure results in an established hESC line, and a variety of other cell types may grow out from isolated embryo cultures. Furthermore, hESCs may be derived at different stages of embryo development (i.e. early or late blastocyst) while still retaining pluripotency, which can alter the subsequent features of their cell population. While cell lines may be superficially similar in these aspects, they often show significant differences in stem cell surface antigen expression, DNA methylation status, X-chromosome inactivation, variation in specific gene expression, cell doubling time, and capacity to differentiate. The cause of this variation between cell lines is largely unknown, but it is likely, in part at least, to be due to the wide genetic background of human donors (mESCs, by contrast, are produced from inbred mouse strains); it also depends on environmental conditions and stresses, which can impart phenotypic changes on cells during derivation and culture. It is therefore essential that hESCs are characterized under a set of criteria which allows for accurate, valid and robust comparisons to be made both within and between laboratories. In this section, we look more closely at the characteristics that currently define hESCs.

1.2.1 hESC morphology

Human ESCs typically form compact flat colonies with defined colony borders (Figure 1.2). This morphology is like that of mouse epiblast stem cells, with which hESCs share most similarity, and in contrast to that of mESCs, which form characteristic discrete domed colonies. The hESC possesses a nucleus with distinctive nucleoli and little cytoplasm when viewed by phase-contrast microscopy. These characteristics, together with colony formation, provide effective initial identification. Although hESCs dissociate readily with a

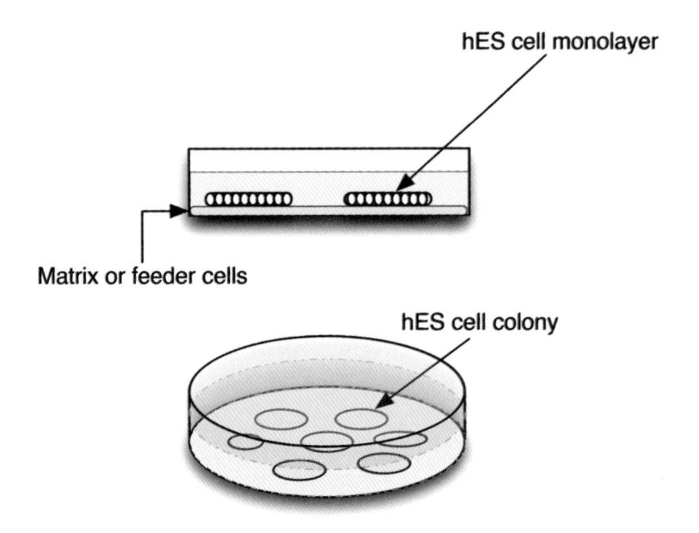

Figure 1.2 Human ESCs grow as flat colonies on a matrix- or feeder cell-coated dish.

variety of enzymes and protocols (i.e. low salt conditions) to disrupt cell–cell adhesion, their survival is poor, with single cell colony-forming capacity often less than 1%. For this reason, most standard passaging involves clumps or sheets of hESCs to limit apoptosis. In contrast, human embryonic germ cells (hEGCs), which are also pluripotent (Shamblott *et al.*, 1998), form spherical colonies, which unlike hESCs are refractory to standard cell dissociation methods.

1.2.2 Stem cell markers

Besides the typical cell/colony morphology, which is a routine check during cell culture, hESCs are characterized mainly by their expression of a variety of specific cell-surface and intracellular protein markers using antibodies (usually monoclonal), often in combination with flow cytometry or high-content image analysis. These cell-surface markers were first identified in the preimplantation mouse embryo or in embryonal carcinoma cells (ECCs; pluripotent cancer cell lines). The phenotypic morphology of a hESC may alter as spontaneous differentiation occurs during cell culture, with cells gradually losing expression of markers associated with pluripotency and upregulating those associated with differentiation; therefore, a panel of markers can rapidly identify subpopulations of cells. If quantitative analysis is used, the stability of a hESC culture can be monitored accurately over time. Surface markers indicative of an undifferentiated hESC state include SSEA-3, SSEA-4 and the high-molecular-weight glycoproteins TRA-1-60 and TRA-1-81 (Thomson *et al.*, 1998). HESCs also express the intracellular markers OCT4, Nanog and REX1 and stain positive for alkaline phosphatase activity (Figure 1.3).

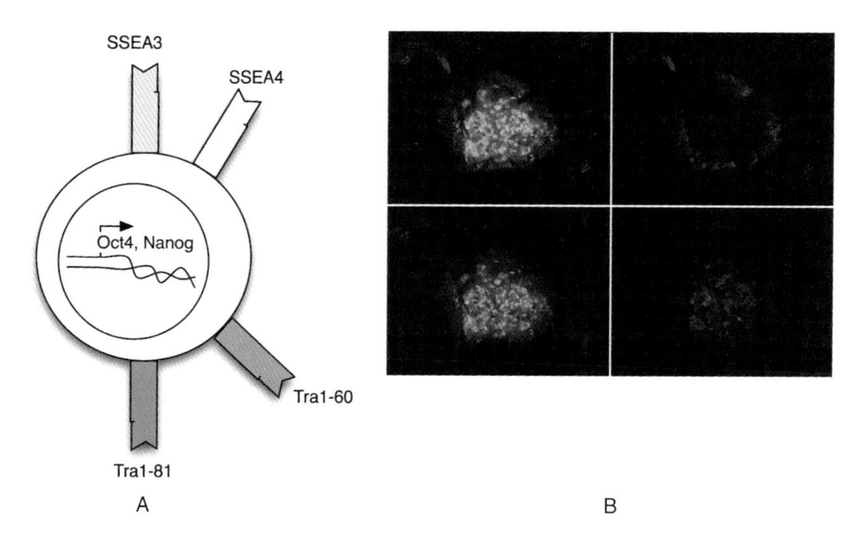

Figure 1.3 (A) Main intracellular and extracellular markers used to identify hESCs. (B) A colony of Shef1 hESCs plated on ECM (Matrigel). Immunofluorescent localization of cell-surface markers Tra-1-60 (green), SSEA3 (blue) and SSEA4 (red). Although all three markers identify pluripotent cells, the expression patterns in the colony differ.

Significantly, mESCs differ in their surface-antigen profile, failing to express SSEA-3 or SSEA-4 but expressing SSEA-1, a cell-surface marker characteristic of differentiated hESCs. The markers display differences in sensitivity to shifts in the differentiation status of the cell, which can be exploited to some extent to forecast developmental changes. For example, SSEA-3 expression is the first to downregulate upon early differentiation while markers such as SSEA-4 and TRA1-60 lag behind (Henderson *et al.*, 2002).

1.2.3 Function characterization: differentiation potential

ESCs are unique in their ability to self-renew and differentiate into all three embryonic germ layers, in principal forming any fully terminally differentiated cell within the body. In the mouse, ESC pluripotency is defined by the ability to generate chimeric offspring and contribute to the germ line. However, for ethical and practical reasons, in humans and some nonhuman primate species, the ability of ESCs to form chimeras is not a testable property, and alternative protocols on which to base functional pluripotency must be used. In the absence of the natural stem cell niche of the embryo, hESCs are in a dynamic balance between cell fates and are highly susceptible to environmental cues, which can induce spontaneous cell differentiation or, in the correct combination, can be employed to drive a more 'directed' cell differentiation. Therefore, pluripotency is measured either *in vitro* by differentiation of cells as aggregates in suspension culture (called embryoid bodies, EBs) or *in vivo* by their formation in the mouse as benign tumours called teratomas.

1.2.3.1 *In vitro*: EBs Human ESCs can be induced to differentiate *in vitro* by the process of EB formation (Figure 1.4). The process involves growing hESCs in suspension to form cell aggregates on a nonadhesive substrate to prevent their dissociation. As the EBs mature, hESCs alter their morphological appearance and acquire molecular markers characteristic of differentiated derivatives. Markers specific to each embryonic lineage can include neurofilament 68Kd (ectoderm), β-globin (mesoderm) and α-fetoprotein (endoderm) (Itskovitz-Eldor *et al*., 2000). However, more markers per germ layer are usually analysed, to illustrate a more global picture of differentiation ability. Initial testing of differentiation capacity is commonly done by spontaneous EB differentiation in medium supplemented with serum. Methods have become more refined, however, using defined number of cells and defined media formulations (Ng *et al*., 2005). An EB formation assay should always be part of the basic hESC characterization, and should clearly show either upregulation of markers from the three germ layers in the EBs or outgrowth from them.

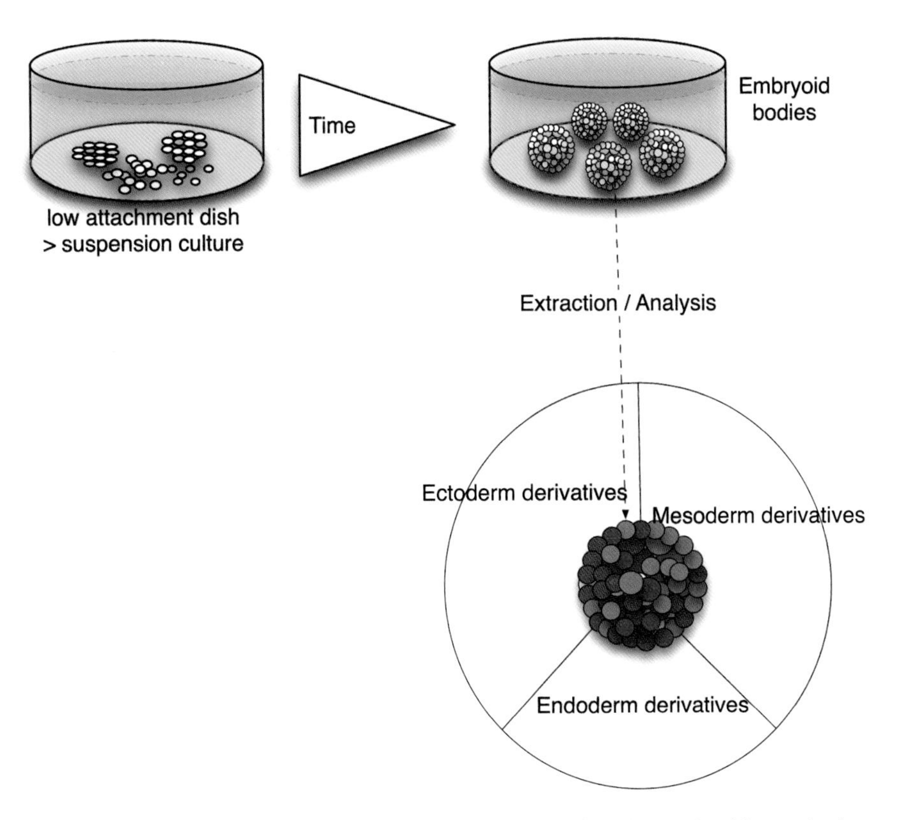

Figure 1.4 Simple overview of EB formation from hESCs. EBs from hESCs should contain tissues derived from all three embryonic germ layers.

1.2.3.2 *In vivo*: teratoma formation The formation of a teratoma is a formal demonstration of pluripotency of hESCs *in vivo*. Teratomas are benign tumours that contain different types of developmental tissue derived from all three germ layers. They are formed after injection of undifferentiated hESCs into the hind leg, testis or kidney capsule of immunocompromised mice (i.e. nonobese diabetic severe combined-immunodeficient, NOD/SCID). They are then usually analysed by histological evaluation of the tumour mass for the presence of representatives of all three germ layers (Figure 1.5)

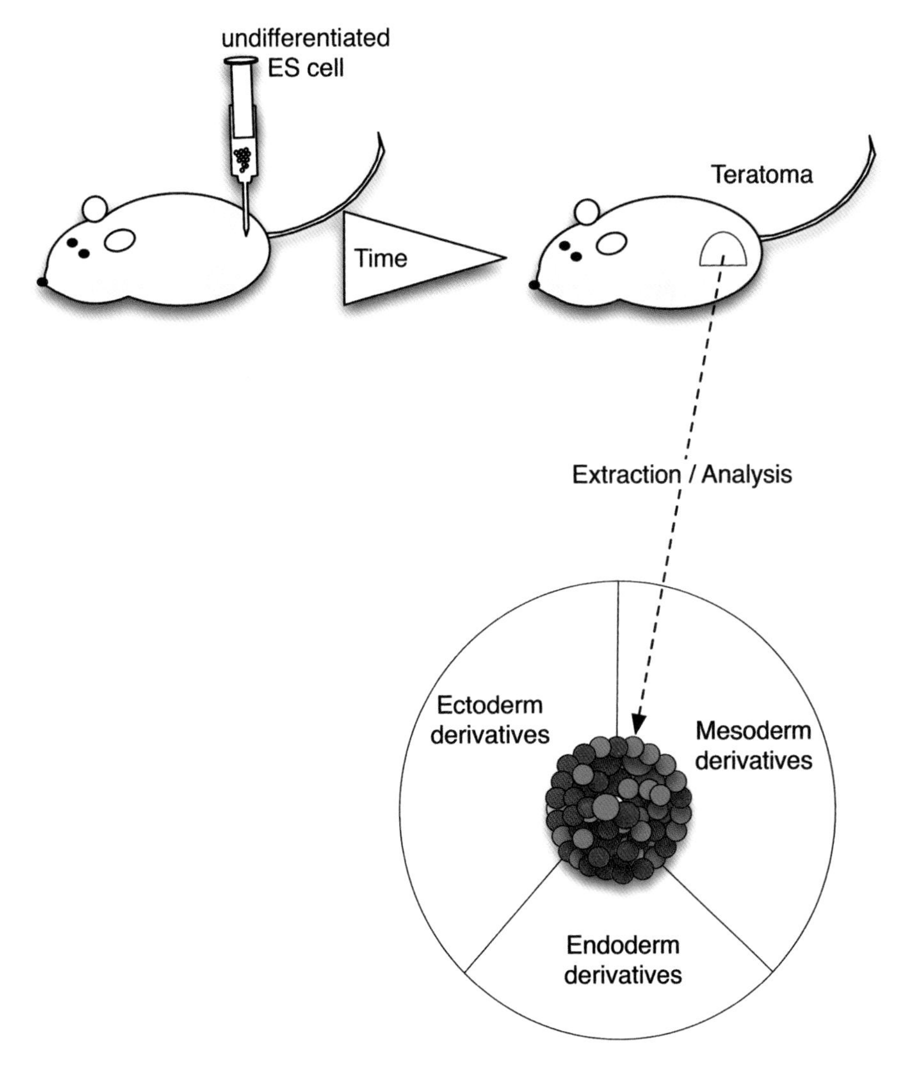

Figure 1.5 Simple overview of teratoma formation in immunocompromised mice. Teratomas from hESCs contain tissues derived from all three embryonic germ layers.

(Thomson *et al.*, 1998). On occasion, after injection, hESCs fail to form teratomas; therefore, injection of more than one mouse is often necessary to account for any variability. This may be due to the abnormal environment in which hESCs are placed, residual immune reactivity, the quality of the hESCs or the scientific methodology. Efficiencies can be improved by adding ECMs such as inactivated mouse embryonic fibroblasts (MEFs) or Matrigel with the hESCs and by using more severely immunocompromised mice (Gropp *et al.*, 2012).

While teratoma formation is an expected part of the basic characterization panel for new hESC lines, a search for less expensive and shorter surrogate assays is ongoing. In particular, more streamlined and time-efficient methods are required for the mass generation of iPSCs (Muller *et al.*, 2011).

1.3 Stem cell quality and culture adaptation with reference to cancer

In the embryo, during normal development, the cells of the ICM usually exist for just a few days before differentiating into more mature cell types to form the three germ layers. During the derivation of human embryonic stem cell lines, cells from the ICM of a blastocyst are transferred to a culture dish and need to adapt to this *in vitro* environment. Prolonged culture of these cells exposes them to various stress factors, which can then lead to further selection of the most adapted cells. Initially, this process of adaptation occurs mostly through epigenetic mechanisms, as *in vitro*-cultured mESCs can convert back to form a normal mouse embryo *in vivo*. However, with extended laboratory culture for months or years (as is possible with pluripotent stem cells), selection of cells that have increased survival may occur, further helping their culture adaptation. This can lead to not only epigenetic but also genomic changes in the cell population.

Any genetic or epigenetic changes that occur in hESCs over extended culture may alter their developmental potential, function or behaviour and should therefore be avoided, if possible. In particular, nonreversible genomic changes need to be tracked and controlled to minimize effects on experimental studies or treatments.

1.3.1 Genomic abnormalities

Genomic abnormalities that have been observed in pluripotent stem cell cultures range from large chromosomal changes to single-nucleotide mutations.

1.3.1.1 Chromosomal aberrations The study of large chromosomal aberrations has been possible since chromosomal banding methods were established in the late 1960s. ‘Karyotyping’, in which metaphase chromosomes are stained with either quinacrine mustard (q-banding) (Caspersson *et al.*, 1970) or Giemsa (g-banding) (Sumner *et al.*, 1971) to give a characteristic banding pattern to each chromosome, is now a routine method. Depending on the chromosomal region, a resolution of 5–10 megabases can be achieved. The detection of aneuploidy in patient cells can be an indicator or marker for disease; for example, trisomy 21 is found in Down syndrome.

Initial studies revealed that hESC lines could maintain a normal diploid set of chromosomes during extended periods in culture (>6 months) (Thomson *et al.*, 1998). However, follow-up studies soon revealed that hESC lines could also acquire chromosomal changes (Draper *et al.*, 2003) and thereby emphasized the need for genome monitoring.

Recurrent large aberrations in hESCs after extended culture are mostly gains of regions in chromosomes 1, 12, 17 and X. Interestingly, the most frequent gain of human chromosome 17 (Figure 1.6) is also syntenic to the distal part of mouse chromosome 11, which is most often gained in mESCs (Ben-David and Benvenisty, 2012). Such changes are nonrandom gains that seem to be selected for by *in vitro* culture systems, and have been seen to occur at a rate of 10–20%. However, the general frequency of changes, including subchromosomal changes, is at a rate of 30–35%; this includes aberrations that are selected against during culture and those that are introduced at derivation or come from the embryo (Amps *et al.*, 2011). The observed frequency of chromosomal abnormalities clearly reiterates the need to monitor cells over time, with karyotyping being the most commonly used method.

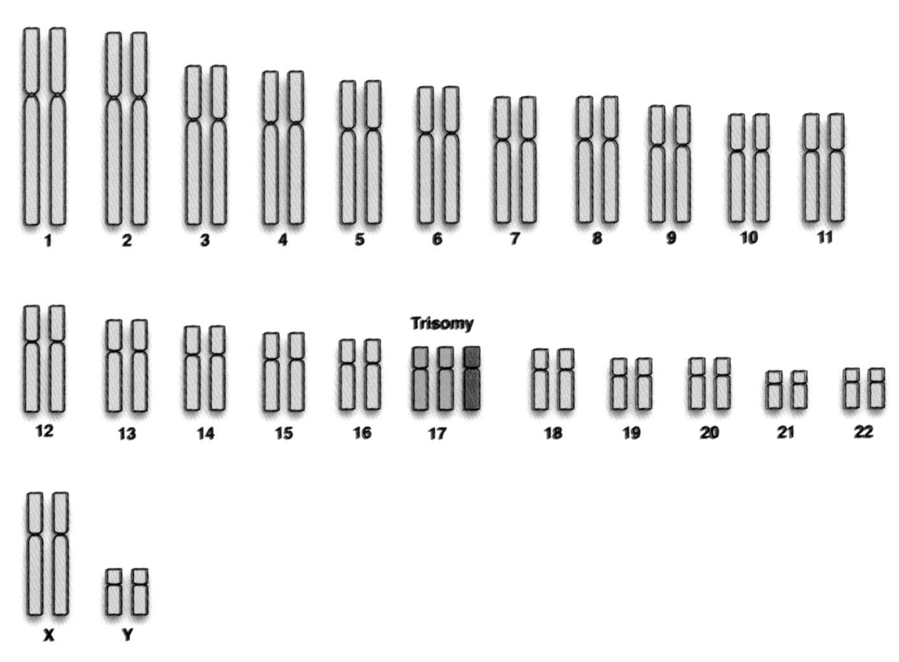

Figure 1.6 Illustration of karyotype with an extra chromosome 17. Trisomy 17 is one of the most common chromosome changes acquired during hESC culture.

1.3.1.2 Copy-number variations While karyotyping initially identified large chromosomal changes, recent application of higher-resolution technologies has both confirmed such large deviations and revealed additional changes on a subchromosomal level. Several studies using single-nucleotide polymorphism (SNP) data have established that all hESC lines exhibit copy-number variations (CNVs) of various sizes, many of which are specific to hESCs (Figure 1.7). At a higher resolution, changes that naturally exist in the human population must be differentiated from changes that have been acquired during *in vitro* culture. Analysis conducted on early and late passage cell populations revealed several regions with gain or loss of heterozygosity (Narva *et al.*, 2010; Hanahan and Weinberg, 2011; Avery *et al.*, 2013). In particular, a minimal amplicon in chromosome 20q11.21 was found in more than 20% of cell lines (Werbowetski-Ogilvie *et al.*, 2009; Amps *et al.*, 2011). Furthermore, it was revealed that the gain of this minimal amplicon introduces a resistance to apoptosis, most likely caused by one specific gene, BCL2L1. A simple genomic quantitative polymerase chain reaction (qPCR)-based approach or fluorescence *in situ* hybridization (FISH) on karyotyping slides should be a good measure to verify that this region has not changed in a particular set of hESC cultures (Avery *et al.*, 2013).

1.3.1.3 Single-nucleotide variations With the advent of whole-genome sequencing, a few studies on iPSCs have been able to increase their resolution

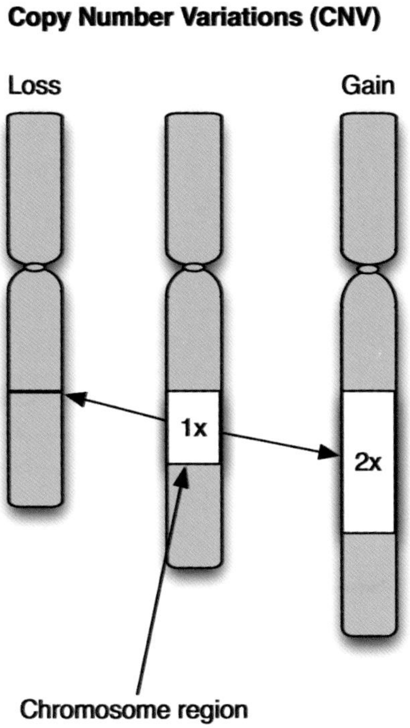

Figure 1.7 Illustration of a possible CNV in hESs.

to the single base pair level and have thus identified single-nucleotide variations (SNVs) (Gore *et al.*, 2011). An average of five to six mutations in coding regions have been reported, but many of these likely derive from the parental cell lines. It is important to bear in mind that only a few complete human genomes have been sequenced to date, so the extent of normal variation amongst our population is unclear. However, over time, and with further advanced sequencing technology, bigger data sets will reveal more answers with regards to genome stability. More scientific studies using whole-genome sequencing on hESC lines will be very interesting and may reveal significant SNVs that cannot be detected with other methods and which impact the quality of hESC lines.

1.3.2 Epigenetics

Epigenetics is the study of changes acting upon but not altering the DNA sequence, namely such mechanisms as imprinting, DNA methylation and histone modification to regulate gene expression. The epigenetic characterization of hESC lines, and other cell types, is much less established than genomic analysis, because of its higher level of complexity. While we have sequenced the whole human genome, we do not yet have the same understanding of our epigenome. There are hundreds of epigenomes for every genome, because every person has hundreds of cell types, each of which has different DNA modifications. Another reason is that epigenetic changes are dynamic: able to adapt to changes in the cellular environment over time. High-throughput methods for the analysis of a full set of methylations are now available, but the technology is still new and expensive, and it still needs to develop reliable references for hESCs.

Epigenetic mechanisms give specific cell types their identity by allowing only a subset of genes to be active. Faulty regulation in early embryonic development can result in embryo mortality or distort differentiation and should be evaluated and deselected from hESC cultures, if necessary.

Human ESCs are derived from an early blastocyst stage – a developmental time point at which cells are fragile because processes like X-chromosome inactivation (XCI) are still ongoing (van den Berg *et al.*, 2009).

1.3.2.1 Imprinting and XCI Expression of a number of genes is necessarily mono-allelic, which means that one of the parental alleles needs to be silenced (in a process termed 'imprinting') for proper development to occur. For example, mono-allelic expression is important for X-linked genes and hence requires XCI in female embryos. With regards to imprinting and XCI, the gene dosage is important, and failure to properly silence one allele can result in lethality or developmental disorders. Several studies have indicated that prolonged hESC culture can affect the XCI pattern of pluripotent cell lines (Lengner *et al.*, 2010; Tchieu *et al.*, 2010; Nazor *et al.*, 2012). Cell culture under low oxygen allows for the derivation of female ESC lines with two active X chromosomes, while normal oxygen will produce mixed cultures, indicating that the cell culture environment has a profound effect on XCI (Lengner *et al.*, 2010). Consequently, if a certain X-linked expression is required for disease modelling, the activation or inactivation should be evaluated during ESC characterization. A PCR for X-inactive specific transcript (XIST) expression, which is expressed from the inactivated X chromosome, will give an initial idea of whether XCI has occurred and is maintained in a particular culture system.

1.3.2.2 Methylation pattern DNA methylation silences promoter regions and prevents gene expression where it is not required. New technologies have now started to evaluate genome-wide DNA methylation patterns and are building a reference map for ESCs. While many promoter regions are equally methylated and demethylated between ESC lines, other genes appear to be variably methylated (Bock *et al.*, 2011). Processes that give rise to variation include underlying human variability, cell culture methods, the time point, the method of derivation and other stress factors. What seems clear is that these changes in methylation patterns are impacting the differentiation capacity of ESCs and could be used to predict their ability to differentiate along certain lineages (Bock *et al.*, 2011). Hence, methylation analysis on promoter regions for genes that are important for lineage-specific differentiation can give important insight into the selection of a cell line for a specific purpose and may be included in the characterization of a line. Established methods such as methylation-specific polymerase chain reaction (MSP), pyrosequencing or array-based methylation analysis, together with a reference map, can give clues as to whether a particular cell line is able to differentiate towards all lineages equally.

1.3.2.3 Histone modifications Histones are proteins that package the DNA in eukaryotic cells and play a role in gene regulation, by rendering DNA active or inactive. They can be highly modified through various modifying enzymes and thereby affect gene regulation. For example, promoters occupied by a histone H3 lysine 4 trimethylation (H3K4me3) or histone H3 lysine 27 trimethylation (H3K27me3) are associated with gene activation and repression, respectively. Histone modifications can be affected by cell culture adaptation, and may lead to higher proliferation and differential expression of tumour suppressor genes with parallels to cancer cells. While analysis of histone modification is not commonly carried out for hESC characterization, they impact many genes that are linked to severe developmental disorders and cancers (Lund *et al.*, 2013).

1.3.3 hESC culture adaptation with reference to cancer (genomic and epigenetic)

Cell culture can induce genomic and epigenetic changes in hESCs and should be controlled for. In fact, most cultured cells will have or acquire changes over time, and it is important to find out whether these changes are in an acceptable range for normal functionality. The hESC field is still relatively new and much remains to be understood before the right conclusions can be drawn from particular changes, making it necessary to screen for such changes in order to increase our knowledge. Genome instability and resistance of cell death through abnormalities are a hallmark of cancer (Hanahan and Weinberg, 2011), so introducing such abnormalities might be a big risk for future clinical applications. Some abnormalities enriched in hESC cultures are also found in tumours and might therefore carry a higher risk of inducing cancer-like changes.

Primordial germ cells and ESCs are closely related cell types as they originate from a similar developmental stage. Their similarity is partly mirrored in the abnormalities they acquire, with germ cell tumours (GCTs) most often amplifying chromosome 12p and gaining material from chromosome 17, much like culture-adapted hESCs (Summersgill *et al.*, 2001); it has therefore been proposed that hESC culture adaptation may be used as a model for GCT malignancy (Harrison *et al.*, 2007). During the malignant evolution of ECCs, differentiation capacity is lost in favour of proliferation proficiency, eventually leading to nullipotent ECCs with a high self-renewal capability. Culture-adapted cells may therefore lose some or all of their differentiation capacity and cause embryonal carcinoma-like tumours if undifferentiated cells are contaminating the differentiated cells used in clinical protocols.

1.3.3.1 Impact of hESC culture-induced genomic and epigenetic changes in differentiated cells Considering that hESCs can differentiate into any cell type found in the human body, there is a real risk that genomic or epigenetic abnormalities in these early stem cells cause more mature cell types to acquire cancer phenotypes. For example, trisomy 12, the most common abnormality in hESCs, is also associated with chronic lymphoid leukaemia (Juliusson *et al.*, 1990), while gain of chromosome 17, particularly the long arm of 17, is strongly associated with neuroblastoma (Plantaz *et al.*, 1997) and CNVs on chromosome locus 20q11 are associated with a variety of cancers (Beroukhim *et al.*, 2010).

Epigenetically, there are many links between hESC abnormalities and neoplasia. The methylation of tumour-suppressor genes or the activation of oncogenes through epigenetic mechanisms might be a prime reason for the transformation of benign cells.

1.4 Future applications in tissue-engineering therapies

Tissue engineering is a concept that evolved from organ transplantation and has existed since the mid 1980s – over a decade before the isolation of the first hESC lines. It aims to maintain or restore function to tissues whose failure is threatening illness. This can be done in three different ways: the support of preexisting tissues to prevent loss of function; the encouragement of damaged tissues to regain lost function; and the replacement of lost or damaged tissue. The main approaches have included treatment with bio-active molecules such as inhibitors and growth factors, the use of structural biomaterials as scaffolds and the introduction of new cells or tissues, as well as various combinations of these methods.

Many approaches to cell or tissue transplantation have met with significant levels of success. One of the best established of these procedures is the autologous transplantation of hematopoietic stem cells to restore blood cell production after chemotherapy-induced bone marrow ablation. Other cell therapies include the implantation of foetal dopaminergic neurons into patients suffering from Parkinson's disease (Ali *et al.*, 2013), the grafting of a retinal pigmented epithelium (RPE)-choroid patch to treat age-related macular degeneration (Buchholz *et al.*, 2013) and transplantation of Islet cells or the whole pancreas to treat diabetes (Pavlakis and Khwaja, 2007). Each of these examples provides good proof of concept for cell-replacement therapy, but restricted levels of source tissue prevent many such therapies from being commonly applied in a clinical setting.

Human pluripotent stem cells, such as hESCs and hiPSCs, have the potential to solve this issue. Theoretically, they can differentiate into virtually any cell type in the human body, allowing a single source of cells to be applied to multiple different clinical uses. Furthermore, unlike many primary cell types, hPSCs can be easily maintained in culture and it is possible to scale up their cell numbers exponentially. This means that they have the potential to solve the problem of tissue supply faced by cell-replacement tissue engineering and even to provide previously unobtainable cell types for regenerative purposes.

So far, very few clinical trials that aim to utilize hPSC-derived tissues for replacement therapies have been announced. The differentiation potential and long-term *in vitro* culture of hPSCs introduces a level of complexity to the engineering process that makes an understanding of early human development, and subsequent control of cell phenotype, an essential key to success. Issues facing the implementation of hPSC-derived replacement therapies are: production and maintenance of high-quality and safe source hPSCs; development of efficient protocols for generating the cell type of interest; acquisition of pure differentiated cells of interest, without contamination of undifferentiated or other unwanted cell types; and circumvention of immune-compatibility issues, to prevent immune-rejection (Figure 1.8).

1.4.1 Efficient differentiation and purification of the cell type of interest

Protocols need to be devised that can recapitulate the embryology of a specific cell type *in vitro*. It is unlikely to be possible to produce just one phenotype, so methods have to be in place to recognize the cell of interest and enrich the population for that particular cell type. This all needs to be done with enough

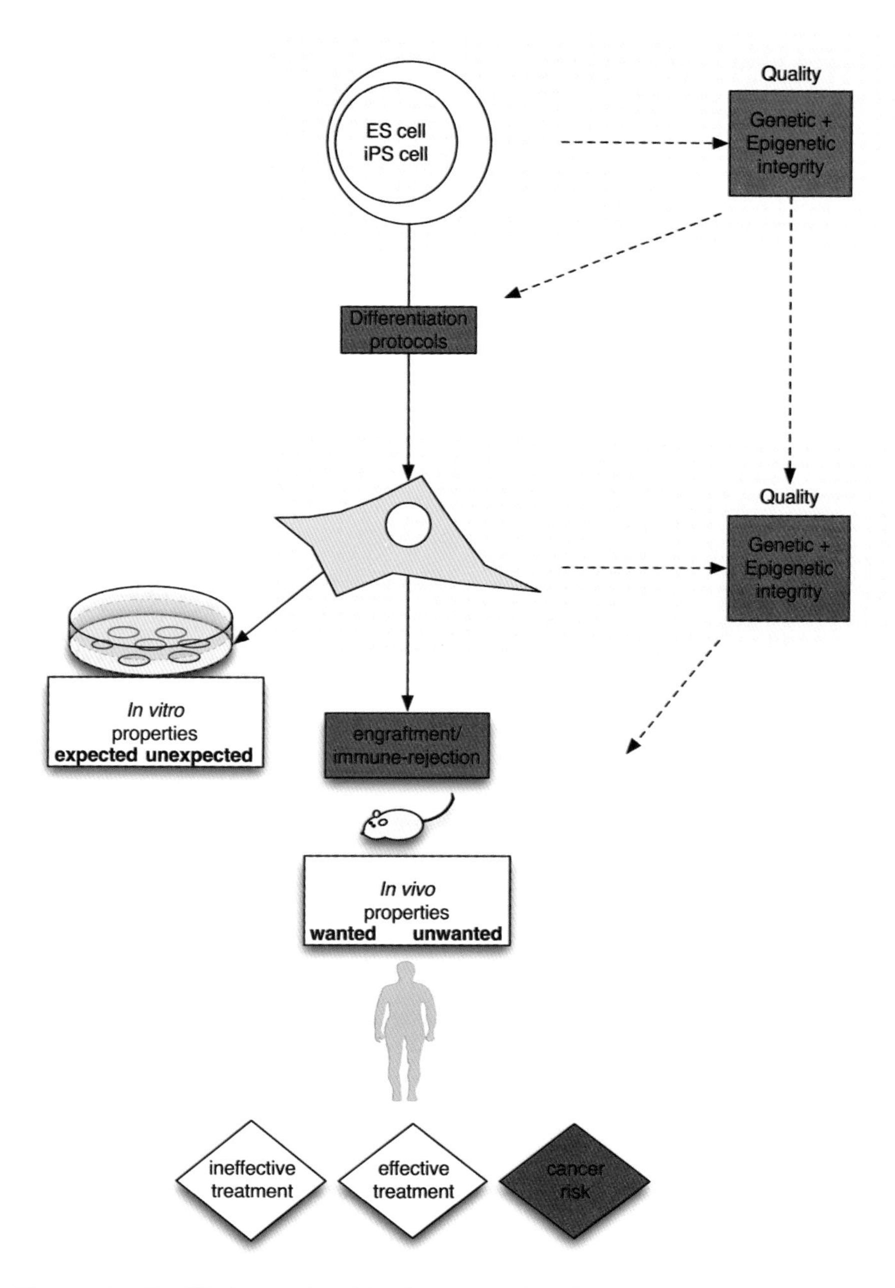

Figure 1.8 Simplified research and development pathway for future therapies from ES/iPSCs to differentiated, potentially therapeutically applicable cells, highlighting factors that can affect cancer risk (in red), as described in this chapter.

cells to allow the required function to be elicited from their transplantation into the patient.

In the developing embryo, differentiation of pluripotent cells is spatially, temporally and chemically regulated. When cellular transplantation from a multipotent or terminally differentiated source is used for clinical purposes, the cells already have the required function and can be identified easily through tissue location and marker expression. When hPSC derivatives are used, they first have to be differentiated *in vitro*. For example, in order to replace foetal dopaminergic neuron treatment of Parkinson's disease with hPSC derivatives, this process has been mimicked with a defined protocol that drives differentiation in the correct direction. Use of small-molecule inhibitors (SMIs) to block or stimulate specific signalling pathways has allowed the development of well-defined culture conditions that do not require co-culture with secondary cell types (Ali *et al.*, 2013).

When intrastriatal transplantation of foetal neurons is carried out, the ventral mesencephalon of the foetus is dissected and dissociated before it is implanted into the courdate and putamen of the patient. When this is done, knowledge of dissection location combined with approximately 15% of cells expressing tyrosine hydroxylase is enough to be confident of cell type (Widner *et al.*, 1992). The range of different derivatives that can be produced from hPSCs raises the possibility that contaminating phenotypes, even with comparable marker expression, may be present in any differentiated culture. For this reason, the dopaminergic differentiation protocol for hPSCs can be used in concert with serum-free culture methods that remove any extrinsic signalling from the system, reducing contaminating cell types and enriching the final cell population for the required phenotype. These cells are able to respond to the axon guidance cues that allow innervation and connection to host neurons (Cord *et al.*, 2010) and have been shown to alleviate symptoms of Parkinson's disease in Rat models (Ali *et al.*, 2013). All of this work has brought us to a stage where there is a real possibility of clinical trials using hPS derived neurons to treat Parkinson's disease.

Some cell types require more than clever combinations of signalling molecules and SMIs for successful differentiation from hPSCs. Production of glucose-sensitive, insulin-producing β cells for use in the treatment of diabetes has been a much sought-after goal ever since the isolation of the first hESCs. Human PSCs can be induced to form definitive endoderm by Activin A/Nodal signalling, or more efficiently by using the SMIs IDE1 and IDE2. Following this, there are a number of different methods for differentiation and enrichment of the culture for pancreatic cells, assessed on the basis of the pancreatic marker PDX1 (Aguayo-Mazzucato and Bonner-Weir, 2010). Insulin-expressing cells have been produced in this manner, paving the way for a cell-replacement therapy for diabetes, but none of these methods appears to be capable of producing cells that secrete insulin in a glucose-sensitive manner. This is a characteristic that they share with foetal β cells, perhaps identifying them as a nascent counterpart to the fully functioning cell required for clinical use. The failure to achieve this last step of β-cell maturation has prevented a cell-replacement therapy for diabetes from reaching clinical trials and is driving research into methods for better recapitulating pancreatic development *in vitro* by spatially organizing the cells into 3D structures, with the hope that polarization of the cells will allow them to mature (Greggio *et al.*, 2013).

Despite the establishment of specific and selective methods for producing a desired cell type, such protocols will rarely be able to exclude all other cell types that might develop from hPSCs. Because of this, most protocols, especially those for clinical applications, will include steps for purification (i.e. antibody-based magnetic-activated cell sorting). Importantly, undifferentiated hPSCs that can form teratomas and cells that might in any way negatively affect the treatment must be removed from a final cell-therapy product.

1.4.2 Genetic stability and tumourigenic potential

Antiapoptotic and proliferative selection pressures subjected to cells in culture can lead to the accumulation of genetic and epigenetic changes, which promote a phenotype analogous to that of a tumour cell. This process of culture adaptation is fully discussed in Section 1.3, but it should be noted that such changes present a significant barrier to the successful application of hPSCs in tissue engineering. If an adapted cell of this type were to be implanted into a person, there would be a significant danger of tumour formation. For this reason, any cell population that is used for such a purpose needs to undergo a thorough genetic analysis of the kind described in Section 1.3.

1.4.3 Immune compatibility

One of the greatest challenges facing any kind of cell-replacement therapy is the response of the host's own immune system to a graft. Immune reactions to grafts and graft-versus-host disease (GVHD) can be so severe as to not only render the transplant useless but also kill the patient. In some cases, immunosuppressant drugs originally developed for organ transplantation can circumvent this, but these can produce their own complications. The intricate play between host and graft has been widely studied since the advent of hematopoietic stem cell transplantations. While engraftment of many cell types can strongly benefit from a limited host immune response, it must be considered that residual host response in allogeneic transplantation might also have applications in therapies.

The advent of cellular reprogramming holds the potential to make allogeneic transplants a thing of the past, circumventing the issue of immune-rejection altogether. By taking a small skin biopsy or blood sample, it is now possible to create patient-specific iPSCs that can be used in place of hESCs in all of the aforementioned examples of tissue engineering. Derivatives of such reprogrammed cells will be a perfectly immune-compatible match to a host system, and considered as the host's 'own' cells. It is of note that this will deprive the host of any means of immune detection, should such cells carry abnormalities that could cause cancer.

1.5 Conclusions

One of the most promising applications to date for stem cell-based tissue engineering is the treatment of the dry form of age-related macular degeneration (AMD) with hPSC-derived RPE. This is the first approach using hPSCs to move to clinical trials, partly because of the ease with which differentiation into RPE cells was first achieved. Human PSCs will spontaneously develop into RPE cells at low efficiencies and can be easily identified via pigmentation. Recent advances in differentiation protocols have increased efficiencies to ~60%, allowing purification of enough cells for transplantation into patients, through either the injection of a single cell suspension or insertion of a 'patch' of cells grown in a monolayer on a scaffold (Buchholz *et al.*, 2013)*. These first clinical trials may successfully cure AMD and push hPSCs into a clinical era. However, they might also uncover unwanted side effects, such as those described with the first attempts to cure severe combined immunodeficiency (SCID-X1) by using gene therapy to genetically modify cells (Hacein-Bey-Abina *et al.*, 2003). An early clinical trial conducted by the Geron Corporation attempted to use hESC derivatives to treat spinal cord injury, but it ended prematurely, partly because of the unanticipated cost of the project. This is an example of the unforeseen hurdles that can still slow advancement to treatments.

The use of hPSCs for tissue engineering could circumvent issues of tissue supply, regress symptoms of neurodegenerative diseases rather than just alleviate them, and provide replacement cells that were previously unobtainable. The approaches described in this chapter represent only those that are closest to clinical use. In the future, it may be possible to use gene therapy to cure genetic defects in the iPSCs of patients before re-implanting them as an autologous cell-replacement therapy, and even to use additive manufacturing to create whole organs from hPSC derivatives (Orlando *et al.*, 2013). Currently, the costs of these therapies are prohibitive, but, although the need for extensive preclinical analysis of cells will always be present, optimization of these procedures alongside differentiation, transplantation and reprogramming will eventually bring costs down.

As our understanding of human embryology expands and we learn to better recapitulate it *in vitro*, we can expect to see the number of applications for hPSCs in tissue engineering rise exponentially.

* Since the time of writing trials have shown improvements in the eyesight of patients taking part in trials for this treatment of AMD (Schwartz *et al.*, 2014).

References

Adewumi, O., Aflatoonian, B., Ahrlund-Richter, L., Amit, M., Andrews, P. W., Beighton, G., *et al.*, 2007. Characterization of human embryonic stem cell lines by the international stem cell initiative. Nat. Biotechnol. 25(7), 803–816.

Aguayo-Mazzucato, C., Bonner-Weir, S., 2010. Stem cell therapy for type 1 diabetes mellitus. Nat. Rev. Endocrinol. 6(3), 139–148.

Ali, F., Stott, S.R.W., Barker, R.A., 2013. Stem cells and the treatment of Parkinson's disease. Exp. Neurol. 260, 3–11.

Amps, K., Andrews, P.W., Anyfantis, G., Armstrong, L., Avery, S., Baharvand, H., *et al.*, 2011. Screening ethnically diverse human embryonic stem cells identifies a chromosome 20 minimal amplicon conferring growth advantage. Nat. Biotechnol. 29(12), 1132–1144.

Avery, S., Hirst, A.J., Baker, D., Lim, C.Y., Alagaratnam, S., Skotheim, R.I., *et al.*, 2013. BCL-XL mediates the strong selective advantage of a 20q11.21 amplification commonly found in human embryonic stem cell cultures. Stem Cell Rep. 1(5), 379–386.

Baker, D.E.C., Harrison, N.J., Maltby, E., Smith, K., Moore, H.D., Shaw, P.J., *et al.*, 2007. Adaptation to culture of human embryonic stem cells and oncogenesis *in vivo*. Nat. Biotechnol. 25(2), 207–215.

Ben-David, U., Benvenisty, N., 2012. High prevalence of evolutionarily conserved and species-specific genomic aberrations in mouse pluripotent stem cells. Stem Cells (Dayton, Ohio) 30(4), 612–622.

Beroukhim, R., Mermel, C.H., Porter, D., Wei, G., Raychaudhuri, S., Donovan, J., *et al.*, 2010. The landscape of somatic copy-number alteration across human cancers. Nature 463(7283), 899–905.

Blasco, M.A., 2007. Telomere length, stem cells and aging. Nat. Chem. Biol. 3(10), 640–649.

Bock, C., Kiskinis, E., Verstappen, G., Gu, H., Boulting, G., Smith, Z.D., *et al.*, 2011. Reference maps of human ES and iPS cell variation enable high-throughput characterization of pluripotent cell lines. Cell 144(3), 439–452.

Brons, I.G.M., Smithers, L.E., Trotter, M.W.B., Rugg-Gunn, P., Sun, B., Chuva de Sousa Lopes, S.M., *et al.*, 2007. Derivation of pluripotent epiblast stem cells from mammalian embryos. Nature 448(7150), 191–195.

Buchholz, D.E., Pennington, B.O., Croze, R.H., Hinman, C.R., Coffey, P.J., Clegg, D.O., 2013. Rapid and efficient directed differentiation of human pluripotent stem cells into retinal pigmented epithelium. Stem Cells Trans. Med. 2(5), 384–393.

Caspersson, T., Zech, L., Johansson, C., 1970. Analysis of human metaphase chromosome set by aid of DNA-binding fluorescent agents. Exp. Cell Res. 62(2–3), 490–492.

Cord, B.J., Li, J., Works, M., McConnell, S.K., Palmer, T., Hynes, M.A., 2010. Characterization of axon guidance cue sensitivity of human embryonic stem cell-derived dopaminergic neurons. Mol. Cell. Neurosci. 45(4), 324–334.

Draper, J.S., Smith, K., Gokhale, P., Moore, H.D., Maltby, E., Johnson, J., *et al.*, 2003. Recurrent gain of chromosomes 17q and 12 in cultured human embryonic stem cells. Nat. Biotechnol., 22(1), 53–54.

Evans, M.J., Kaufman, M.H., 1981. Establishment in culture of pluripotential cells from mouse embryos. Nature 292(5819), 154–156.

Gore, A., Li, Z., Fung, H.-L., Young, J.E., Agarwal, S., Antosiewicz-Bourget, J., *et al.*, 2011. Somatic coding mutations in human induced pluripotent stem cells. Nature 470(7336), 63–67.

Greggio, C., De Franceschi, F., Figueiredo-Larsen, M., Gobaa, S., Ranga, A., Semb, H., *et al.*, 2013. Artificial three-dimensional niches deconstruct pancreas development *in vitro*. Development 140(21), 4452–4462.

Gropp, M., Shilo, V., Vainer, G., Gov, M., Gil, Y., Khaner, H., *et al.*, 2012. Standardization of the teratoma assay for analysis of pluripotency of human ES cells and biosafety of their differentiated progeny. PLoS One 7(9), e45532.

Hacein-Bey-Abina, S., von Kalle, C., Schmidt, M., Le Deist, F., Wulffraat, N., McIntyre, E., *et al.*, 2003. A serious adverse event after successful gene therapy for X-linked severe combined immunodeficiency. N. Engl. J. Med. 348(3), 255–256.

Hanahan, D., Weinberg, R.A., 2011. Hallmarks of cancer: the next generation. Cell 144(5), 646–674.

Harrison, N.J., Baker, D., Andrews, P.W., 2007. Culture adaptation of embryonic stem cells echoes germ cell malignancy. Int. J. Androl. 30(4), 275–281, disc. 281.

Hasegawa, K., Pomeroy, J.E., Pera, M.F., 2010. Current technology for the derivation of pluripotent stem cell lines from human embryos. Cell Stem Cell, 6(6), 521–531.

Henderson, J.K., Draper, J.S., Baillie, H.S., Fishel, S., Thomson, J.A., Moore, H., *et al.*, 2002. Preimplantation human embryos and embryonic stem cells show comparable expression of stage-specific embryonic antigens. Stem Cells (Dayton, Ohio) 20(4), 329–337.

Itskovitz-Eldor, J., Schuldiner, M., Karsenti, D., Eden, A., Yanuka, O., Amit, M., *et al.*, 2000. Differentiation of human embryonic stem cells into embryoid bodies compromising the three embryonic germ layers. Mol. Med. 6(2), 88–95.

Juliusson, G., Oscier, D.G., Fitchett, M., Ross, F.M., Stockdill, G., Mackie, M.J., *et al.*, 1990. Prognostic subgroups in B-cell chronic lymphocytic leukemia defined by specific chromosomal abnormalities. N. Engl. J. Med. 323(11), 720–724.

Lengner, C.J., Gimelbrant, A.A., Erwin, J.A., Cheng, A.W., Guenther, M.G., Welstead, G.G., *et al.*, 2010. Derivation of pre-X inactivation human embryonic stem cells under physiological oxygen concentrations. Cell 141(5), 872–883.

Lund, R.J., Emani, M.R., Barbaric, I., Kivinen, V., Jones, M., Baker, D., *et al.*, 2013. Karyotypically abnormal human ESCs are sensitive to HDAC inhibitors and show altered regulation of genes linked to cancers and neurological diseases. Stem Cell Res. 11(3), 1022–1036.

Martin, G.R., 1981. Isolation of a pluripotent cell line from early mouse embryos cultured in medium conditioned by teratocarcinoma stem cells. Proc. Nat. Acad. Sci. USA 78(12), 7634–7638.

Muller, F.-J., Schuldt, B.M., Williams, R., Mason, D., Altun, G., Papapetrou, E.P., *et al.*, 2011. A bioinformatic assay for pluripotency in human cells. Nat. Meth. 8, 315–317.

Narva, E., Autio, R., Rahkonen, N., Kong, L., Harrison, N., Kitsberg, D., *et al.*, 2010. High-resolution DNA analysis of human embryonic stem cell lines reveals culture-induced copy number changes and loss of heterozygosity. Nat. Biotechnol. 28(4), 371–377.

Nazor, K.L., Altun, G., Lynch, C., Tran, H., Harness, J.V., Slavin, I., *et al.*, 2012. Recurrent variations in DNA methylation in human pluripotent stem cells and their differentiated derivatives. Cell Stem Cell 10(5), 620–634.

Ng, E.S., Davis, R.P., Azzola, L., Stanley, E.G., Elefanty, A.G., 2005. Forced aggregation of defined numbers of human embryonic stem cells into embryoid bodies fosters robust, reproducible hematopoietic differentiation. Blood 106(5), 1601–1603.

Orlando, G., Soker, S., Stratta, R.J., Atala, A., 2013. Will regenerative medicine replace transplantation? Cold Spring Harb. Perspect. Med. 3(8).

Pavlakis, M., Khwaja, K., 2007. Pancreas and islet cell transplantation in diabetes. Curr. Opin. Endocrinol. Diabetes Obes 14(2), 146–150.

Plantaz, D., Mohapatra, G., Matthay, K.K., Pellarin, M., Seeger, R.C., Feuerstein, B.G., 1997. Gain of chromosome 17 is the most frequent abnormality detected in neuroblastoma by comparative genomic hybridization. Am. J. Pathol. 150(1), 81–89.

Schwartz, S.D., Regillo, C.D., Lam, B.L., Eliott, D., Rosenfeld, P.J., Gregori, N.Z., *et al.*, 2014. Human embryonic stem cell-derived retinal pigment epithelium in patients with age-related macular degeneration and Stargardt's macular dystrophy: follow-up of two open-label phase 1/2 studies. The Lancet. doi:10.1016/S0140-6736(14)61376-3

Shamblott, M.J., Axelman, J., Wang, S., Bugg, E.M., Littlefield, J.W., Donovan, P.J., *et al.*, 1998. Derivation of pluripotent stem cells from cultured human primordial germ cells. Proc. Nat. Acad. Sci. USA 95(23), 13 726–13 731.

Silva, J., Smith, A., 2008. Capturing pluripotency. Cell 132(4), 532–536.

Smith, A.G., Heath, J.K., Donaldson, D.D., Wong, G.G., Moreau, J., Stahl, M., *et al.*, 1988. Inhibition of pluripotential embryonic stem cell differentiation by purified polypeptides. Nature 336(6200), 688–690.

Summersgill, B., Osin, P., Lu, Y.J., Huddart, R., Shipley, J., 2001. Chromosomal imbalances associated with carcinoma in situ and associated testicular germ cell tumours of adolescents and adults. Brit. J. Cancer 85(2), 213–220.

Sumner, A.T., Evans, H.J., Buckland, R.A., 1971. New technique for distinguishing between human chromosomes. Nature: New Biol. 232(27), 31–32.

Takahashi, K., Tanabe, K., Ohnuki, M., Narita, M., Ichisaka, T., Tomoda, K., *et al.*, 2007. Induction of pluripotent stem cells from adult human fibroblasts by defined factors. Nat. Protoc. 131(12), 3081–3089.

Tchieu, J., Kuoy, E., Chin, M.H., Trinh, H., Patterson, M., Sherman, S.P., *et al.*, 2010. Female human iPSCs retain an inactive X chromosome. Cell Stem Cell 7(3), 329–342.

Thomson, J.A., Itskovitz-Eldor, J., Shapiro, S.S., Waknitz, M.A., Swiergiel, J.J., Marshall, V.S., *et al.*, 1998. Embryonic stem cell lines derived from human blastocysts. Science 282(5391), 1145–1147.

van den Berg, I.M., Laven, J.S.E., Stevens, M., Jonkers, I., Galjaard, R.-J., Gribnau, J., *et al.*, 2009. X chromosome inactivation is initiated in human preimplantation embryos. Am. J. Hum. Genet. 84(6), 771–779.

Werbowetski-Ogilvie, T.E., Bosse, M., Stewart, M., Schnerch, A., Ramos-Mejía, V., Rouleau, A., *et al.*, 2009. Characterization of human embryonic stem cells with features of neoplastic progression. Nature Biotechnol. 27(1), 91–97.

Widner, H., Tetrud, J., Rehncrona, S., Snow, B., Brundin, P., Gustavii, B., *et al.*, 1992. Bilateral fetal mesencephalic grafting in two patients with parkinsonism induced by 1-methyl-4-phenyl-1,2,3,6-tetrahydropyridine (MPTP). N. Engl. J. Med. 327(22), 1556–1563.

Yamanaka, S., 2012. Induced pluripotent stem cells: past, present, and future. Cell Stem Cell 10(6), 678–684.

2 Epigenetics, Stem Cell Pluripotency and Differentiation

Maria J. Barrero
CNIO-LILLY Epigenetics Laboratory, Spanish National Cancer Research Center (CNIO), Madrid, Spain

2.1 Introduction

'Epigenetics' refers to particular features that can be inherited from cell to cell that are not part of the DNA sequence and that, unlike mutations, are reversible. These most commonly consist of chemical modifications of histones or DNA that have profound effects on the regulation of gene expression and are critical for establishing and maintaining cell identity. Establishment of aberrant epigenetic marks during the *in vitro* differentiation of embryonic stem cells (ESCs) can result in cells that are different from their *in vivo* counterparts, compromising their use for clinical applications, since they may be prone to reverting to more proliferative phenotypes once implanted into the body.

It has been known for some decades that DNA is wrapped around histones to form the nucleosome; however, the tremendous impact that the organization of DNA into chromatin has on the regulation of gene expression was not fully realized until recently. In general terms, areas of the genome immersed in compact chromatin will have a low chance of being accessed by transcription factors and RNA polymerase II (Pol II). However, genes located in less condensed areas will be more likely to recruit Pol II and be actively transcribed.

Levels of chromatin condensation and gene expression have been correlated with the presence of certain histone or DNA modifications. DNA can

Principles of Stem Cell Biology and Cancer: Future Applications and Therapeutics, First Edition.
Edited by Tarik Regad, Thomas J. Sayers and Robert C. Rees.

be methylated at cytosine residues located at CpG and non-CpG sites by the action of DNA methyltransferases (DNMTs). In both cases, methylation is associated with transcriptional repression, but methylation at non-CpG sites seems far less frequent than CpG methylation (Guo *et al.*, 2013) and remains less explored. Additionally, 5-methylcytosine (5-mC) can be oxidized to 5-hydroxymethylcytosine (5-hmC) by the action of the 2-oxoglutarate-(2-OG) and Fe(II)-dependant dioxygenase Tet proteins (Tahiliani *et al.*, 2009). While hydroxymethylation has been proposed to be an intermediary in the pathway of demethylation (Guo *et al.*, 2011), it might also have other specific functions, since several proteins can specifically bind to hydroxymethylated DNA (Spruijt *et al.*, 2013).

Histones can suffer virtually any described post-translational modification, including acetylation, methylation, phosphorylation and others. It is more likely that these modifications will take place at the N-terminal tails, since they tend to protrude out of the nucleosome and are more available to modifying enzymes. Histone modifications can be correlated with either transcriptional activation or transcriptional repression, depending on the modification and the residue that becomes modified, and are dynamically regulated by enzymes capable of writing and erasing the marks. The collection of large amounts of data concerning levels of gene expression and genome-wide maps of histone modifications has allowed the establishment of clear correlations between certain modifications and gene expression. Among others, acetylation usually correlates with transcriptional activation, likely because this modification introduces a positive charge in the nucleosome believed to render chromatin less condensed. Trimethylation of histone H3 at lysine 4 (H3K4me3) is usually present at the transcription start site (TSS) of actively transcribed genes. Trimethylation of histone H3 at lysine 27 (H3K27me3) is commonly located at repressed genes. Certain modifications, such as H3K4me3 and H3K9me3, seem to be mutually exclusive and rarely mark the same genomic locations. Others, such as H3K4me1 and H3K27 acetylation, co-localize at certain genomic elements, such as enhancers, and allow the prediction of genetic elements according to their co-presence.

DNA and histone modifications, with the exception of acetylation, do not change the structural properties of chromatin. Instead, modified residues can serve as docking sites for other factors, which are referred to as 'readers'. In many cases, readers are subunits of remodelling complexes that use ATP to mediate the condensation or relaxation of chromatin and therefore prevent or facilitate the expression of genes.

The environment is believed to have a deep impact on the cell epigenome, although how the complexity of the signal transduction pathways converge on histone and DNA modifications remains poorly understood. Among other effects, environmental cues can impact on the balance of writers, erasers and readers, and therefore have important consequences for the expression of critical genes.

2.2 Epigenetic regulation of the pluripotent state

ESCs are derived from the inner cell mass (ICM) of the blastocyst. During the last 2 decades, a large effort has been dedicated to identifying the environmental cues that could allow researchers to freeze and preserve the two main properties of the ICM in a Petri dish; this would involve keeping cells in a self-renewing condition while retaining their ability to differentiate into virtually any cell type of the adult organism, a property known as pluripotency. In this way, ESCs have emerged as an unprecedentedly useful tool for the study of mechanisms of differentiation, as well as an endless source of differentiated tissues for regenerative medicine.

The identity of ESCs is maintained by a complex network of transcription factors, with Oct4, Sox2 and Nanog the most studied. These transcription factors maintain high levels of expression of self-renewal genes, at the same time as participating in repressing developmental regulators. The molecular basis of this duality is not fully understood. Self-renewal genes have very high rates of transcription, which seem to be supported by the presence of large enhancers called super-enhancers (which in some instances can span over 1 Mb of DNA) (Hnisz *et al.*, 2013). Super-enhancers are fully loaded with pluripotency-related transcription factors like Oct4, Sox2 and Nanog, and strongly occupied by the Mediator complex, which facilitates the physical interaction between enhancers and promoters and the efficient recruitment of Pol II to support high rates of transcription (Whyte *et al.*, 2013). They display high levels of H3K27 acetylation, which are dynamically regulated, since both histone acetyltransferases and histone deacetylases occupy these sites. Eventually, this dynamic regulation renders super-enhancer-regulated genes extremely vulnerable to perturbations of its components, providing a highly sensitive mechanism that will rapidly turn off the expression of these genes in response to differentiation signals.

The basis of pluripotency consists in critical developmental genes remaining silent but being ready for expression when differentiation signals arrive. Differentiated cells have lost this property, and therefore cannot change their identity in response to environmental signals. It has been proposed that a particular chromatin conformation is responsible for keeping developmental genes poised for activation in ESCs. This chromatin conformation consists in the simultaneous presence of both H3K27me3 and H3K4me3 at the regulatory regions of developmental genes and has been called 'bivalent domains' (Figure 2.1) (Bernstein *et al.*, 2006). These antagonistic modifications are unlikely to occur in the same histone tail, but seem to co-exist asymmetrically in nucleosomes; that is, on opposite H3 tails (Voigt *et al.*, 2012). It has been proposed that during differentiation, bivalent domains resolve into H3K4me3 only in genes that become activated and into H3K27me3 only in genes that become repressed (Bernstein *et al.*, 2006).

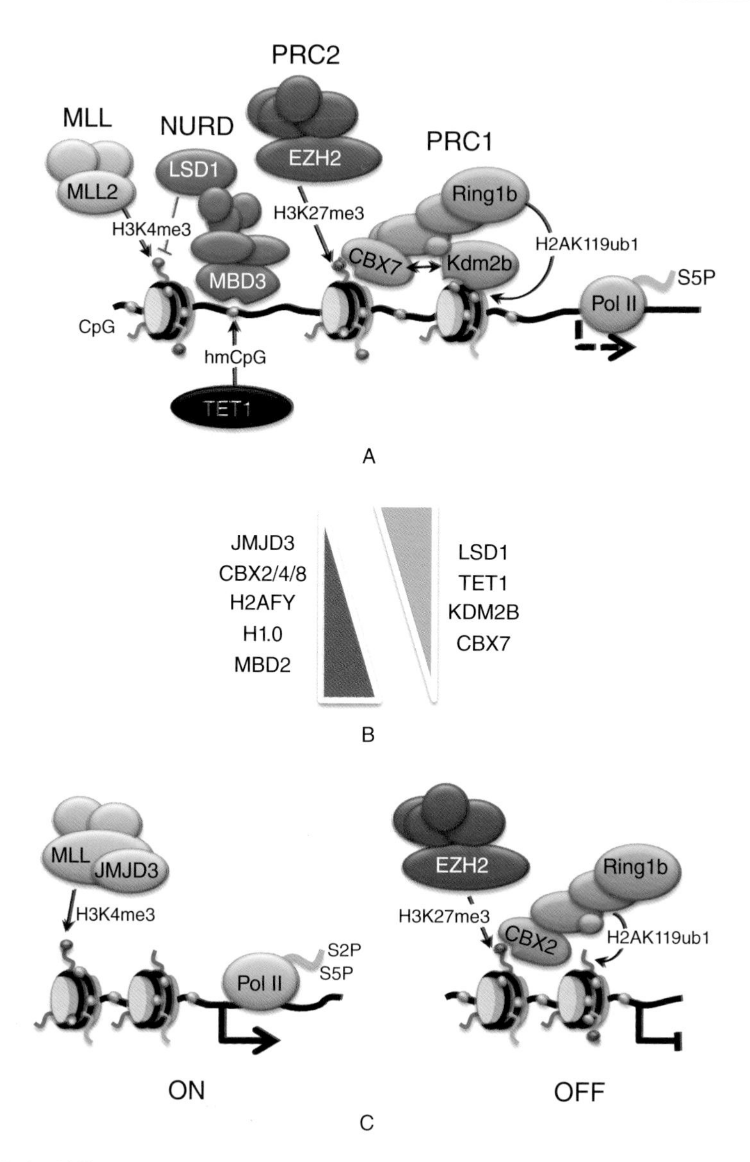

Figure 2.1 Different complexes are involved in maintaining bivalent domains in ESCs. (A) Bivalent domain-containing genes in ESCs are CpG-rich and are hypomethylated or hydroxymethylated by Tet1. Hydroxymethylated CpG islands provide binding sites for the NURD complex through the MBD3 subunit. LSD1 is recruited as part of the NURD complex and counteracts the methylation of H3K4 mediated by MLL2. Different PRC1 complexes are recruited to these regions through the binding of Kdm2b to unmethylated CpG islands or the recognition of Ezh2-mediated H3K27me3 by CBX7. (B) Induction (red) or repression (green) of the expression of different chromatin factors and subunits contributes to the resolution of bivalent domains during differentiation. (C) Bivalent domains resolved after differentiation.

Different complexes mediate the modifications found at bivalent domains. Methylation of H3K27 is catalysed by the Polycomb group of proteins (PcG), which refers to two distinct complexes. Polycomb repressive complex 2 (PRC2) contains four core components, including the SET domain-containing EZH2 subunit capable of trimethylating H3K27. Polycomb repressive complex 1 (PRC1) is represented in mammals by several complexes with different subunit compositions. All PRC1 complexes contain the ubiquitin ligase Ring1b but they are recruited to chromatin through different mechanisms, depending on the presence of other subunits (Figure 2.1). For example, CBX-containing complexes are recruited through chromodomain-mediated methylated H3K27 interaction (Cao *et al.*, 2002), while Kdm2b-containing complexes are recruited through Kdm2b-mediated recognition of nonmethylated CpG-rich areas (Farcas *et al.*, 2012; He *et al.*, 2013; Wu *et al.*, 2013). Methylation of H3K4 in mammals can be catalyzed by at least six different methyltransferases, Set1A and Set1B and MLL1 to MLL4. MLL2 has been recently suggested to be responsible for the H3K4me3 found at bivalent promoters (Hu *et al.*, 2013; Denissov *et al.*, 2014).

Although the resolution of bivalent domains during differentiation seems clearly correlated with changes in gene expression at different stages of cellular commitment (Xie *et al.*, 2013a), the role of bivalency in poising developmental genes remains controversial. Genes containing bivalent domains have also been located in differentiated cells, although in lower numbers (Mikkelsen *et al.*, 2007). Treatment of mouse ESCs (mESCs) with '2i' (inhibitors of Mek and GSK3 kinases) renders cells more naïve, with a lower expression of developmental regulators that correlates with reduced levels of H3K27me3 at their promoters, while Pol II appears sharply restrained at their TSSs (Marks *et al.*, 2012). This argues against a major role of H3K27me3 in poising developmental genes and is coincident with the observation that, following the maternal–zygotic transition in zebrafish, a subset of developmental genes are kept repressed in the presence of H3K4me3 only (Vastenhouw *et al.*, 2010). Additionally, mESCs can be derived from PRC2-deficient blastocysts and maintained in culture for many generations, although they show defects in differentiation (Pasini *et al.*, 2007). All together, these data suggest that the presence of H3K27me3 at developmental genes might reflect a transitional state towards final resolution during differentiation. The role of H3K4me3 in poising genes also remains obscure, since mESCs null for MLL2 loose H3K4me3 at many bivalent genes, but this does not prevent their induction by retinoic acid, suggesting that this modification is not critical to priming genes for activation (Hu *et al.*, 2013; Denissov *et al.*, 2014).

The ultimate event that holds the key to pluripotency is the presence of Pol II restrained at the TSSs of relevant developmental regulators, ready to respond to differentiation signals and engage in productive rounds of transcription. Pol II stalled at bivalent gene promoters in ESCs is preferentially phosphorylated on Ser-5 residues (Stock *et al.*, 2007), but its conformation appears to be different from that of the conventional paused Pol II found in differentiated cells, since it is confined to regions extremely close to the TSS (Min *et al.*, 2011). This unique Pol II conformation has yet to be fully characterized, but it is usually referred to as 'poised' (Stock *et al.*, 2007). Ablation of Ring1B, the PRC1 subunit that mediates ubiquitination of histone H2A, results in loss of ubiquitinated H2A at bivalent genes and de-repression of these genes without changes in H3K27me3 (Stock *et al.*, 2007), suggesting that H2A ubiquitination is involved in restraining Pol II at these promoters. More recent data show that, in ESCs, the kinases Erk1/2 mediate the phosphorylation of Pol II at Ser-5, contributing to holding it back at the TSSs of developmental genes (Tee *et al.*, 2014). Further analysis will be needed to elucidate how this very particular Pol II conformation is achieved at critical promoters in ESCs.

The epigenetic events that regulate the expression of developmental genes in ESCs appear to be linked to genetic features at the regulatory regions of such genes. Most of these genes contain CpG-rich regions, called 'islands', next to their TSSs. These CpG islands remain hypomethylated and facilitate the recruitment of both PRC1 and PRC2 (Wu *et al.*, 2011, 2013). Occupancy by Tet1, which is highly expressed in ESCs, may both protect these sites against DNA methylation and mediate their hydroxymethylation (Wu *et al.*, 2011). Hydroxythylated CpG islands are likely to provide binding sites for MBD3, a subunit of the NuRD repressive complex (Yildirim *et al.*, 2011). The histone demethylase LSD1 is also recruited to these genes as part of this complex, where it finely regulates the levels of H3K4me2/3 (Adamo *et al.*, 2011). Importantly, the recruitment of both activating and repressing complexes to these domains may keep them in a dynamic equilibrium that allows for rapid response during differentiation, merely by tipping the balance of these complexes (Figure 2.1).

2.3 Epigenetic changes during differentiation

Differentiation cues have tremendous effects on the transcriptional profiles of ESCs. Pluripotency-related genes become repressed as developmental regulators become activated in a germ layer-specific manner. How environmental signals drive these coordinated changes and what mechanism triggers the differentiation of cells into one particular embryonic layer is largely unknown. Most likely, differentiation signals activate the expression of early transcription factors capable of both inducing the expression of tissue-specific genes and repressing the expression of alternative-lineage and pluripotency-related genes. For example, CDX2 plays a key role in silencing Oct4 expression in the trophectoderm lineage (Niwa *et al.*, 2005), as well as in inducing differentiation genes (Nishiyama *et al.*, 2009). The transcription factors that are among the first to be induced and can access tissue-specific loci at early stages of development have been termed 'pioneer factors'. These factors have been found to be able to enter compact chromatin regions in order to activate gene expression during development (Zaret and Carroll, 2011).

Several studies have shown that the Polycomb complex participates in the silencing of the pluripotency network during differentiation (Pasini *et al.*, 2007; Shen *et al.*, 2008), while the methyltransferase G9a has been suggested to contribute to the silencing of OCT4 expression by mediating the deposition of the heterochromatin mark H3K9me2/3 at its regulatory regions (Feldman *et al.*, 2006). However, genes like OCT4 and NANOG do not display either H3K27me3 or H3K9me3 marks in human differentiated cells, but show DNA methylation at dispersed CpG sites (Figure 2.2). Nonetheless, Polycomb and G9a might contribute to establishing a transitional chromatin state

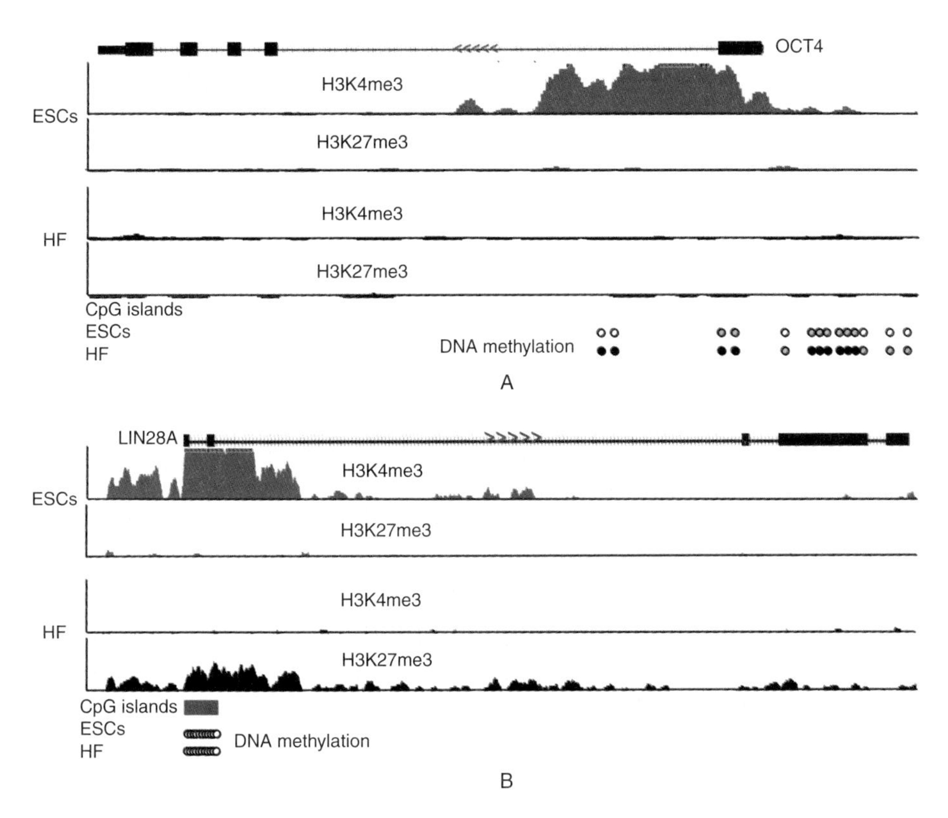

Figure 2.2 Various silencing mechanisms of pluripotency-related genes in differentiated cells according to CpG densitiy, comparing the histone- and DNA-modification profiles of pluripotency genes in ESCs and human fibroblasts (HFs) in two pluripotency-related genes. (A) OCT4 contains dispersed CpGs that become hypermethylated in differentiated cells. (B) LIN28A contains a CpG island that is hypomethylated in both cell types. Repression of this gene in differentiated cells correlates with presence of H3K27me3. Black, grey and white dots represent hypermethylated, partially methylated and nonmethylated CpG sites. Histone modifications from ENCODE publically available data (ENCODE Project Consortium, 2011) generated in the Bernstein laboratory and displayed in the UCSC genome browser are shown.

that favours the final silencing of some pluripotency-related genes through DNA methylation (Epsztejn-Litman *et al.*, 2008). Additionally, recruitment of linker histone H1 (Terme *et al.*, 2011) and histone variant macroH2A (Barrero *et al.*, 2013b) to pluripotency-related genes during differentiation contributes to the repression of these genes.

Developmental specification implies a transition from a permissive chromatin state to a restrictive state with prevalent Polycomb repression. In pluripotent cells, the H3K27me3 mark is confined to peaks at bivalent GC-rich promoters, but it becomes more broadly distributed in differentiated cells and tissues, contributing to maintenance of the silencing of genes with functions in alternate lineages (Zhu *et al.*, 2013). Although DNA methylation of certain CpG islands correlates with repression of genes during differentiation, most CpG islands remain hypomethylated in differentiated cells, with H3K27me3 the most common mechanism of repression of CpG island-containing genes.

Promoters of developmental regulators that become active in early developmental stages resolve their bivalency into H3K4me3 only, a process that entails loss of H3K27me3. This can be achieved by both passive and active mechanisms. Passively, changes in PcG subunit composition during differentiation can help to relocate complexes from genes that become active to genes that need to be repressed, such as the pluripotency network (Morey *et al.*, 2012). More actively, the H3K27me3 demethylases UTX and JMJD3 can be recruited to developmental genes to facilitate activation (Agger *et al.*, 2007; Burgold *et al.*, 2008). Importantly, UTX and JMJD3 associate with different MLL complexes, which likely determines their recruitment to distinct target genes and suggests a coordinated mechanism for the removal of H3K27me3 and the deposition of H3K4me3.

Interestingly, not all bivalent domains are resolved during differentiation. However, these domains are likely to be different in differentiated cells than in pluripotent cells. The presence of additional histone modifications, a different balance of H3K4me2/3 versus H3K27me3 (Figure 2.3) or occupancy by the H2A variant macroH2A (Barrero *et al.*, 2013a) may all contribute to making these domains less permissive in differentiated cells. The role of these domains in differentiated cells remains unclear, but it might reflect a compromise between the plasticity required for some adult cells, such as adult stem cells and germ cells, and the irreversibility of these domains in terminally differentiated cells.

Unlike in developmental regulators, the regulatory regions of tissue-specific genes are often CpG-poor and become activated preferentially at later stages of differentiation, likely due to the binding of developmental transcription factors. Interestingly, loss of DNA methylation at these CpG-poor genes seems to be accompanied by a gain of H3K4me1 or H3K27me3 during early stages of differentiation. These transitions occur without significant change in gene expression and likely function to prime these genes for later induction in terminally differentiated cells (Gifford *et al.*, 2013). Those tissue-specific genes

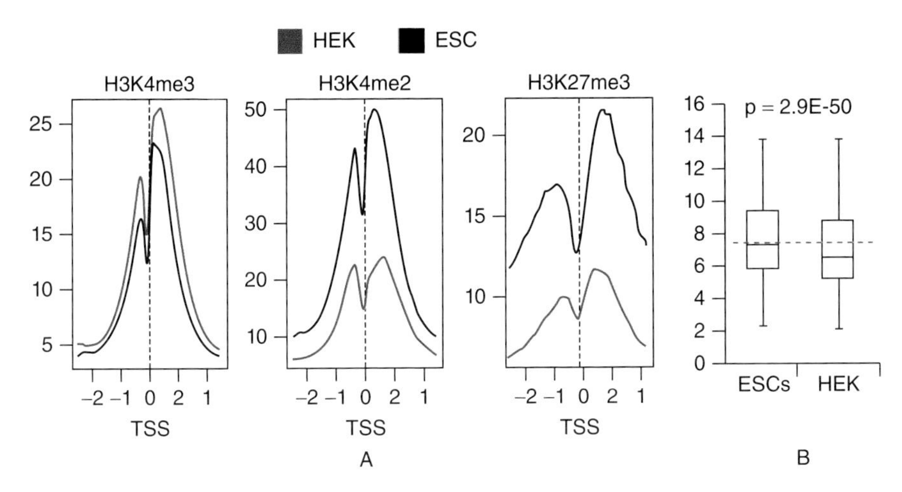

Figure 2.3 Bivalent domains contain different ratios of modifications in pluripotent and differentiated cells. (A) Comparison of the levels of H3K4me3, H3K4me2 and H3K27me3 between human ESCs and keratinocytes (HEKs) around the TSSs of genes that are bivalent in both cell types. (B) Box plot of mRNA levels in genes marked with bivalent domains in both ESCs and HEKs. Keratinocytes show lower levels of H3K4me2 and H3k27me3 at bivalent domains than ESCs, which might contribute to the different transcriptional statuses found in different cell lines. In both cell lines, bivalent genes are expressed at low levels, but expression is more permissive in ESCs than in HEKs. ENCODE publically available data (ENCODE Project Consortium, 2011) generated in the Bernstein laboratory are shown. Aggregation plots were drawn using SitePro, from the CEAS package (Shin *et al.*, 2009). Expression data have been previously published (Aasen *et al.*, 2008).

that remain silenced in a layer-specific manner usually retain DNA methylation at their sparse CpG sites (Xie *et al.*, 2013b).

Broad changes in the heterochromatin mark H3K9me2/3 have been proposed to play a role in differentiation. Initial reports suggested that differentiation of ESCs causes the expansion of H3K9me2 to large megabase-sized domains, termed 'large organized chromatin K9-modifications' (LOCKs), which are maintained by the histone methyltransferase G9a (Wen *et al.*, 2009). These domains are likely to contribute to the establishment of heterochromatin in differentiated cells. However, others have reported no significant differences in the coverage of these domains in pluripotent versus differentiated cells (Filion and van Steensel, 2010). A recent report suggests that increased H3K9me3 signal is found in primary cell cultures compared to tissues and ESCs, and that this is likely triggered by the culture environment (Zhu *et al.*, 2013).

Recent work has revealed dynamic changes in enhancer–promoter interactions during differentiation. The changing enhancer landscape in mammalian development results from the recruitment of lineage-determining factors, which associate with enhancers anchored not only to tissue-specific promoters but also to constitutively active ones (Kieffer-Kwon *et al.*,

2013). This reorganization of interactions also involves changes in histone modifications. Most cell-specific enhancers are CpG-poor and appear to be depleted of H3K27me3 in the cell types in which they are active. A subset of enhancers becomes weakly marked with H3K27me3 in differentiated cells, suggesting that the expansion of the H3K27me3 domains that arise during differentiation may contribute to repression of enhancers that are active in other lineages. However, DNA methylation seems to be a more common mechanism in the inactivation of enhancers during differentiation (Xie *et al.*, 2013b).

Despite the highly dynamic exchange of promoter–enhancer interactions during differentiation, recent data suggest that such interactions are restricted to large yet defined megabase-sized chromatin domains, which have been termed 'topologically associated domains' (TADs) (Dixon *et al.*, 2012; Nora *et al.*, 2012). These regions are separated by narrow segments devoid of chromatin interactions that appear to function as boundaries or insulators which block the interactions between two adjacent domains (Figure 2.4). Interestingly, TADs align with several domain-wide features of the epigenome, such as H3K27me3 or H3K9me2 blocks and lamina-associated domains, suggesting a common epigenetic environment for the genes located within one domain and a potential role for the boundaries in preventing the spreading of certain histone modifications. Accordingly, genes located within one particular TAD are coordinately expressed during development (Nora *et al.*, 2012). Although TADs can switch their association with particular chromatin landmarks during differentiation (Nora *et al.*, 2012), their boundaries are stable across different cell types and are highly conserved across species (Xie *et al.*, 2013b), indicating that TADs are a fundamental property of mammalian genomes, organized into broad regulatory domains.

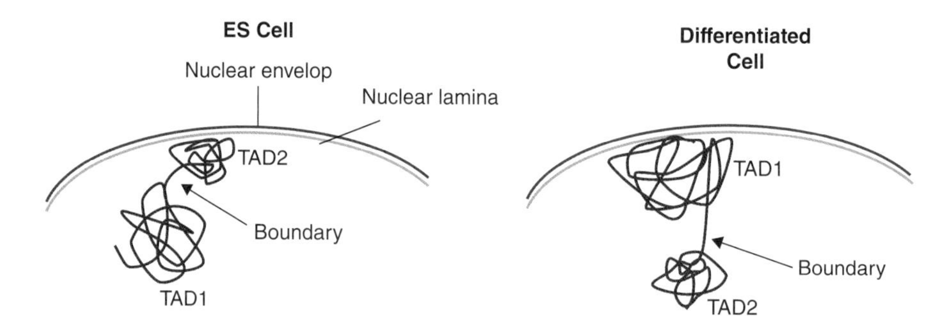

Figure 2.4 Changes in the association of TADs with the nuclear lamina during differentiation. Two different domains are shown (TAD1 and TAD2). Genes located in TAD1 are highly expressed in ES Cell, interact within the same domain and are separated from TAD2 by a boundary region that restricts interactions between the two. Most genes located in TAD2 are silent in ES Cell and interact with the nuclear lamina. Differentiation causes silencing and association with the nuclear lamina of genes located in TAD1, and expression of genes located in TAD2. Despite the changes in expression and interaction with the nuclear lamina, boundaries are conserved, restricting interactions within one domain.

2.4 Reprogramming of somatic cells to pluripotency: reversing the epigenetic landscape

It is a fundamental belief that pluripotent and multipotent cells differentiate into terminal irreversible states. However, in animals that are able to regenerate certain organs, differentiated cells can revert to more undifferentiated phenotypes, allowing proliferation and re-differentiation to repair or even rebuild the new organ. Mammals seem to have lost this property, raising the question of whether differentiated states could be reversed. Nuclear transfer experiments 2 decades ago showed for the first time that it is possible to revert the differentiated phenotype (Wilmut *et al.*, 1997), and, more recently, it has been shown that it is possible to reprogramme somatic cells to pluripotency using the transcription factors Oct4, Sox2, Klf4 and Myc (Takahashi and Yamanaka, 2006).

The process of reprogramming using ectopic expression of factors is very inefficient, likely due to the impossibility of hitting condensed chromatin with these factors. The use of small molecules such as inhibitors of histone deacetylases and DNA methyltransferases to enhance preprogramming suggests that chromatin acts as a barrier in this process (Feng *et al.*, 2009). Chromatin remodelling factors like CHD1 (Gaspar-Maia *et al.*, 2009) and components of the BAF complex (Singhal *et al.*, 2010) have also been shown to facilitate reprogramming, by contributing to the opening up of chromatin at genomic sites that are bound by the reprogramming factors (Figure 2.5).

Methylation of H3K27 needs to be properly regulated during reprogramming. While the H3K27me3 demethylase UTX is needed to remove the H3K27me3 mark at pluripotency-related genes (Mansour *et al.*, 2013), cells that show defective PRC2 activity fail to silence developmental regulators and reprogramme into pluripotency (Fragola *et al.*, 2013). Removal of DNA methylation at self-renewal genes is another critical step that requires the action of TET2 and TET1 (Doege *et al.*, 2012; Costa *et al.*, 2013; Gao *et al.*, 2013) (Figure 2.5). Additionally, gain of H3K4me3/2 at self-renewal genes during reprogramming requires the action of MLL complexes (Ang *et al.*, 2011).

In accordance with the role of H3K9 methylation in the establishment of heterochromatin, depletion of the H3K9 methyltransferases Ehmt1, Ehmt2 and Setdb1 and of the methyl H3K9-binding protein Cbx3 facilitates reprogramming (Sridharan *et al.*, 2013), while the presence of H3K9 demethylases promotes reprogramming (Chen *et al.*, 2013). These findings are in agreement with the discovery of megabase-sized blocks marked with H3K9me3 in differentiated cells that are refractory to binding by the reprogramming factors during the early stages of reprogramming (Soufi *et al.*, 2012).

During the reprogramming process, the changes in gene expression appear ordered in time: the silencing of the cell-specific programmes occurs first, followed by activation of the pluripotency network (Sridharan *et al.*, 2009). However, most early change in histone modification consist in the gain of H3K4me2 at pluripotency-related and developmentally regulated

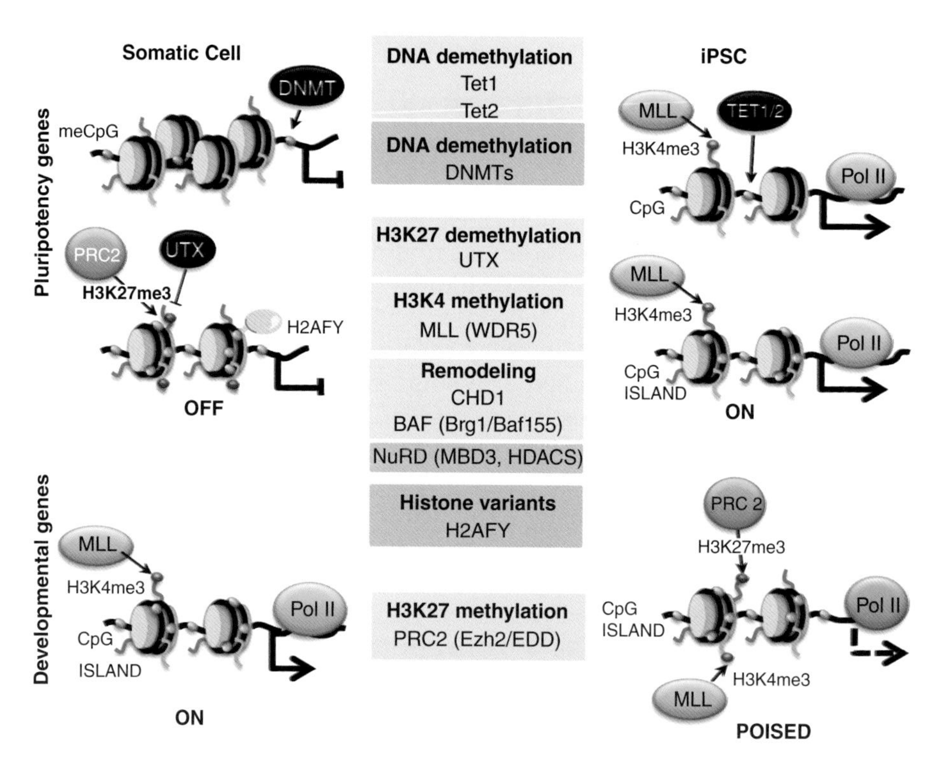

Figure 2.5 Chromatin factors involved in the process of somatic cell reprogramming to pluripotency. Middle panel shows factors that have been shown to facilitate (green) or block (red) reprogramming to induced pluripotent stem cells (iPSC). These factors are involved in critical events such as DNA demethylation or H3K27me3 removal and gain of H3K4me2/3 at pluripotency genes and silencing of developmental genes through gain of H3K27me3.

gene promoters and enhancers. This change, which precedes changes in H3K27me3, might act as a priming event, since it precedes transcriptional changes within the corresponding locus (Koche *et al.*, 2011). Epigenetic barriers that prevent the gain of H3K4me2 at this locus, such as the presence of macroH2A (Barrero *et al.*, 2013a), are likely to contribute to the low efficiency of the reprogramming process.

2.5 Epigenetics, stem cells and cancer

Although malignant transformation has typically been associated with mutations and translocations, aberrant patterns of histone and DNA modification can also account for driver events in tumorigenesis. Altered expression and mutation of chromatin modifiers have been largely described in tumours (Rodriguez-Paredes and Esteller, 2011). The fact that DNA methyltransferases and histone deacetylases inhibitors are successfully used to treat cancer shows the relevance of the epigenetic machinery in this process (Esteller, 2008). Moreover, cancer-dependent epigenetic signatures can be reverted during reprogramming to pluripotency, highlighting the reversibility of aberrant epigenetic marks (Ron-Bigger *et al.*, 2010).

Several studies have proposed potential links between chromatin features in ESCs and cancer cells. Promoters marked with bivalent domains in ESCs, which are commonly DNA hypomethylated in both ESCs and differentiated cells, have a tendency to become hypermethylated in cancer (Widschwendter *et al.*, 2007; Doi *et al.*, 2009). The acquisition of DNA methylation at bivalent genes in self-renewing adult stem cells can impair their induction during differentiation and lock stem cells in an immature proliferating state (Trowbridge *et al.*, 2012). Similarly, activating mutations of Ezh2 found in transformed B cells can lead to aberrant silencing of bivalent differentiation genes, with similar consequences (Béguelin *et al.*, 2013).

2.6 Conclusions

ESCs provide a unique system in which to study changes in gene expression and chromatin regulation during differentiation. Understanding these mechanisms will not only be relevant to improving the *in vitro* generation of cell types and tissues useful for regenerative medicine but also provides insights into regulatory pathways that are frequently altered in pathological situations such as cancer. Importantly, the reversibility of the epigenetic marks can offer multiple possibilities for therapeutic intervention.

Acknowledgements

I would like to apologize to those authors whose work could not be cited due to space limitations.

References

Aasen, T., Raya, A., Barrero, M.J., Garreta, E., Consiglio, A., Gonzalez, F., *et al.*, 2008. Efficient and rapid generation of induced pluripotent stem cells from human keratinocytes. Nat. Biotechnol. 26, 1276–1284.

Adamo, A., Sese, B., Boue, S., Castano, J., Paramonov, I., Barrero, M.J., Izpisua Belmonte, J.C. 2011. LSD1 regulates the balance between self-renewal and differentiation in human embryonic stem cells. Nat. Cell Biol. 13, 652–659.

Agger, K., Cloos, P.A., Christensen, J., Pasini, D., Rose, S., Rappsilber, J., *et al.*, 2007. UTX and JMJD3 are histone H3K27 demethylases involved in HOX gene regulation and development. Nature 449, 731–734.

Ang, Y.S., Tsai, S.Y., Lee, D.F., Monk, J., Su, J., Ratnakumar, K., *et al.*, 2011. Wdr5 mediates self-renewal and reprogramming via the embryonic stem cell core transcriptional network. Cell 145, 183–197.

Barrero, M.J., Sese, B., Kuebler, B., Bilic, J., Boue, S., Marti, M., Izpisua Belmonte, J.C. 2013a, Macrohistone variants preserve cell identity by preventing the gain of H3K4me2 during reprogramming to pluripotency. Cell Rep. 3, 1005–1011.

Barrero, M.J., Sese, B., Marti, M., Izpisua Belmonte, J.C., 2013b. Macro histone variants are critical for the differentiation of human pluripotent cells. J. Biol. Chem. 288, 16 110–16 116.

Béguelin, W., Popovic, R., Teater, M., Jiang, Y., Bunting, K.L., Rosen, M., *et al.*, 2013. EZH2 is required for germinal center formation and somatic EZH2 mutations promote lymphoid transformation. Cancer Cell 23, 677–692.

Bernstein, B.E., Mikkelsen, T.S., Xie, X., Kamal, M., Huebert, D.J., Cuff, J., *et al.*, 2006. A bivalent chromatin structure marks key developmental genes in embryonic stem cells. Cell 125, 315–326.

Burgold, T., Spreafico, F., De Santa, F., Totaro, M.G., Prosperini, E., Natoli, G., Testa, G., 2008. The histone H3 lysine 27-specific demethylase Jmjd3 is required for neural commitment. PLoS One 3, e3034.

Cao, R., Wang, L., Wang, H., Xia, L., Erdjument-Bromage, H., Tempst, P., *et al.*, 2002. Role of histone H3 lysine 27 methylation in Polycomb-group silencing. Science 298, 1039–1043.

Costa, Y., Ding, J., Theunissen, T.W., Faiola, F., Hore, T.A., Shliaha, P.V., *et al.*, 2013. NANOG-dependent function of TET1 and TET2 in establishment of pluripotency. Nature 495, 370–374.

Chen, J., Liu, H., Liu, J., Qi, J., Wei, B., Yang, J., *et al.*, 2013. H3K9 methylation is a barrier during somatic cell reprogramming into iPSCs. Nat. Genet. 45, 34–42.

Denissov, S., Hofemeister, H., Marks, H., Kranz, A., Ciotta, G., Singh, S., *et al.*, 2014. Mll2 is required for H3K4 trimethylation on bivalent promoters in embryonic stem cells, whereas Mll1 is redundant. Development 141, 526–537.

Dixon, J.R., Selvaraj, S., Yue, F., Kim, A., Li, Y., Shen, Y., *et al.*, 2012. Topological domains in mammalian genomes identified by analysis of chromatin interactions. Nature 485, 376–380.

Doege, C.A., Inoue, K., Yamashita, T., Rhee, D.B., Travis, S., Fujita, R., *et al.*, 2012. Early-stage epigenetic modification during somatic cell reprogramming by Parp1 and Tet2. Nature 488, 652–655.

Doi, A., Park, I.H., Wen, B., Murakami, P., Aryee, M.J., Irizarry, R., *et al.*, 2009. Differential methylation of tissue- and cancer-specific CpG island shores distinguishes human induced pluripotent stem cells, embryonic stem cells and fibroblasts. Nat. Genet. 41, 1350–1353.

ENCODE Project Consortium. 2011. A user's guide to the encyclopedia of DNA elements (ENCODE). PLoS Biol. 9, e1001046.

Epsztejn-Litman, S., Feldman, N., Abu-Remaileh, M., Shufaro, Y., Gerson, A., Ueda, J., *et al.*, 2008. De novo DNA methylation promoted by G9a prevents reprogramming of embryonically silenced genes. Nat. Struct. Mol. Biol. 15, 1176–1183.

Esteller, M. 2008. Epigenetics in cancer. N. Engl. J. Med. 358, 1148–1159.

Farcas, A.M., Blackledge, N.P., Sudbery, I., Long, H.K., Mcgouran, J.F., Rose, N.R., *et al.*, 2012. KDM2B links the Polycomb Repressive Complex 1 (PRC1) to recognition of CpG islands. eLife, 1, e00205.

Feldman, N., Gerson, A., Fang, J., Li, E., Zhang, Y., Shinkai, Y., *et al.*, 2006. G9a-mediated irreversible epigenetic inactivation of Oct-3/4 during early embryogenesis. Nat. Cell Biol. 8, 188–194.

Feng, B., Ng, J.H., Heng, J.C., Ng, H.H. 2009. Molecules that promote or enhance reprogramming of somatic cells to induced pluripotent stem cells. Cell Stem Cell, 4, 301–312.

Filion, G.J., Van Steensel, B. 2010. Reassessing the abundance of H3K9me2 chromatin domains in embryonic stem cells. Nat. Genet. 42, 4; author reply 5–6.

Fragola, G., Germain, P.L., Laise, P., Cuomo, A., Blasimme, A., Gross, F., *et al.*, 2013. Cell reprogramming requires silencing of a core subset of polycomb targets. PLoS Genet. 9, e1003292.

Gao, Y., Chen, J., Li, K., Wu, T., Huang, B., Liu, W., *et al.*, 2013. Replacement of Oct4 by Tet1 during iPSC induction reveals an important role of DNA methylation and hydroxymethylation in reprogramming. Cell Stem Cell 12, 453–469.

Gaspar-Maia, A., Alajem, A., Polesso, F., Sridharan, R., Mason, M.J., Heidersbach, A., *et al.*, 2009. Chd1 regulates open chromatin and pluripotency of embryonic stem cells. Nature 460, 863–868.

Gifford, C.A., Ziller, M.J., Gu, H., Trapnell, C., Donaghey, J., Tsankov, A., *et al.*, 2013. Transcriptional and epigenetic dynamics during specification of human embryonic stem cells. Cell 153, 1149–1163.

Guo, J.U., Su, Y., Zhong, C., Ming, G.-L., Song, H. 2011. Hydroxylation of 5-methylcytosine by TET1 promotes active DNA demethylation in the adult brain. Cell 145, 423–434.

Guo, J.U., Su, Y., Shin, J.H., Shin, J., Li, H., Xie, B., *et al.*, 2013. Distribution, recognition and regulation of non-CpG methylation in the adult mammalian brain. Nat. Neurosci. 17, 215–222.

He, J., Shen, L., Wan, M., Taranova, O., Wu, H., Zhang, Y. 2013. Kdm2b maintains murine embryonic stem cell status by recruiting PRC1 complex to CpG islands of developmental genes. Nat. Cell Biol. 15, 373–384.

Hnisz, D., Abraham, B.J., Lee, T.I., Lau, A., Saint-André, V., Sigova, A.A., *et al.*, 2013. Super-enhancers in the control of cell identity and disease. Cell 155, 934–947.

Hu, D., Garruss, A.S., Gao, X., Morgan, M.A., Cook, M., Smith, E.R., Shilatifard, A. 2013, The Mll2 branch of the COMPASS family regulates bivalent promoters in mouse embryonic stem cells. Nat. Struct. Mol. Biol. 20, 1093–1097.

Kieffer-Kwon, K.-R., Tang, Z., Mathe, E., Qian, J., Sung, M.-H., Li, G., *et al.*, 2013. Interactome maps of mouse gene regulatory domains reveal basic principles of transcriptional regulation. Cell 155, 1507–1520.

Koche, R.P., Smith, Z.D., Adli, M., Gu, H., Ku, M., Gnirke, A., *et al.*, 2011. Reprogramming factor expression initiates widespread targeted chromatin remodeling. Cell Stem Cell 8, 96–105.

Mansour, A.A., Gafni, O., Weinberger, L., Zviran, A., Ayyash, M., Rais, Y., *et al.*, 2013. The H3K27 demethylase Utx regulates somatic and germ cell epigenetic reprogramming. Nature 488, 409–413.

Marks, H., Kalkan, T., Menafra, R., Denissov, S., Jones, K., Hofemeister, H., *et al.*, 2012. The transcriptional and epigenomic foundations of ground state pluripotency. Cell 149, 590–604.

Mikkelsen, T.S., Ku, M., Jaffe, D.B., Issac, B., Lieberman, E., Giannoukos, G., *et al.*, 2007. Genome-wide maps of chromatin state in pluripotent and lineage-committed cells. Nature 448, 553–560.

Min, I.M., Waterfall, J.J., Core, L.J., Munroe, R.J., Schimenti, J., Lis, J.T. 2011. Regulating RNA polymerase pausing and transcription elongation in embryonic stem cells. Genes Dev. 25, 742–754.

Morey, L., Pascual, G., Cozzuto, L., Roma, G., Wutz, A., Benitah, S.A., Di Croce, L. 2012. Nonoverlapping functions of the Polycomb group Cbx family of proteins in embryonic stem cells. Cell Stem Cell 10, 47–62.

Nishiyama, A., Xin, L., Sharov, A.A., Thomas, M., Mowrer, G., Meyers, E., *et al.*, 2009. Uncovering early response of gene regulatory networks in ESCs by systematic induction of transcription factors. Cell Stem Cell 5, 420–433.

Niwa, H., Toyooka, Y., Shimosato, D., Strumpf, D., Takahashi, K., Yagi, R., Rossant, J. 2005. Interaction between Oct3/4 and Cdx2 determines trophectoderm differentiation. Cell 123, 917–929.

Nora, E.P., Lajoie, B.R., Schulz, E.G., Giorgetti, L., Okamoto, I., Servant, N., *et al.*, 2012. Spatial partitioning of the regulatory landscape of the X-inactivation centre. Nature 485, 381–385.

Pasini, D., Bracken, A.P., Hansen, J.B., Capillo, M., Helin, K. 2007. The polycomb group protein Suz12 is required for embryonic stem cell differentiation. Mol. Cell Biol. 27, 3769–3779.

Rodriguez-Paredes, M., Esteller, M. 2011. Cancer epigenetics reaches mainstream oncology. Nat. Med. 17, 330–339.

Ron-Bigger, S., Bar-Nur, O., Isaac, S., Bocker, M., Lyko, F., Eden, A. 2010. Aberrant epigenetic silencing of tumor suppressor genes is reversed by direct reprogramming. Stem Cells 28, 1349–1354.

Shen, X., Liu, Y., Hsu, Y.J., Fujiwara, Y., Kim, J., Mao, X., *et al.*, 2008. EZH1 mediates methylation on histone H3 lysine 27 and complements EZH2 in maintaining stem cell identity and executing pluripotency. Mol. Cell 32, 491–502.

Shin, H., Liu, T., Manrai, A.K., Liu, X.S. 2009. CEAS: cis-regulatory element annotation system. Bioinformatics 25, 2605–2606.

Singhal, N., Graumann, J., Wu, G., Arauzo-Bravo, M.J., Han, D.W., Greber, B., *et al.*, 2010. Chromatin-remodeling components of the BAF complex facilitate reprogramming. Cell 141, 943–955.

Soufi, A., Donahue, G., Zaret, K.S. 2012. Facilitators and impediments of the pluripotency reprogramming factors' initial engagement with the genome. Cell 151, 994–1004.

Spruijt, C.G., Gnerlich, F., Smits, A.H., Pfaffeneder, T., Jansen, P.W., Bauer, C., *et al.*, 2013. Dynamic readers for 5-(hydroxy)methylcytosine and its oxidized derivatives. Cell 152, 1146–1159.

Sridharan, R., Tchieu, J., Mason, M.J., Yachechko, R., Kuoy, E., Horvath, S., *et al.*, 2009. Role of the murine reprogramming factors in the induction of pluripotency. Cell 136, 364–377.

Sridharan, R., Gonzales-Cope, M., Chronis, C., Bonora, G., Mckee, R., Huang, C., *et al.*, 2013. Proteomic and genomic approaches reveal critical functions of H3K9 methylation and heterochromatin protein-1gamma in reprogramming to pluripotency. Nat. Cell Biol. 15, 872–882.

Stock, J.K., Giadrossi, S., Casanova, M., Brookes, E., Vidal, M., Koseki, H., *et al.*, 2007. Ring1-mediated ubiquitination of H2A restrains poised RNA polymerase II at bivalent genes in mouse ES cells. Nat. Cell Biol. 9, 1428–1435.

Tahiliani, M., Koh, K.P., Shen, Y., Pastor, W.A., Bandukwala, H., Brudno, Y., *et al.*, 2009. Conversion of 5-methylcytosine to 5-hydroxymethylcytosine in mammalian DNA by MLL partner TET1. Science 324, 930–935.

Takahashi, K., Yamanaka, S. 2006. Induction of pluripotent stem cells from mouse embryonic and adult fibroblast cultures by defined factors. Cell 126, 663–676.

Tee, W.-W., Shen, S.S., Oksuz, O., Narendra, V., Reinberg, D. 2014. Erk1/2 activity promotes chromatin features and RNAPII phosphorylation at developmental promoters in mouse ESCs. Cell 156, 678–690.

Terme, J.M., Sese, B., Millan-Arino, L., Mayor, R., Belmonte, J.C., Barrero, M.J., Jordan, A. 2011. Histone H1 variants are differentially expressed and incorporated into chromatin during differentiation and reprogramming to pluripotency. J. Biol. Chem. 286, 35 347–35 357.

Trowbridge, J.J., Sinha, A.U., Zhu, N., Li, M., Armstrong, S.A., Orkin, S.H. 2012. Haploinsufficiency of Dnmt1 impairs leukemia stem cell function through derepression of bivalent chromatin domains. Genes Dev. 26, 344–349.

Vastenhouw, N.L., Zhang, Y., Woods, I.G., Imam, F., Regev, A., Liu, X.S., *et al.*, 2010. Chromatin signature of embryonic pluripotency is established during genome activation. Nature 464, 922–926.

Voigt, P., Leroy, G., Drury, W.J., Zee, B.M., Son, J., Beck, D.B., *et al.*, 2012. Asymmetrically modified nucleosomes. Cell 151, 181–193.

Wen, B., Wu, H., Shinkai, Y., Irizarry, R.A., Feinberg, A.P. 2009. Large histone H3 lysine 9 dimethylated chromatin blocks distinguish differentiated from embryonic stem cells. Nat. Genet. 41, 246–250.

Whyte, W.A., Orlando, D.A., Hnisz, D., Abraham, B.J., Lin, C.Y., Kagey, M.H., *et al.*, 2013. Master transcription factors and mediator establish super-enhancers at key cell identity genes. Cell 153, 307–319.

Widschwendter, M., Fiegl, H., Egle, D., Mueller-Holzner, E., Spizzo, G., Marth, C., *et al.*, 2007. Epigenetic stem cell signature in cancer. Nat. Genet. 39, 157–158.

Wilmut, I., Schnieke, A.E., Mcwhir, J., Kind, A.J., Campbell, K.H. 1997. Viable offspring derived from fetal and adult mammalian cells. Nature 385, 810–813.

Wu, H., D'Alessio, A.C., Ito, S., Xia, K., Wang, Z., Cui, K., *et al.*, 2011. Dual functions of Tet1 in transcriptional regulation in mouse embryonic stem cells. Nature 473, 389–393.

Wu, X., Johansen, J.V., Helin, K. 2013. Fbxl10/Kdm2b recruits polycomb repressive complex 1 to CpG islands and regulates H2A ubiquitylation. Mol. Cell 49, 1134–1146.

Xie, R., Everett, L.J., Lim, H.W., Patel, N.A., Schug, J., Kroon, E., *et al.*, 2013a. Dynamic chromatin remodeling mediated by polycomb proteins orchestrates pancreatic differentiation of human embryonic stem cells. Cell Stem Cell 12, 224–237.

Xie, W., Schultz, M.D., Lister, R., Hou, Z., Rajagopal, N., Ray, P., *et al.*, 2013b. Epigenomic analysis of multilineage differentiation of human embryonic stem cells. Cell 153, 1134–1148.

Yildirim, O., Li, R., Hung, J.H., Chen, P.B., Dong, X., Ee, L.S., *et al.*, 2011. Mbd3/NURD complex regulates expression of 5-hydroxymethylcytosine marked genes in embryonic stem cells. Cell 147, 1498–1510.

Zaret, K.S., Carroll, J.S. 2011. Pioneer transcription factors: establishing competence for gene expression. Genes Dev. 25, 2227–2241.

Zhu, J., Adli, M., Zou, J.Y., Verstappen, G., Coyne, M., Zhang, X., *et al.*, 2013. Genome-wide chromatin state transitions associated with developmental and environmental cues. Cell 152, 642–654.

3 Stem Cell Niche and Microenvironment

Marcio Alvarez-Silva
Laboratory of Stem Cell and Bioengineering, Department of Cell Biology, Embryology and Genetics, Federal University of Santa Catarina, Florianópolis, Brazil

3.1 Concept of the stem cell niche

In normal adult tissues, stem cells are responsible for renewal, repair, and remodelling through their self-renewal capacity and multipotency. They govern tissue homeostasis under diverse physiological (ageing) and pathological (injury or disease) conditions throughout the lifetime of the organism. Such homeostasis is dependent on the ability of the stem cells to maintain themselves over prolonged periods of time (self-renewal) and to differentiate into multiple, but limited, cell types (multipotency). By adjusting cell divisions to meet the needs of resident tissues, stem cells enable lifelong optimization of tissue and organ formation and function (O'Brien and Bilder, 2013). The balance between self-renewal and differentiation is the basis for tissue homeostasis, as well as for its regeneration. This can be achieved by combining different extracellular components that act as regulators of these two distinct processes. These extracellular components are mainly composed of the extracellular matrix (ECM) – which functions in cell adhesion, cell–cell communication and differentiation – and of cells and growth factors (Abedin and King, 2010; Brizzi *et al.*, 2012). In 1977, it was demonstrated that many different cells, including endothelial cells and macrophages, could interact with haematopoietic stem cells (HSCs), to support their survival and differentiation (Dexter *et al.*, 1977). Other studies have shown that many other cells are able to function as a supportive stroma (stromal cells) for HSCs, including adventitial reticular cells (Weiss, 1976), osteoblasts (Calvi *et al.*, 2003), glial cells (Yamazaki *et al.*, 2011), adipocytes (Naveiras *et al.*,

Principles of Stem Cell Biology and Cancer: Future Applications and Therapeutics, First Edition.
Edited by Tarik Regad, Thomas J. Sayers and Robert C. Rees.

2009), fibroblasts and mesenchymal stem cells (MSCs) (Bianco, 2011). These cells are able to produce a broad combination of ECM and growth factors and to express cell membrane molecules that can interact with and control the self-renewal and differentiation of HSCs (Kunisaki *et al.*, 2013; Morrison and Scadden, 2014). The association of ECM, cells and growth factors organizes the bone microenvironment, which ultimately regulates HSCs. Although complex, the microenvironment influences haematopoiesis in a coordinated manner to maintain homeostasis of the organism. The ability of HSCs to sense and respond to organismal needs derives, in large part, from their intimate association with this microenvironment.

Schofield (1978) proposed the existence of a niche for HSCs that could dynamically regulate stem cell behaviour, maintaining equilibrium amongst quiescence, self-renewal and differentiation. The stem cell niche was understood as a physiological microenvironment consisting of specialized cells within fixed compartments that would act as stromal cells to HSCs and provide the growth factors and ECM required to maintain HSC properties. This niche can be viewed as specific areas of a tissue with local and specialized microenvironments consisting of soluble and surface-bound signalling factors, cell–cell contacts, stem-cell-niche support cells and the ECM, all able to maintain stem cell functions (Figure 3.1). In sum, the stem cell niche provides a protective environment that regulates proliferation, differentiation and apoptosis to control stem cell reserves. Thus, maintaining a balance between stem cell quiescence and activity is a hallmark of a functional niche (Moore and Lemischka, 2006).

Although the idea of a stem cell niche originates from studies on mammalian HSCs, a detailed description of a stem cell compartment at the cellular level was first achieved in the *Drosophila melanogaster* ovary (Xie and Spradling, 2000; Losick *et al.*, 2011). Since then, the concept of the stem cell niche has been extended to include various stem cell types, including the progenitors of mammalian gut and hair cells.

3.2 Organization of the niche

Various populations of stem cells reside in the body, undergoing continuous self-renewal throughout the organism's lifetime. The complex milieu comprising many types of cells and the ECM, as well as the signalling molecules associated with each population of stem cells, is collectively termed the 'stem cell niche'. Since the stem cell niche controls the fate of stem cells in different tissues, its cellular and molecular organization is very important.

The cellular composition of supportive stromal cells varies in different types of stem cell niches. The HSC niche contains different cell types, including osteoblasts, osteoclasts, reticular cells, vascular cells, MSCs and neuron-Schwann cells (Adams and Scadden, 2006; Wang and Wagers, 2011). The intestinal stem cell (ISC) niche is less complex, in that the epithelial monolayer comprises one type of absorptive cell (enterocyte) and four

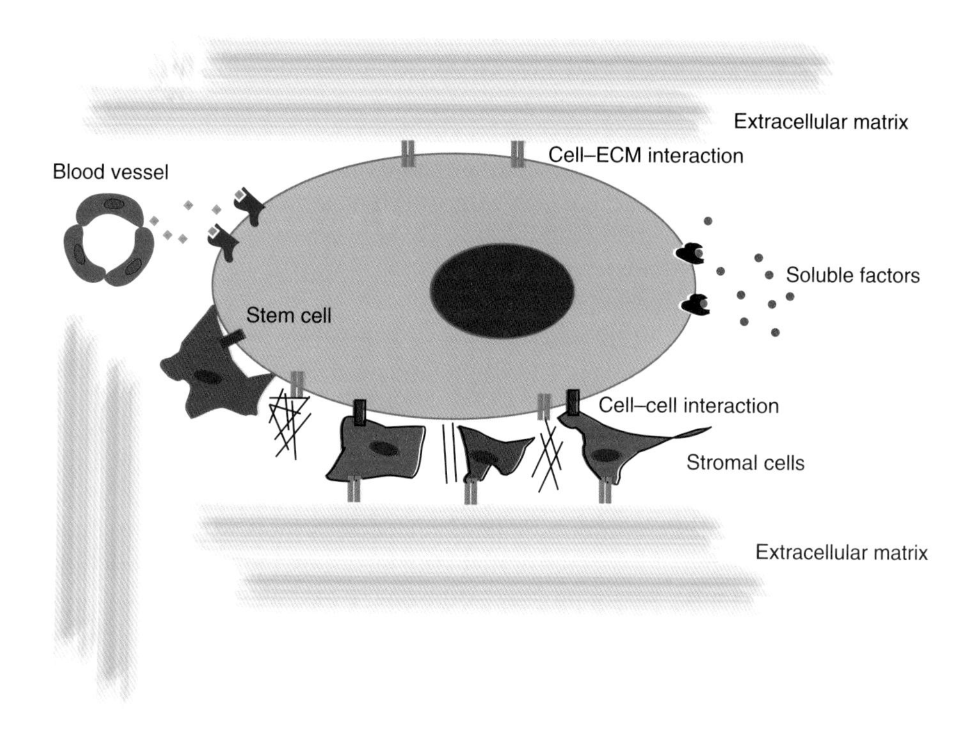

Figure 3.1 The stem cell niche is a specialized and dynamic microenvironment in which a number of inputs regulate stem cell behaviour. This schematic illustration summarizes the components of a typical stem cell niche: the stem cell itself, stromal supportive cells, soluble factors, ECM, vascular network and cell adhesion components (cell–cell and cell–ECM). Although many stem cell niche constituents are conserved in tissues, it is unlikely that every stem cell niche in every tissue includes all of the components listed. Different niches, for different types of stem cells, can combine a selection of these components for proper self-renewal and differentiation.

types of secretory cells (Goblet, Paneth, enteroendocrine and Tuft cells). The mucosal surface forms crypts, which project downward, and villi, which project upward. Stem cells have been identified in the crypt of the small intestine, which consists of the duodenum, jejunum and ileum (Ema and Suda, 2012).

The niche for muscle stem cells, comprising satellite cells, is simpler. Satellite cells (about 5% of adult muscle nuclei) are quiescent muscle precursors, forming a small population beneath the basal lamina of each myofibre (Mauro, 1961). The satellite cell niche contains a basement membrane and muscle fibres and/or endothelial cells in close proximity, which can function as supporting stromal cells (Dhawan and Rando, 2005; Christov *et al.*, 2007).

ECM components make up the constitutive part of the niche. The ECM is a complex assembly of many proteins and polysaccharides, forming an elaborate meshwork in a three-dimensional organization within tissues (Hay, 1991). Accordingly, the diversity of ECM components includes both large secreted glycoproteins and small secreted protein factors, either diffusing or associated with the ECM or cell membrane. In general, ECM proteins are highly asymmetric in shape, and when they are combined, a functional network results (Hay, 1991; Mecham, 2011).

The ECM comprises several secreted proteins, including collagens, fibronectin, elastin and fibrillins. Although the ECM is formed by a large variety of proteins and polysaccharides with different structures and functions, some common features are evident. Many ECM proteins are very large, their size often contributing to extensive glycosylation. In particular, proteoglycans (PGs) contain long, charged glycosaminoglycan (GAG) chains covalently attached to serines or threonines of the core protein. Some GAG chains are also found unconnected to a protein, such as hyaluronan, a major non-proteinaceous GAG component of ECM. Hydration of these carbohydrate-rich components exerts a swelling pressure against the surrounding fibrous network, providing tissue turgidity and compressibility and facilitating molecular transport (Hubmacher and Apte, 2013).

All ECM molecules are multidomain elements, in which different or equal domains are arranged in a specific domain organization. Domains are defined as homologous units, the homology following from sugar or amino acid sequence comparisons. Even homologous domains may have large sequential and structural differences, which present rather different functions. The combination of different domains leads virtually all ECM components being multifunctional. Commonly, several GAG and protein domains act in a coordinated fashion. Domains also interact such that the resultant multidomain elements are able to form new functional entities. This multifunctionality and the resultant expanded shapes provide the potential for lateral interactions, favouring the formation of fibres and other supramolecular assemblies of ECM components (Mecham, 2011).

Growth factors are key components of the stem cell niche. It is well known that secreted growth factors play a key role in coordinating many biological functions, such as stem cell growth, division, differentiation, apoptosis and signalling. Some signals mediate communication between direct neighbours (juxtacrine) or over several cell diameters (paracrine), whereas others act on distant tissues or organs (endocrine). Many individual growth factors are themselves pleiotropic, exerting multiple actions in different cell types, and redundant where many different growth factors have overlapping actions. Growth factor pleiotropy and redundancy can be respectively explained, at least partially, by the ability to signal via more than one type of receptor complex and by the sharing of an individual receptor component (Ozaki and Leonard, 2002; Sizonenko *et al.*, 2007).

Studies in flies indicate that supportive stromal cells secrete growth factors, such as paracrine signals, that are required for maintenance of stem cell identity and for specification of stem cell self-renewal (Xie and Spradling, 2000; Gonzalez-Reyes, 2003; Losick *et al.*, 2011). Extracellular signals, such as Notch, Wnt and Sonic hedgehog (Shh), have been associated with self-renewal and maintenance of HSCs (Maillard *et al.*, 2003; Reya and Clevers, 2005). However, rather than being expressed by surrounding stromal cells in bone marrow (paracrine signalling), Wnt may be secreted from the HSCs themselves and thus might act through autocrine signalling to control stem cell fate (Van Den Berg *et al.*, 1998; Malhotra and Kincade, 2009).

Extracellular growth factors signal from both soluble growth factors and ECM-bound factors. The ECM contains growth factor-binding GAGs, heparan sulfate (HS) and chondroitin sulfate (CS). Perlecan is a ubiquitous heparan sulfate proteoglycan (HSPG) found in basement membranes and cartilage that binds multiple regulatory factors, including fibroblast growth factor (FGF), vascular endothelial growth factor (VEGF), platelet-derived growth factor (PDGF) and interleukin 2 (IL-2), to concentrate them and form morphogen gradients. Perlecan knockout mice have revealed that cartilage perlecan is essential for vascularization of the perichondrium. Cartilage perlecan promotes the activation of VEGF/vascular endothelial growth factor receptor (VEGFR) by binding to the VEGFR of endothelial cells (Yayon *et al.*, 1991; Ishijima *et al.*, 2012), providing gradients of soluble factors that influence cell growth and the differentiation, motility and viability of stem cells.

Syndecans are type I transmembrane cell-surface HSPGs. Four syndecan family members occur in vertebrates: syndecan-1, -2, -3 and -4. Syndecans are expressed at up to 1 million copies per cell in every nucleated cell of the organism. Most cell types, with the exception of erythrocytes, express at least one syndecan family member, and a few can express all four. Syndecan-1 is expressed in epithelial cells and plasma cells; syndecan-2 is mainly expressed in fibroblasts, endothelial cells, neurons and smooth-muscle cells; syndecan-3 is primarily found in neuronal tissues, but it is also important for chondrocyte proliferation; while syndecan-4 is nearly ubiquitous (Woods *et al.*, 1998; Couchman, 2003; Choi *et al.*, 2011). Syndecans are expressed by stem cells and their supportive stroma, as well as cancer cells (Chen *et al.*, 2004; Drzeniek *et al.*, 1997). The protein cores of syndecans consist of a highly conserved C-terminal cytoplasmic domain, a single-pass transmembrane domain and a large N-terminal extracellular domain (Bernfield *et al.*, 1999; Couchman, 2010). GAG chains are present on syndecan ectodomains, which carry up to five GAG chains. Syndecan-1s from different tissues display different GAG types, comprising HS and CS of varying lengths and fine structures (Sanderson and Bernfield, 1988; Bernfield *et al.*, 1992).

Cell-surface HSPGs bind extracellular proteins and form signalling complexes with receptors for growth factors (Yayon *et al.*, 1991; Bernfield *et al.*, 1999). This ligand binding is a means of regulating the occupancy and response of the specific receptor (Harmer, 2006). Outcomes seem to depend on whether the ligand is soluble (i.e. a growth factor) or insoluble (i.e. a cell or ECM component), whether the ligand also interacts with a signalling receptor and whether the ligand binds to the HS chains on HSPGs or their core proteins. HSPG ectodomains are shed from the cell surface, resulting in soluble PGs that, presumably, retain all the binding properties of their parent molecules. Shedding can rapidly reduce HSPGs at the cell surface, thereby enabling the shed ectodomains to compete for ligands with their cell-surface counterparts. Therefore, shedding provides a means of regulating all HSPG–ligand interactions. Because the soluble ectodomains can interact with any extracellular molecule, they can have activities other than those shown by the cell-surface PGs (Gallagher, 2001; Harmer, 2006). It has been proposed that HS binds soluble protein ligands from three-dimensional space, converting them into a two-dimensional array that enhances further molecular encounters (Lander, 1998). These proteins might also be protected from degradation by binding to GAG, preventing oligomerization and/or conformational changes that facilitate further interactions. One example is the interaction between FGF2 and HS that is required for binding of FGF2 to its high-affinity tyrosine-kinase receptor (Yayon *et al.*, 1991). A ternary complex of growth factor, HS and cell-surface receptor seems to maximize signalling potential (Couchman, 2003). HSPGs are considered the most common and widely acting low-affinity receptors on the cell surface, and they play a central role in the reception and modulation of a wide range of growth factors (Tamm *et al.*, 2012).

Growth factors also bind to specific domains of ECM proteins. For example, fibronectin and vitronectin bind to hepatocyte growth factor (HGF) and modulate its biological activity (Rahman *et al.*, 2005). Fibronectin also binds to VEGF, enhancing specific cellular responses to it (Wijelath *et al.*, 2006). Finally, fibronectin has been found to bind platelet-derived growth factor-BB (PDGF-BB), a key factor for the survival of mesenchymal cells (Lin *et al.*, 2011). Type IV collagens bind to bone morphogenetic protein (BMP) and regulate its activity, enhancing signalling in target cells (Wang *et al.*, 2008). The PG agrin binds BMP2, BMP4 and transforming growth factor-β1 (TGFβ1) with relatively high affinity. This inhibits the activity of BMP2 and BMP4, but enhances the activity of TGFβ1 (Banyai *et al.*, 2010).

The biological activity of many growth factors can be intrinsically related to their interaction with ECM elements of the niche. The degree of regulation of growth factor activity is dependent on adequate levels of ECM components, which can vary in different tissues. Such activity can promote niche specificity in certain tissues, such as bone marrow and brain, as well as support specific SC differentiations, including, for example, into HSCs and neuronal stem cells.

Two essential mechanisms are closely related during embryogenesis: patterning and growth. Since the ECM can act as a reservoir of growth factors (e.g. VEGF, Wnt, BMP and FGF), such factors form gradients that control pattern formation during developmental processes. ECM binding of these factors can influence these gradients. Morphogens (secreted signalling molecules, including growth factors, that regulate the size, shape and patterning of animal tissues and organs) are conserved molecules. It seems reasonable that morphogen gradients incorporate ECM binding as part of their regulation (Wartlick *et al.*, 2009; Yan and Lin, 2009). Morphogen binding to molecules in the extracellular space affects signal movement. The diffusion of a particle that is interacting with binding partners in this manner is referred to as ‘effective diffusion’ (Crank, 1979). Interactions with binding partners can modify ligand dispersal and activity in at least four ways:

(i) altering the mobility/diffusivity of a signal, (ii) concentrating ligand at the surface of a cell, (iii) promoting or hindering ligand–receptor interactions, and (iv) influencing the extracellular stability of a ligand (Müller and Schier, 2011). Moreover, the niche can generate high or low local gradients of growth factors, in which the ECM molecules elicit specific biological responses by the juxtaposition of growth factors and their cell-surface receptors in a confined spatial environment, resulting in short-range paracrine or autocrine growth factor signalling. This can lead to different stem cell responses, depending on the local level of growth factor.

3.3 Relationship between niche and stem cells

Morphogens are required for morphogenesis, as the term itself indicates. Therefore, it is expected that adult tissue regeneration may also be controlled by morphogens. Classical morphogens are expressed in the adult organism and play a central role in the regeneration of tissues. BMP2 associated with ECM is effective for differentiation of stem cells to osteoblasts. This controls osteoblast-specific gene-encoding proteins, such as osteocalcin, that are selectively expressed in bone (Seib *et al.*, 2010), a central issue in the repair of bone fractures. Wnt and Shh signalling are involved in liver regeneration after partial hepatectomy (Omenetti and Diehl, 2008). Extracellular ligands are a critical component of the regulatory niche necessary for neural stem cell (NSC) maintenance of self-renewal, differentiation, cell adhesion and migration. Some of the many extracellular factors implicated in cell specification and signalling include FGF, epidermal growth factor (EGF), CXCL12 (stromal cell-derived factor-1, SDF-1), Shh, BMP, and Notch (Wade *et al.*, 2014).

Neurogenesis, the process of generating functionally integrated neurons from stem cells, was once believed to occur only during embryonic stages in the mammalian central nervous system (CNS). However, new neurons are now known to be added in distinct regions of the mammalian CNS system in the adult brain, suggesting a degree of inherent plasticity (Lie *et al.*, 2004). In most mammals, active neurogenesis occurs throughout life in the subventricular zone (SVZ) of the lateral ventricle and in the subgranular zone (SGZ) of the dentate gyrus in the hippocampus. Neurogenesis outside these two regions appears to be extremely limited or nonexistent in the intact adult mammalian CNS. Following pathological stimulation, such as by brain insults, adult neurogenesis appears to occur in regions otherwise considered to be nonneurogenic (Gage, 2000; Lie *et al.*, 2004). The niche plays an extremely crucial role in orchestrating the complex cascade of events involving proliferation and fate specification of NSCs to migration, nerve guidance, neuronal maturation and synaptic integration of newborn neurons in the adult CNS environment. The highly specialized architecture within this niche implicates both cell–cell interactions and soluble factors as important regulators of NSC behaviour. Shh has been shown to regulate both SVZ and SGZ NSCs. Overexpression of Shh near the dentate gyrus increases proliferation and neurogenesis of SGZ cells and maintains proliferation of adult hippocampal neuronal progenitors (Lai *et al.*, 2003). Interestingly, in the adult SVZ, both BMP and Shh signalling are intermixed within the same region. BMP signalling during development promotes astrocyte differentiation of SVZ precursors at the expense of oligodendrogliogenesis and neurogenesis. Adult SVZ cells themselves produce BMPs and their receptors (Alvarez-Buylla and Lim, 2004). EGF, FGF and PDGF are important mitogenic regulators for multipotent stem cells in the SVZ (Gonzalez-Perez and Alvarez-Buylla, 2011). ECM structures termed fractones, which resemble basement membranes with components including laminin, collagen IV, nidogen and HSPG, have been identified in neurogenic niches located in the SVZ of the human brain. Binding of FGF-2 next to these fractones stimulates the proliferation of stem cells (Kerever *et al.*, 2007).

Within the bone marrow, HSCs reside at, or near, the endosteum. The endosteal surface is covered by a protective layer of bone-lining cells that can differentiate into bone-forming osteoblasts. Endosteal cells secrete regulatory factors that promote HSC maintenance (Li and Xie, 2005; Adams and Scadden, 2006). This generates a complex three-dimensional stem cell niche comprising many cell types, growth factors and ECM proteins, including hyaluronic acid, fibronectin, collagen types I, III and IV, laminin, thrompospondin, hemonectin and PGs (Keating and Gordon, 1988; Nilsson *et al.*, 2003; Sagar *et al.*, 2006; Can, 2008). The homing, proliferation and differentiation of HSCs after grafting is strictly dependent on adequate endosteal niche composition, with the expression of ECM elements and growth factors, such as SDF-1, steel factor (SLF), interleukin 11 (IL-11), interleukin 6 (IL-6), thrombopoietin (TPO), granulocyte-macrophage colony-stimulating factor (GM-CSF), granulocyte colony-stimulating factor (G-CSF) and macrophage colony-stimulating factor (M-CSF) (Marquez-Curtis *et al.*, 2011; Sharma *et al.*, 2011; Suárez-Álvarez *et al.*, 2012; Khurana *et al.*, 2013). The elements of the niche are essential for reconstitution of haematopoiesis in clinical haematopoietic regeneration in transplantation by HSC adhesion and differentiation.

3.4 Instructive niche effect

Stem cell behaviour is regulated by coordinated environmental signals and intrinsic programmes. Environmental signals are supported by the niche. Currently, proposed niche functions include regulation of stem cell number, cell cycle, fate determination and motility. For example, stem cell number is thought to maintain niche constancy under physiological conditions. The niche provides positive and negative factors controlling proliferation, and the maintenance of stem cell quiescence is a common niche feature. The complex nature of the stem cell niche allows the formation of a distinct and specialized microenvironment for different stem cell types in the same organism. For most stem cell types, the simultaneous activation of several pathways is needed for continuous stem cell self-renewal. For example, the FGF, BDNF and Shh signalling pathways are needed for NSC self-renewal (Zhao *et al.*, 2008). Since specific signals or combinations of signals are needed by different niches, stem cells must stay inside the niche in order to maintain self-renewal. The most consistent method of accomplishing this is to anchor stem cells in their niche using adhesion molecules. Thus, it can be said that niches (i) create a signalling microenvironment that regulates stemness (the capacity for self-renewal and development into more differentiated cell types) and (ii) use adhesive molecules to maintain stem cells within signal-receiving range.

Cadherins, whose main function is cell–cell adhesion, play an important role in cell signalling during critical processes, such as cell migration, gene regulation through catenins (especially the β-catenin/Wnt signalling pathway) and epithelial–mesenchymal transition (EMT). EMT has biological relevance during early embryonic development and later organogenesis, and it can also be activated during wound healing in fibrotic or cancer tissues (Hay, 1995). Cadherins are transmembrane proteins and promote cell adhesion, especially epithelial cadherin (E-cadherin). The family of catenins includes cytoskeletal proteins (α-, β-, γ-, δ-catenins) that are important to the formation of a junction between cells, through their ability to link the actin filaments of the cytoskeleton to cadherins.

Integrins are a large family of cell-surface receptors that can bind ECM components, soluble extracellular ligands and other membrane-bound cell-surface molecules (Prowse *et al.*, 2011). Integrins and other adhesion molecules are involved in the regulation of cell–cell interactions and other cellular events, such as cell signalling and cell polarity. Integrins are heterodimers of noncovalently associated α- and β-subunits. In vertebrates, there are 18 α- and 8 β-subunits, which can assemble into 24 different receptors with different binding properties and different tissue distributions (Hynes, 2002; Campbell and Humphries, 2011). Many stem cell types express integrin molecules and directly contact the ECM or the ECM-rich basal membrane of the niche, such as collagen, fibronectin, laminin and vitronectin, as well as members of the SIBLING family (Small Integrin Binding Ligand, N-Linked Glycoproteins; e.g. osteopontin and bone sialoprotein), by their extracellular domain. Integrins can also associate with receptors on the surfaces of other cells through direct interaction with membrane-bound proteins, such as vascular or intracellular cell adhesion molecules (VCAMs and ICAMs) (Calderwood, 2004; Humphries *et al.*, 2006).

In the SVZ of mouse brain, stem cells attach to the capillary endothelial basement membrane by the laminin receptor α6β1 integrin (Shen *et al.*, 2008). Human stem cells with neurogenic potential express high levels of α6β1 integrin chains and can be used as a marker to isolate SC-enriched populations (Hall *et al.*, 2006). Stem cells in the mouse testis also express high levels of α6β1 integrin (Shinohara *et al.*, 1999). HSCs express high levels of α4, α6, α7, α9 and β1 integrins, which are fundamental for HSC niche homing (Potocnik *et al.*, 2000; Grassinger *et al.*, 2009; Schreiber *et al.*, 2009). In the skin, stem cells attach to the basal membrane via α6, β1 and β4 integrins (Watt, 2002), while in skeletal muscle, α7β1 integrin is localized on the side of satellite cells facing the basal membrane component of the stem cell niche (Kuang *et al.*, 2008).

The ECM exerts control over cells through interaction with the integrins that mediate mechanical and chemical signals. The signals regulate the activities of cytoplasmic kinases, growth factor receptors and ion channels and control the organization of the intracellular actin cytoskeleton. This regulates cell migration, cell survival and growth (Giancotti and Ruoslahti, 1999). Integrins are bidirectional signalling receptors that are involved in outside-in and inside-out signalling. Outside-in signalling of integrin receptors through their interactions with the ECM generates clustering of integrin heterodimers called focal adhesion sites. Inside-out signalling mainly acts to bring the integrin into active conformation (Barczyk *et al.*, 2010).

Interaction of stem cells with niche elements can alter gene expression. Binding with extracellular ligands can induce conformational changes in integrins, which can result in the separation of the α- and β-subunits. This changes the cytoplasmic tails, which may contribute to outside-in signalling by favouring recruitment of cytoplasmic proteins, since complex conformational rearrangements governing affinity for ECM proteins will be directed to the cytoplasmic tail associated with cytoskeletal proteins (Cary and Guan, 1999; Arnaout *et al.*, 2007). This can lead to specific tyrosine phosphorylation cascade of a limited number of protein substrates, which will then participate in regulating cytoskeletal organization and gene expression. Osf2 (also called Cbfa1/AML3/PEBP2αA), an osteoblast-specific transcription factor, is a key regulator of the osteoblast phenotype and is necessary for osteoblast-specific expression of the osteocalcin gene. It has been shown that the α2 integrin associated with collagen increases binding of Osf2 to DNA, resulting in activation of the osteocalcin promoter, followed by osteoblast differentiation (Xiao *et al.*, 1998). Interaction of stem cells with niche elements can alter gene expression.

The human salivary gland (HSG) epithelial cell line is an undifferentiated population with a phenotype similar to intercalated duct. The adhesion of HSG cells to fibronectin and collagen rapidly upregulates 32 different genes (Lafrenie and Yamada, 1998), indicating the role of niche–integrin interaction in stem cell regulation and fate. The varied integrin expression of stem cells of different tissue origins, as well as integrin interaction with niche elements, results in integrin-mediated signalling mechanisms that can alter intracellular phosphorylation of target proteins. This results in altered gene expression, which regulates cell motility, as well as proliferation and differentiation.

Stem cells can respond to specific microenvironments and differentiate to a phenotype different from that of their origin. The induction of genes that regulate stem cell differentiation seems to be regulated by specific niche conditions. When placenta-derived MSCs interact with the brain microenvironment, they differentiate to the neural phenotype (Martini *et al.*, 2013). Neurite outgrowth is dependent on contact between MSCs and the astrocyte cell membrane in the neural niche. This reveals the instructive effect of the niche, perhaps sequestering neurogenic growth factors through PGs shed from the cell surface or deposited from ECM (Kerever *et al.*, 2007). Cell surface-shed or ECM-deposited PGs can exert two different biological effects: activation of growth factors or deactivation of biological activity (Alvarez-Silva and Borojevic, 1996; Delehedde *et al.*, 2001; Simons and Horowitz, 2001). This balance in the activity of growth factors may influence the differentiation of local stem cells, resulting in changes in their phenotype. Neuritogenesis is dependent on multiple regulatory molecules, including hormones, growth factors, gangliosides and extracellular molecules, acting as both positive and negative signals (Skaper, 2005; Trentin, 2006). It is possible that the neurogenic growth factors associated with cell membrane molecules and ECM elements are able to properly control neuritogenesis in the course of MSC differentiation. The different types of stem cells are complex systems that often respond differently in different cells, even to the same stimuli.

3.5 Cancer stem cell niche

NSCs must possess two qualities to perform their natural function: self-renewal and differentiation. According to the cancer stem cell (CSC) theory, a defined subset of cancer cells has the exclusive properties of stem cells, driving the growth and spread of the tumour. These CSCs can give rise to cancer cell progeny that are more differentiated and are destined to stop proliferating or die, because they have limited or no ability to undergo further mitotic divisions (Lobo *et al.*, 2007). Thus, the CSC theory considers some elements of the cellular hierarchy seen in normal tissues to be retained in many tumours. The concept of CSCs has been demonstrated in several human cancers, including leukaemia, brain and thyroid tumours, breast cancer, prostate cancer, lung cancer, pancreatic cancer and colon cancer (Lin, 2011; Fazilaty *et al.*, 2013; Yi *et al.*, 2013). Although CSCs represent a minority of tumour masses that retain self-renewal capacity, they possess the aggressive characteristics needed for metastatic development, including motility, invasiveness and apoptotic resistance (Pang *et al.*, 2010).

Three key observations classically define the existence of a CSC population. First, only a few cancer cells within each tumour have tumourigenic potential when transplanted into nonobese diabetic severe combined-immunodeficient (NOD/SCID) mice. Second, tumourigenic cancer cells are characterized by a distinctive profile of surface markers and can be differentially and reproducibly isolated from nontumourigenic surface markers by means of flow cytometry or other immunoselection procedures. Third, tumours grown from tumourigenic cells contain mixed populations of tumourigenic and non-tumourigenic cancer cells, thus recreating the full phenotypic heterogeneity of the parent tumour (Dalerba *et al.*, 2007).

As discussed in Section 3.2, the fate of stem cells is controlled by a particular microenvironment known as the stem cell niche. It has been reported that CSCs also depend on a similar niche, called the 'CSC niche', which controls their differentiation and proliferation (Li and Neaves, 2006; Visvader and Lindeman, 2008). At least two possibilities for generating CSC niches exist: either they are created as nascent domains by tumour cells or they adopt existing tissue-specific stem cell niches (LaBarge, 2010).

The evidence that CSCs are effectively located in a well-defined niche within tumours, close to NSCs, was observed in some brain tumours (Calabrese *et al.*, 2007). NSCs usually reside in perivascular niches located mainly in the hippocampus and SVZ, where rates of cell proliferation are quite low. Starting from these observations on NSCs, several studies have analysed whether the same situation is found in diverse brain tumours. The brain CSCs (Nestin$^+$/CD133$^+$) were found to be strictly associated with perivascular endothelial cells in the tumours, enabling the maintenance of self-renewal and thus affording protection from radiation damage. It was demonstrated that glioblastoma CSCs (CD133$^+$/CD144$^-$) could differentiate into endothelial cells and participate in blood vessel formation. Glioblastoma CSCs differentiated into endothelial progenitors that gave rise to tumour and endothelial cells *in vivo* (Ricci-Vitiani *et al.*, 2010; Wang *et al.*, 2010). When CD133$^+$/CD144$^-$ cells were killed, tumour growth was greatly suppressed, indicating that such CSCs directly contribute to tumour neoangiogenesis, at least in glioblastoma, regulating the generation of cellular elements of the niche.

It must be considered that regulatory factors produced by endothelial cells can control glioma CSC biology. Molecular interaction with the perivascular niche can control glioma CSC maintenance. Notch signalling appears to play a critical role in this process. Glioma CSCs (Nestin$^+$/CD133$^+$) express Notch1 and Notch2 receptors and show elevated levels of activity, as evidenced by upregulated expression of the Notch target gene *Hes5*. Endothelial cells, on the other hand, express the membrane-bound Notch ligands Delta-like 4 (Dll4) and Jagged-1, and knockdown of these ligands in brain microvascular endothelial cells reduces tumour growth (Zhu *et al.*, 2011). Glioma CSC can stimulate endothelial cells through the production of VEGF (Bao *et al.*, 2006), and the VEGF signalling on endothelial cells can increase *Notch* and *Dll4* gene expression (Bao *et al.*, 2006). Extracellular signals from the niche are crucial for the maintenance of tumour bulk, since the differentiation of CSCs into cancer stromal elements seems to be dependent on niche–CSC interaction.

The interaction of leukaemia stem cells (LSCs) with the marrow niche for their malignant self-renewal and quiescence has been described. Like normal HSCs, LSCs are enriched in CD34$^+$/CD38$^-$ cells (Bonnet and Dick, 1997; Hope *et al.*, 2004). Perturbing the adhesion between LSCs and the marrow niche might therefore 'mobilize' LSCs from their protective environment (Lutz *et al.*, 2013). Since LSCs possess prerequisites for interaction with the bone marrow niche, targeting this association might be an effective therapeutic approach to potentially inhibiting their proliferation or stimulating their apoptosis. *In vivo* blocking of the binding of CD44 to the niche alters LSC fate by transplanting human acute myeloid leukaemia (AML) LSCs into NOD-SCID mice and thus selectively eradicating AML, indicating that LSCs are niche-dependent (Jin *et al.*, 2006). The chemokine stromal cell-derived factor-1 (SDF-1/CXCL12) and its receptor, CXCR4, are involved in the interaction of HSCs with the bone marrow niche. Higher expression of CXCR4 in AML LSCs increases the homing of neoplastic cells, defining, in turn, an unfavourable prognosis of AML (Rombouts *et al.*, 2004). The expression of CXCR4 on LSCs, as well as in many CSCs, implies that the CXCL12/CXCR4 axis may play a critical role in directing the migration/metastasis of CXCR4$^+$ tumour cells to organs that express CXCL12, such as lymph nodes, lungs, liver and bones (Teicher and Fricker, 2010). Several CXCR4$^+$ cancer cells metastasize to the bones and lymph nodes in a CXCL12-dependent manner, and the bone marrow in particular can provide a protective environment for tumour cells (Meads *et al.*, 2008). These data suggest that interventions targeting the CXCL12/CXCR4 axis might significantly perturb CSC trafficking, retention and survival, and could be important therapeutic tools in clinical trials.

The CSC hypothesis holds that the CSC niche in tumours is capable of unlimited self-renewal and that eradication of these cells will ultimately halt neoplastic expansion. Consequently, more differentiated cells have limited mitogenic capacity and will not contribute to long-term tumour growth. Niche elements, as discussed in this review, can represent central counterparts for tumour growth, since their interaction with CSCs seems to be essential for proper CSC self-renewal.

3.6 Conclusions

The stem cell, whether normal or cancerous, is intrinsically associated with its niche, the strong interaction between the two resulting in the niche's control over stem cell fate. Therefore, understanding the complex interactions between CSCs and their niche microenvironments will contribute to the development of effective new treatment strategies that eliminate both the bulk of tumour cells and CSCs. The concept that the niche microenvironment might be a therapeutic target is a consistent alternative to the direct targeting of the tumour lesion.

Therapies that target CSCs may have unique properties compared to those that target the bulk of a tumour. Assuming that CSCs represent only a small proportion of an entire tumour, eradicating them might have little impact on the size of the tumour. Over time, however, the tumour can be expected to exhaust itself and dissipate, because it will have lost the capacity for long-term self-renewal (Yang and Wechsler-Reya, 2007). Therefore, successful therapeutic strategies will target the functions that all CSC niches have in common: mediating self-renewal and maintaining an undifferentiated state and CSC activity.

From a clinical perspective, it remains to be established whether such therapies are feasible. Antibodies against the fibronectin receptor VLA-4 ($\alpha 4\beta 1$ integrin) and disruption of VEGFR1$^+$ cellular cluster formation could block the metastasis of well-established tumours (Kaplan *et al.*, 2005). Interestingly, high VLA-4 expression on leukaemia cells reduces their chemosensitivity through interaction with fibronectin in the bone marrow, resulting in a poor induction of remission and minimal residual disease (MRD) in the bone marrow, ultimately leading to recurrence and short-term survival (Matsunaga *et al.*, 2003). Thus, the combination of CSC targeting with the use of conventional agents against tumour mass should be considered. Indeed, combinations of antiangiogenic drugs and conventional chemotherapies have proven to be more effective than either mode of therapy alone (Tozer *et al.*, 2005), particularly because they can prevent metastasis by disrupting the interaction between CSCs and their niche.

The theory that tumours depend for their long-term growth and propagation on the interaction between a population of CSCs and their niche has profound implications for our understanding and treatment of cancer. The latest drugs targeting CSCs or their microenvironments are expected to minimize complications and, at the same time, improve patient quality of life. Thus, from both scientific and clinical perspectives, the biology of stem cells and their respective niches is one of the most promising fields of research for understanding and targeting cancer.

References

Abedin, M., King, N. 2010. Diverse evolutionary paths to cell adhesion. Trend. Cell Biol. 20, 734–742.

Adams, G.B., Scadden, D.T. 2006. The hematopoietic stem cell in its place. Nat. Immunol. 7, 333–337.

Alvarez-Buylla, A., Lim, D.A. 2004. For the long run: maintaining germinal niches in the adult brain. Neuron. 41, 683–686.

Alvarez-Silva, M., Borojevic, R. 1996. GM-CSF and IL-3 activities in schistosomal liver granulomas are controlled by stroma-associated heparan sulfate proteoglycans. J. Leuk. Biol. 59, 435–441.

Arnaout, M.A., Goodman, S.L., Xiong, J.-P. 2007. Structure and mechanics of integrin-based cell adhesion. Curr. Opin. Cell Biol. 19, 495–507.

Banyai, L., Sonderegger, P., Patthy, L. 2010. Agrin binds BMP2, BMP4 and TGFbeta1. PLoS One 5, 0010758.

Bao, S., Wu, Q., Sathornsumetee, S., Hao, Y., Li, Z., Hjelmeland, A.B., Shi, Q., *et al.*, 2006. Stem cell-like glioma cells promote tumor angiogenesis through vascular endothelial growth factor. Cancer Res. 66, 7843–7848.

Barczyk, M., Carracedo, S., Gullberg, D. 2010. Integrins. Cell Tiss. Res. 339, 269–280.

Bernfield, M., Kokenyesi, R., Kato, M., Hinkes, M.T., Spring, J., Gallo, R.L., Lose, E.J. 1992. Biology of the syndecans: a family of transmembrane heparan sulfate proteoglycans. Ann. Rev. Cell Biol. 8, 365–393.

Bernfield, M., Gotte, M., Park, P.W., Reizes, O., Fitzgerald, M.L., Lincecum, J., Zako, M. 1999. Functions of cell surface heparan sulfate proteoglycans. Ann. Rev. Biochem. 68, 729–777.

Bianco, P. 2011. Bone and the hematopoietic niche: a tale of two stem cells. Blood 117, 5281–5288.

Bonnet, D., Dick, J.E. 1997. Human acute myeloid leukemia is organized as a hierarchy that originates from a primitive hematopoietic cell. Nat. Med. 3, 730–737.

Brizzi, M.F., Tarone, G., Defilippi, P. 2012. Extracellular matrix, integrins, and growth factors as tailors of the stem cell niche. Curr. Opin. Cell Biol. 24, 645–651.

Calabrese, C., Poppleton, H., Kocak, M., Hogg, T.L., Fuller, C., Hamner, B., *et al.*, 2007. A perivascular niche for brain tumor stem cells. Cancer Cell 11, 69–82.

Calderwood, D.A. 2004. Integrin activation. J. Cell Sci. 117, 657–666.

Calvi, L.M., Adams, G.B., Weibrecht, K.W., Weber, J.M., Olson, D.P., Knight, M.C., *et al.*, 2003. Osteoblastic cells regulate the haematopoietic stem cell niche. Nature 425, 841–846.

Campbell, I.D., Humphries, M.J. 2011. Integrin structure, activation, and interactions. Cold Spring Harb. Perspect. Biol., 3.

Can, A. 2008. Haematopoietic stem cells niches: interrelations between structure and function. Transfus. Apheresis Sci. 38, 261–268.

Cary, L.A., Guan, J.L. 1999. Focal adhesion kinase in integrin-mediated signaling. Frontiers Biosci. 4, D102–D113.

Chen, E., Hermanson, S., Ekker, S.C. 2004. Syndecan-2 is essential for angiogenic sprouting during zebrafish development. Blood 103, 1710–1719.

Choi, Y., Chung, H., Jung, H., Couchman, J.R., Oh, E.-S. 2011. Syndecans as cell surface receptors: unique structure equates with functional diversity. Matrix Biol. 30, 93–99.

Christov, C., Chretien, F., Abou-Khalil, R., Bassez, G., Vallet, G., Authier, F.J., *et al.*, 2007. Muscle satellite cells and endothelial cells: close neighbors and privileged partners. Mol. Biol. Cell, 18, 1397–1409.

Couchman, J.R. 2003. Syndecans: proteoglycan regulators of cell-surface microdomains? Nat. Rev. Mol. Cell Biol. 4, 926–937.

Couchman, J.R. 2010. Transmembrane signaling proteoglycans. Ann. Rev. Cell Dev. Biol. 26, 89–114.

Crank, J. 1979. The Mathematics of Diffusion. Oxford: Clarendon Press.

Dalerba, P., Cho, R.W., Clarke, M.F. 2007. Cancer stem cells: models and concepts. Ann. Rev. Med. 58, 267–284.

Delehedde, M., Lyon, M., Sergeant, N., Rahmoune, H., Fernig, D.G. 2001. Proteoglycans: pericellular and cell surface multireceptors that integrate external stimuli in the mammary gland. J. Mamm. Gland Biol. Neoplasia. 6, 253–273.

Dexter, T.M., Allen, T.D., Lajtha, L.G. 1977. Conditions controlling the proliferation of haemopoietic stem cells in vitro. J. Cell. Physiol. 91, 335–344.

Dhawan, J., Rando, T.A. 2005. Stem cells in postnatal myogenesis: molecular mechanisms of satellite cell quiescence, activation and replenishment. Trends Cell Biol. 15, 666–673.

Drzeniek, Z., Siebertz, B., Stocker, G., Just, U., Ostertag, W., Greiling, H., Haubeck, H.D. 1997. Proteoglycan synthesis in haematopoietic cells: isolation and characterization of heparan sulphate proteoglycans expressed by the bone-marrow stromal cell line MS-5. Biochem. J. 327(Pt 2), 473–480.

Ema, H., Suda, T. 2012. Two anatomically distinct niches regulate stem cell activity. Blood 120, 2174–2181.

Fazilaty, H., Gardaneh, M., Bahrami, T., Salmaninejad, A., Behnam, B. 2013. Crosstalk between breast cancer stem cells and metastatic niche: emerging molecular metastasis pathway? Tumor Biol. 34, 2019–2030.

Gage, F.H. 2000. Mammalian neural stem cells. Science 287, 1433–1438.

Gallagher, J.T. 2001. Heparan sulfate: growth control with a restricted sequence menu. J. Clin. Invest. 108, 357–361.

Giancotti, F.G., Ruoslahti, E. 1999. Integrin signaling. Science 285, 1028–1032.

Gonzalez-Perez, O., Alvarez-Buylla, A. 2011. Oligodendrogenesis in the subventricular zone and the role of epidermal growth factor. Brain Res. Rev. 67, 147–156.

Gonzalez-Reyes, A. 2003. Stem cells, niches and cadherins: a view from Drosophila. J. Cell Sci. 116, 949–954.

Grassinger, J., Haylock, D.N., Storan, M.J., Haines, G.O., Williams, B., Whitty, G.A., *et al.*, 2009. Thrombin-cleaved osteopontin regulates hemopoietic stem and progenitor cell functions through interactions with α9β1 and α4β1 integrins. Blood 114, 49–59.

Hall, P.E., Lathia, J.D., Miller, N.G.A., Caldwell, M.A., Ffrench-Constant, C. 2006. Integrins are markers of human neural stem cells. Stem Cells 24, 2078–2084.

Harmer, N.J. 2006. Insights into the role of heparan sulphate in fibroblast growth factor signalling. Biochem. Soc. Trans. 34, 442–445.

Hay, E.D. 1991. Cell Biology of Extracellular Matrix. New York: Plenum Press.

Hay, E.D. 1995. An overview of epithelio-mesenchymal transformation. Cells Tissues Organs 154, 8–20.

Hope, K.J., Jin, L., Dick, J.E. 2004. Acute myeloid leukemia originates from a hierarchy of leukemic stem cell classes that differ in self-renewal capacity. Nat. Immunol. 5, 738–743.

Hubmacher, D., Apte, S.S. 2013. The biology of the extracellular matrix: novel insights. Curr. Opin. Rheumatol. 25, 65–70.

Humphries, J.D., Byron, A., Humphries, M.J. 2006. Integrin ligands at a glance. J. Cell Sci. 119, 3901–3903.

Hynes, R.O. 2002. Integrins: bidirectional, allosteric signaling machines. Cell 110, 673–687.

Ishijima, M., Suzuki, N., Hozumi, K., Matsunobu, T., Kosaki, K., Kaneko, H., *et al.*, 2012. Perlecan modulates VEGF signaling and is essential for vascularization in endochondral bone formation. Matrix Biol. 31, 234–245.

Jin, L., Hope, K.J., Zhai, Q., Smadja-Joffe, F., Dick, J.E. 2006. Targeting of CD44 eradicates human acute myeloid leukemic stem cells. Nat. Med. 12, 1167–1174.

Kaplan, R.N., Riba, R.D., Zacharoulis, S., Bramley, A.H., Vincent, L., Costa, C., *et al.*, 2005. VEGFR1-positive haematopoietic bone marrow progenitors initiate the pre-metastatic niche. Nature 438, 820–827.

Keating, A., Gordon, M.Y. 1988. Hierarchical organization of hematopoietic microenvironments: role of proteoglycans. Leukemia 2, 766–9.

Kerever, A., Schnack, J., Vellinga, D., Ichikawa, N., Moon, C., Arikawa-Hirasawa, E., *et al.*, 2007. Novel extracellular matrix structures in the neural stem cell niche capture the neurogenic factor fibroblast growth factor 2 from the extracellular milieu. Stem Cells 25, 2146–2157.

Khurana, S., Margamuljana, L., Joseph, C., Schouteden, S., Buckley, S.M., Verfaillie, C.M. 2013. Glypican-3-mediated inhibition of CD26 by TFPI: a novel mechanism in hematopoietic stem cell homing and maintenance. Blood 121, 2587–2595.

Kuang, S., Gillespie, M.A., Rudnicki, M.A. 2008. Niche regulation of muscle satellite cell self-renewal and differentiation. Cell Stem Cell 2, 22–31.

Kunisaki, Y., Bruns, I., Scheiermann, C., Ahmed, J., Pinho, S., Zhang, D., *et al.*, 2013. Arteriolar niches maintain haematopoietic stem cell quiescence. Nature 502, 637–643.

LaBarge, M.A. 2010. The difficulty of targeting cancer stem cell niches. Clin. Cancer Res. 16, 3121–3129.

Lafrenie, R.M., Yamada, K.M. 1998. Integrins and matrix molecules in salivary gland cell adhesion, signaling, and gene expression. Ann. NY Acad. Sci. 842, 42–48.

Lai, K., Kaspar, B.K., Gage, F.H., Schaffer, D.V. 2003. Sonic hedgehog regulates adult neural progenitor proliferation *in vitro* and *in vivo*. Nat. Neurosci. 6, 21–27.

Lander, A.D. 1998. Proteoglycans: master regulators of molecular encounter? Matrix Biol. 17, 465–472.

Li, L., Neaves, W.B. 2006. Normal stem cells and cancer stem cells: the niche matters. Cancer Res. 66, 4553–4557.

Li, L., Xie, T. 2005. Stem cell niche: structure and function. Ann. Rev. Cell Dev. Biol. 21, 605–631.

Lie, D.C., Song, H., Colamarino, S.A., Ming, G.L., Gage, F.H. 2004. Neurogenesis in the adult brain: new strategies for central nervous system diseases. Ann. Rev. Pharmacol. Toxicol. 44, 399–421.

Lin, R.Y. 2011. Thyroid cancer stem cells. Nat. Rev. Endocrinol. 7, 609–616.

Lin, F., Ren, X.D., Pan, Z., Macri, L., Zong, W.X., Tonnesen, M.G., *et al.*, 2011. Fibronectin growth factor-binding domains are required for fibroblast survival. J. Invest. Dermatol., 131, 84–98.

Lobo, N.A., Shimono, Y., Qian, D., Clarke, M.F. 2007. The biology of cancer stem cells. Ann. Rev. Cell Dev. Biol. 23, 675–699.

Losick, V.P., Morris, L.X., Fox, D.T., Spradling, A. 2011. Drosophila stem cell niches: a decade of discovery suggests a unified view of stem cell regulation. Dev. Cell 21, 159–171.

Lutz, C., Hoang, V.T., Buss, E., Ho, A.D. 2013. Identifying leukemia stem cells – is it feasible and does it matter? Cancer Lett. 338, 10–14.

Maillard, I., Adler, S.H., Pear, W.S. 2003. Notch and the immune system. Immunity 19, 781–791.

Malhotra, S., Kincade, P.W. 2009. Wnt-related molecules and signaling pathway equilibrium in hematopoiesis. Cell Stem Cell 4, 27–36.

Marquez-Curtis, L.A., Turner, A.R., Sridharan, S., Ratajczak, M.Z., Janowska-Wieczorek, A. 2011. The ins and outs of hematopoietic stem cells: studies to improve transplantation outcomes. Stem Cell Rev 7, 590–607.

Martini, M.M., Jeremias Talita Da, S., Kohler, M.C., Marostica, L.L., Trentin, A.G., Alvarez-Silva, M. 2013. Human placenta-derived mesenchymal stem cells acquire neural phenotype under the appropriate niche conditions. DNA Cell Biol. 32, 58–65.

Matsunaga, T., Takemoto, N., Sato, T., Takimoto, R., Tanaka, I., Fujimi, A., *et al.*, 2003. Interaction between leukemic-cell VLA-4 and stromal fibronectin is a decisive factor for minimal residual disease of acute myelogenous leukemia. Nat. Med. 9, 1158–1165.

Mauro, A. 1961. Satellite cell of skeletal muscle fibers. J. Biophys. Biochem. Cytol. 9, 493–495.

Meads, M.B., Hazlehurst, L.A., Dalton, W.S. 2008. The bone marrow microenvironment as a tumor sanctuary and contributor to drug resistance. Clin.Cancer Res. 14, 2519–2526.

Mecham, R.P. 2011. The Extracellular Matrix: An Overview. Berlin: Springer.

Moore, K.A., Lemischka, I.R. 2006. Stem cells and their niches. Science 311, 1880–1885.

Morrison, S.J., Scadden, D.T. 2014. The bone marrow niche for haematopoietic stem cells. Nature 505, 327–334.

Müller, P., Schier, A.F. 2011. Extracellular movement of signaling molecules. Dev.Cell 21, 145–158.

Naveiras, O., Nardi, V., Wenzel, P.L., Hauschka, P.V., Fahey, F., Daley, G.Q. 2009. Bone-marrow adipocytes as negative regulators of the haematopoietic microenvironment. Nature 460, 259–263.

Nilsson, S.K., Haylock, D.N., Johnston, H.M., Occhiodoro, T., Brown, T.J., Simmons, P.J. 2003. Hyaluronan is synthesized by primitive hemopoietic cells, participates in their lodgment at the endosteum following transplantation, and is involved in the regulation of their proliferation and differentiation *in vitro*. Blood 101, 856–862.

O'Brien, L.E., Bilder, D. 2013. Beyond the niche: tissue-level coordination of stem cell dynamics. Ann. Rev. Cell Dev. Biol. 29, 107–136.

Omenetti, A., Diehl, A.M. 2008. The adventures of sonic hedgehog in development and repair. II. Sonic hedgehog and liver development, inflammation, and cancer. Am. J. Physiol. Gastrointest. Liver Physiol. 294, 24.

Ozaki, K., Leonard, W.J. 2002. Cytokine and cytokine receptor pleiotropy and redundancy. J. Biol. Chem. 277, 29 355–29 358.

Pang, R., Law, W.L., Chu, A.C., Poon, J.T., Lam, C.S., Chow, A.K., *et al.*, 2010. A subpopulation of $CD26^+$ cancer stem cells with metastatic capacity in human colorectal cancer. Cell Stem Cell 6(6), 603–615.

Potocnik, A.J., Brakebusch, C., Fassler, R. 2000. Fetal and adult hematopoietic stem cells require beta1 integrin function for colonizing fetal liver, spleen, and bone marrow. Immunity 12, 653–663.

Prowse, A.B.J., Chong, F., Gray, P.P., Munro, T.P. 2011. Stem cell integrins: implications for ex-vivo culture and cellular therapies. Stem Cell Res. 6, 1–12.

Rahman, S., Patel, Y., Murray, J., Patel, K.V., Sumathipala, R., Sobel, M., Wijelath, E.S. 2005. Novel hepatocyte growth factor (HGF) binding domains on fibronectin and vitronectin coordinate a distinct and amplified Met-integrin induced signalling pathway in endothelial cells. BMC Cell Biol. 6, 8.

Reya, T., Clevers, H. 2005. Wnt signalling in stem cells and cancer. Nature 434, 843–850.

Ricci-Vitiani, L., Pallini, R., Biffoni, M., Todaro, M., Invernici, G., Cenci, T., *et al.*, 2010. Tumour vascularization via endothelial differentiation of glioblastoma stem-like cells. Nature 468, 824–828.

Rombouts, E.J.C., Pavic, B., Löwenberg, B., Ploemacher, R.E. 2004. Relation between CXCR-4 expression, Flt3 mutations, and unfavorable prognosis of adult acute myeloid leukemia. Blood 104, 550–557.

Sagar, B.M., Rentala, S., Gopal, P.N., Sharma, S., Mukhopadhyay, A. 2006. Fibronectin and laminin enhance engraftibility of cultured hematopoietic stem cells. Biochem. Biophys. Res. Commun. 350, 1000–1005.

Sanderson, R.D., Bernfield, M. 1988. Molecular polymorphism of a cell surface proteoglycan: distinct structures on simple and stratified epithelia. Proc. Nat. Acad. Sci. USA 85, 9562–9566.

Schofield, R. 1978. The relationship between the spleen colony-forming cell and the haemopoietic stem cell. Blood Cells 4(1–2), 7–25.

Schreiber, T.D., Steinl, C., Essl, M., Abele, H., Geiger, K., Müller, C.A., *et al.*, 2009. The integrin α9β1 on hematopoietic stem and progenitor cells: involvement in cell adhesion, proliferation and differentiation. Haematologica 94, 1493–1501.

Seib, F.P., Lanfer, B., Bornhäuser, M., Werner, C. 2010. Biological activity of extracellular matrix-associated BMP-2. J. Tissue Eng. Regen. Med. 4, 324–327.

Sharma, M., Afrin, F., Satija, N., Tripathi, R.P., Gangenahalli, G.U. 2011. Stromal-derived factor-1/CXCR4 signaling: indispensable role in homing and engraftment of hematopoietic stem cells in bone marrow. Stem Cell. Dev. 20, 933–946.

Shen, Q., Wang, Y., Kokovay, E., Lin, G., Chuang, S.-M., Goderie, S.K., *et al.*, 2008. Adult SVZ stem cells lie in a vascular niche: a quantitative analysis of niche cell–cell interactions. Cell Stem Cell 3, 289–300.

Shinohara, T., Avarbock, M.R., Brinster, R.L. 1999. β1- and α6-integrin are surface markers on mouse spermatogonial stem cells. Proc. Nat. Acad. Sci. 96, 5504–5509.

Simons, M., Horowitz, A. 2001. Syndecan-4-mediated signalling. Cell Signal 13, 855–862.

Sizonenko, S.V., Bednarek, N., Gressens, P. 2007. Growth factors and plasticity. Semin. Fetal Neonatal Med. 12, 241–249.

Skaper, S.D. 2005. Neuronal growth-promoting and inhibitory cues in neuroprotection and neuroregeneration. Ann. NY Acad. Sci. 1053, 376–385.

Suárez-Álvarez, B., López-Vázquez, A., López-Larrea, C. 2012. Mobilization and homing of hematopoietic stem cells. In: López-Larrea, C., López-Vázquez, A., Suárez-Álvarez, B. (eds.) Stem Cell Transplantation. New York: Springer.

Tamm, C., Kjellen, L., Li, J.P. 2012. Heparan sulfate biosynthesis enzymes in embryonic stem cell biology. J. Histochem. Cytochem. 60, 943–949.

Teicher, B.A., Fricker, S.P. 2010. CXCL12 (SDF-1)/CXCR4 pathway in cancer. Clin. Cancer Res. 16, 2927–2931.

Tozer, G.M., Kanthou, C., Baguley, B.C. 2005. Disrupting tumour blood vessels. Nat. Rev. Cancer 5, 423–435.

Trentin, A.G. 2006. Thyroid hormone and astrocyte morphogenesis. J. Endocrinol. 189, 189–197.

Van Den Berg, D.J., Sharma, A.K., Bruno, E., Hoffman, R. 1998. Role of members of the Wnt gene family in human hematopoiesis. Blood 92, 3189–3202.

Visvader, J.E., Lindeman, G.J. 2008. Cancer stem cells in solid tumours: accumulating evidence and unresolved questions. Nat. Rev. Cancer, 8, 755–768.

Wade, A., Mckinney, A., Phillips, J.J. 2014. Matrix regulators in neural stem cell functions. Biochim. Biophys. Acta, 1840(8), 2520–2525.

Wang, L.D., Wagers, A.J. 2011. Dynamic niches in the origination and differentiation of haematopoietic stem cells. Nat. Rev. Mol. Cell Biol. 12, 643–655.

Wang, X., Harris, R.E., Bayston, L.J., Ashe, H.L. 2008. Type IV collagens regulate BMP signalling in Drosophila. Nature 455, 72–77.

Wang, R., Chadalavada, K., Wilshire, J., Kowalik, U., Hovinga, K.E., Geber, A., *et al.*, 2010. Glioblastoma stem-like cells give rise to tumour endothelium. Nature 468, 829–833.

Wartlick, O., Kicheva, A., González-Gaitán, M. 2009. Morphogen gradient formation. Cold Spring Harb. Perspect. Biol. 1.

Watt, F.M. 2002. Role of integrins in regulating epidermal adhesion, growth and differentiation. EMBO J. 21, 3919–3926.

Weiss, L. 1976. The hematopoietic microenvironment of the bone marrow: an ultrastructural study of the stroma in rats. Anatom. Rec. 186, 161–184.

Wijelath, E.S., Rahman, S., Namekata, M., Murray, J., Nishimura, T., Mostafavi-Pour, Z., *et al.*, 2006. Heparin-II domain of fibronectin is a vascular endothelial growth factor-binding domain: enhancement of VEGF biological activity by a singular growth factor/matrix protein synergism. Circulat. Res. 99, 853–860.

Woods, A., Oh, E.-S., Couchman, J.R. 1998. Syndecan proteoglycans and cell adhesion. Matrix Biol. 17, 477–483.

Xiao, G., Wang, D., Benson, M.D., Karsenty, G., Franceschi, R.T. 1998. Role of the α2-integrin in osteoblast-specific gene expression and activation of the Osf2 transcription factor. J. Biol. Chem. 273, 32 988–32 994.

Xie, T., Spradling, A.C. 2000. A niche maintaining germ line stem cells in the drosophila ovary. Science 290, 328–330.

Yamazaki, S., Ema, H., Karlsson, G., Yamaguchi, T., Miyoshi, H., Shioda, S., *et al.*, 2011. Nonmyelinating Schwann cells maintain hematopoietic stem cell hibernation in the bone marrow niche. Cell 147, 1146–1158.

Yan, D., Lin, X. 2009. Shaping morphogen gradients by proteoglycans. Cold Spring Harb. Perspect. Biol. 1.

Yang, Z.-J., Wechsler-Reya, R.J. 2007. Hit 'em where they live: targeting the cancer stem cell niche. Cancer Cell 11, 3–5.

Yayon, A., Klagsbrun, M., Esko, J.D., Leder, P., Ornitz, D.M. 1991. Cell surface, heparin-like molecules are required for binding of basic fibroblast growth factor to its high affinity receptor. Cell 64, 841–848.

Yi, S.-Y., Hao, Y.-B., Nan, K.-J., Fan, T.-L. 2013. Cancer stem cells niche: a target for novel cancer therapeutics. Cancer Treat. Rev. 39, 290–296.

Zhao, C., Deng, W., Gage, F.H. 2008. Mechanisms and functional implications of adult neurogenesis. Cell 132, 645–660.

Zhu, T.S., Costello, M.A., Talsma, C.E., Flack, C.G., Crowley, J.G., Hamm, L.L., *et al.*, 2011. Endothelial cells create a stem cell niche in glioblastoma by providing NOTCH ligands that nurture self-renewal of cancer stem-like cells. Cancer Res. 71, 6061–6072.

4
Haematopoietic Stem Cells in Therapy

Jos Domen[1] and Jignesh Dalal[2]

[1]*Cardiac Transplant Research Laboratory, Section of Cardiac Surgery, The Children's Mercy Hospital, Kansas City, MO, USA*
[2]*Cardiac Transplant Research Laboratory, Section of Bone Marrow Transplantation, The Children's Mercy Hospital, Kansas City, MO, USA*

4.1 Introduction

4.1.1 Discovery

The discovery of haematopoietic stem cells (HSCs) followed from experiments designed to elucidate the compromised haematopoietic system in humans exposed to irradiation after the atomic bomb explosions in World War II. The first quantitative assay for HSCs, developed by Till and McCullogh, was the Colony Forming Units – Spleen (CFU-S) assay, in which visible colonies of haematopoietic cells form in an irradiated, mostly empty, mouse spleen. This provided early quantitative data (although mostly of progenitors, rather than HSCs), but also allowed for the unequivocal demonstration, using cells with chromosomal aberrations, that all the cells in a colony, independent of lineage, are derived from a single cell. Labelling and reconstitution assays developed since (Spooncer *et al.*, 1985; Spangrude *et al.*, 1988) have greatly improved our ability to study these cells. The haematopoietic system is similarly sensitive to chemotherapeutic agents used to treat cancer.

HSCs have found widespread application in the clinic following pioneering work by people such as Thomas and Mathé in the late 1950s (Thomas *et al.*, 1957; Mathé *et al.*, 1959). Mimicking the original observation that bone marrow could rescue the haematopoietic system in an otherwise lethally irradiated recipient, bone marrow transplantation has become the standard of care for cancer patients undergoing high-dose chemo and/or radiation therapy, or

Principles of Stem Cell Biology and Cancer: Future Applications and Therapeutics, First Edition.
Edited by Tarik Regad, Thomas J. Sayers and Robert C. Rees.

suffering from haematopoietic malignancies. Globally, each year more than 50 000 people receive bone marrow transplantations (Gratwohl *et al.*, 2010); approximately 21 000 of these are allogeneic transplants, while the remainder are autologous. In addition to bone marrow, mobilized peripheral blood or umbilical cord blood (UCB) is increasingly used as the source.

4.1.2 What are HSCs?

Certain tissues, such as skin, gut and, as discussed here, the haematopoietic system, have a high turnover of cells and need to be regenerated continuously. To accomplish this, the bone marrow contains cells called HSCs that are capable of regenerating all the components of the haematopoietic system. The essential property of HSCs (and other stem cells) is the ability to choose between self-renewal and differentiation into one of several lineages. Self-renewal, according to the strictest definition, means that cells have the ability to remain present as productive stem cells indefinitely. While indefinite renewal is difficult to establish experimentally, HSCs are present and function for the lifetime of an organism. Progenitor cells, on the other hand, may retain the potential for significant expansion, and may have the ability to choose from one of several cell fates, but are unable to remain present as productive cells for very long; that is, they can't self-renew. The essential cellular component in haematopoietic cell transplantation in the form of bone marrow, mobilized peripheral blood or UCB transplantation is HSCs, as they alone have the potential to engraft and produce all the haematopoietic cells needed for life. While other cells in the graft can play important roles, only HSCs can produce all of the cells needed to allow the haematopoietic system to function long-term.

4.1.3 How are HSCs obtained for clinical use?

HSC therapy has been studied in preclinical settings using HSCs from different sources and with different degrees of purity, up to highly purified, fluorescence-activated cell sorting (FACS)-sorted cells. Figure 4.1 illustrates the different types of 'stem cells' used. Bone marrow, mobilized peripheral blood or UCB can be used either as is or following bulk enrichment/depletion for a single parameter, using either magnetic beads (in the past Isolex, currently e.g. Clinimacs) or depleting antibodies. These methods can result in considerable enrichment, but contaminants remain. In the case of dangerous contaminants, such as malignant cells, this can contribute to an eventual relapse. Multiparameter enrichment, such as FACS, yields highly purified HSCs (orders of magnitude cleaner than those produced using other methods), but has rarely been used clinically, due to technical limitations. Small trials do suggest that highly purified HSCs can have a better outcome in autologous haematopoietic stem cell transplantation (HSCT) in breast cancer patients (Muller *et al.*, 2012). Multiparameter FACS purification is used routinely to study HSCs in research settings.

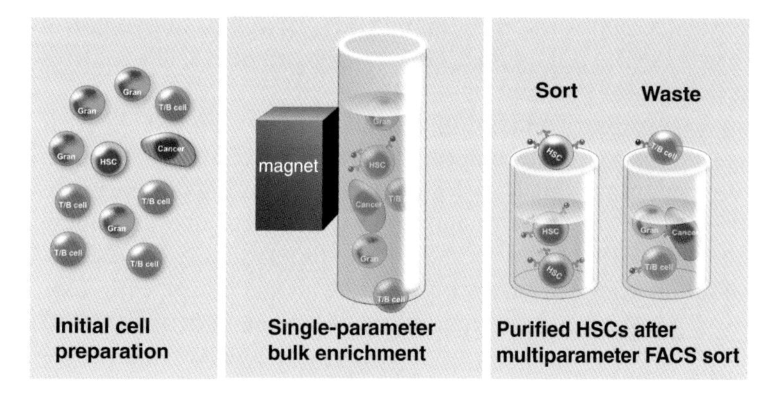

Figure 4.1 Purification of HSCs. Left panel: HSC preparations directly after harvest consist of a mixture of cells, including various types of myeloid and lymphoid cells. Other cells, including tumour cells, can also be present in the preparation. The types and amounts of cells depend on the source. Middle panel: HSCs can be enriched from the mixture using bulk enrichment for one parameter, typically CD34 expression. In this approach, in which very large numbers of cells can be processed rapidly in a closed system, antibodies with attached magnetic beads bind the HSCs and allow the external magnet to bind and eventually release them, separating them from the cells not recognized by the antibody. While this type of enrichment can be quite impressive, measurable quantities of all sorts of contaminants remain present. Clinical systems based on this approach have been in use for several decades. Right panel: Multiparameter FACS can sort cells based on the presence or absence of 10 or more markers. However, each additional marker, while improving the purity, reduces the yield. In practice, clinical sorts have used two markers, such as CD34 and CD90.

Clinically, it is important to consider the nature of the cell preparation used. While HSCs provide the long-term engraftment, other cells also play important roles. Cell components can adversely affect the transplant outcome, such as malignant cells in autologous grafts and graft-versus-host disease (GVHD)-inducing lymphocytes in allogeneic grafts. Examples of positive roles include graft-versus-leukaemia (GVL) responses and improved engraftment by allogeneic lymphocytes. Donor lymphocyte infusions (DLIs) are sometimes used to aid a failing graft.

4.2 Established clinical use of HSCs: common indications of HSCT in children

In this section, we will briefly discuss how HSCs and HSCT (see Section 4.1) are currently used in cancer therapy, focusing on paediatric use.

4.2.1 Leukaemias

4.2.1.1 Acute myeloid leukaemia Despite intensive chemotherapy, less than half of all patients with acute myeloid leukaemia (AML) will survive in the long term (Gibson *et al.*, 2005). Disease recurrence and treatment-related toxicities still account for approximately 40% of deaths in paediatric patients with *de novo* AML. Allogeneic HSCT may provide a GVL effect in such patients (Table 4.1).

Current indicators of HSCT in AML include: relapsed AML in second complete remission (CR2); relapsed AML in first complete remission (CR1) with high-risk features such as FLT3/ITD$^+$ with a high allelic ratio of >0.4 (HR FLT3/ITD$^+$); presence of monosomy 7, monosomy 5 or del5q; lack of evidence of low-risk cytogenetics; evidence of residual AML (MRD $\geq$ 0.1%) at end of induction; and lack of low-risk cytogenetics with radiographic evidence of progressive extramedullary AML at end of induction (Kelly *et al.*, 2014). Currently, low-risk AML (defined as an inversion (16)/t(16;16) or t(8,21),NPM1 mutation, Down syndrome or acute promyelocytic leukaemia (APL) in CR1) is not considered for transplantation, as survival is >60% with just chemotherapy.

4.2.1.2 Acute lymphoblastic leukaemia Acute lymphoblastic leukaemia (ALL) in children is a heterogeneous disease with different molecular and chromosomal abnormalities. Current indicators of HSCT in ALL in children include ultra-high-risk ALL in CR1, in which estimated survival with chemotherapy is <50%. The ultra-high-risk group includes: hypodiploidy with <45 chromosomes; T cell precursor ALL; very-high-risk infants with myelomonocytic leukaemia (MML) rearrangement at age < 6 months and either poor response to glucocorticoid treatment or initial leukocyte count $\geq 300 \times 10^9$/l; and poor early responders with $\geq$1% minimal residual disease at the end of 46 days of induction therapy (Pulsipher *et al.*, 2011).

Table 4.1 Common indications for allogeneic HSCT in children.

Category	Diseases
Malignant disorders	ALL in CR1 ultra-high-risk; ALL with medullary relapse within 30 months of diagnosis; multiple-relapsed extramedullary disease; relapsed T-cell ALL AML in CR1 with high-risk features; telapsed AML Primary or secondary myelodysplastic syndrome (MDS) Recurrent Hodgkin's lymphoma and NHL
Haematological disorders	Idiopathic aplastic anaemia with matched sibling or immunotherapy resistance without sibling; other bone marrow failure syndromes, including Fanconi anaemia, dyskeratosis congenita, haemoglobinopathies sickle cell disease, beta thalassemia
Immunodeficiency disorders	SCIDS; CD40 ligand Def.; Wiskott–Aldrich syndrome; chronic granulomatous disease; congenital neutropenia; familial HLH
Metabolic disorders	Hurler syndrome; Maroteaux–Lamy syndrome (MPS VI); adrenal leukodystrophy (ALD); Krabbe disease; metachromatic leucodystrophy; congenital erythropoietic porphyria (CEP); malignant osteopetrosis
Autoimmune disorders	IPEX; infantile-onset Crohn's disease

HSCT is a treatment option for relapsed ALL based on the sites of relapse and on the duration of first remission. Indications for treatment include relapsed T cell ALL, early bone marrow relapse within 30 months of diagnosis and multiple extramedullay relapses. Estimated long-term survival after transplantation is currently around 60% (Fagioli *et al.*, 2013).

4.2.1.3 Juvenile myelomonocytic leukaemia and chronic myeloid leukaemia Juvenile myelomonocytic leukaemia (JMML) is a rare type of leukaemia that occurs in young children. Allogenic HSCT is the only curative option. Pretransplant chemotherapy does not change outcome, so current practice is to undertake transplantation as soon as an appropriate donor is available (Yoshida *et al.*, 2012). Long-term survival after transplantation is around 70% and relapse is a significant problem (Loh, 2011).

Currently, HSCT in children with chronic myeloid leukaemia (CML) in the chronic phase is reserved to those who fail tyrosine kinase inhibitor therapy. Two-year overall survival in children remains high, at 80%, after the HSCT chronic phase (Hamidieh *et al.*, 2013).

4.2.2 Lymphomas

4.2.2.1 Hodgkin's lymphoma Autologous HSCT is the standard of care for patients with Hodgkin's lymphoma in their first chemosensitive relapse or CR2 (Table 4.2) (Schmitz *et al.*, 2002). Allogeneic transplantation with a reduced-intensity conditioning (RIC) regimen is reserved for relapse after autologous transplant or refractory Hodgkin's lymphoma. Transplant-related mortality with RIC remains low and long-term survival is around 50% (Thomson *et al.*, 2008).

4.2.2.2 Non-Hodgkin's lymphoma Five-year event-free survival (EFS) is similar for allogeneic and autologous transplants in recurrent, diffuse large B cell lymphoma (50 vs 52%), Burkitt lymphoma (31 vs 27%), and anaplastic

Table 4.2 Common indications for autologous HSCT in children.

	Indication of autologous transplant	
Diseases	Single	Tandem
High risk of chemosensitive relapse	Standard of care	Experimental
High risk of Ewing's sarcoma or chemosensitive relapse	May be useful	
Brain tumour (medulloblastoma, high-grade glioma, atypical teratoid rhabdoid tumour)		May be useful, especially in young children (<3 years) in whom radiation therapy is not an option
Relapsed Hodgkin's lymphoma	Standard of care	
Relapsed non-Hodgkin's lymphoma	Standard of care	
Autoimmune disorder	Experimental	

large-cell lymphoma (46 vs 35%). However, a higher EFS has been reported for relapsed lymphoblastic lymphoma following allogeneic transplantation versus autologous transplantation (40 vs 4%) by a recent study published by Children's Oncology Group investigators (Gross *et al.*, 2010).

4.3 Emerging and experimental therapeutic uses for HSCs

While therapeutic use of HSCs has mostly, and successfully, focused on cancer – and continues to do so – these cells have other applications, too (Figure 4.2). Some diseases, such as severe anaemias, have been treated with HSCT for decades. Treatment options for autoimmune disease and metabolic diseases increasingly include HSCs. Other uses, such as tolerance induction for solid organ transplantation and combination with gene therapy, are under active investigation, but not in routine use. Emerging improvements in the treatment of cancer may focus on more highly purified preparations, especially for autologous transplantation and specific adjuvants,

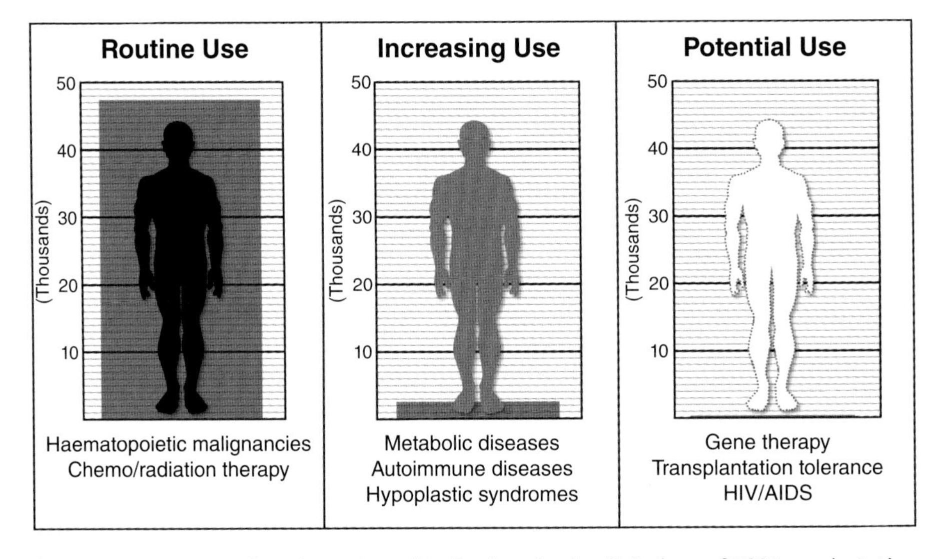

Figure 4.2 Overview of routine and novel indications for the clinical use of HSC transplantation. Current use is heavily focused on the treatment of haematopoietic and other malignancies (left). Grey boxes depict the ratio of transplants performed in each of the three categories. The middle column lists various indications for which HSCT has been used successfully for a longer period of time, but these still represent only a small proportion of the HSCT performed annually: slightly less than 2600 out of more than 50 000 transplants performed worldwide in 2006 (Gratwohl et al., 2010). The right column shows emerging potential uses, for which clinical experience is very limited – sometimes to only a few patients. Similar data have been reported for transplants in the United States in 2010; approximately 5% of the almost 18 000 HSCT performed have been for aplastic anaemia and other nonmalignant diseases (Pasquini and Wang, 2012).

such as T cells aimed at the tumour cells. We will briefly discuss some of these indications and the role of HSC in this section, as many of the future developments of HSC use will focus here.

4.3.1 Autoimmunity

Autoimmune diseases, which affect 3–5% of the general population (Rabusin *et al.*, 2008), are characterized by the recognition and destruction of specific tissues by the immune system. Standard therapy consists of immunosuppressants, which control but do not cure; to achieve a cure, the immune system needs to be eradicated and replaced using an HSCT (Sullivan *et al.*, 2010; Tyndall, 2012). Autologous cells have fewer complications for this purpose, but if a predisposition to a specific autoimmune disease is present, the disease may return. Nevertheless, in some instances, autologous cells can work (Reiff *et al.*, 2011).

To remove the underlying susceptibility to a disease, allogeneic HSCT is necessary. This replaces the susceptible immune system with a nonsusceptible one. However, the morbidity and mortality associated with allogeneic HSCT are higher than for autologous HSCT. Treatment-related complications that are potentially beneficial in cancer treatment, such as low-degree GVHD, are problematic in the treatment of autoimmune disease. It is not easy to justify the benefits of treatment modalities with 2–4% short-term mortality, such as HSCT, for diseases with mostly long-term associated mortality. With the advent of next-generation sequencing, it seems increasingly likely that autologous HSCT combined with gene therapy to remove underlying susceptibility from transplanted cells may become a realistic option in the future (Coleman and Steptoe, 2012).

4.3.2 Metabolic diseases

HSC therapy can also be used to treat certain inherited metabolic disorders caused by single gene mutations. Untreated, these can be devastating. As an example, in Hurler syndrome (lack of the lysosomal enzyme alpha-L-iduronidase (Boelens *et al.*, 2009)), patients are unable to break down glycosaminoglycans. This results in a progressive disease that is fatal in early childhood, with symptoms including psychomotor retardation, skeletal abnormalities and life-threatening cardiac and pulmonary problems. Available treatment options are limited. Cells can take up externally provided enzyme, but enzyme replacement therapy is limited by cost and, even more importantly, by the inability of the enzyme to cross the blood–brain barrier, making it impossible to treat symptoms that involve the central nervous system (CNS). The only treatment option currently available for diseases that involve the CNS is bone marrow or cord blood transplantation. The engrafting haematopoietic cells that produce the missing enzyme are present throughout the body, including – in the form of macrophages and microglial cells – the CNS (Prasad and Kurtzberg, 2010a, 2010b; Valayannopoulos and Wijburg, 2011).

Over 2000 patients with various inherited metabolic disorders have been treated this way in the past 30 years. UCB has become the preferred source, because patients have to be treated as early as possible, and UCB is available from tissue banks, engrafts better and results in higher enzyme levels. Improvement – especially developmental – can be quite dramatic. Where necessary, supplemental treatment with enzyme replacement therapy can be initiated at any time post-transplant (Valayannopoulos *et al.*, 2010). The associated musculoskeletal defects, however, have proven unresponsive to this mode of treatment, and require other (surgical) intervention.

4.3.3 Gene therapy

HSCs are a favourite target for gene therapy, due to the fact that they are well characterized and are routinely transplanted in clinical settings. Gene therapy – the ability to change the genetic makeup of specific cells, either by introduction of genes or, more recently, by the removal (inactivation) or alteration of genes – has developed through numerous clinical trials during the past 40 years (Gillet *et al.*, 2009). Much progress has been made during that time, but setbacks have slowed the process and continue to drive the need for further improvements. The field continues to show great potential, however (Kay, 2011). The large strides made in vector design and gene control will enable much of the use currently anticipated. Technical developments continue to increase our options. Especially important in this respect is the increasingly routine use of whole-genome sequencing techniques, which allow for exact identification of mutated genes or control regions. The ongoing development of advanced techniques such as nucleotide editing at the genome level using oligonucleotides or CRISPR/Cas (Aarts and te Riele, 2011; Wang *et al.*, 2013; Wei *et al.*, 2013) allows for far more precise approaches to gene therapy. Developments like these mean that gene therapy, even after 40 years of work, continues to hold great promise for the expansion of the clinical use of HSCs, both in treating cancers and in other indications.

4.3.4 Tolerance

HSC transplantation can also be used to induce tolerance for a subsequently transplanted solid organ. 'Tolerance' here would be defined as the ability to accept a transplanted organ as self, while retaining (near) normal immunological competence. This principle has been recognized for over half a century, following pioneering work by Peter Medawar and others. Implementation has lagged due to the morbidity and mortality associated with allogeneic bone marrow transplantation, however.

Organ transplantation is currently dependent on the continued use of immunosuppressants to prevent rejection. While successful in the short term, long-term survival is limited. Ten-year survival ranges from approximately 60% of transplanted organs (heart, liver) to barely 25% (lungs, small bowels) (HHS/HRSA/HSB/DOT, 2009). In addition, continued immunosuppression is associated with complications, including renal and neurotoxicity and opportunistic infections (Mueller *et al.*, 1994; Cattaneo *et al.*, 2004; Kalil *et al.*, 2007; Vajdic and Van Leeuwen, 2009). Multiple methods of inducing tolerance have been proposed and tested, mostly with only incidental success (Table 4.3) (for reviews, see Gandy *et al.*, 2007; Sykes, 2009; Bluestone, 2011; Strober *et al.*, 2011; Domen *et al.*, 2012; and Leventhal *et al.*, 2013b). These include costimulatory blockade, microchimerism and T cell depletion. More successful have been approaches involving HSCs, several of which continue to be developed and will be discussed in more detail in this section.

Experimentally, tolerance can be induced by different types of HSC-containing preparations, including both mixed populations and rigorously purified HSCs. Other haematopoietic cells, including regulatory T cells, macrophages and dendritic cells, have also been used to protect grafts experimentally. Clinically, the complications associated with allogeneic HSCT continue to be a major hurdle. Several strategies for overcoming this are under active development. The combination of total lymphoid irradiation (TLI) and HSCT has shown promise in clinical trials (Scandling *et al.*, 2012). Sublethal preconditioning has also been pursued systematically (Sachs *et al.*, 2011), while the use of so-called 'facilitator cells' – cells that can facilitate allogeneic engraftments (Gandy *et al.*, 1999) – has recently shown clinical promise in tolerance induction protocols (Leventhal *et al.*, 2013a). The use of progenitor cells may be promising in combining some of the reduced toxicity of more mature cell therapy with the ability to address several different tolerance induction mechanisms of HSC therapy (Domen *et al.*, 2011). However this approach, while promising, is still strictly preclinical.

Table 4.3 Comparison of standard immunosuppressant therapy with therapies aimed at inducing tolerance.

Therapy	Use	Outcome
Immunosuppressants	Standard therapy	Effective in short term, less so in the long term; significant side effects
Costimulatory blockade, T-cell depletion	Trials	Limited effectiveness alone, may be good in combination therapy
HSC transplant (BMT, MPB, UCB)	Trials	Very effective; works at multiple levels and through many mechanisms; severe comorbidity
Progenitor cell therapy, e.g. myeloid progenitors.	Experimental	Effective; may work through different mechanisms
Other cellular therapies (Treg, macrophages, dendritic cells)	Trials	Promising but addresses fewer issues (single cell type)

4.3.5 Graft engineering

Graft engineering encompasses the many approaches currently under development to improve on the outcome for patients given cellular grafts. It is limited for the purpose of this chapter to haematopoietic grafts. Graft engineering can involve the use of gene therapy, as discussed in Section 4.3.3. Grafts can also be altered by the addition of specific cells to the graft to improve outcome, either by facilitating engraftment of the haematopoietic (stem) cells, or by providing additional functionality, such as anti-tumour activity or improved immunocompetence in the post-engraftment period.

Cells that can facilitate the engraftment of HSCs, especially across allogeneic barriers, have been studied for years. DLIs have been used extensively to bolster failing allogeneic HSCT, and can often do so quite well (Tomblyn and Lazarus, 2008; Deol and Lum, 2010; Bar *et al.*, 2013; Chang and Huang, 2013). However, the term 'facilitating cells' is more typically used for cell preparations that accompany the original graft and whose function is to improve engraftment. Their use is especially interesting in the context of highly purified HSCs, or HSCs with strict T cell depletion. Facilitating cells can reduce the number of allogeneic HSCs needed for successful engraftment. Several cell types that have this activity have been described, including $CD8^+$ cells that lack a functional T cell receptor. Dendritic cell precursors make up an important facilitator compartment (Gandy *et al.*, 1999; Leventhal *et al.*, 2013b). Recently, facilitator cells have been used successfully clinically in HSCT aimed at inducing tolerance for subsequent organ transplantation (Leventhal *et al.*, 2012).

Other cells may be of interest as adjuvants to HSCs. These include mesenchymal stem cells (MSCs), various types of T cells and myeloid progenitor cells. MSCs, cells that attach and form fibroblast-like colonies when bone marrow is plated on plastic, can be expanded *in vitro* to generate large numbers of cells. They are of interest for several reasons. They can differentiate into mesenchymal lineage cells of the mesoderm, such as osteocytes, adipocytes and chondrocytes, and can provide support for the growth and differentiation of haematopoietic progenitor cells in the bone marrow environment (Dalal *et al.*, 2012; Leatherman, 2013; Miura *et al.*, 2013). It has also been found, somewhat unexpectedly, that they can inhibit lymphocyte proliferation. They have anti-inflammatory, antiproliferative and immunosuppressive properties, most likely through the release of soluble molecules (Dalal *et al.*, 2012; Miura *et al.*, 2013). While certainly not completely exempt from recognition and rejection by immune cells, MSCs are now used in attempts to modulate undesired immune responses. This includes facilitating engraftment by allogeneic HSCs and reducing GVHD. Long-term systemic engraftment with MSCs has proven elusive, but local administration seems to result in higher engraftment levels.

Lymphocytes, especially T lymphocytes, are of interest in engineered grafts. A full discussion of all the various options and their advantages and disadvantages is well outside the scope of this chapter, but we will list some examples.

One option is complete infusion of donor lymphocytes, harvested from the HSC donor, which, as discussed already, can help in allogeneic engraftment, and can be used to rescue a failing graft. But there are also more defined lymphocyte populations that are of interest. These include regulatory T cells (Treg), classically described by the phenotype $CD4^+CD25^+FoxP3^+$, which are capable of suppressing immune responses. The development of these cells for clinical use has attracted significant interest, as reviewed in Michael *et al.* (2013). Another experimental approach that has proven spectacularly successful in a limited number of cases to date in cancer patients is the use of chimeric antigen receptors (CARs), which allows the targeting of large numbers of T cells against a tumour target (solid tumour or haematopoietic malignancy) (Louis *et al.*, 2011; Davila *et al.*, 2014). These cells, however, are typically not given in the context of an HSCT. Cytotoxic T lymphocytes, expanded as reactive to two human viruses, have been successfully used in allogeneic transplants (Leen *et al.*, 2009). Also of potential interest is the addition of non-naïve T cells to an engineered HSCT graft. Non-naïve T cells have responded and been shown to recognize a target other than host tissue (which wasn't present in the donor). These cells normally do not cause GVHD, but, interestingly, seem to retain GVL activity (Chen *et al.*, 2004, 2006). Of interest, but with no clear path to obtaining clinical useful numbers of cells, is the addition of lymphoid progenitor cells. These cells, which mature and undergo selection in the host, do not cause GVHD and in preclinical models have been shown to be effective in reducing viral infections (Arber *et al.*, 2003).

Myeloid progenitor cells, progenitors capable of differentiating into the myelo-erythroid lineages, are being developed and are undergoing clinical trials as adjuvants. These cells, which can be obtained in clinically useful numbers in short-term defined cultures, will expand significantly and rapidly differentiate into functional myeloid cells (Akashi *et al.*, 2000; Manz *et al.*, 2002). In preclinical models, these cells have been shown to be effective in reducing bacterial and fungal infectious complications in neutropenic animals, and to be able to do so without any need for haplotype matching (Arber *et al.*, 2005; BitMansour *et al.*, 2002). In preclinical models, these cells are also effective in treating acute radiation syndrome (Singh *et al.*, 2012). Clinical trials with these cells in HSCT settings, aimed at reducing infectious complications, are ongoing (Cellerant Therapeutics, 2012).

4.3.6 Future developments

The clinical use of HSCs continues to improve and expand. Current limitations include the toxicity of preconditioning, GVHD for allogeneic transplants, contaminating cancer cells for autologous transplants and graft failure due to insufficient cells. Mostly outside of the scope of this overview, treatment of the underlying cancer is another limitation. Some areas in which improvements can be expected include the preconditioning of the patient, the purity of the stem cell graft, the addition of defined cells to the stem cell graft and, potentially, the origin of the HSCs used. These developments will also help expand the use of HSCs beyond current indications, which are still mostly focused on the treatment of cancer.

The morbidity and mortality associated with HSC transplantation have been reduced through many modifications. The development of reduced-intensity conditioning regimens (so-called 'minitransplants') has been very important in reducing transplant-related complications. Graft-versus-host activity is both good and bad in the treatment of cancer. The response against the host cancer cells (GVL) can be decisive in achieving a cure. However, the uncontrolled rejection of healthy host tissue can be life threatening. In treating conditions other than cancer, the rejection of host tissue in GVHD is bad and not acceptable. The stem cell source and preconditioning can help reduce the incidence. The use of purified HSCs (without any contaminating T cells) can completely prevent GVHD, but these require higher cell doses to ensure engraftment. Adjuvant populations such as facilitator cells can help engraftment without increasing the risk of GVHD. Other cells can be used to reduce the incidence of infections in the (immediate) post-transplant period. For example, myeloid progenitor cells, which can be cryopreserved and used without matching, provide rapid and mostly transient engraftment with myeloid cells that can protect from fungal and bacterial infections (Arber *et al.*, 2005). They are currently undergoing clinical trials (Cellerant Therapeutics, 2012). While enormous progress has been made in reducing the toxic effects of preconditioning, further improvements may well be possible. More targeted therapy can use antibodies to prepare the host to receive a stem cell graft. Preclinical studies have indicated that antibodies blocking c-Kit can be used to eliminate HSCs and prepare a host for transplant (Czechowicz *et al.*, 2007), based on the fact that HSCs are dependent on two signals for survival: one involving the Bcl-2 pathway and one involving signalling through c-Kit (Domen and Weissman, 2000). Combined with antibodies or small molecules targeting specific immune cells, this could be one path towards designing a targeted approach with limited toxicity. Additional approaches to this end may arise from ongoing work on defining factors involved in stem cell specification and interaction (Clements and Traver, 2013; Leatherman, 2013).

A final piece of the puzzle – the ability to generate stem cells – is also changing rapidly (Figure 4.3). Classically, HSCs are harvested from the bone marrow, either directly, after mobilization into circulation or from UCB. While extremely successful, this approach has limitations. Most importantly, the stem cell numbers that can be obtained are limited and, in autologous harvests, often contaminated. One of the most exciting developments in stem cell biology in the last decade has been the discovery that a pluripotent stem cell phenotype can be induced in many different types of somatic cells following transduction with a limited set of genes (Takahashi *et al.*, 2007; Yu *et al.*, 2007). These so-called 'induced pluripotent stem cells' (iPSCs) are of obvious interest, as they can be patient-specific and have the potential to differentiate

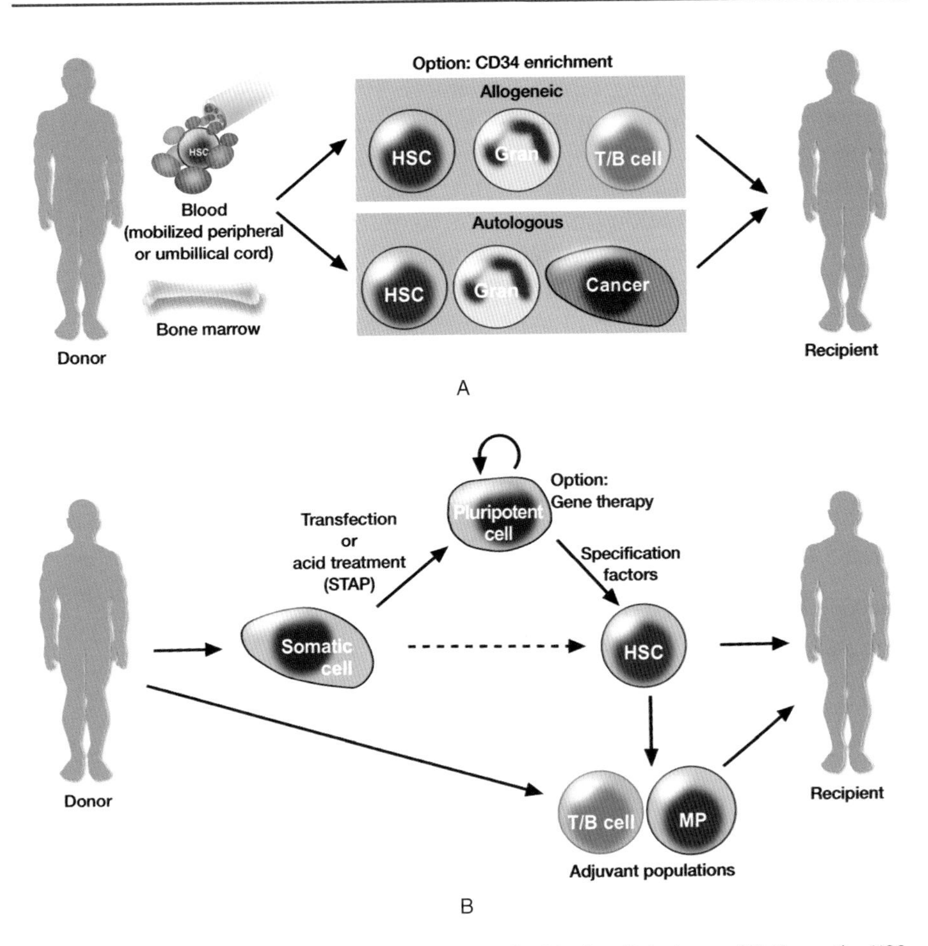

Figure 4.3 Current and potential future sources of HSCs for clinical use. (A) Currently, HSCs are harvested directly from the donor, in the form of either bone marrow or blood (as pheresed mobilized peripheral blood in adults or UCB at birth). The preparation can undergo limited manipulation, such as enrichment for $CD34^+$ cells. Some of the major limitations are the number of cells that can be obtained, the presence of alloreactive lymphocytes in allogeneic preparations and the presence of cancer cells in autologous preparations. (B) Going forward, it may become possible to harvest somatic cells (possibly blood cells) from the donor, and either to derive pluripotent cells through transfection or stress treatment, expand these and rederive committed HSCs, or, possibly, if conditions can be defined, to directly reprogramme the harvested somatic cells into HSCs. The HSCs can then be administered to the recipient or can be used to derive specific adjuvant populations, such as myeloid progenitor cells. Other adjuvant populations, such as lymphocytes, may be harvested directly from the donor. The pluripotent cell phase will allow for manipulations such as gene therapy, and the expansion at this stage may go a long way towards ensuring a sufficient numbers of cells. HSCs derived through this process should be free of host reactive lymphocytes and contaminating cancer cells. Due to the expansion step, it may eventually be possible to bank these cells based on haplotype.

into many different lineages. The transduction process, however, raises questions about potential long-term risks, especially in therapeutic settings. Reprogramming of cells from one fate to another without their passing through a pluripotent intermediary (e.g. fibroblasts into neurons) has also been demonstrated recently (Chambers and Studer, 2011). While it is in early stages, this approach may open additional avenues to obtaining genetically defined cells for research and therapy. The recent observation that it is possible to reprogramme somatic cells into pluripotent stem cells simply by stressing them through incubation in a mild acid environment for 30 minutes (STAP protocol) (Obokata *et al.*, 2014), provided it can be confirmed, opens up what may be the most exciting novel method for obtaining cells, including HSCs, for clinical use, as it does not require transfection. If the potential suggested by these initial reports can be realized – something that will require a lot of additional work – it may in the long term transform many of the aspects of the therapeutic use of HSCs and other stem cells discussed in this chapter.

4.4 Conclusions

Bone marrow transplantation is an incredible success story in the therapeutic use of stem cells in treating cancer. It has an exciting future, as continued improvements in the procedures will widen the therapeutic applications for which HSCs can be administered routinely. However, for many of these indications, routine clinical use is still far from reality. Continued development and properly designed clinical trials will be needed to bring novel indications to routine and effective clinical use. Care has to be taken with shortcuts, however tempting these may be. Only methodical approaches can delineate definitively useful approaches from ones that aren't.

References

Aarts, M., Te Riele, H. 2011. Progress and prospects: oligonucleotide-directed gene modification in mouse embryonic stem cells: a route to therapeutic application. Gene Ther. 18, 213–219.

Akashi, K., Traver, D., Miyamoto, T., Weissman, I.L. 2000. A clonogenic common myeloid progenitor that gives rise to all myeloid lineages. Nature 404, 193–197.

Arber, C., BitMansour, A., Sparer, T.E., Higgins, J.P., Mocarski, E.S., Weissman, I.L., *et al.*, 2003. Common lymphoid progenitors rapidly engraft and protect against lethal murine cytomegalovirus infection after hematopoietic stem cell transplantation. Blood 102, 421–428.

Arber, C., Bitmansour, A., Shashidhar, S., Wang, S., Tseng, B., Brown, J.M. 2005. Protection against lethal *Aspergillus fumigatus* infection in mice by allogeneic myeloid progenitors is not major histocompatibility complex restricted. J. Infect. Dis. 192, 1666–1671.

Bar, M., Sandmaier, B.M., Inamoto, Y., Bruno, B., Hari, P., Chauncey, T., *et al.*, 2013. Donor lymphocyte infusion for relapsed hematological malignancies after allogeneic hematopoietic cell transplantation: prognostic relevance of the initial CD3+ T cell dose. Biol. Blood Marrow Trans. 19, 949–957.

BitMansour, A., Burns, S.M., Traver, D., Akashi, K., Contag, C.H., Weissman, I.L., Brown, J.M. 2002. Myeloid progenitors protect against invasive aspergillosis and Pseudomonas aeruginosa infection following hematopoietic stem cell transplantation. Blood 100, 4660–4667.

Bluestone, J.A. 2011. Mechanisms of tolerance. Immunol. Rev. 241, 5–19.

Boelens, J.J., Rocha, V., Aldenhoven, M., Wynn, R., O'Meara, A., Michel, G., *et al.*, 2009. Risk factor analysis of outcomes after unrelated cord blood transplantation in patients with hurler syndrome. Biol. Blood Marrow Trans. 15, 618–625.

Cattaneo, D., Perico, N., Gaspari, F., Remuzzi, G. 2004. Nephrotoxic aspects of cyclosporine. Transplant Proc. 36, 234S–239S.

Cellerant Therapeutics. 2012. Cellerant Therapeutics initiates a Phase I/II clinical trial of CLT-008 for chemotherapy induced neutropenia in acute leukemia patients. Available from: http://www.cellerant.com/pr_032211.html (last accessed 28 November, 2014).

Chambers, S.M., Studer, L. 2011. Cell fate plug and play: direct reprogramming and induced pluripotency. Cell 145, 827–830.

Chang, Y.J., Huang, X.J. 2013. Donor lymphocyte infusions for relapse after allogeneic transplantation: when, if and for whom? Blood Rev. 27, 55–62.

Chen, B.J., Cui, X., Sempowski, G.D., Liu, C., Chao, N.J. 2004. Transfer of allogeneic CD62L-memory T cells without graft-versus-host disease. Blood 103, 1534–1541.

Chen, B.J., Deoliveira, D., Cui, X., Le, N.T., Son, J., Whitesides, J.F., Chao, N.J. 2006. Inability of memory T cells to induce graft-versus-host disease is a result of an abortive alloresponse. Blood 109(7), 3115–3123.

Clements, W.K., Traver, D. 2013. Signalling pathways that control vertebrate haematopoietic stem cell specification. Nat. Rev. Immunol. 13, 336–348.

Coleman, M.A., Steptoe, R.J. 2012. Induction of antigen-specific tolerance through hematopoietic stem cell-mediated gene therapy: the future for therapy of autoimmune disease? Autoimmun. Rev. 12, 195–203.

Czechowicz, A., Kraft, D., Weissman, I.L., Bhattacharya, D. 2007. Efficient transplantation via antibody-based clearance of hematopoietic stem cell niches. Science 318, 1296–1299.

Dalal, J., Gandy, K., Domen, J. 2012. Role of mesenchymal stem cell therapy in Crohn's disease. Pediatr. Res. 71, 445–451.

Davila, M.L., Riviere, I., Wang, X., Bartido, S., Park, J., Curran, K., *et al.*, 2014. Efficacy and toxicity management of 19-28z CAR T cell therapy in B cell acute lymphoblastic leukemia. Sci. Transl. Med. 6, 224ra25.

Deol, A., Lum, L.G. 2010. Role of donor lymphocyte infusions in relapsed hematological malignancies after stem cell transplantation revisited. Cancer Treat. Rev. 36, 528–538.

Domen, J., Weissman, I.L. 2000. Hematopoietic stem cells need two signals to prevent apoptosis; BCL-2 can provide one of these, Kitl/c-Kit signaling the other. J. Exp. Med. 192, 1707–1718.

Domen, J., Sun, L., Trapp, K., Maghami, N., Inagaki, E., Li, Y., *et al.*, 2011. Tolerance induction by hematopoietic cell transplantation: combined use of stem cells and progenitor cells. J. Heart Lung Transplant, 30, 507–514.

Domen, J., Gandy, K., Dalal, J. 2012. Emerging uses for pediatric hematopoietic stem cells. Pediatr. Res. 71, 411–417.

Fagioli, F., Quarello, P., Zecca, M., Lanino, E., Rognoni, C., Balduzzi, A., *et al.*, 2013. Hematopoietic stem cell transplantation for children with high-risk acute lymphoblastic leukemia in first complete remission: a report from the AIEOP registry. Haematologica 98, 1273–1281.

Gandy, K.L., Domen, J., Aguila, H., Weissman, I.L. 1999. CD8+TCR+ and CD8+TCR− cells in whole bone marrow facilitate the engraftment of hematopoietic stem cells across allogeneic barriers. Immunity 11, 579–590.

Gandy, K., Domen, J., Copeland, J. 2007. Tolerance in heart transplantation: current and future role. In: Watson, R.R., Larson, D.F. (eds.) Immune Dysfunction and Immunotherapy in Heart Disease. Oxford: Blackwell.

Gibson, B.E., Wheatley, K., Hann, I.M., Stevens, R.F., Webb, D., Hills, R.K., *et al.*, 2005. Treatment strategy and long-term results in paediatric patients treated in consecutive UK AML trials. Leukemia 19, 2130–2138.

Gillet, J.P., Macadangdang, B., Fathke, R.L., Gottesman, M.M., Kimchi-Sarfaty, C. 2009. The development of gene therapy: from monogenic recessive disorders to complex diseases such as cancer. Methods Mol. Biol. 542, 5–54.

Gratwohl, A., Baldomero, H., Aljurf, M., Pasquini, M.C., Bouzas, L.F., Yoshimi, A., *et al.*, 2010. Hematopoietic stem cell transplantation: a global perspective. JAMA 303, 1617–1624.

Gross, T.G., Hale, G.A., He, W., Camitta, B.M., Sanders, J.E., Cairo, M.S., *et al.*, 2010. Hematopoietic stem cell transplantation for refractory or recurrent non-Hodgkin lymphoma in children and adolescents. Biol. Blood Marrow Trans. 16, 223–230.

Hamidieh, A.A., Ansari, S., Darbandi, B., Soroush, A., Arjmandi Rafsanjani, K., Alimoghaddam, K., *et al.*, 2013. The treatment of children suffering from chronic myelogenous leukemia: a comparison of the result of treatment with imatinib mesylate and allogeneic hematopoietic stem cell transplantation. Pediatr. Trans. 17, 380–386.

HHS/HRSA/HSB/DOT. 2009. OPTN/SRTR annual report 1999–2008. Available from: http://www.ustransplant.org/annual_reports/current/ (last accessed 28 November 2014).

Kalil, A.C., Dakroub, H., Freifeld, A.G. 2007. Sepsis and solid organ transplantation. Curr. Drug Targets 8, 533–541.

Kay, M.A. 2011. State-of-the-art gene-based therapies: the road ahead. Nat. Rev. Genet. 12, 316–328.

Kelly, M.J., Horan, J.T., Alonzo, T.A., Eapen, M., Gerbing, R.B., He, W., *et al.*, 2014. Comparable survival for pediatric acute myeloid leukemia with poor-risk cytogenetics following chemotherapy, matched related donor, or unrelated donor transplantation. Pediatr. Blood Cancer 61, 269–275.

Leatherman, J. 2013. Stem cells supporting other stem cells. Front Genet. 4, 257.

Leen, A.M., Christin, A., Myers, G.D., Liu, H., Cruz, C.R., Hanley, P.J., *et al.*, 2009. Cytotoxic T lymphocyte therapy with donor T cells prevents and treats adenovirus and Epstein-Barr virus infections after haploidentical and matched unrelated stem cell transplantation. Blood 114, 4283–4292.

Leventhal, J., Abecassis, M., Miller, J., Gallon, L., Ravindra, K., Tollerud, D.J., *et al.*, 2012. Chimerism and tolerance without GVHD or engraftment syndrome in

HLA-mismatched combined kidney and hematopoietic stem cell transplantation. Sci. Transl. Med. 4, 124ra28.

Leventhal, J., Abecassis, M., Miller, J., Gallon, L., Tollerud, D., Elliott, M.J., *et al.*, 2013a. Tolerance induction in HLA disparate living donor kidney transplantation by donor stem cell infusion: durable chimerism predicts outcome. Transplantation 95, 169–176.

Leventhal, J., Miller, J., Abecassis, M., Tollerud, D.J., Ildstad, S.T. 2013b. Evolving approaches of hematopoietic stem cell-based therapies to induce tolerance to organ transplants: the long road to tolerance. Clin. Pharmacol. Ther. 93, 36–45.

Loh, M.L. 2011. Recent advances in the pathogenesis and treatment of juvenile myelomonocytic leukaemia. Br. J. Haematol. 152, 677–687.

Louis, C.U., Savoldo, B., Dotti, G., Pule, M., Yvon, E., Myers, G.D., *et al.*, 2011. Anti-tumor activity and long-term fate of chimeric antigen receptor-positive T cells in patients with neuroblastoma. Blood 118, 6050–6056.

Manz, M.G., Miyamoto, T., Akashi, K., Weissman, I.L. 2002. Prospective isolation of human clonogenic common myeloid progenitors. Proc. Nat. Acad. Sci. USA 99, 11 872–11 877.

Mathé, G., Jammet, H., Pendic, B., Schwarzenberg, L., Duplan, J.F., Maupin, B., *et al.*, 1959. [Transfusions and grafts of homologous bone marrow in humans after accidental high dosage irradiation.] Rev. Fr. Etud. Clin. Biol. 4, 226–238.

Michael, M., Shimoni, A., Nagler, A. 2013. Regulatory T cells in allogeneic stem cell transplantation. Clin. Dev. Immunol. 2013, 608951.

Miura, Y., Yoshioka, S., Yao, H., Takaori-Kondo, A., Maekawa, T., Ichinohe, T. 2013. Chimerism of bone marrow mesenchymal stem/stromal cells in allogeneic hematopoietic cell transplantation: is it clinically relevant? Chimerism, 4, 78–83.

Mueller, A.R., Platz, K.P., Schattenfroh, N., Bechstein, W.O., Christe, W., Neuhaus, P. 1994. Neurotoxicity after orthotopic liver transplantation in cyclosporin A- and FK 506-treated patients. Transpl. Int. 7(Suppl. 1), S37–S42.

Muller, A.M., Kohrt, H.E., Cha, S., Laport, G., Klein, J., Guardino, A.E., *et al.*, 2012. Long-term outcome of patients with metastatic breast cancer treated with high-dose chemotherapy and transplantation of purified autologous hematopoietic stem cells. Biol. Blood Marrow Trans. 18, 125–133.

Obokata, H., Wakayama, T., Sasai, Y., Kojima, K., Vacanti, M.P., Niwa, H., *et al.*, 2014. Stimulus-triggered fate conversion of somatic cells into pluripotency. Nature 505, 641–7.

Pasquini, M.C., Wang, Z. 2012. Current use and outcome of hematopoietic stem cell transplantation: CIBMTR summary slides. Available from: http://www.cibmtr.org (last accessed 28 November 2014).

Prasad, V.K., Kurtzberg, J. 2010a. Cord blood and bone marrow transplantation in inherited metabolic diseases: scientific basis, current status and future directions. Br. J. Haematol. 148, 356–372.

Prasad, V.K., Kurtzberg, J. 2010b. Transplant outcomes in mucopolysaccharidoses. Semin. Hematol. 47, 59–69.

Pulsipher, M.A., Peters, C., Pui, C.H. 2011. High-risk pediatric acute lymphoblastic leukemia: to transplant or not to transplant? Biol. Blood Marrow Trans. 17, S137–S148.

Rabusin, M., Andolina, M., Maximova, N. 2008. Haematopoietic SCT in autoimmune diseases in children: rationale and new perspectives. Bone Marrow Trans. 41(Suppl. 2), S96–S99.

Reiff, A., Shaham, B., Weinberg, K.I., Crooks, G.M., Parkman, R. 2011. Anti-CD52 antibody-mediated immune ablation with autologous immune recovery for the treatment of refractory juvenile polymyositis. J. Clin. Immunol. 31, 615–622.

Sachs, D.H., Sykes, M., Kawai, T., Cosimi, A.B. 2011. Immuno-intervention for the induction of transplantation tolerance through mixed chimerism. Semin. Immunol. 23, 165–173.

Scandling, J.D., Busque, S., Dejbakhsh-Jones, S., Benike, C., Sarwal, M., Millan, M.T., *et al.*, 2012. Tolerance and withdrawal of immunosuppressive drugs in patients given kidney and hematopoietic cell transplants. Am. J. Transplant. 12, 1133–1145.

Schmitz, N., Pfistner, B., Sextro, M., Sieber, M., Carella, A.M., Haenel, M., *et al.*, 2002. Aggressive conventional chemotherapy compared with high-dose chemotherapy with autologous haemopoietic stem-cell transplantation for relapsed chemosensitive Hodgkin's disease: a randomised trial. Lancet 359, 2065–2071.

Singh, V.K., Christensen, J., Fatanmi, O.O., Gille, D., Ducey, E.J., Wise, S.Y., *et al.*, 2012. Myeloid progenitors: a radiation countermeasure that is effective when initiated days after irradiation. Radiat. Res. 177, 781–791.

Spangrude, G.J., Heimfeld, S., Weissman, I.L. 1988. Purification and characterization of mouse hematopoietic stem cells. Science 241, 58–62.

Spooncer, E., Lord, B.I., Dexter, T.M. 1985. Defective ability to self-renew in vitro of highly purified primitive haematopoietic cells. Nature 316, 62–64.

Strober, S., Spitzer, T.R., Lowsky, R., Sykes, M. 2011. Translational studies in hematopoietic cell transplantation: treatment of hematologic malignancies as a stepping stone to tolerance induction. Semin. Immunol. 23, 273–281.

Sullivan, K.M., Muraro, P., Tyndall, A. 2010. Hematopoietic cell transplantation for autoimmune disease: updates from Europe and the United States. Biol. Blood Marrow Trans. 16, S48–S56.

Sykes, M. 2009. Hematopoietic cell transplantation for tolerance induction: animal models to clinical trials. Transplantation 87, 309–316.

Takahashi, K., Tanabe, K., Ohnuki, M., Narita, M., Ichisaka, T., Tomoda, K., Yamanaka, S. 2007. Induction of pluripotent stem cells from adult human fibroblasts by defined factors. Cell 131, 861–872.

Thomas, E.D., Lochte, H.L. Jr, Lu, W.C., Ferrebee, J.W. 1957. Intravenous infusion of bone marrow in patients receiving radiation and chemotherapy. N. Engl. J. Med. 257, 491–496.

Thomson, K.J., Peggs, K.S., Smith, P., Cavet, J., Hunter, A., Parker, A., *et al.*, 2008. Superiority of reduced-intensity allogeneic transplantation over conventional treatment for relapse of Hodgkin's lymphoma following autologous stem cell transplantation. Bone Marrow Trans. 41, 765–770.

Tomblyn, M., Lazarus, H.M. 2008. Donor lymphocyte infusions: the long and winding road: how should it be traveled? Bone Marrow Trans. 42, 569–579.

Tyndall, A. 2012. Application of autologous stem cell transplantation in various adult and pediatric rheumatic diseases. Pediatr. Res. 71, 433–438.

Vajdic, C.M., Van Leeuwen, M.T. 2009. Cancer incidence and risk factors after solid organ transplantation. Int. J. Cancer 125, 1747–1754.

Valayannopoulos, V., Wijburg, F.A. 2011. Therapy for the mucopolysaccharidoses. Rheumatology (Oxford), 50(Suppl. 5), v49–v59.

Valayannopoulos, V., De Blic, J., Mahlaoui, N., Stos, B., Jaubert, F., Bonnet, D., *et al.*, 2010. Laronidase for cardiopulmonary disease in Hurler syndrome 12 years after bone marrow transplantation. Pediatrics, 126, e1242-e1247.

Wang, H., Yang, H., Shivalila, C.S., Dawlaty, M.M., Cheng, A.W., Zhang, F., Jaenisch, R. 2013. One-step generation of mice carrying mutations in multiple genes by CRISPR/Cas-mediated genome engineering. Cell 153, 910–918.

Wei, C., Liu, J., Yu, Z., Zhang, B., Gao, G., Jiao, R. 2013. TALEN or Cas9 – rapid, efficient and specific choices for genome modifications. J. Genet. Genomics 40, 281–289.

Yoshida, N., Doisaki, S., Kojima, S. 2012. Current management of juvenile myelomonocytic leukemia and the impact of RAS mutations. Paediatr. Drugs 14, 157–163.

Yu, J., Vodyanik, M.A., Smuga-Otto, K., Antosiewicz-Bourget, J., Frane, J.L., Tian, S., *et al.*, 2007. Induced pluripotent stem cell lines derived from human somatic cells. Science 318, 1917–1920.

5

Isolation and Identification of Neural Stem/Progenitor Cells

Loic P. Deleyrolle and Brent A. Reynolds

McKnight Brain Institute, Department of Neurosurgery, University of Florida, Gainesville, FL, USA

5.1 Introduction

In general, during evolution, the capacity to regenerate organs, tissue and body parts is diminished as a species' complexity increases and it moves higher up the evolutionary ladder. This phenomenon may reflect an increased pressure to protect rather than regenerate, or it could be the result of dependence on specialized cells to perform sophisticated functions. However, it is clear that the degree of plasticity seen in lower vertebrates is diminished in most mammals. The concept of regeneration has intrigued scientists since 1712, when the French mathematician René-Antoine Ferchault de Réaumur presented a study to the French Academy on the regeneration of crayfish legs (Barroux, 2003; Ratcliff, 2005). In the 1740s, the Swiss naturalist Abraham Trembley uncovered the tremendous regenerative ability of freshwater polyps. In an effort to determine whether the species were plant or animal, Trembley divided the polyps transversely, with the expectation that animals would die and plants would regenerate. To his surprise, each part grew a new polyp, with the head growing a tail and the tail growing a head (Trembley, 1744; Barroux, 2003). In subsequent experiments involving repeated splitting of new heads, he generated a multiheaded animal that was eventually termed a 'hydra', after the mythological creature. This capacity to regenerate is characteristic of adult simple organisms and amphibians, which either retain clusters of stem cells (planaria) or possess other cells near the site of injury

Principles of Stem Cell Biology and Cancer: Future Applications and Therapeutics, First Edition.
Edited by Tarik Regad, Thomas J. Sayers and Robert C. Rees.

that de-differentiate into a stem cell state and recapitulate developmental tissue growth. This is not to say that humans have lost the ability to repair and regenerate altogether, but this capability appears to be limited to specific circumstances and organs.

5.2 Stem cell concept

While scientists have been studying the outcomes of stem cell-based repair for several centuries, it was not until the late 1800s that the term 'stem cell' was coined with the definition used today (Ramalho-Santos and Willenbring, 2007). The first definitive evidence of adult stem cells was provided by Till, McCulloch and colleagues at the University of Toronto in the 1960s (Till and McCulloch, 1961; Siminovitch *et al.*, 1963; McCulloch *et al.*, 1965). In a series of papers from 1961 to 1965, using an *in vivo* spleen colony-forming-cell assay, this group demonstrated and defined the characteristics of an adult stem cell. Their assay involved the implantation of donor bone marrow cells (BMCs) into recipient animals that had been previously irradiated, followed by the analysis of clonally derived colonies in the host's spleen (Figure 5.1). From these studies, they surmised that the colony-forming cells were present in haematopoietic tissue in relatively small numbers (1 : 10 000 cells) and that they had no distinguishing features other than their ability to give rise

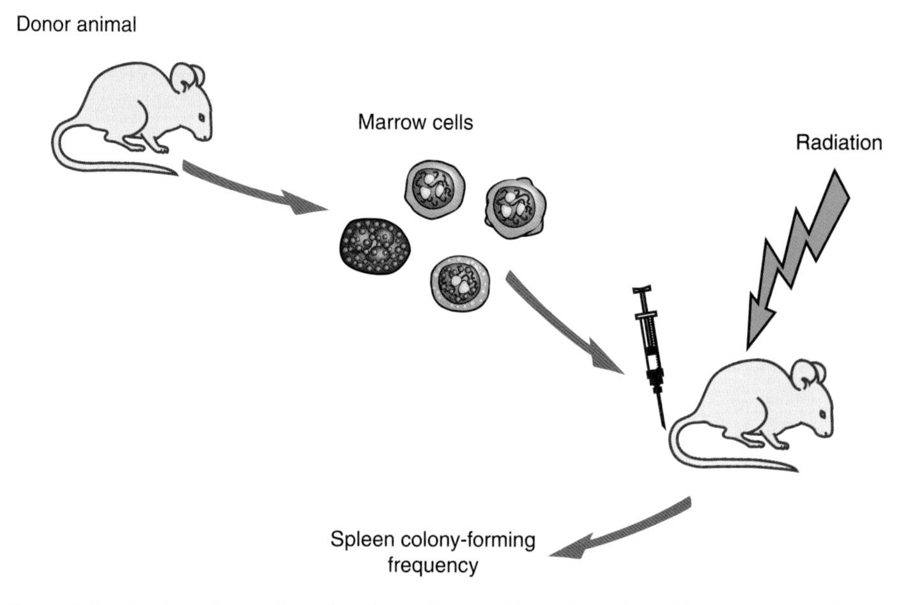

Figure 5.1 *In vivo* spleen colony-forming-cell assay. Transplantation of bone marrow cells into an irradiated recipient mouse induces spleen colony formation at a frequency of 1 : 104. This assay provides a retrospective analysis of stem cell properties based on a functional definition (colony-forming ability).

to spleen colonies. Based on this analysis, the group concluded that the spleen-forming stem cell properties must be deduced by investigation of the colonies they produced, or that stem cells are studied based on a functional definition and therefore in a retrospective manner. Till and McCulloch defined the stem cell properties of the colony-forming cells as:

1. Expression of an extensive proliferative ability.
2. Production of cells that are capable of differentiation.
3. Capability for self-renewal, such that a colony has the ability to form secondary colonies.

These tenets have formed the foundation for the study of all adult stem cells and have defined the key elements of the functional features of a stem cell. Since this seminal finding, stem cells have been extensively studied in other tissues, such as blood, which exhibits a high degree of cell turnover, and the skin and small intestines, both of which contain a relatively rare population of cells that are ultimately responsible for generating the millions of cells produced by these organ systems on a daily basis. Potten and Loeffler (1990) published a thoughtful and elegant review in which they considered many of the problems and concepts used to define solid tissue stem cells. Following on from the work of Till and McCulloch, they defined stem cells as undifferentiated cells capable of:

1. proliferation;
2. self-maintenance;
3. production of a large number of differentiated functional progeny;
4. regeneration of tissue after injury; and
5. flexibility in the use of these options.

While a stem cell should ideally be able to satisfy all of these criteria, they are not an absolute requirement, and are rather weighted towards exhibiting certain features, such as self-maintenance over the lifespan of the animal, the generation of a large number of functional progeny and regeneration of tissue after injury. One of the problems created by a functional definition is that one must be in a position to observe a stem cell acting, or one must force it to act, in order to define it as a stem cell. As there is no universally accepted or defining morphological or antigenic profile for solid tissue stem cells, the functional criteria are still the gold standard used for measurement, and while controversy exists over the accuracy of such an approach, it has in general served the stem cell community well.

Up until the early 1990s, adult stem cells were thought to exist only in tissues that exhibited a significant degree of cell turnover, such as the blood, skin and intestines, and it was never expected that they would be found in what was thought to be the most static tissue of the body: the brain.

5.3 Neural stem cells

Neural regenerative medicine was dominated for the better part of the last century by a doctrine originally established in the early 20th century by the famous and widely influential Spanish histologist Ramon y Cajal. This eminent biologist stated that 'once the development was ended, the founts of growth and regeneration of the axons and dendrites dried up irrevocably. In the adult centers, the nerve paths are something fixed, ended, and immutable. Everything may die, nothing may be regenerated' (Ramon y Cajal, 1913). This became known as the 'no new neuron dogma' and stifled any hopes of the central nervous system (CNS) being capable of regeneration. It naturally followed that, since stem cells were located in tissues in which cell turnover was a normal occurrence, the brain, which exhibited no capacity for regeneration, would not contain any stem cells. In early 1960s, however, Joseph Altman published several articles challenging the 'central dogma of neurology', demonstrating cell proliferation in the adult CNS (Altman, 1962, 1963; Altman and Das, 1965). These findings were reproduced by Kaplan, using additional techniques such as electron microscopy, who provided a strong indication that neuron genesis was occurring in the rat olfactory bulb and the dentate gyrus of the hippocampus (Kaplan and Hinds, 1977). Meanwhile, Privat and colleagues demonstrated the generation of new glial cells in postnatal mammalian brain (Privat and Leblond, 1972; Privat, 1975) and Fernando Nottebohm published a paper demonstrating neuronal replacement in birds during mating season, adding to the indication of adult CNS plasticity and neurogenesis (Nottebohm, 1985). The field denied the evidence of cell genesis in the adult nervous system until the early 1990s, however, when a laboratory in Canada demonstrated that neural stem cells (NSCs) could be isolated from adult mammal brain and expanded *in vitro* using growth factors (Reynolds and Weiss, 1992). This study contributed to the demise of the 'no new cell genesis dogma'. A few years later, the presence of stem cells was also demonstrated in the human embryonic and adult nervous systems (Eriksson *et al.*, 1998; Kukekov *et al.*, 1999; Vescovi *et al.*, 1999). NSCs have been identified lining the ventricular system throughout the neuroaxis in the adult nervous system, and are at the origin of neuron genesis in specific brain areas, including the subventricular zone (SVZ) of the forebrain lateral ventricles and the subgranular zone (SGZ) in the dentate gyrus of the hippocampus (Figure 5.2) (Cameron *et al.*, 1993; Kuhn *et al.*, 1996; Weiss *et al.*, 1996). Hence, adult neurogenesis persists in the majority of mammalian species, including humans and nonhuman primates (Eriksson *et al.*, 1998; Gould *et al.*, 1998; Kornack and Rakic, 1999; Pencea *et al.*, 2001; Bedard and Parent, 2004). In the SVZ, it is believed that new neurons arise from a glial cell (type-B cell), located in the subependymal zone. This type-B cell gives rise to progenitors called 'transient amplifying type-C cells', which in turn produce the direct precursors of neurons (type-A cells), which migrate along the rostral migratory stream (RMS) into the olfactory bulb to differentiate in interneurons

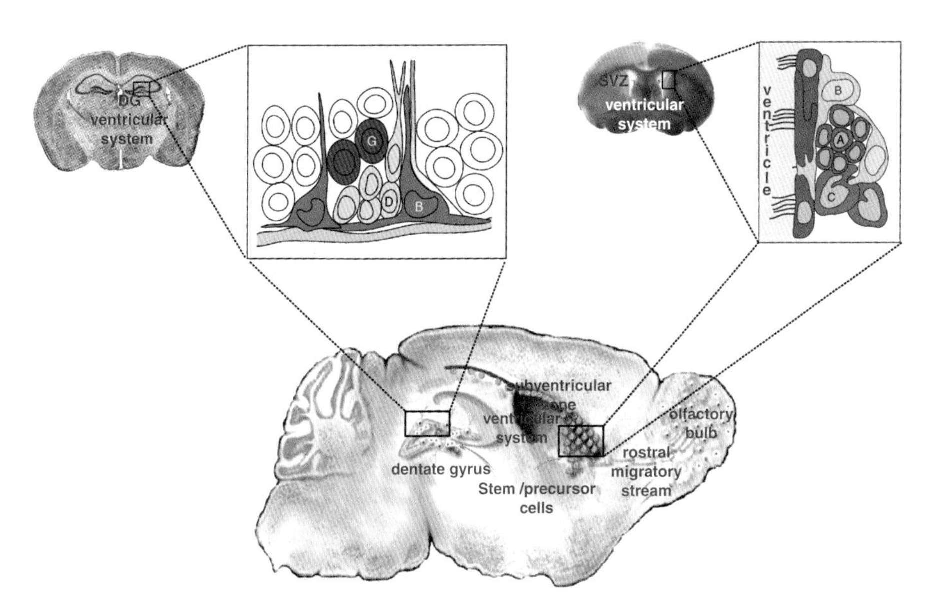

Figure 5.2 Adult neurogenic regions: The SVZ of the lateral ventricles and the hippocampal dentate gyrus (DG). Stem and precursor cells in the striatal SVZ (type-B cells) proliferate and give rise to transient amplifying cells (type-C cells), from which neuroblasts (type-A cells) are derived; these migrate tangentially in the RMS to the olfactory bulb, where they differentiate into granule and periglomerular neurons. Neural precursors (type B) in the dentate gyrus proliferate and generate type-D cells, which produce granule neurons (type G) throughout life.

(Lois and Alvarez-Buylla, 1994; Doetsch, 2003; Doetsch *et al.*, 1999). Growing indication promotes a role of the newly generated neurons in olfactory memory and discrimination (Gheusi *et al.*, 2000; Petreanu and Alvarez-Buylla, 2002; Rochefort *et al.*, 2002; Enwere *et al.*, 2004). In the SGZ, astrocyte-like cells (type-B cells) divide and give rise to type-D cells, which generate granule neurons (type-G cells). Several studies have demonstrated that approximately 9000 new cells are generated every day in the young adult rodent dentate gyrus, and that a portion of these cells integrate into the neuronal circuitry by receiving synaptic input and sending axonal projections to appropriate targets (Stanfield and Trice, 1988; Hastings and Gould, 1999; Markakis and Gage, 1999; Cameron and McKay, 2001; Carlen *et al.*, 2002). Depletion of cell genesis in the hippocampus alters the formation of some types of memory, implicating dentate gyrus neurogenesis in particular hippocampus-dependent learning and memory mechanisms (Shors *et al.*, 2001, 2002).

Prior works suggest that olfactory bulb and dentate gyrus neurogenesis are independently regulated by distinct physiological processes. For instance, pregnancy and mating stimulate olfactory bulb neurogenesis but do not affect dentate gyrus neuron formation (Shingo *et al.*, 2003); conversely, external cues like enriched environment, learning and running induce an increase in dentate gyrus neurogenesis but leave olfactory bulb cell genesis unchanged

(Kempermann *et al.*, 1997; Gould *et al.*, 1999; van Praag *et al.*, 1999; Brown *et al.*, 2003). A recent study proposed that an association exists between neurogenesis in both areas: Mak *et al.* (2007) established that dominant male pheromones regulate female reproductive success by controlling neurogenesis in both the olfactory bulb and the dentate gyrus. For the first time, the authors demonstrated interplay between olfactory-dependent reproductive behaviours, hippocampal-dependent memories produced for social identification and neurogenesis in both the olfactory bulb and the hippocampus.

Production of new neurons has also been described in other regions of the mature brain, such as the adult striatum, substantia nigra, amygdala and hypothalamus (Bedard *et al.*, 2002; Bernier *et al.*, 2002; Zhao *et al.*, 2003; Dayer *et al.*, 2005; Kokoeva *et al.*, 2007). Although the number of reports supporting neurogenic potential from stem/precursor cells in other areas of the adult brain is growing, a consensus on this topic has not been reached.

5.4 The neurosphere assay: a gold-standard assay

Due to the lack of unique cell surface markers and the absence of a definite and evident morphological phenotype, NSCs are usually defined and studied by virtue of their functional criteria. A general definition of a stem cell is: an undifferentiated cell which retains the capacity to (i) self-renew over an extended period of time, (ii) generate a large number of progeny, and (iii) exhibit a multilineage differentiation potential. Reynolds and Weiss (1992) determined optimal culture conditions for the expansion of NSCs and progenitor cells from the mammalian CNS, thus providing for the first time a simple and robust assay by which to investigate the activity and regulation of the cardinal properties of an NSC (as listed in Section 5.2). The neurosphere assay (NSA) is thus currently the most frequently used method for isolating, enriching and expanding NSCs. Cells from the primary tissue are dissociated into a single cell suspension under serum-free conditions in the presence of cytokines such as epidermal growth factor (EGF) and basic fibroblast growth factor (bFGF). Under these minimalistic growth conditions, the only cells that survive are those that divide in response to the mitogenic stimulation (Figure 5.3). This culture system allows a small population of growth factor-responsive cells to enter a period of active proliferation in which they form clusters of undifferentiated cells referred to as 'neurospheres'. These in turn can: (i) be dissociated to form numerous secondary spheres or (ii) be induced to differentiate, generating the three major cell types of the CNS: neurons, astrocytes and oligodendrocytes (Figure 5.3). In theory, under these conditions, stem cells can be passed indefinitely (Foroni *et al.*, 2007). When the number of cells generated at each passage is graphed as a function of time (or passage number), the growth curve presented in Figure 5.3 is obtained. Demonstration of extensive renewal and generation of a large number of differentiated progeny using the NSA reveal that this assay represents a *bona fide* methodology for the isolation and propagation of NSCs.

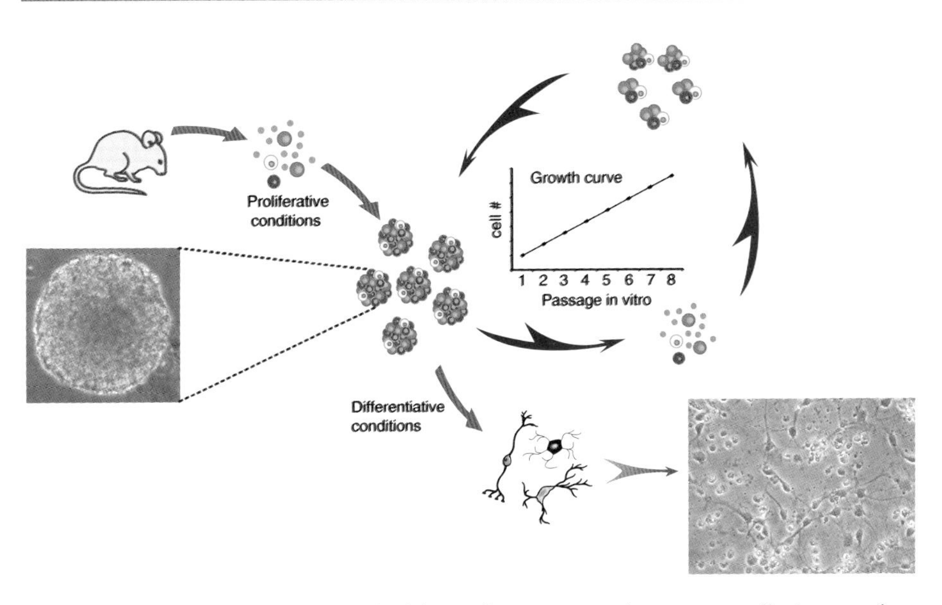

Figure 5.3 NSA features. This methodology allows stem and precursor cells to grow in a serum-free media containing EGF and bFGF. Under these conditions, growth factor-responsive cells proliferate and form spheres called 'neurospheres', which can be dissociated to form secondary spheres, which in turn can be passed and generate tertiary spheres. Theoretically, under these conditions, stem cells can be maintained indefinitely. Growth curves are obtained by plotting the accumulation of the cell number quantified at each passage. Growth factor removal and serum addition provide differentiative conditions that induce generation of the three major cell types of the CNS: neurons, astrocytes and oligodendrocytes.

5.5 Limitations of the NSA

Although the NSA allows detection of stem cell activity, the generation of an individual neurosphere does not reflect the presence of a single stem cell. Hence, the concept that a one-to-one correlation exists between neurospheres and stem cells is mistaken. Indeed, there is no experimental evidence to support this association, and the number of studies arguing that the presence of neurospheres is not fundamentally associated with the presence of actual stem cells is increasing (Seaberg and van der Kooy, 2002; Bull and Bartlett, 2005; Marshall *et al.*, 2006). Reynolds and Rietze (2005) and Louis *et al.* (2008) challenged the central tenet of the NSA that all neurospheres are derived from an NSC and showed that the majority of brain-derived spheres are not formed from stem cells but from progenitor cells with limited capacity for self-renewal. Accordingly, the sphere-forming frequency overrates the NSC frequency and quantification of the number of spheres is not an accurate measure of stem cell number or self-renewal properties (for review, see Deleyrolle *et al.*, 2008). The NSA in and of itself cannot be used as a read-out of NSC frequency, but represents by far the best assay by which to assess regulation of progenitor cell activity. Clearly, novel culture conditions are required, or else new assays need to be designed so as to distinguish between stem and precursor activity. Also, studies that have used the NSA readout for strategies aimed at purifying and enriching NSC need to be revised. Indeed, accurate and reliable measures are critical: errors in the readout lead to erroneous or irrelevant NSC populations.

5.6 The neural colony-forming cell assay: enumeration of actual NSCs

In order to address the need for an assay that can meaningfully detect specific changes in NSC activity, Louis *et al.* (2008) have developed a new method, the neural colony-forming cell assay (NCFCA), which allows discrimination between NSCs and progenitors according to the size of colony they produce (i.e. their proliferative potential). The assay is based on the assumption that progenitor cells exhibit limited proliferative capacity when compared to stem cells, which suggests that the size (diameter) of the colony can be used to distinguish its initiator cell type. Although the NCFCA uses a defined serum-free medium like the NSA, its specific culture conditions, utilizing a semisolid collagen matrix, ensure clonality of the colonies. The conditions also allow stem or progenitor cells to grow for 3–4 weeks, enabling the cells to exhibit their full proliferative capacity, which is not feasible under the NSA, as spheres are typically passed after 5–7 days *in vitro*, beyond which they become unhealthy and die. As a result, NSC and neural progenitor cell frequency can be distinguished based on colony size. As colonies are generated with a diverse size distribution, due to the differences in the proliferative potential of the original colony-forming cell, four categories of colonies are quantified, according to their diameter (0.5, 0.5–1.0, 1.0–2.0, and >2.0 mm). Using this description, only large colonies (>2.0 mm) are able to exhibit stem cell features (extensive self-renewal, generation of a large number of progeny and multilineage differentiation potential), while smaller-sized colonies are not (they show limited self-renewal capacity and restricted differentiation capability). Hence, the NCFCA is a meaningful method by which to analyse NSC biology and represents an essential tool in the study of neural development and regeneration.

5.7 Purification strategies: the quest for a cellular identity

Since the first characterization of haematopoietic stem cells (HSCs) over half a century ago, when it was revealed that injection of blood marrow cells could rescue animals from lethal marrow aplastic anaemia caused by irradiation (see Figure 5.1) (Jacobson *et al.*, 1951; Lorenz *et al.*, 1951), the field of HSC biology has elaborated consistent methodologies for the characterization, isolation and study of HSCs *in vitro* and *in vivo*. To date, the HSC represents the best characterized adult stem cell, and HSC-based therapies are routinely used in the clinic. Neurobiologists have sought inspiration from the haematopoietic field in designing methods for the identification, characterization and isolation of stem cells and developing NSC-based therapies. Hence, common strategies for the prospective isolation of putative NSCs are often based on the expression (or lack of expression) of specific cell-surface epitopes. Fluorescence-activated cell sorting (FACS) represents the most popular approach to the separation and purification of specific populations. Isolating stem cells from neural tissue is difficult due to the paucity of rigorously characterized markers. However, NSCs share cellular and molecular properties with their haematopoietic equivalents that may serve as surrogate identifiers of NSCs. Hence, several extracellular markers have been used to separate putative NSCs using positive and negative sorting strategies. Using a multiple-selection step based on cell size and low expression of peanut agglutinin (PNA) and heat-stable antigen (HSA, mCD24a) binding, Rietze *et al.* (2001) isolated a population of cells enriched in sphere-forming ability from the adult subependymal zone that accounted for 63% of the total sphere-forming activity. Capela and colleagues identified the carbohydrate Lewis X (LeX, also known as stage-specific embryonic antigen 1, SSEA-1, or leucocyte cluster of differentiation 15, CD15), which is expressed by embryonic pluripotent stem cells, as a potential marker for adult NSCs, based on LeX expression on cells located in the subependymal zone and their ability to form neurospheres (Gooi *et al.*, 1981; Capela and Temple, 2002). The subpopulation of LeX-expressing cells represented 4% of the SVZ cells and exhibited BrdU uptake and retention characteristics expected in adult NSCs. In addition, CD133 (prominin-1) has been identified in a number of tissue stem cells, including those derived from myogenic and haematopoietic lineages (Miraglia *et al.*, 1997). Identification of CD133 based on antibody binding allowed the separation of stem cells from human foetal brain cells; the authors of this study used flow cytometry to isolate a $CD133^+$ cell capable of neurosphere initiation 140 times greater than the negative population (Uchida *et al.*, 2000). Prominin expression has also been used to identify and isolate murine NSCs from the developing and adult brain (Corti *et al.*, 2007), revealing that CD133 antigen expression is represented on 3.2% of adult brain-derived cells, located near the ventricle zone and expressing several markers used to identify immature neural cells, such as Sox2, Nestin and Musashi1. The $CD133^+$ population presented a significantly higher sphere-forming capacity than the $CD133^-$ portion (Corti *et al.*, 2007). More recently, Coskun *et al.* (2008) demonstrated that a subpopulation of postnatal ependymal cells displayed CD133 expression in combination with stem cell characteristics. *In vivo*, these relatively quiescent ependymal $CD133^+$ cells retained cell-division ability and became activated upon lesion. Further, after transplantation in postnatal brain, the $CD133^+$ ependymal population generated new neuronal cells destined to integrate into the olfactory bulb (Coskun *et al.*, 2008). These data support the hypothesis that ependymal cells contain a population of stem-like cells and support CD133 as a potential marker for identifying putative NSCs.

Rather than focusing stem cell isolation procedures on cell-surface marker expression, some studies have described enzymatic activity-based sorting. As a result, using Aldefluor, a fluorescent-conjugated aldehyde dehydrogenase (ALDH) substrate, HSCs have been separated from mixed cell populations on the basis of their high ALDH expression and activity (Kastan *et al.*, 1990; Storms *et al.*, 1999) – shared characteristics between HSCs and NSCs. Consequently, Aldefluor labelling intensity has been used to isolate NSCs (Corti *et al.*, 2006). According to neurosphere-forming frequency, Corti *et al.* (2006) described stem cell enrichment by 39 and 30 in a low-side-scatter/bright-ALDH ($SSC^{lo}ALDH^{br}$) population derived from embryonic and adult neurospheres, respectively, compared to the negative fraction. Similar stem cell enrichments in the $SSC^{lo}ALDH^{br}$ population have been observed from embryonic and adult tissue (61 and 53 times, respectively). Hoechst 33342 is also commonly used to separate HSCs from non-HSCs in the mixed cell population, on the basis of low fluorescence of the dye (Goodell *et al.*, 1996). Primitive HSCs have been isolated using flow cytometry to collect the Hoechst negative/low side population (SP) of cells, which exhibited lower fluorescence due to the rapid dye efflux through specific transporters like ATP binding cassette (ABC)-transporter proteins. It has been suggested that stem cells in general have the ability to pump out toxins. This assumption has been tested in the neural tissue by performing SP analysis on a mixed population of embryonic or adult NSCs (Kim and Morshead, 2003). By comparing the neuropshere forming frequency between the two populations (SP and non-SP), this study demonstrated that embryonic and adult SPs were enriched by 278 and 7.5 in stem cells, and contained >99% and ~90% of total NSC activity, respectively (Kim and Morshead, 2003).

All of the studies described in this section relied on the NSA as a measure of stem cell frequency. As mentioned in Section 5.5, the reliability and accuracy of the NSA in evaluating a particular stem cell enrichment strategy are open to criticism. Therefore, the conclusions of these studies, or at least the stem cell purification outputs from these results, are questionable. Louis *et al.* (2008) have recently used the NCFCA to re-evaluate the stem cell content of a supposedly pure population of adult stem cells (1 : 1.28) obtained by sorting $PNA^{lo}HSA^{lo}$ cells with a diameter >12 μm (Figure 5.4) (Rietze *et al.*, 2001). This multiple-selection protocol allowed 236-fold enrichment in sphere-forming frequency relative to controls (78%, $PNA^{lo}HSA^{lo}$ > 12 μm cells vs 0.33%, controls). However, this sorting strategy gave a 15-fold increase in large (>2 mm diameter) colony-forming frequency, which reflected actual stem cell frequency (15%, $PNA^{lo}HSA^{lo}$ > 12 μm cells vs 1%, controls). Hence, the conclusions drawn from the NSA overestimated the stem cell purification by 15 times. These results highlight the danger of using neurosphere-forming frequency as a readout of stem cell frequency and show the relevance of using the NCFCA.

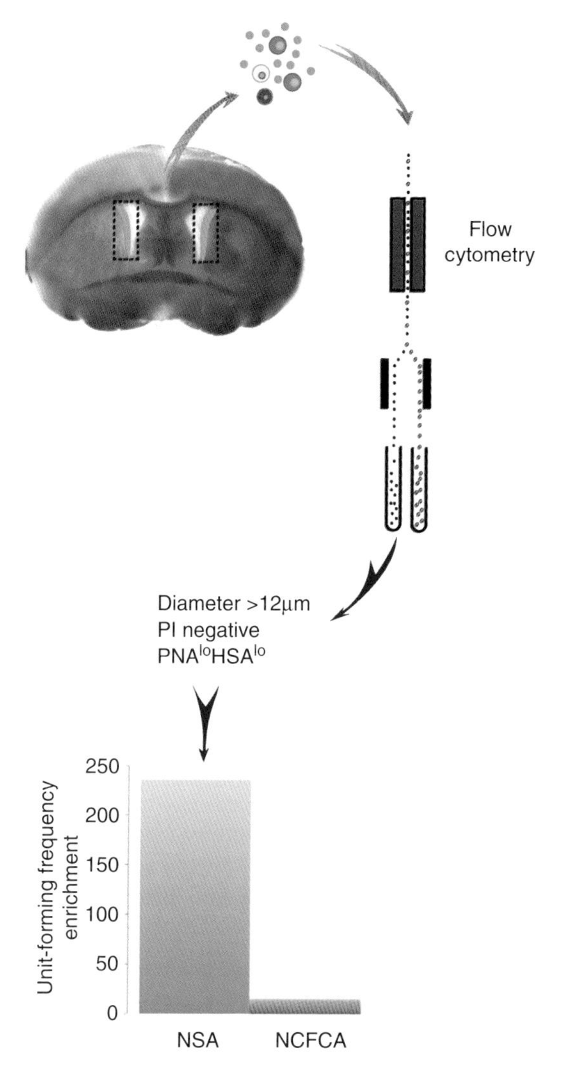

Figure 5.4 Sorting of stem cells from the adult rostral periventricular region. Cells from the adult subventricular area were harvested and sorted by flow cytometry. The sort used multiple selection steps. Viable cells, exhibiting a diameter >12 μm, were selected based on negative labelling for propidium iodide. From this population, $PNA^{lo}HSA^{lo}$ cells were isolated. This sort strategy greatly enriched neurosphere-forming cells (236-fold), but the colony-forming frequency (more accurate in stem cell activity assessment than the NSA) was only increased 15 times.

5.8 Conclusions

The discovery of NSCs has opened new horizons for neurobiology research. Although no consensus has been reached on positive or negative selection criteria for NSCs, considerable progress has been made in their characterization. Improvement and multiplication of methodologies like the NSA and the NCFCA, in combination with flow cytometry and *in vivo* transplantation assays, offer the opportunity to study and uncover NSC markers and functions that can contribute to a better understanding of adult brain function and dysfunction.

Our understanding of NSC biology reveals the therapeutic potential of these cells in regenerative medicine. The persistence of stem cells in the adult brain and their implication in its homeostasis provide the opportunity to modulate the cellular composition of the injured or diseased adult brain and therefore represent a recruitable endogenous source for repair. Cell-replacement therapy using stem cells also offer the possibility of restoring aged, injured or lost cells and thus regenerating and/or repairing damaged tissues and organs. Finally, these cells can be used as delivery vehicles for therapeutic molecules in the treatment of lesions and genetic or degenerative disorders.

Although the promise of stem cell research is vast, the barriers of translational research are high. Further investigation is needed to ensure effective and safe clinical trials if the young stem cell therapy field is to avoid the fate of the gene therapy field, which became stalled and mistrusted following the tragedies of early trials. Nonetheless, although caution is required, translation to the clinic is essential.

References

Altman, J. 1962. Are new neurons formed in the brains of adult mammals? Science 135, 1127–1128.

Altman, J. 1963. Autoradiographic investigation of cell proliferation in the brains of rats and cats. Anat. Rec. 145, 573–591.

Altman, J., Das, G.D. 1965. Autoradiographic and histological evidence of postnatal hippocampal neurogenesis in rats. J. Comp. Neurol. 124, 319–335.

Barroux, G. 2003. Lorsque Trembley et Réamur parlaient de 'régénération'. Medecine/Science 19, 761–762.

Bedard, A., Parent, A. 2004. Evidence of newly generated neurons in the human olfactory bulb. Brain Res. Dev. Brain Res. 151, 159–168.

Bedard, A., Cossette, M., Levesque, M., Parent, A. 2002. Proliferating cells can differentiate into neurons in the striatum of normal adult monkey. Neurosci. Lett. 328, 213–216.

Bernier, P.J., Bedard, A., Vinet, J., Levesque, M., Parent, A. 2002. Newly generated neurons in the amygdala and adjoining cortex of adult primates. Proc. Nat. Acad. Sci. USA 99, 11 464–11 469.

Brown, J., Cooper-Kuhn, C.M., Kempermann, G., Van Praag, H., Winkler, J., Gage, F.H., Kuhn, H.G. 2003. Enriched environment and physical activity stimulate hippocampal but not olfactory bulb neurogenesis. Eur. J. Neurosci. 17, 2042–2046.

Bull, N.D., Bartlett, P.F. 2005. The adult mouse hippocampal progenitor is neurogenic but not a stem cell. J. Neurosci. 25, 10 815–10 821.

Cameron, H.A., McKay, R.D. 2001. Adult neurogenesis produces a large pool of new granule cells in the dentate gyrus. J. Comp. Neurol. 435, 406–417.

Cameron, H.A., Woolley, C.S., McEwen, B.S., Gould, E. 1993. Differentiation of newly born neurons and glia in the dentate gyrus of the adult rat. Neuroscience 56, 337–344.

Capela, A., Temple, S. 2002. LeX/ssea-1 is expressed by adult mouse CNS stem cells, identifying them as nonependymal. Neuron 35, 865–875.

Carlen, M., Cassidy, R.M., Brismar, H., Smith, G.A., Enquist, L.W., Frisen, J. 2002. Functional integration of adult-born neurons. Curr. Biol. 12, 606–608.

Corti, S., Locatelli, F., Papadimitriou, D., Donadoni, C., Salani, S., *et al.*, 2006. Identification of a primitive brain-derived neural stem cell population based on aldehyde dehydrogenase activity. Stem Cells 24, 975–985.

Corti, S., Nizzardo, M., Nardini, M., Donadoni, C., Locatelli, F., Papadimitriou, D., *et al.*, 2007. Isolation and characterization of murine neural stem/progenitor cells based on Prominin-1 expression. Exp. Neurol. 205, 547–562.

Coskun, V., Wu, H., Blanchi, B., Tsao, S., Kim, K., Zhao, J., *et al.*, 2008. CD133+ neural stem cells in the ependyma of mammalian postnatal forebrain. Proc. Nat. Acad. Sci. USA 105(3), 1026–1031.

Dayer, A.G., Cleaver, K.M., Abouantoun, T., Cameron, H.A. 2005. New GABAergic interneurons in the adult neocortex and striatum are generated from different precursors. J. Cell. Biol. 168, 415–427.

Deleyrolle, L., Rietze, R.L., Reynolds, B.A. 2008. The neurosphere assay, a method under scrutiny. Acta Neuropsych. 20(1), 2–8.

Doetsch, F. 2003. The glial identity of neural stem cells. Nat. Neurosci. 6, 1127–1134.

Doetsch, F., Caille, I., Lim, D.A., Garcia-Verdugo, J.M., Alvarez-Buylla, A. 1999. Subventricular zone astrocytes are neural stem cells in the adult mammalian brain. Cell 97, 703–716.

Enwere, E., Shingo, T., Gregg, C., Fujikawa, H., Ohta, S., Weiss, S. 2004. Aging results in reduced epidermal growth factor receptor signaling, diminished olfactory neurogenesis, and deficits in fine olfactory discrimination. J. Neurosci. 24, 8354–8365.

Eriksson, P.S., Perfilieva, E., Bjork-Eriksson, T., Alborn, A.M., Nordborg, C., Peterson, D.A., Gage, F.H. 1998. Neurogenesis in the adult human hippocampus. Nat. Med. 4, 1313–1317.

Foroni, C., Galli, R., Cipelletti, B., Caumo, A., Alberti, S., Fiocco, R., Vescovi, A. 2007. Resilience to transformation and inherent genetic and functional stability of adult neural stem cells *ex vivo*. Cancer Res. 67(8), 3725–3733.

Gheusi, G., Cremer, H., McLean, H., Chazal, G., Vincent, J.D., Lledo, P.M. 2000. Importance of newly generated neurons in the adult olfactory bulb for odor discrimination. Proc. Nat. Acad. Sci. USA 97, 1823–1828.

Goodell, M.A., Brose, K., Paradis, G., Conner, A.S., Mulligan, R.C. 1996. Isolation and functional properties of murine hematopoietic stem cells that are replicating *in vivo*. J. Exp. Med. 183, 1797–1806.

Gooi, H.C., Feizi, T., Kapadia, A., Knowles, B.B., Solter, D., Evans, M.J. 1981. Stage-specific embryonic antigen involves alpha 1 goes to 3 fucosylated type 2 blood group chains. Nature 292, 156–158.

Gould, E., Tanapat, P., McEwen, B.S., Flugge, G., Fuchs, E. 1998. Proliferation of granule cell precursors in the dentate gyrus of adult monkeys is diminished by stress. Proc. Nat. Acad. Sci. USA 95, 3168–3171.

Gould, E., Beylin, A., Tanapat, P., Reeves, A., Shors, T.J. 1999. Learning enhances adult neurogenesis in the hippocampal formation. Nat. Neurosci. 2, 260–265.

Hastings, N.B., Gould, E. 1999. Rapid extension of axons into the CA3 region by adult-generated granule cells. J. Comp. Neurol. 413, 146–154.

Jacobson, L.O., Simmons, E.L., Marks, E.K., Eldredge, J.H. 1951. Recovery from radiation injury. Science 113, 510–511.

Kaplan, M.S., Hinds, J.W. 1977. Neurogenesis in the adult rat: electron microscopic analysis of light radioautographs. Science 197, 1092–1094.

Kastan, M.B., Schlaffer, E., Russo, J.E., Colvin, O.M., Civin, C.I., Hilton, J. 1990. Direct demonstration of elevated aldehyde dehydrogenase in human hematopoietic progenitor cells. Blood 75, 1947–1950.

Kempermann, G., Kuhn, H.G., Gage, F.H. 1997. More hippocampal neurons in adult mice living in an enriched environment. Nature 386, 493–495.

Kim, M., Morshead, C.M. 2003. Distinct populations of forebrain neural stem and progenitor cells can be isolated using side-population analysis. J. Neurosci. 23, 10 703–10 709.

Kokoeva, M.V., Yin, H., Flier, J.S. 2007. Evidence for constitutive neural cell proliferation in the adult murine hypothalamus. J. Comp. Neurol. 505, 209–220.

Kornack, D.R., Rakic, P. 1999. Continuation of neurogenesis in the hippocampus of the adult macaque monkey. Proc. Nat. Acad. Sci. USA 96, 5768–5773.

Kuhn, H.G., Dickinson-Anson, H., Gage, F.H. 1996. Neurogenesis in the dentate gyrus of the adult rat: age-related decrease of neuronal progenitor proliferation. J. Neurosci. 16, 2027–2033.

Kukekov, V.G., Laywell, E.D., Suslov, O., Davies, K., Scheffler, B., Thomas, L.B., *et al.*, 1999. Multipotent stem/progenitor cells with similar properties arise from two neurogenic regions of adult human brain. Exp. Neurol. 156, 333–344.

Lois, C., Alvarez-Buylla, A. 1994. Long-distance neuronal migration in the adult mammalian brain. Science 264, 1145–1148.

Lorenz, E., Uphoff, D., Reid, T.R., Shelton, E. 1951. Modification of irradiation injury in mice and guinea pigs by bone marrow injections. J. Nat. Cancer Inst. 12, 197–201.

Louis, S.A., Rietze, R.L., Deleyrolle, L., Wagey, R.E., Thomas, T.E., Eaves, A.C., Reynolds, B.A. 2008. Enumeration of neural stem and progenitor cells in the neural colony-forming cell assay. Stem Cells 26, 988–996.

Mak, G.K., Enwere, E.K., Gregg, C., Pakarainen, T., Poutanen, M., Huhtaniemi, I., Weiss, S. 2007. Male pheromone-stimulated neurogenesis in the adult female brain: possible role in mating behavior. Nat. Neurosci. 10, 1003–1011.

Markakis, E.A., Gage, F.H. 1999. Adult-generated neurons in the dentate gyrus send axonal projections to field CA3 and are surrounded by synaptic vesicles. J. Comp. Neurol. 406, 449–460.

Marshall, G.P. 2nd, Laywell, E.D., Zheng, T., Steindler, D.A., Scott, E.W. 2006. *In vitro*-derived 'neural stem cells' function as neural progenitors without the capacity for self-renewal. Stem Cells 24, 731–738.

McCulloch, E.A., Till, J.E., Siminovitch, L. 1965. The role of independent and dependent stem cells in the control of hemopoietic and immunologic responses. Wistar Inst. Symp. Monogr. 4, 61–68.

Miraglia, S., Godfrey, W., Yin, A.H., Atkins, K., Warnke, R., Holden, J.T., *et al.*, 1997. A novel five-transmembrane hematopoietic stem cell antigen: isolation, characterization, and molecular cloning. Blood 90, 5013–5021.

Nottebohm, F. 1985. Neuronal replacement in adulthood. Ann, NY Acad. Sci. 457, 143–161.

Pencea, V., Bingaman, K.D., Freedman, L.J., Luskin, M.B. 2001. Neurogenesis in the subventricular zone and rostral migratory stream of the neonatal and adult primate forebrain. Exp. Neurol. 172, 1–16.

Petreanu, L., Alvarez-Buylla, A. 2002. Maturation and death of adult-born olfactory bulb granule neurons: role of olfaction. J. Neurosci. 22, 6106–6113.

Potten, C.S., Loeffler, M. 1990. Stem cells: attributes, cycles, spirals, pitfalls and uncertainties. Lessons for and from the crypt. Development 110(4), 1001–1020.

Privat, A. 1975. Postnatal gliogenesis in the mammalian brain. Int. Rev. Cytol. 40, 281–323.

Privat, A., Leblond, C.P. 1972. The subependymal layer and neighboring region in the brain of the young rat. J. Comp. Neurol. 146, 277–302.

Ramalho-Santos, M., Willenbring, H. 2007. On the origin of the term 'stem cell'. Cell Stem Cell 1, 35–38.

Ramon y Cajal, S. 1913. Degeneration and Regeneration of the Nervous System. Oxford: Oxford University Press.

Ratcliff, M.J. 2005. Experimentation, communication and patronage: a perspective on Rene-Antoine Ferchault de Reaumur 1683–1757). Biol Cell 97, 231–233.

Reynolds, B.A., Weiss, S. 1992. Generation of neurons and astrocytes from isolated cells of the adult mammalian central nervous system. Science 255, 1707–1710.

Reynolds, B.A., Rietze, R.L. 2005. Neural stem cells and neurospheres – re-evaluating the relationship. Nat. Methods 2, 333–336.

Rietze, R.L., Valcanis, H., Brooker, G.F., Thomas, T., Voss, A.K., Bartlett, P.F. 2001. Purification of a pluripotent neural stem cell from the adult mouse brain. Nature 412, 736–739.

Rochefort, C., Gheusi, G., Vincent, J.D., Lledo, P.M. 2002. Enriched odor exposure increases the number of newborn neurons in the adult olfactory bulb and improves odor memory. J. Neurosci. 22, 2679–2689.

Seaberg, R.M., van der Kooy, D. 2002. Adult rodent neurogenic regions: the ventricular subependyma contains neural stem cells, but the dentate gyrus contains restricted progenitors. J. Neurosci. 22, 1784–1793.

Shingo, T., Gregg, C., Enwere, E., Fujikawa, H., Hassam, R., Geary, C., *et al.*, 2003. Pregnancy-stimulated neurogenesis in the adult female forebrain mediated by prolactin. Science 299, 117–120.

Shors, T.J., Miesegaes, G., Beylin, A., Zhao, M., Rydel, T., Gould, E. 2001. Neurogenesis in the adult is involved in the formation of trace memories. Nature 410, 372–376.

Shors, T.J., Townsend, D.A., Zhao, M., Kozorovitskiy, Y., Gould, E. 2002. Neurogenesis may relate to some but not all types of hippocampal-dependent learning. Hippocampus 12, 578–584.

Siminovitch, L., McCulloch, E.A., Till, J.E. 1963. The distribution of colony-forming cells among spleen colonies. J. Cell Physiol. 62, 327–336.

Stanfield, B.B., Trice, J.E. 1988. Evidence that granule cells generated in the denate gyrus of adult rats extend axonal projections. Exp. Brain Res. 72, 399–406.

Storms, R.W., Trujillo, A.P., Springer, J.B., Shah, L., Colvin, O.M., Ludeman, S.M., Smith, C. 1999. Isolation of primitive human hematopoietic progenitors on the basis of aldehyde dehydrogenase activity. Proc. Nat. Acad. Sci. USA 96, 9118–9123.

Till, J.E., McCulloch, E.A. 1961. A direct measurement of the radiation sensitivity of normal mouse bone marrow cells. Radiation Res. 14, 213–222.

Trembley, A. 1744. Mémoire pour servir à l'histoire d'un genre de polypes d'eau douce. A Leide: Chez Jean & Herman Verbeek.

Uchida, N., Buck, D.W., He, D., Reitsma, M.J., Masek, M., Phan, T.V., *et al.*, 2000. Direct isolation of human central nervous system stem cells. Proc. Nat. Acad. Sci. USA 97, 14 720–14 725.

van Praag, H., Kempermann, G., Gage, F.H. 1999. Running increases cell proliferation and neurogenesis in the adult mouse dentate gyrus. Nat. Neurosci. 2, 266–270.

Vescovi, A.L., Parati, E.A., Gritti, A., Poulin, P., Ferrario, M., Wanke, E., *et al.*, 1999. Isolation and cloning of multipotential stem cells from the embryonic human CNS and establishment of transplantable human neural stem cell lines by epigenetic stimulation. Exp. Neurol. 156, 71–83.

Weiss, S., Dunne, C., Hewson, J., Wohl, C., Wheatley, M., Peterson, A.C., Reynolds, B.A. 1996. Multipotent CNS stem cells are present in the adult mammalian spinal cord and ventricular neuroaxis. J. Neurosci. 16, 7599–7609.

Zhao, M., Momma, S., Delfani, K., Carlen, M., Cassidy, R.M., Johansson, C.B., *et al.*, 2003. Evidence for neurogenesis in the adult mammalian substantia nigra. Proc. Nat. Acad. Sci. USA 100, 7925–7930.

Part II

Cancer Stem Cells

6 The Role of Epithelial–Mesenchymal Transition in Cancer Metastasis

Paul C. McDonald[1] **and Shoukat Dedhar**[1,2]

[1]*Department of Integrative Oncology, British Columbia Cancer Research Centre, Vancouver, BC, Canada*

[2]*Department of Biochemistry and Molecular Biology, University of British Columbia, Vancouver, BC, Canada*

6.1 Introduction

The epithelial–mesenchymal transition (EMT) is a highly conserved cellular programme that is central to the normal development of multicellular organisms and that, when engaged, allows stationary epithelial cells to acquire mesenchymal characteristics and gain the ability to migrate and invade (Kalluri and Weinberg, 2009; Nieto, 2013). The EMT programme comprises several key events that are considered hallmarks of the process. These include the loss of epithelial cell–cell junctions, the loss of apical–basal polarity concomitant with the acquisition of front–rear polarity, the coordinated downregulation of an epithelial gene signature – particularly repression of E-cadherin – and the upregulation/activation of genes that define the mesenchymal phenotype, such as vimentin, N-cadherin and smooth-muscle actin (SMA) (Nieto, 2013; Tam and Weinberg, 2013; Lamouille *et al.*, 2014). These phenotypic changes are associated with substantive alterations in cell behaviour, notably the ability to degrade surrounding extracellular matrix (ECM) and basement membrane components, increase motility and invasive properties and resist apoptosis (Nieto, 2013; Lamouille *et al.*, 2014), traits that allow mesenchymal cells to move from one location to another. Importantly, upon reaching their destination, mesenchymal cells revert to an epithelial

Principles of Stem Cell Biology and Cancer: Future Applications and Therapeutics, First Edition.
Edited by Tarik Regad, Thomas J. Sayers and Robert C. Rees.

phenotype, through a process called mesenchymal–epithelial transition (MET), allowing them to establish, proliferate and differentiate, resulting in the formation of complex structures and organs and demonstrating the high degree of cellular plasticity (i.e. the ability of cells to change phenotype in a reversible manner) inherent within the EMT programme (Thiery, 2002; Nieto, 2013; Tam and Weinberg, 2013). In addition to the well-established role of EMT in normal development, it is now known that this core cellular programme, which is silenced in adult tissues, can be activated in a variety of pathologies. This has resulted in the recognition of three distinct types of EMT: developmental EMTs (Type 1), EMTs related to wound-healing fibrosis and tissue regeneration (Type 2) and EMTs associated with cancer progression and metastasis (Type 3) (Kalluri and Weinberg, 2009; Nieto, 2013). Indeed, research over the past decade has demonstrated that epithelial tumour cells can co-opt the EMT programme in order to promote cancer progression and, in particular, metastasis.

Cancer metastasis is an inherently inefficient process and is generally viewed as a cascade of several rate-limiting stages or steps (invasion, intravasation, systemic transport, extravasation and colonization) (Nguyen *et al.*, 2009; Chaffer and Weinberg, 2011). Initial invasion at the primary tumour site requires that epithelial tumour cells degrade the underlying basement membrane and ECM, acquire migratory properties and invade the adjacent tissue stroma. Invading tumour cells must then penetrate the endothelial lining of adjacent vasculature, in a process called intravasation. Successful intravasation results in the entrance of cells into the systemic circulation, allowing tumour cells that have adapted for survival in this environment to travel to distant locations. Ultimately, select circulating tumour cells (CTCs) may arrest in the microvessels and capillaries of distant organs, penetrate the endothelial barrier once more (in a process called extravasation) and enter the stromal environment of the organs. In the final step of metastasis, called colonization, a small subset of these cells establish in the new environment and eventually proliferate to form clinically detectable macrometastases. EMT represents a vital programme required by epithelial cancer cells for successful completion of these steps (Chaffer and Weinberg, 2011). Indeed, EMT in cancer cells shares many of the hallmarks of EMT in development, and ultimately leads to cell dissociation and motility, two key steps in metastasis (Nieto, 2013; Tam and Weinberg, 2013).

An appreciation of the role of EMT in cancer progression and metastasis is evolving rapidly. Many studies using *in vitro* and *in vivo* models of cancer have clearly shown the importance of EMT in cancer progression (Yang and Weinberg, 2008; Thiery *et al.*, 2009; Scheel and Weinberg, 2012; Nieto, 2013). However, the clinical relevance of EMT in the progression and metastasis of human cancer has been debated. While epithelial cancer cells in primary human tumours and CTCs exhibit hallmarks of EMT (Wan *et al.*, 2013; Krebs *et al.*, 2014), cells comprising distant macrometastases generally show an epithelial phenotype (Nieto, 2013; Tam and Weinberg, 2013), suggesting that mesenchymal disseminating tumour cells must revert to an epithelial phenotype (the MET) in order to successfully establish macrometastases. In addition, cancer cells that undergo EMT share many morphological and functional attributes with stromal cells, such as cancer-associated fibroblasts (CAFs), making the identification of tumour cells that have undergone EMT in complex tissues a particular challenge. The relevance of EMT to human cancer metastasis is highlighted by studies demonstrating that the expression of master regulators of EMT, including Twist1, Zeb1, Snail1 and Snail2, in primary tumours is linked to an increased risk of metastasis (De Craene and Berx, 2013). Also, breast and prostate cancer patients with a poor prognosis have been shown to have CTCs that have undergone EMT. Thus, the EMT programme must be very dynamic, and a model of EMT incorporating the concept of reversibility is key to understanding its role in cancer metastasis.

In this chapter, we will provide an overview of the basic components of EMT activation as they relate to metastasis, including the core inducers, transcriptional regulators and effector molecules. We will examine the role of EMT in the various steps of the metastatic cascade and discuss the important concept of epithelial–mesenchymal plasticity as it relates to the EMT programme in metastasis. We will then turn our attention to cancer stem cells (CSCs) and the role of EMT in their biology. We will also discuss the contribution of hypoxia to EMT and CSCs. Finally, we will highlight the role of EMT in the resistance of tumour cells to cancer treatment and discuss ways of interfering with EMT and CSC stemness to specifically target these treatment-resistant cells.

6.2 Fundamental components of the cellular and molecular programme of EMT in cancer

Several recent reviews have provided comprehensive details of the molecular pathways regulating EMT and its counterpart, MET (Lee *et al.*, 2006; Kalluri and Weinberg, 2009; De Craene and Berx, 2013; Lamouille *et al.*, 2013, 2014). The purpose of this section is to provide an overview of the major components involved in the activation of the EMT programme and to provide a framework for further discussions relating to the role of EMT in metastasis. Integration of multiple molecular signalling pathways and regulators is required in order to activate the programme and to mediate the phenotypic and functional changes observed in EMT. These components can be placed into one of three broad categories, based on their functional attributes (Tsai and Yang, 2013): the extracellular microenvironmental stimuli that activate EMT signalling pathways are known as 'EMT inducers', the transcription factors that orchestrate and control EMT are called the 'core regulators' and the molecules that execute the programme are termed the 'effectors' (Tsai and Yang, 2013).

6.2.1 EMT inducers

The induction in epithelial cancer cells of the EMT programme occurs, in large part, through extracellular stimuli present within the tumour microenvironment, including cellular constituents such as CAFs and myeloid-derived suppressor cells (MDSCs), soluble mediators such as growth factors and cytokines and environmental conditions such as hypoxia and acidosis (Gao *et al.*, 2012; Hanahan and Coussens, 2012; Kang and Pantel, 2013; Tam and Weinberg, 2013; Pattabiraman and Weinberg, 2014) (Figure 6.1). Importantly, the heterogeneity inherent to the tumour microenvironment defines the context and tissue-type and cell-type dependence of EMT inducers (Tam and Weinberg, 2013; Tsai and Yang, 2013). Thus, the specific inducers of the EMT programme in cancer are dependent on tissue context and may vary from tumour to tumour, but all induce a common set of signal pathways that

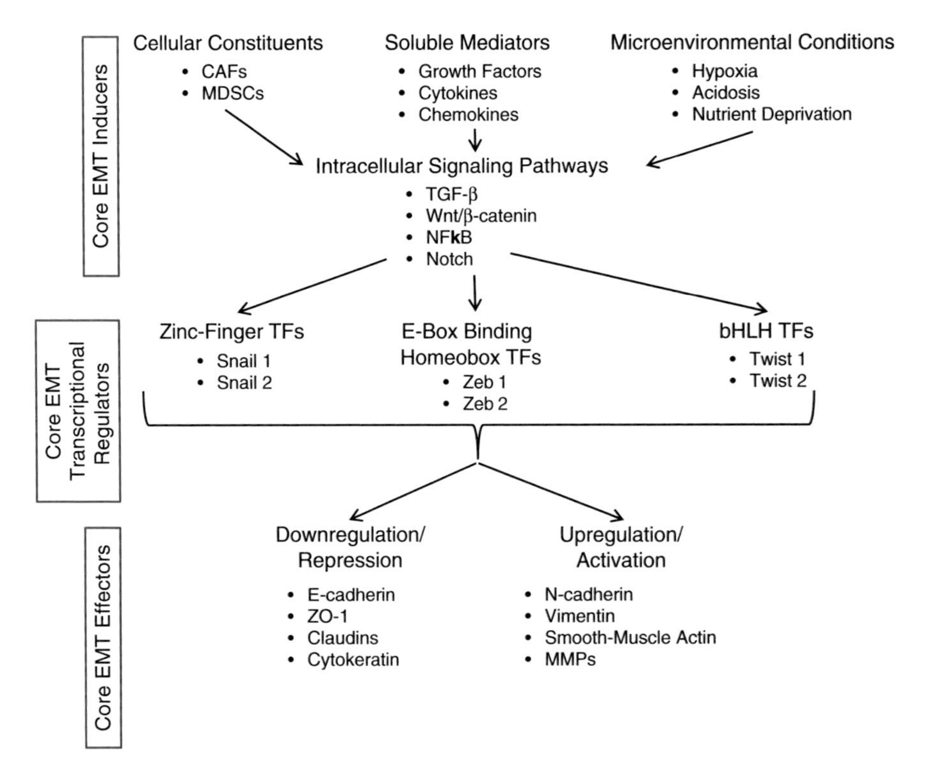

Figure 6.1 Core components of the activation of the EMT programme in cancer cells. The cells, soluble mediators and microenvironmental conditions that make up the heterogeneous tumour microenvironment serve to induce EMT through the coordinated activation and regulation of multiple intracellular signalling pathways. The activation of these pathways results in the expression and stimulation of several core transcriptional regulators. Once activated, these EMT TFs modulate the differential regulation of several effector molecules of EMT, including the downregulation and repression of epithelial junctional proteins, especially E-cadherin, and the upregulation and activation of mesenchymal proteins such as vimentin and SMA.

make up the core EMT programme. These inducers activate EMT through the coordinated regulation of several developmental signalling pathways, including TGF-β, Wnt, Notch and NF-κB signalling cascades (Figure 6.1). The prototypic inducer of EMT, TGF-β, initiates the programme through stimulation of the TGF-β pathway, in association with the Wnt/β-catenin/LEF-1 and Ras kinase pathways (Nieto, 2013; Tam and Weinberg, 2013). Signalling through the TGF-β pathway results in the formation of a transcription factor complex that migrates to the nucleus and induces a transcription programme that mediates acquisition of mesenchymal traits and suppression of epithelial properties (Tsai and Yang, 2013). Similarly, inflammatory cytokines such as TNF-α can activate EMT through the NF-κB pathway. Hypoxia, through the stabilization and activation of HIF-1α, can also activate and promote EMT (see Section 6.6) (Marie-Egyptienne *et al.*, 2013; Tsai and Yang, 2013). Induction of EMT through cooperative regulation of multiple signalling pathways culminates in the expression and activation of the core transcriptional regulators of EMT.

6.2.2 Core regulators

In epithelial cancers, the induction of EMT by extracellular stimuli and subsequent activation of signalling cascades results in the expression and stimulation of transcription factors that mediate the regulation of a myriad of genes involved in the processes of cell adhesion, mesenchymal differentiation, migration and invasion (Kalluri and Weinberg, 2009; Scheel and Weinberg, 2012; Nieto, 2013; Tsai and Yang, 2013). The core transcriptional regulators of EMT, known collectively as EMT transcription factors (EMT TFs), incorporate three separate protein families: the zinc-finger family transcription factors Snail1 and Snail2, the E-box-binding homoeobox-family transcription factors Zeb1 and Zeb2 and the basic helix-loop-helix (bHLH)-family transcription factors Twist1 and Twist2 (Kalluri and Weinberg, 2009; Tsai and Yang, 2013) (Figure 6.1). Importantly, these transcription factors may induce EMT alone or in cooperation with one another other. All of them serve to downregulate cell junctional protein expression, especially E-cadherin expression. Snail and Zeb also suppress claudins and ZO-1 junctional proteins, while Twist activates processes critical for cell invasion, demonstrating that these transcription factors are robust mediators of EMT, directing multiple components of the programme (Scheel and Weinberg, 2012; Tam and Weinberg, 2013). These core transcriptional regulators function to modulate the expression of the effector molecules involved in EMT.

6.2.3 Effectors

The key effector molecules of EMT, including E-cadherin, N-cadherin, vimentin, cytokeratins and SMA, are involved in the conversion from an epithelial phenotype to a mesenchymal phenotype, as well as the promotion of cell migration and invasion (Tsai and Yang, 2013) (Figure 6.1). Specifically, epithelial cancer cells must downregulate junctional proteins, the most critical of which is E-cadherin, the protein regarded as the seminal gatekeeper of the epithelial state (Nieto, 2013; Tam and Weinberg, 2013). In fact, E-cadherin is subject to multiple levels of regulation, including transcriptional repression, promoter methylation and protein degradation. In conjunction with repression of E-cadherin expression, cells undergoing EMT increase their levels of the intermediate filament protein vimentin, as well as SMA. The differential regulation of these key genes allows for the phenotypic conversion of epithelial cells to mesenchymal cells and the acquisition of the migratory and invasive behaviours critical for metastasis.

6.3 The role of EMT in metastasis

The majority of the carcinoma cells present within a growing tumour exhibit primarily epithelial characteristics and properties, including robust expression of E-cadherin, high levels of proliferation and limited motility (Scheel and Weinberg, 2012; Tam and Weinberg, 2013). However, some cancer cells develop the ability to metastasize, resulting in the growth of tumours at locations distant from the primary tumour. For these cells to invade local tissue, disseminate to distant organs and tissues through the systemic circulation and eventually give rise to metastatic colonies, they must undergo a phenotypic and functional shift towards a more motile, invasive mesenchymal cell (Chaffer and Weinberg, 2011; Scheel and Weinberg, 2012). Activation of EMT is the mechanism by which cancer cells acquire these attributes. Strong concordance exists between the mesenchymal phenotype of cancer cells and the phenotypic characteristics required by cancer cells for metastasis. During EMT, epithelial cancer cells lose their differentiated epithelial characteristics, including intercellular adhesion, polarity and nonmotility, and acquire mesenchymal traits, including motility, invasiveness, resistance to apoptosis and resistance to treatment (Nieto, 2013; Tam and Weinberg, 2013). Prototypic features of EMT in the context of cancer are similar to those critical for EMT in development: loss of epithelial cell polarity and cell–cell contacts; loss of the epithelial marker E-cadherin; acquisition of mesenchymal markers such as vimentin, N-cadherin and SMA; upregulation of cell migration and invasion; and resistance to treatment (Tam and Weinberg, 2013; Tsai and Yang, 2013). These changes result in a more aggressive, metastatic phenotype. Thus, EMT can provide epithelial cells with the critical traits required for metastasis (Nieto, 2013). This section discusses the role of EMT and MET in the various steps of metastasis.

6.3.1 Mesenchymal conversion and local invasion

Local invasion of the surrounding ECM is one of the earliest steps in the process of metastasis and is an important determinant of the metastatic potential of cancer cells. In order to invade local tissue, epithelial cancer cells must undergo adaptive structural changes that allow them to simultaneously loosen cell–cell adhesions, degrade and remodel the ECM and become motile (Nieto, 2013; Tsai and Yang, 2013) (Figures 6.2 and 6.3). EMT core regulators such as Snail1 repress the expression of E-cadherin and other junctional proteins and simultaneously induce the expression of soluble and membrane-tethered (MT) matrix metalloproteases (e.g. MT1-MMP, MT2-MMP, MMP2, MMP9) to digest or degrade the basement membrane (Tsai and Yang, 2013). In addition to their degradative properties, these proteases may also be involved in regulating the disruption of cell–cell junctional complexes, suggesting that there may be coordinated induction of proteases and loss of cellular junctions in the promotion of the invasion process (Tsai and Yang, 2013).

In addition to MMP activation, emerging evidence suggests that the activation of EMT TFs by inducers of EMT can result in the formation of invadopodia, specialized actin-based cellular protrusions capable of recruiting proteases to sites of cell–ECM contact in order to focus matrix breakdown at specific locations (Murphy and Courtneidge, 2011; Tsai and Yang, 2013) (Figure 6.2). For example, growth factor signalling coupled with the activation of Twist is reported to promote the formation of invadopodia in cultured cancer cells, as is Zeppo1, a novel promoter of metastasis that represses E-cadherin expression (Tsai and Yang, 2013). Collectively, the

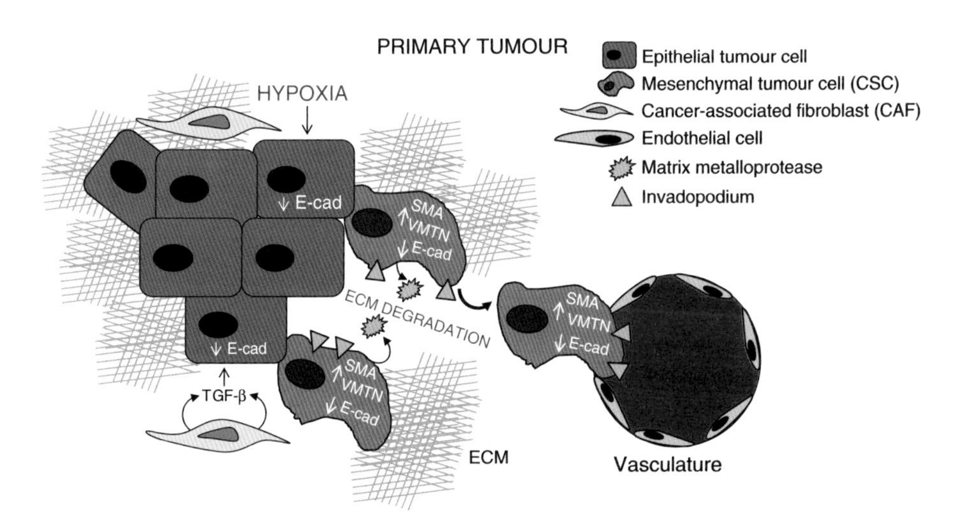

Figure 6.2 Role of the EMT programme in local invasion. Stimuli present in the tumour microenvironment, including TGF-β produced by CAFs and hypoxia, activate EMT in epithelial cancer cells. Activation of EMT results in the repression of E-cadherin (a hallmark of the cellular programme), acquisition of mesenchymal phenotypic traits and activation of ECM degradative enzymes. EMT activation also induces the formation of invadopodia, to allow for focused matrix breakdown. These events culminate in increased migration and invasion of tumour cells, allowing them to invade the surrounding tissue and intravasate into the vasculature.

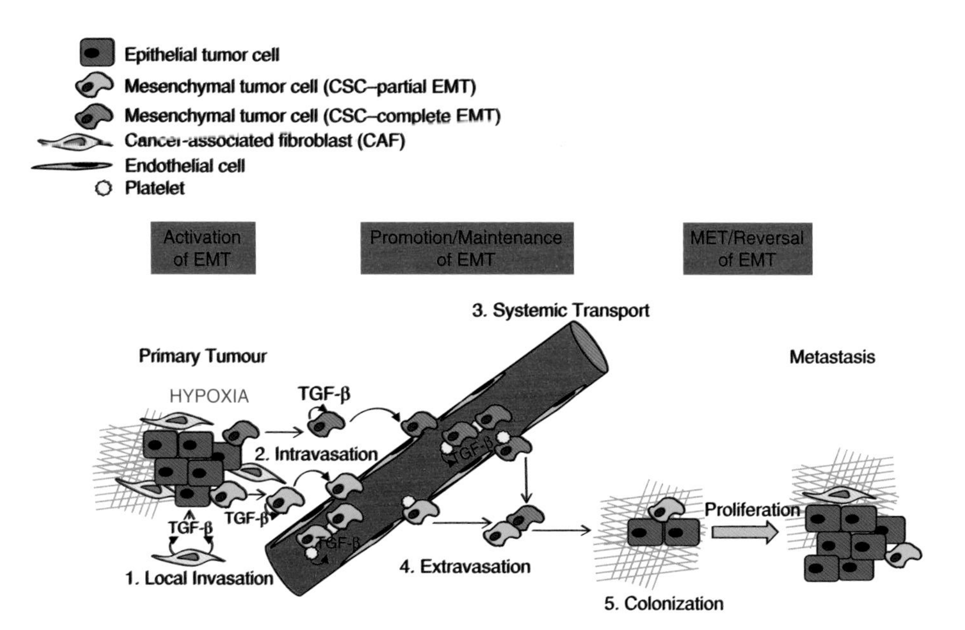

Figure 6.3 Role of EMT in metastasis cascade, epithelial–mesenchymal plasticity and CSCs. Activation of EMT by paracrine signals present in the tumour microenvironment induces a phenotypic and behavioural shift in epithelial cancer cells towards a more mesenchymal cell, including increased migratory, invasive and survival properties that allow for local invasion. Individual cells advance to different points along the EMT pathway, with some cells undergoing 'partial EMT' and some undergoing 'complete EMT', yielding cells with different metastatic propensities. Cancer cells that have undergone EMT exhibit hallmarks of CSCs and maintain their mesenchymal phenotype through both paracrine and autocrine signalling. Mesenchymal CSCs invade local tissue and intravasate into vessels to disseminate into the systemic circulation. Survival of these CTCs is critical for metastasis and is mediated, in part, through maintenance of the EMT programme by platelet-derived paracrine signalling and adhesion. Successful CTCs may then extravasate at distant sites and establish in the tissue stroma, a process that is influenced by cellular plasticity and requires the ability of mesenchymal cells (CSCs) to revert to an epithelial state through MET. Proliferation of the established tumour cells results in the formation of clinically relevant macrometastases.

repression of epithelial traits and upregulation of mesenchymal properties results in cells being permissive for invasion.

6.3.2 Tumour cell intravasation and transport within the systemic circulation

Subsequent to local invasion, tumour cells must intravasate into the vasculature to allow for systemic dissemination by the vascular system (Chaffer and Weinberg, 2011) (Figures 6.2 and 6.3). The modulation of properties acquired by cancer cells during the activation of EMT, especially those involved in migration and invasion, likely plays a role in the transit of invasive tumour cells across the endothelial cell barrier. However, the detailed mechanisms involved in the process of intravasation by tumour cells that have undergone EMT are still being delineated (Tsai and Yang, 2013). Recent studies have shown that the expression of Zeb1 in PC-3 prostate cancer cells is required for increased migration through the endothelial cell barrier, while the overexpression of Snail1 has been reported to promote intravasation through activation of membrane-bound but not secreted MMPs, providing initial evidence of a role for EMT in this process (Tsai and Yang, 2013). Indeed, EMT may be an enabling force, providing cells with the tools required to intravasate.

Invading tumour cells that successfully intravasate and enter the systemic circulation are termed 'CTCs' (Ksiazkiewicz *et al.*, 2012; Kang and Pantel, 2013; Krebs *et al.*, 2014). Their presence in the circulation represents the mechanism for the transport of cancer cells with metastatic potential to distant sites, where they may disseminate and eventually form metastases. A critical requirement of CTCs is their ability to resist anoikis and survive in the circulation (Krebs *et al.*, 2014). Maintenance of the EMT programme by CTCs is thought to help provide this capacity.

Several recent studies have demonstrated the presence of mesenchymal markers on CTCs, both in preclinical models of cancer and in human cancer patients (Tsai and Yang, 2013), strongly suggesting that CTCs have undergone EMT. Clinical studies have shown that the presence of increased numbers of CTCs is an indicator of poor prognosis and of distant relapse (Tsai and Yang, 2013; Krebs *et al.*, 2014). The presence of mesenchymal CTCs has also been correlated with resistance to treatment. Importantly, many of the technologies presently available for the detection and isolation of CTCs rely on epithelial markers, thus missing the subset of mesenchymal cells – a fact that has resulted in the development of assays capable of detecting mesenchymal CTCs (Krebs *et al.*, 2014). Recent interrogation of CTCs in patients, particularly when through non-antigen-dependent techniques, has demonstrated clear heterogeneity with respect to EMT markers, further suggesting the existence of a continuum of EMT phenotypes (Krebs *et al.*, 2014).

The maintenance of a mesenchymal state by CTCs in the systemic circulation may promote survival, an enabling characteristic of tumour cell dissemination. In particular, EMT may be important in preventing single tumour cells from suffering from detachment-induced anoikis by promoting CTC aggregation to leukocytes and platelets in the circulation (Tsai and Yang, 2013; Krebs *et al.*, 2014). Indeed, platelets may be important for the maintenance of the mesenchymal phenotype of CTCs (Figure 6.3). In addition to providing CTCs with physical protection in the circulation, platelets may promote EMT in these cells through TGF-β and NF-κB signalling and thus prime them for metastasis (Tsai and Yang, 2013; Krebs *et al.*, 2014). Breast cancer CTCs enriched in TGF-β-pathway components have been found clustered to platelets, a major source of the EMT inducer TGF-β, and depletion of platlet-derived TGF-β reduces distant metastasis, suggesting that maintenance of a mesenchymal state by CTCs may be important for metastasis.

6.3.3 Tumour cell extravasation

The exit of tumour cells from the systemic circulation through the endothelial barrier, a process called extravasation, is a critical step in the metastatic cascade (Figure 6.3). While one can imagine that this step, like that of intravasation, would require substantial modulation of migration and invasion processes, the precise role of EMT in extravasation remains to be extensively investigated. A major challenge to the study of EMT in this context, especially *in vivo*, is the paucity of experimental models that faithfully recapitulate physiologically relevant levels of tumour cell dissemination and extravasation, which are highly inefficient (Tsai and Yang, 2013). Most studies have relied on experimental metastasis assays, such as tail vein injection or intracardiac injection, to interrogate the mechanisms of extravasation, but these methods involve injecting very large numbers of tumour cells directly into the circulation, resulting in intravessel growth in the lung microvasculature (Tsai and Yang, 2013). Recent studies have demonstrated that tumour cells arriving at distant metastatic sites develop integrin-containing filopodium-like protrusions (FLPs) in order to interact with the surrounding ECM. FLPs can be induced by Twist1 and Snail1 expression. Their generation is correlated with the mesenchymal state of the tumour cells and their formation has been shown to be essential for successful metastasis, suggesting a role for EMT in the process of extravasation (Scheel and Weinberg, 2012; Tam and Weinberg, 2013).

6.3.4 MET and metastatic colonization

Subsequent to extravasation, disseminated tumour cells must survive in their new environment and proliferate to form first the micrometastases and then the macrometastases that are ultimately identified as clinically relevant metastatic disease (Chaffer and Weinberg, 2011; Scheel and Weinberg, 2012). The development of micro- and macrometastases at distant organ sites highlights the dynamic plasticity of the EMT programme (Figure 6.3). Specifically, while locally invading cells and CTCs demonstrate many of the cellular and molecular markers of EMT, cells comprising metastatic foci are largely of epithelial phenotype. These observations have suggested that, once metastatic cells have 'set up shop' at a secondary site, they undergo MET and revert to epithelial cells to expand the metastasis (Scheel and Weinberg, 2012; Nieto, 2013). Recent experiments suggest the presence of epithelial–mesenchymal plasticity during metastasis. In an inducible Twist1 mouse skin tumour model, activation of EMT promoted invasion, intravasation and extravasation, while subsequent loss of the EMT-inducing signal was required for cell proliferation and formation of metastases (Scheel and Weinberg, 2012). In addition, lung metastasis formation following tail vein injection of BT-549 breast cancer cells required the downregulation of a novel EMT inducer, Prrx-1, and reversion of EMT following initial Prrx1–Twist1 cooperation in EMT promotion (Scheel and Weinberg, 2012). The reversion to the epithelial state by tumour cells may be required for the re-engagement of the proliferative programmes necessary for the development of metastases. EMT regulators such as Snail1, Zeb2 and Twist1 suppress cell division, although the signalling pathways that couple EMT to proliferation remain to be elucidated.

The signals and switches required for the reversion of cancer cells from a mesenchymal phenotype to an epithelial state are not yet understood (Scheel and Weinberg, 2012; Tsai and Yang, 2013). For example, it is possible that the absence of an EMT-inducing signal may be sufficient for the reversion process. It is equally possible that specific MET-inducing signals may be necessary to actively engage MET. Alternatively, the balance between the removal of EMT-inducing signals coupled with the onset of MET-inducing signals may dictate the phenotypic direction of the cell. It is likely that signal inputs from the metastatic microenvironment may also help regulate the EMT–MET conversion, although how this occurs remains to be investigated.

6.4 Epithelial–mesenchymal plasticity

Epithelial–mesenchymal plasticity is an important concept that is critical to understanding EMT as it relates to cancer progression and metastasis. In contrast to the complete loss of epithelial markers concomitant with the acquisition of well-defined mesenchymal markers – migratory and invasive properties typical of cells that have undergone a 'complete EMT' (Kalluri, 2009; Nieto, 2013) – it is now evident that cancer cells that initiate an EMT programme often reside in a state in which newly acquired mesenchymal traits are expressed simultaneously with existing epithelial traits: a state often termed a 'partial EMT' (Nieto, 2013; Tam and Weinberg, 2013; Pattabiraman and Weinberg, 2014). Indeed, activation of the EMT programme in cancer usually results in tumour cells displaying a spectrum of phenotypes intermediate between the fully differentiated epithelial state and the completely mesenchymal state, the two opposing extremes (Tam and Weinberg, 2013). Thus, cancer cells can advance to different points along the EMT pathway (Figure 6.3). In cancer, these phenotypic shifts are initiated by a multitude of contextual signals present in the tumour microenvironment. Those cells capable of a high degree of epithelial–mesenchymal plasticity and which undergo partial EMT are the most likely to metastasize effectively, as they maintain the capacity to profoundly and reversibly change their phenotype and behaviour in order to successfully complete the various steps in the metastatic cascade. In contrast, cancer cells that undergo a complete EMT may lose the phenotypic plasticity required for tumour-initiating properties (Pattabiraman and Weinberg, 2014). Therefore, the progression of epithelial cancer cells through an EMT programme, together with the reversion of such cells to an epithelial state through MET, must be viewed as a highly plastic, dynamic process that ultimately provides epithelial cancer cells with the biological properties required for invasion and metastasis (Scheel and Weinberg, 2012).

The functional aspects of epithelial–mesenchymal plasticity rely, at least in part, on the recently elucidated connection between paracrine signals from the tumour stroma and autocrine signals produced by the cancer cells themselves (Pattabiraman and Weinberg, 2014) (Figure 6.3). Paracrine signals produced by the stroma, including ECM components, soluble factors such as TGF-β and the Wnt proteins and environmental conditions such as hypoxia and acidosis, serve to trigger the EMT programme. Autocrine signals are then produced by the tumour cells, perpetuating the presence of mesenchymal traits in the absence of further signals from the stromal compartment. The introduction of a cell-autonomous mechanism for the maintenance of mesenchymal traits provides a way for individual invading tumour cells to disseminate and produce metastases (Pattabiraman and Weinberg, 2014). Furthermore, it has been suggested that autocrine signalling loops may be important to preventing epithelial cells from activating EMT through the inhibition of EMT-promoting mediators, including TGF-β and Wnt, and that repression of these inhibitors enables pro-EMT autocrine signalling (Pattabiraman and Weinberg, 2014). The complex interplay between paracrine and autocrine signalling creates a vast array of signalling possibilities, which may dictate the extent to which cells move along the EMT continuum (Pattabiraman and Weinberg, 2014).

6.5 Association between EMT and CSCs

CSCs are important contributors to tumour recurrence and metastasis (Figure 6.3). They are functionally defined as a subset of tumour cells that are capable of self-renewal and that, when seeded at limiting dilutions in immunocompromised mice, will form new tumours with a cellular heterogeneity similar to that of the primary tumour (Chang and Mani, 2013; Marie-Egyptienne *et al.*, 2013). CSCs are generally identified using characteristic cell-surface markers, the most common of which are CD44 (often in combination with CD24), CD133 and EpCAM (Hill *et al.*, 2009; Scheel and Weinberg, 2012; Marie-Egyptienne *et al.*, 2013). It is important to recognize, however, that universal CSC cell-surface markers do not exist, the functional relevance of these markers to the biology of CSCs is not understood and the precise markers used to label CSCs are tissue type-dependent (Chang and Mani, 2013; Marie-Egyptienne *et al.*, 2013). Despite these complexities, cell-surface markers have been used to isolate CSCs from many solid tumours, including breast, brain, prostate, lung, pancreas and ovarian cancers (Marie-Egyptienne *et al.*, 2013). The CSC phenotype has several characteristics, including the capacity for self-renewal, expression of stem cell markers, increased tumourigenicity *in vivo*, enhanced survival and resistance to chemotherapy and radiotherapy. Furthermore, it is now well recognized that CSCs exhibit many of the characteristics associated with cancer cells that have undergone EMT, providing a strong biological link between EMT, CSCs and metastasis (Scheel and Weinberg, 2012).

It is becoming increasingly clear that strong connections exist between the CSC phenotype and cancer cells that have undergone EMT (Scheel and Weinberg, 2012). Several studies have shown that the activation of EMT by the inducer TGF-β and the transcriptional activators Snail1, Twist1 and Zeb1 in normal epithelial cells results in the acquisition of CSC traits, including a stem cell surface-marker profile, a mesenchymal phenotype, self-renewal capacity and increased tumourigenicity (Chang and Mani, 2013; Tam and Weinberg, 2013). A number of molecular components common to the EMT programme and the CSC state have been identified. For example, Zeb1 inhibits the expression of miR-200 family members, which are involved in the suppression of genes involved in stemness and EMT, including the polycomb protein Bmi1 and the histone-modifying enzyme Suz12. Furthermore, Zeb1 and miR-200 family members negatively regulate each other through a double-feedback loop, suggesting that the inhibition of epithelial differentiation and the acquisition of CSC traits are interconnected (Scheel and Weinberg, 2012). Furthermore, the Wnt-β-catenin signalling pathway, which is a prototypic component of the EMT programme, is active in CSCs in several types of cancer, including breast and colon cancer. The Wnt-β-catenin signalling pathway is involved in regulating the activity of Snail, which in turn represses E-cadherin expression, resulting in a feedback loop that reinforces Wnt-β-catenin activity. It is a critical inducer of EMT in carcinomas, and is also important for the maintenance of CSCs, indicating a common mechanistic connection between CSCs and cells that have undergone EMT.

An inherent consequence of the EMT process is that it imparts epithelial cancer cells with traits that are associated with increasingly malignant cell behaviours, including motility, invasiveness and resistance to apoptosis, the combination of which allows for metastatic dissemination. In addition, passage through an EMT imparts tumour-initiating properties to cancer cells – traits that may be critical for the ability of cancer cells to seed distant sites, resulting in the growth of clinically significant macrometastases (Tam and Weinberg, 2013).

Like the concept of epithelial–mesenchymal plasticity (see Section 6.4), it is evident that stemness may also be a flexible or plastic state, and that the microenvironment may be an important regulator of this flexibility. Certainly, studies have linked the CSC phenotype with cancer cells undergoing EMT (Scheel and Weinberg, 2012). The question of whether bulk tumour cells can become CSCs remains hotly debated, but CSCs that arise from the bulk tumour population may get there via EMT (Scheel and Weinberg, 2012). Current evidence points to a model whereby cancer cells co-opt an already established cell biological programme present in normal epithelial cells to provide a source of CSCs by enabling dedifferentiation of cancer epithelial cells within carcinomas, suggesting that the stemness state of CSCs may exhibit considerable plasticity. However, the extent to which CSCs derived from progenitor stem cells and those derived from differentiated cells through an EMT are similar or different remains to be rigorously investigated (Nieto, 2013).

6.6 Role of hypoxia in the induction of EMT and in supporting metastasis of cells that have undergone EMT

Contextual cues present in the tumour microenvironment are key inducers of EMT in epithelial cancer cells. In addition to the presence of numerous cell types (e.g. CAFs, vascular cells, immune cells such as TAMs), ECM and soluble mediators, the tumour microenvironment is also characterized by a modified metabolism (switch to glycolysis) and a disorganized vasculature compared to normal tissue, resulting in regions of acidic pHe, low nutrient levels and hypoxia (Hanahan and Coussens, 2012; Marie-Egyptienne *et al.*, 2013; Gatenby and Gillies, 2004). Hypoxia is a prominent element of the microenvironment of many solid tumours, including cancers of the breast, lung, colon and pancreas, and is well recognized as a critical mediator of metastasis and drug resistance. Tumour cells adapt to hypoxia by activating hypoxia-inducible factor 1 (HIF-1α), the master regulator of hypoxia-induced gene expression, and modifying critical molecular pathways that allow for survival, proliferation and invasion (Lendahl *et al.*, 2009). Hypoxia gradients in tumours are very dynamic (fluctuating levels of hypoxia due to changes in blood flow) and result in the activation of adaptive mechanisms that promote stemness, metastasis and treatment resistance, including the induction of EMT (Marie-Egyptienne *et al.*, 2013).

Hypoxia has been shown to regulate many of the core components of the EMT programme. Through the stabilization and activation of HIF-1α, it is a potent inducer of EMT regulators, including Snail, Slug and Twist (Hill *et al.*, 2009; Marie-Egyptienne *et al.*, 2013). It also results in increased expression of TGF-β, a potent inducer of EMT, and activates several intracellular signalling pathways of importance to the activation of EMT, including TGF-β, Wnt-β-catenin, Notch, NF-κB and Hedgehog (Hh) (Bao *et al.*, 2012). Hypoxia also induces the downregulation of E-cadherin and concomitant expression of N-cadherin in epithelial cells, and stimulates directed MMP production, resulting in the activation of cell motility and invasion (Philip *et al.*, 2013). The level of chronicity of hypoxia is reported to have an effect on the reversibility of EMT, with the presence of acute hypoxia resulting in Twist-mediated, reversible EMT, and long-term, chronic hypoxia mediating a less transient ZEB2-dependent EMT (Philip *et al.*, 2013). Therefore, hypoxia can serve to drive the EMT programme in cancer cells.

In addition to its direct impact on the regulation of the EMT programme, hypoxia results in the induction of a glycolytic switch by tumour cells, leading to increased production of acidic metabolites, including lactic acid, protons and CO_2 (Parks *et al.*, 2011). In order to avoid prolonged intracellular acidosis, tumour cells activate a network of proteins and buffer systems that function to maintain pH homeostasis (Gatenby and Gillies, 2008; Neri and Supuran, 2011). The membrane-bound, exofacial carbonic anhydrases, especially HIF-1α-inducible, tumour-assoicated carbonic anhydrase IX and XII (CAIX and CAXII), are critical components of this pH regulatory system (McDonald *et al.*, 2012). By catalysing the reversible hydration of carbon dioxide (CO_2) to bicarbonate (HCO_3^-) and protons (H^+), CAIX enables the maintenance of an intracellular pH favourable to cancer cell survival and growth, and simultaneously participates in extracellular acidification, facilitating tumour cell migration, invasion and metastasis (McDonald *et al.*, 2012). In the context of EMT and metastasis, recent studies have demonstrated the importance of CAIX expression and activity for the growth and metastasis of hypoxic tumours (Chiche *et al.*, 2009; Lou *et al.*, 2011; McIntyre *et al.*, 2012). Furthermore, inhibition of CAIX results in depletion of CSCs, inhibition of EMT and inhibition of tumour cell invasion in hypoxia in models of breast cancer (Lock *et al.*, 2013), indicating the importance of hypoxia and of CAIX, EMT and metastasis.

Tumour hypoxia may also support metastasis of cancer cells that have undergone EMT, particularly those with CSC properties, by acting as a component of the microenvironment or niche favourable for CSC survival (Marie-Egyptienne *et al.*, 2013). CSCs are thought to exist in microenvironments or niches that make them more resistant to chemo- and radiotherapy, and hypoxia may represent a major part of such environments (Marie-Egyptienne *et al.*, 2013). Indeed, hypoxia may provide tumour cells with cues for the maintenance of a stem-like state.

6.7 EMT and resistance to cancer treatment

The acquisition of drug resistance by cancer cells subjected to chemotherapy, targeted therapies and radiotherapy leads to tumour recurrence and metastasis and is a formidable clinical challenge for effective treatment of cancer patients (Holohan *et al.*, 2013). It is becoming increasingly clear that EMT plays an important role in such acquisition. In cancer cell lines exposed to cytotoxic chemotherapy, cells that emerge following treatment express increased levels of mesenchymal markers, indicating that they have undergone EMT (Tsai and Yang, 2013). In human patients with colorectal cancer or non-small-cell lung cancer (NSCLC), comparison of biopsied tumour tissue before and after chemotherapy has demonstrated an increase in mesenchymal markers subsequent to treatment (Tsai and Yang, 2013). Recurrent tumours in a subset of patients exhibit an EMT gene signature coupled with reduced disease-free survival (Tsai and Yang, 2013). Resistance to chemotherapy has also been linked to expression of EMT TFs, and TGF-β-induced EMT activation by chemotherapy may increase metastasis

(Chang and Mani, 2013). Recent studies have also indicated a role of EMT in the acquisition of resistance to targeted therapies. Examination of NSCLC cell lines has revealed that those expressing markers of EMT are resistant to treatment with EGF-R inhibitors such as gefitinib and erlotinib (Holohan *et al.*, 2013; Tsai and Yang, 2013). Furthermore, markers of EMT have been observed in tumour biopsies from patients who developed resistance to EGF-R inhibitors, demonstrating that EMT-associated resistance to drug therapy occurs clinically (Holohan *et al.*, 2013). A further example of acquired drug resistance involving EMT is the failure of antiangiogenic therapy in the treatment of metastatic tumours. Recent studies have established that the use of antiangiogenic therapy results in the development of hypoxia and activation of the hypoxia cascade, leading to the activation of EMT.

There are several potential mechanisms that may account for the acquisition of drug resistance via EMT. One consequence of EMT is a reduction in cell proliferation, resulting in the resistance of cells to conventional chemotherapy, which primarily targets rapidly dividing cells (Tsai and Yang, 2013). Cells that have undergone EMT also exhibit increased resistance to apoptosis, providing a potential survival advantage during a therapeutic assault (Chang and Mani, 2013; Tsai and Yang, 2013). In addition, epithelial cancer cells that have undergone EMT and acquired CSC properties exhibit increased resistance to cancer therapy and the fraction of CSCs in tumours often increases after chemotherapy, suggesting a link between CSCs derived through EMT and resistance to treatment (Chang and Mani, 2013).

Since EMT is involved in the acquisition of drug and radiation resistance by cancer cells and plays a critical role in metastatic disease, targeted killing of cancer cells that have undergone EMT will be required if durable clinical responses are to be realized. However, because these dangerous cells are, by definition, resistant to therapy, current therapeutic modalities have been relatively ineffective in eliminating them. Several challenges exist in targeting cells that have undergone EMT. In addition to accounting for their inherent drug resistance, identification of appropriate targets is critical. EMT TFs are technically challenging to target (Tsai and Yang, 2013) and therapeutic strategies that interfere with the induction of EMT or the functional consequences of EMT may be more effective (Nieto, 2013; Tsai and Yang, 2013). The plasticity of EMT also requires the appropriate temporal window for effective treatment to be identified (Tsai and Yang, 2013). For example, inhibiting EMT prior to tumour cell dissemination may be beneficial in delaying or preventing metastasis, while doing so after cells have disseminated may result in reversion of EMT at distant sites, potentially promoting metastatic spread (Nieto, 2013; Tsai and Yang, 2013). Furthermore, the potential for differentiation of epithelial cells present in the bulk tumour cell population towards a CSC state requires that the bulk tumour cells be targeted together with novel agents against CSCs (Holohan *et al.*, 2013). Therefore, the combination of cytotoxic chemotherapy and therapeutic agents targeting EMT may overcome drug resistance in disseminated tumour cells (Tsai and Yang, 2013).

Potential therapeutic strategies for targeting CSCs that arise via EMT include selectively killing mesenchymal cells by blocking EMT pathways and forcing epithelial differentiation to convert resistant mesenchymal cells into sensitive epithelial cells (Chang and Mani, 2013). Interestingly, restoration of E-cadherin expression and inhibition of Twist or Zeb1 have been shown to restore chemosensitivity in cancer cell lines, suggesting that shifting cells back towards the epithelial state can induce sensitivity (Chang and Mani, 2013). Resistance to EGF-R inhibitors resulting from EMT may be overcome through the use of inhibitors against the receptor tyrosine kinase AXL, suggesting that the use of multiple targeted therapies may be beneficial in eliminating cells with elevated metastatic potential (Holohan *et al.*, 2013).

6.8 Conclusions

EMT is a critical cellular programme in normal development that is co-opted by epithelial cancer cells and is required for progression along the metastasis cascade. It provides cancer cells with mesenchymal traits critical for metastatic processes, including invasion, dissemination and colonization. The activation of EMT in epithelial cancers requires induction by stimuli present within the tumour microenvironment, subsequent activation of intracellular signalling networks and the stimulation of core transcription factors that orchestrate the regulation of many effector molecules involved in the processes of cell adhesion, mesenchymal differentiation, migration and invasion. The concept of epithelial–mesenchymal plasticity encompasses the adaptive changes that cancer cells can undergo along the EMT–MET axis and defines the high degree of flexibility that cancer cells exhibit between the epithelial and mesenchymal states. This cellular plasticity is evident in the strong association between cells that have undergone EMT and CSCs, a subpopulation of cancer cells capable of self-renewal that is an important contributor to tumour recurrence and metastasis. Hypoxia, a microenvironmental condition prominent in many solid tumours, induces EMT and is important for the regulation of CSCs and cancer cell metastasis. Finally, EMT plays an important role in the acquisition of treatment resistance, a situation that leads to tumour recurrence and metastasis and involves cancer cells with CSC properties. Therefore, targeting of cancer cells that have undergone EMT represents an important therapeutic avenue for the control and potential elimination of metatstatic disease.

References

Bao, B., Azmi, A.S., Ali, S., Ahmad, A., Li, Y., Banerjee, S., *et al.*, 2012. The biological kinship of hypoxia with CSC and EMT and their relationship with deregulated expression of miRNAs and tumor aggressiveness. Biochim Biophys Acta 1826, 272–296.

Chaffer, C.L., Weinberg, R.A. 2011. A perspective on cancer cell metastasis. Science 331, 1559–1564.

Chang, J.T, Mani, S.A. 2013. Sheep, wolf, or werewolf: cancer stem cells and the epithelial-to-mesenchymal transition. Cancer Lett 341, 16–23.

Chiche, J., Ilc, K., Laferrière, J., Trottier, E., Dayan, F., Mazure, N.M., *et al.*, 2009. Hypoxia-inducible carbonic anhydrase IX and XII promote tumor cell growth by counteracting acidosis through the regulation of the intracellular pH. Cancer Res. 69, 358–368.

De Craene, B., Berx, G. 2013. Regulatory networks defining EMT during cancer initiation and progression. Nat. Rev. Cancer 13, 97–110.

Gao, D., Vahdat, L.T., Wong, S., Chang, J.C., Mittal, V. 2012. Microenvironmental regulation of epithelial-mesenchymal transitions in cancer. Cancer Res. 72, 4883–4889.

Gatenby, R.A, Gillies, R.J. 2004. Why do cancers have high aerobic glycolysis? Nat. Rev. Cancer 4, 891–899.

Gatenby, R.A, Gillies, R.J. 2008. A microenvironmental model of carcinogenesis. Nat. Rev. Cancer 8, 56–61.

Hanahan, D., Coussens, L.M. 2012. Accessories to the crime: functions of cells recruited to the tumor microenvironment. Cancer Cell 21, 309–322.

Hill, R.P., Marie-Egyptienne, D.T, Hedley, D.W. 2009. Cancer stem cells, hypoxia and metastasis. Semin. Radiat. Oncol. 19, 106–111.

Holohan, C., Van Schaeybroeck, S., Longley, D.B, Johnston, P.G. 2013. Cancer drug resistance: an evolving paradigm. Nat. Rev. Cancer 13, 714–726.

Kalluri, R. 2009. EMT: when epithelial cells decide to become mesenchymal-like cells. J. Clin. Invest. 119, 1417–1419.

Kalluri, R., Weinberg, R.A. 2009. The basics of epithelial-mesenchymal transition. J. Clin. Invest. 119, 1420–1428.

Kang, Y., Pantel, K. 2013. Tumor cell dissemination: emerging biological insights from animal models and cancer patients. Cancer Cell 23, 573–581.

Krebs, M.G., Metcalf, R.L., Carter, L., Brady, G., Blackhall, F.H., Dive, C. 2014. Molecular analysis of circulating tumour cells-biology and biomarkers. Nat. Rev. Clin. Oncol. 11, 129–144.

Ksiazkiewicz, M., Markiewicz, A., Zaczek, A.J. 2012. Epithelial-mesenchymal transition: a hallmark in metastasis formation linking circulating tumor cells and cancer stem cells. Pathobiology 79, 195–208.

Lamouille, S., Subramanyam, D., Blelloch, R., Derynck, R. 2013. Regulation of epithelial-mesenchymal and mesenchymal-epithelial transitions by microRNAs. Curr. Opin. Cell Biol. 25, 200–207.

Lamouille, S., Xu, J., Derynck, R. 2014. Molecular mechanisms of epithelial-mesenchymal transition. Nat. Rev. Mol. Cell Biol. 15, 178–196.

Lee, J.M., Dedhar, S., Kalluri, R., Thompson, E.W. 2006. The epithelial-mesenchymal transition: new insights in signaling, development, and disease. J. Cell Biol. 172, 973–981.

Lendahl, U., Lee, K.L., Yang, H., Poellinger, L. 2009. Generating specificity and diversity in the transcriptional response to hypoxia. Nat. Rev. Genet. 10, 821–832.

Lock, F.E., McDonald, P.C., Lou, Y., Serrano, I., Chafe, S.C., Ostlund, C., *et al.*, 2013. Targeting carbonic anhydrase IX depletes breast cancer stem cells within the hypoxic niche. Oncogene 32, 5210–5219.

Lou, Y., McDonald, P.C., Oloumi, A., Chia, S., Ostlund, C., Ahmadi, A., *et al.*, 2011. Targeting tumor hypoxia: suppression of breast tumor growth and metastasis by novel carbonic anhydrase IX inhibitors. Cancer Res. 71, 3364–3376.

Marie-Egyptienne, D.T., Lohse, I., Hill, R.P. 2013. Cancer stem cells, the epithelial to mesenchymal transition (EMT) and radioresistance: potential role of hypoxia. Cancer Lett. 341, 63–72.

McDonald, P.C., Winum, J.Y., Supuran, C.T, Dedhar, S. 2012. Recent developments in targeting carbonic anhydrase IX for cancer therapeutics. Oncotarget. 3, 84–97.

McIntyre, A., Patiar, S., Wigfield, S., Li, J.L., Ledaki, I., Turley, H. *et al.* 2012. Carbonic anhydrase IX promotes tumor growth and necrosis in vivo and inhibition enhances anti-VEGF therapy. Clin. Cancer Res. 18, 3100–3111.

Murphy, D.A, Courtneidge, S.A. 2011. The 'ins' and 'outs' of podosomes and invadopodia: characteristics, formation and function. Nat. Rev. Mol. Cell Biol. 12, 413–426.

Neri, D., Supuran, C.T. 2011. Interfering with pH regulation in tumours as a therapeutic strategy. Nat. Rev. Drug Discov. 10, 767–777.

Nguyen, D.X., Bos, P.D, Massague, J. 2009. Metastasis: from dissemination to organ-specific colonization. Nat. Rev. Cancer 9, 274–284.

Nieto, M.A. 2013. Epithelial plasticity: a common theme in embryonic and cancer cells. Science 342, 1234850.

Parks, S.K., Chiche, J., Pouyssegur, J. 2011. pH control mechanisms of tumor survival and growth. J. Cell Physiol. 226, 299–308.

Pattabiraman, D.R., Weinberg, R.A. 2014. Tackling the cancer stem cells – what challenges do they pose? Nat. Rev. Drug Discov. 13, 497–512.

Philip, B., Ito, K., Moreno-Sanchez, R., Ralph, S.J. 2013. HIF expression and the role of hypoxic microenvironments within primary tumours as protective sites driving cancer stem cell renewal and metastatic progression. Carcinogenesis 34, 1699–1707.

Scheel, C., Weinberg, R.A. 2012. Cancer stem cells and epithelial-mesenchymal transition: concepts and molecular links. Semin. Cancer Biol. 22, 396–403.

Tam, W.L., Weinberg, R.A. 2013. The epigenetics of epithelial-mesenchymal plasticity in cancer. Nat. Med. 19, 1438–1449.

Thiery, J.P. 2002. Epithelial-mesenchymal transitions in tumour progression. Nat. Rev. Cancer 2, 442–454.

Thiery, J.P., Acloque, H., Huang, R.Y., Nieto, M.A. 2009. Epithelial-mesenchymal transitions in development and disease. Cell 139, 871–890.

Tsai, J.H., Yang, J. 2013. Epithelial-mesenchymal plasticity in carcinoma metastasis. Genes Dev. 27, 2192–2206.

Wan, L., Pantel, K., Kang, Y. 2013. Tumor metastasis: moving new biological insights into the clinic. Nat. Med. 19, 1450–1464.

Yang, J., Weinberg, R.A. 2008. Epithelial-mesenchymal transition: at the crossroads of development and tumor metastasis. Dev. Cell 14, 818–829.

7 Regulation of Breast Cancer Stem Cells by Mesenchymal Stem Cells in the Metastatic Niche

Fayaz Malik[1], Hasan Korkaya[1,2], Shawn G. Clouthier[1] and Max S. Wicha[1]

[1]*Comprehensive Cancer Center, Department of Internal Medicine, University of Michigan, Ann Arbor, MI, USA*

[2]*Department of Biochemistry and Molecular Biology, GRU Cancer Center, Georgia Regents University, Augusta, GA, USA*

7.1 Introduction

Experimental identification of cancer stem cells (CSCs) in acute myeloid leukaemia (Lapidot *et al.*, 1994) was followed by the isolation and functional characterization of CSCs in solid tumours, including breast (Al-Hajj *et al.*, 2003), brain (Hemmati *et al.*, 2003), colon (O'Brien *et al.*, 2007), melanoma (Fang *et al.*, 2005), pancreatic (Hermann *et al.*, 2007), prostate (Collins *et al.*, 2005), ovarian (Bapat *et al.*, 2005), lung (Ho *et al.*, 2007) and gastric (Fukuda *et al.*, 2009) cancers. Thus, accumulating evidence over the past decade suggests that many human cancers are driven by CSCs, which share many properties with their normal counterparts. Like normal stem cells, CSCs retain properties of self-renewal and lineage differentiation, contributing to tumour heterogeneity. According to the CSC hypothesis, tumours are organized in a hierarchical fashion, whereby self-renewing CSCs drive tumourigenesis or differentiated cells constitute the tumour bulk (O'Brien *et al.*, 2010). Although the existence of CSCs in multiple human tumours has been firmly established, the functional and clinical significance of these cells are currently under intensive investigation (Shipitsin and Polyak, 2008). CSCs are also

Principles of Stem Cell Biology and Cancer: Future Applications and Therapeutics, First Edition.
Edited by Tarik Regad, Thomas J. Sayers and Robert C. Rees.

regulated by various extrinsic and intrinsic signalling pathways, which govern their self-renewal and lineage differentiation (Charafe-Jauffret *et al.*, 2008). Alterations in these pathways may result in stem cell expansion, an event implicated in malignant transformation. The intricate crosstalk between the CSCs and their microenvironment has been an area of interest in recent years. Investigations have led to new insights into the nature of these bidirectional interactions, which play a critical role in tumour development and progression.

7.2 Tumour microenvironment and CSCs

Normal tissue-specific stem cells crossreact with their microenvironment in a bidirectional manner during normal development, as well as in responses to pathological insults. A developing tumour also recruits a multitude of cell types to make up the diverse microenvironment via neovascularization through interactions between the tumour cells and their niche (Polyak *et al.*, 2009). In addition to genetic mutations, recent studies have implicated epigenetic alterations driven by signals from the tumour microenvironment in tumour development and progression (Baylin and Ohm, 2006; Korkaya *et al.*, 2011a). The co-evolution of tumours and the niche may result from genetic and epigenetic changes caused by the reciprocal interactions within this microenvironment (Ma *et al.*, 2009; Polyak *et al.*, 2009).

In recent years, functional characterization of CSCs (in multiple solid tumours) has indeed reinvigorated the 'seed and soil' concept first proposed by Paget, which suggests that tumour seeds must first find favourable soil in order to establish a viable metastasis (Paget *et al.*, 1989). This concept is well supported by many studies exploring the significance of the microenvironment in cancer progression (Littlepage *et al.*, 2005; Albini and Sporn, 2007). By definition, a CSC must be capable of initiating tumours when inoculated into experimental animals at limiting dilutions (Clarke *et al.*, 2006; Burness and Sipkins, 2010). CSCs, like their normal counterparts, are also influenced by bidirectional interactions with their niche, which comprises a diversity of cell types and associated signalling pathways (Calabrese *et al.*, 2007; Vermeulen *et al.*, 2008). Thus, identification of signals originating from the microenvironment may provide an attractive molecular target for CSC-specific therapeutics (Korkaya *et al.*, 2011a).

The breast cancer stroma, like that of other solid tumours, contains an array of cell types, including mesenchymal cells, endothelial cells, lymphocytes, macrophages, neutrophils, fibroblasts and pericytes, which interact with tumour cells via chemokine networks (Jones and Wagers, 2008). The role of these cells in the development of the tumour has been extensively studied, and recently the importance of multipotent mesenchymal cells resident in the tumour microenvironment known as mesenchymal stem cells (MSCs) has been widely recognized (Bhowmick *et al.*, 2004; Arendt *et al.*, 2010; Liu *et al.*, 2011; Yan *et al.*, 2012). MSCs have been defined as multipotent cells that have the ability to differentiate into multiple lineages, including bone, cartilage, stroma, adipose tissue, connective tissue, muscle and tendon (Pittenger *et al.*, 1999; Deans and Moseley, 2000). MSCs can secrete various factors and function in a wide range of physiological responses, such as homing to the site of injury, suppressing immune reactions and repairing and regenerating damaged tissues (Pittenger *et al.*, 2002; Chamberlain *et al.*, 2007; Xia *et al.*, 2012). MSCs are considered to be integral components of the cancer stroma in experimental as well as clinical settings (Chamberlain *et al.*, 2007; Karnoub *et al.*, 2007). Interestingly, disseminated tumour cells from the primary tumour home to microanatomical areas surrounded by MSCs – a process that may contribute to the metastatic process (Trumpp and Wiestler, 2008). These micrometastatic CSCs might recapitulate tumours in distant organs by utilizing microenvironmental signals (Croker and Allan, 2008). A recent study has reported that the homing and engraftment of both the normal HSCs and leukaemic stem cells require specialized microenvironments of bone marrow (i.e. the periendosteal region) (Sipkins *et al.*, 2005). On the whole, the importance of the CSC niche in regulating the fate of CSC supports the concept that targeting the unique microenvironment of CSCs may represent a compelling therapeutic strategy.

7.3 Isolation and identification of MSCs

MSCs are multipotent progenitor cells, first discovered in bone marrow by Friedenstein *et al.* (1970). Since they possess the capacity to self-renew and differentiate, they are considered stem cells (Caplan, 1991). They reside primarily in the bone marrow but are also found in a variety of other tissues throughout the body, including adult bone marrow, compact bone, peripheral blood, adipose tissue, cord blood, amniotic fluid, foetal liver and other tissues (Erices *et al.*, 2000; Young *et al.*, 2001; De Ugarte *et al.*, 2003). They have been defined as nonhaematopoietic, multipotent, stem-like cells capable of differentiating into both mesenchymal and nonmesenchymal lineages (Beyer Nardi and da Silva Meirelles, 2006).

MSCs are able to differentiate into osteoblasts, adipocytes and chondroblasts under standard *in vitro* differentiating conditions (Zhang *et al.*, 2008). To date, there are no unique markers that can unequivocally identify MSCs and distinguish them from other cell types. However, it has been reported that MSCs express the surface antigens CD29, CD44, CD49, CD73,CD90, CD105, CD106, CD140b, CD166 and STRO-1 but are devoid of haematopoietic markers such as CD11b, CD14, CD19,CD31, CD34, CD45 and CD133 (Dominici *et al.*, 2006; Roobrouck *et al.*, 2008). Noel *et al.* (2002) described how MSCs in bone and cartilage can be identified by the absence of expression of CD34 and CD45 haematopoietic cell markers and showed they are positive for expression of CD90 and CD105. MSCs also express adhesion molecules, including VCAM (CD 106), ICAM (CD54) and LFA-3 (Majumdar *et al.*, 2003), and are known to secrete interleukins such as IL-6, IL-7, IL-11, IL-12, IL-14, IL-15, leukaemia inhibitory factor (LIF), macrophage colony-stimulating factor (M-CSF), stem cell factor (SCF) and FLT3 ligand (Majumdar *et al.*, 1998). Expression by MSCs of several chemokines has been reported, including the receptor for SDF-1, CXCR4 (Honczarenko *et al.*, 2006; Ponte *et al.*, 2007; Sordi *et al.*, 2005).

MSCs have been isolated from tissues in several pathological conditions, such as rheumatoid arthritis, peripheral blood of acute burns patients, obstructive apneas and bone sarcomas (Mansilla *et al.*, 2006; Bian *et al.*, 2009; Carreras *et al.*, 2009). These studies suggest that MSCs isolated from different tissues have unique properties. We previously demonstrated that MSCs from bone marrow contain at least two subpopulations – one that expresses and one that lacks the stem cell marker aldehyde dehydrogenase 1 (ALDH1) – which exhibit different functional capacities (Korkaya *et al.*, 2011b). Based on an aldefluor expression assay (STEM CELL technologies, BC, Canada), bone marrow-derived MSCs contain 5–6% aldefluor-positive cells, which are capable of generating both an aldefluor-positive and an aldefluor-negative MSC population. Aldefluor-positive MSCs show both adipogenic and osteogenic differentiation, but aldefluor-negative MSCs fail to do so under the same conditions. When co-cultured with breast cancer cells, an aldefluor-positive MSC population is able to regulate the CSC population (Liu *et al.*, 2011). A recent study reported that MSCs can be polarized by downstream toll-like receptor (TLR) signalling into two homogenous phenotypes: MSC1 and MSC2 (Waterman *et al.*, 2010). TLR4-primed MSC1 mostly elaborates proinflammatory mediators, while TLR3-primed MSC2 expresses mostly immunosuppressive responses. Furthermore, MSC1 and MSC2 have divergent effects on cancer growth and metastasis in *in vitro* assays and xenograft models (Waterman *et al.*, 2012), with MSC1 primarily having an antitumour effect and MSC2 promoting tumour growth and metastases.

7.4 Plasticity and the differentiation potential of MSCs

MSCs propagated in established *in vitro* culture systems have been extensively studied. The classic osteogenic differentiation of MSCs in different tissues requires distinct signals (Jaiswal *et al.*, 1997; Pittenger *et al.*, 1999). It has been demonstrated that only one-third of initial adherent bone marrow-derived MSC colonies are pluripotent and capable of differentiating into osteoblastic, chondrocytic and adipocytic cell lineages, while the majority of MSCs exhibits a bilineage (osteo/chondro) or unilineage (osteo) potential. Subsequent studies have revealed the presence of a potential hierarchical model of differentiation, with human bone marrow clonal MSCs readily differentiating into the three lineages, followed by sequential loss of lineage potential, with the osteogenic precursors as residual cells (Muraglia *et al.*, 2000). In addition, MSCs have been reported to exhibit greater differentiation potential than was originally recognized (Eslaminejad *et al.*, 2010). They can give rise to various cells, including neurons, keratinocytes and parenchymal cells of the lung and intestines (Chapel *et al.*, 2003; Sugaya, 2003).

7.5 The role of MSCs in breast cancer

MSCs have been increasingly recognized as an important population of cells within the tumour microenvironment, where they modulate tumour progression and drug sensitivity (Albini and Sporn, 2007). Several recent studies suggest that the tumour microenvironment can modify the proliferation, survival, polarity, differentiation, invasiveness and metastatic capacity of cancer cells (Frisch and Francis, 1994; Weaver *et al.*, 1996; Aboseif, 1999). Furthermore, the proximity of MSCs and CSCs in biopsies obtained from cancer patients has been confirmed by immunohistochemical analysis, raising the possibility that MSCs may influence the clinical course of human malignancies by regulating CSC functions (Corre *et al.*, 2012). Proinflammatory cytokines, chemokines and other mediators secreted by tumours actively recruit various cells, including bone marrow-derived MSCs, into the tumour microenvironment (Dvorak, 1986). These cells respond to signals and factors produced by the tumour cells and support CSC self-renewal, invasion and metastasis (Hu and Polyak, 2008; Korkaya *et al.*, 2011a). The homing capacity of MSCs has been demonstrated in a number of human cancers, including lung cancer (Loebinger *et al.*, 2009), malignant glioma (Sonabend *et al.*, 2008; Sasportas *et al.*, 2009; Yang *et al.*, 2009), Kaposi's sarcomas (Khakoo *et al.*, 2006), breast cancer (Kidd *et al.*, 2009; Patel *et al.*, 2010), colon carcinoma (Menon *et al.*, 2007), melanoma (Studeny *et al.*, 2002) and ovarian cancer (Kidd *et al.*, 2009).

Accumulating evidence indicates that MSCs play a critical role in supporting breast tumour development (El-Haibi *et al.*, 2012). The formation of breast carcinomas is often accompanied by responses that closely resemble a wound that won't heal, and signals from these malignant cells may attract MSCs into the tumour stroma (El-Haibi and Karnoub, 2010). Breast cancer cells have been observed to have the capacity to chemo-attract MSCs in *in vitro* and *in vivo* experimental settings (Goldstein *et al.*, 2010). These cells, like bone marrow-derived MSCs, migrate and become incorporated into the stroma of developing breast carcinomas (Karnoub *et al.*, 2007; Goldstein *et al.*, 2010). MSCs secrete a number of chemokines, including CCL5/RANTES, IL-17B, IL-6 and IL-8 (Karnoub *et al.*, 2007; Goldstein *et al.*, 2010; Liu *et al.*, 2011), which may act on breast CSCs to mediate invasion, metastasis and therapeutic resistance (El-Haibi *et al.*, 2012). Hypoxia-inducible factors (HIFs) mediate paracrine signalling between breast cancer cells and MSCs to promote metastasis. Co-culture of breast cancer cells and MSCs results in the increased expression of genes encoding proteins that promote metastasis by mediating tissue invasion and premetastatic niche formation (Chaturvedi *et al.*, 2013).

Use of adipose-derived MSCs in *in vitro* experiments suggests that MSCs contribute to the progression of basal-like breast cancers by stimulating growth and invasion (Zhao *et al.*, 2012). It has been demonstrated that MSCs promote the ability of breast CSCs to form spheres, as well as inducing a more mesenchymal phenotype. Furthermore, MSC-originating signals also decrease E-cadherin expression in oestrogen receptor-positive luminal breast cancers (Klopp *et al.*, 2010). This has been convincingly demonstrated by co-culture with MSCs: the MSC-conditioned media promoted sphere formation of normal mammary epithelial cells and MCF7 and SUM149 breast cancer cells, suggesting that MSC-derived factors induce CSC self-renewal. In contrast to their tumour-promoting properties, there are few reports suggesting that MSCs play a role in tumour suppression of breast cancer cells. Co-culture of MDA-MB231 with human MSCs induced apoptosis in MDA-MB-231 cancer cells. The effect of human-derived MSCs on tumourigenesis *in vivo* was assessed by injection of MSCs into nonobese diabetic/severe combined immune-deficient mice following tumour establishment with MDA-MB-231, and it was found that MSCs were able to suppress breast cancer tumourigenesis in mice (Chao *et al.*, 2012).

7.6 Induction of EMT and expansion of breast CSCs by MSCs

Epithelial–mesenchymal transition (EMT) is a form of cell plasticity in which epithelial cells acquire mesenchymal phenotypes. This process is an important component of embryogenesis. During EMT, cells lose their epithelial characteristics, cell–cell adhesion and their basal–apical polarity and experience simultaneous remodelling of their cytoskeletal structures. Furthermore, the expression of epithelial markers such as keratin switches to vimentin-type intermediate filaments and becomes motile and resistant to anoikis (Valdes *et al.*, 2002; Klymkowsky and Savagner, 2009). A number of antiapoptotic transcription factors, such as Twist and Snail, and mesenchymal protein markers, such as vimentin and N-cadherin, have been consistently associated with EMT (Mani *et al.*, 2008), and it has been demonstrated that other transcription factors, such as Snail, Slug and E47, are concomitantly expressed in nascent branches of mammary ducts and activate the EMT programme during morphogenesis and pathogenesis (Lee *et al.*, 2011). Decreased expression of E-cadherin, and the resultant cellular dissociation, is another hallmark of EMT in breast cancer cells (Hombauer and Minguell, 2000; Fierro *et al.*, 2004; Thiery and Sleeman, 2006; Martin *et al.*, 2010). It has been reported that EMT, like invasive basal cancer subtypes, is characterized by vimentin upregulation and has been positively correlated with poor prognosis in breast cancer patients (Yamashita *et al.*, 2013).

The EMT programme has been reported to endow cancer cells with certain stem cell-like properties thought to be necessary for metastatic process (Chaffer and Weinberg, 2011). Breast cancer cells were prompted to undergo EMT in order to acquire certain stem cell characteristics, such as the ability to form mammospheres and an increased capacity to form tumours upon serial transplantation into immune-suppressed mice (Morel *et al.*, 2008). The tumour-promoting effects of the MSCs were also attributed to the induction of EMT, in which MSCs play a central role. The first evidence of the role of MSCs in mesenchymal transition was demonstrated in breast cancer (Martin *et al.*, 2010). Breast cancer cells expressed elevated levels of oncogenes (NCOA4, FOS), proto-oncogenes (FYN, JUN), genes associated with invasion (MMP11), angiogenesis (VEGF) and antiapoptosis (IGF1R, BCL2), as well as downregulation of genes associated with proliferation (Ki67, MYBL2), following direct co-culture with MSCs. At the same time, significant upregulation of EMT specific markers (N-cadherin, vimentin, Twist and Snail) was also observed following co-culture with MSCs, with a reciprocal downregulation in the expression of E-cadherin protein. These changes were predominantly cell contact-mediated and regulated by MSCs.

MCSs have been shown to localize to breast carcinomas, where they integrate into the tumour-associated stroma and trigger EMT-mediated metastasis. It has been shown that bone marrow-derived human MSCs, when mixed with human breast carcinoma cells with low metastatic potential, cause the cancer cells to increase their metastatic potency (Karnoub *et al.*, 2007). Studies suggest that breast cancer cells stimulate *de novo* secretion of the chemokine CCL5 (RANTES) from MSCs, which then acts in a paracrine fashion on the cancer cells to enhance their motility, invasion and metastasis. ALDH1-positive breast cancer cells are enriched for CSCs. The enzymatic activity of ALDH1 is used for the identification and isolation of stem cells from a number of malignancies (Ginestier *et al.*, 2007; Douville *et al.*, 2009). We have recently shown that MSCs are able to promote a fourfold increase in the ALDH1-positive CSC population when co-cultured with MSCs (Korkaya *et al.*, 2011b). Similarly, with MDA-MB-231 and MCF7/Ras cells, MSCs cause a multifold increase in the ALDH1 positivity in breast CSCs when co-cultured with a phenotype mediated by MSCs through contact-dependent mechanisms (El-Haibi *et al.*, 2012). Our study also shows that breast CSCs co-cultured with MSCs exhibit a two- to twelvefold increase in primary and secondary mammosphere-forming capacities, which is consistent with the acquisition of CSC properties (Korkaya *et al.*, 2011b). Bone marrow-derived human MSCs induce the production of lysyl oxidase when co-cultured with human breast carcinoma cells, which in turn enhances metastasis to the lung and bone. Lysyl oxidase produced by MSCs stimulates twist transcription in breast cancer cells via paracrine action, thereby mediating MSC-triggered EMT and metastasis (El-Haibi *et al.*, 2012).

7.7 EMT and MET CSC plasticity in migration and metastasis

In order to generate metastasies, cancer cells at the primary tumour and distant organ sites must undergo transient activation of reversible embryonic programmes of EMT and mesenchymal–epithelial transition (MET), respectively. In contrast to EMT, in which the loss of epithelial characteristics is followed by the acquisition of a mesenchymal phenotype, MET is characterized by E-cadherin expression and the acquisition of cell–cell contacts and cellular polarity. Activation of EMT in cells at primary tumour sites promotes their motility, invasive capacity and ability to survive in the circulation. These EMT-like CSCs form micrometastasis at distant sites, but may remain dormant until they undergo an MET conversion, which is associated with self-renewal as well as differentiation. This generates metastases which display a hierarchical organization, resembling that found in the primary tumour. Recent studies have shown that the processes of EMT and MET are both required driving forces for tumour metastasis. In the case of colorectal cancer, a characteristic phenotypic change in the cells at the invasive tumour front and the metastatic site have been reported (Brabletz *et al.*, 2001). Brabletz *et al.* (2001) showed that cells at the invasive front of the primary tumour site are often associated with the loss of epithelial differentiation and acquisition of a mesenchyme-like phenotype, which later grow and colonize as epithelial-like tumours at the site of metastasis. They showed invasive EMT stem-like cells to be growth-arrested while MET cells were highly proliferative, generating tumours at metastatic sites (Brabletz *et al.*, 2001). While several studies have revealed an association between EMT and the acquisition of CSC properties (Mani *et al.*, 2008; Wellner *et al.*, 2009), a few recent reports have described the gain of stem cell characteristics in MET cells, which are requisite for the initiation of secondary tumours (Chaffer *et al.*, 2006; Brabletz, 2012; Celia-Terrassa *et al.*, 2012). A recent study has shown that breast CSCs exist in alternative EMT and MET states, characterized by the expression of different CSC markers (Liu *et al.*, 2014). Furthermore, high concordance was found in the gene expression profile of $CD24^{-}CD44^{+}$ EMT CSCs and $ALDH^{+}$ MET CSCs isolated from genetically distinct tumour subtypes, suggesting shared CSC properties across the molecular subtypes of breast cancer. Based on the gene expression profiles, we found that mesenchymal-like breast CSCs resemble basal stem cells, whereas the profiles of epithelial-like breast CSCs resemble those of luminal stem cells in the normal breast. Furthermore, we found that breast CSC plasticity enables CSCs to transition between these EMT and MET states, and that this transition may be regulated by the tumour microenvironment. Activation of several stem cell pathways, including Hedgehog (Hh), Wnt and transforming growth factor beta (TGF-β), has been shown to be involved in the regulation of EMT (Shin *et al.*, 2010; Takebe *et al.*, 2011; Yoo *et al.*, 2011). Other stem cell-regulating pathways, including bone morphogenetic protein (BMP) and human epidermal growth factor receptor 2 (HER2) signalling, have also been shown to promote MET. Recently studies have interrogated the role of transcription factors and other signaling molecules involved in the switching of EMT/MET states during metastasis. Ocana *et al.* (2012) showed that the loss of the homeobox factor *PRRX1* plays an essential role in the stemness of MET cells, favouring metastatic progression, while Stankic *et al.* (2013) demonstrated an important role of Id1, along with TGF-β, in the transition of EMT to MET CSCs during lung metastatic colonization. The latter group also demonstrated that TGF-β-induced Id1 acts on EMT CSCs to promote their transition into MET CSCs during lung colonization. MSCs found in secondary tumour sites are known to play a vital role in the promotion of metastasis through cell–cell contact and the secretion of several growth factors and cytokines. A functional interaction between human MSCs and lung adenocarcinoma (LAC) cells shows that MSCs effectively inhibit the migration and invasion of several LAC cell lines and enhance the MET phenotype (Wang *et al.*, 2012). The role of MSCs in MET plasticity is not well characterized. The interaction of MSCs with breast CSCs may depend on the expression of receptors, the degree of differentiation and the type of tumour cells with which they are interacting.

7.8 Signalling crosstalk between MSCs and breast CSCs

Tumours produce a wide range of chemokines and cytokines, which function as ligands for MSC surface receptors (Dwyer *et al.*, 2007). These cytokines and their corresponding receptors include SDF-1/CXCR4, SCbF/c-Kit, HGF/c-Met, VEGF/VEGF receptor, MCP/CCR2 and HMGB1/RAGE (Imitola *et al.*, 2004; Schmidt *et al.*, 2005; Son *et al.*, 2006). In addition, MSCs produce a number of angiogenic growth factors that promote the growth of primary and metastatic tumours by inducing the formation of new blood vessels (De Luca *et al.*, 2011). The tumour cells also secrete inflammatory cytokines and growth factors, such as IL-6, IL-8, neurotrophin-3, TGF-ß, IL-1ß, TNF-α, PDGF and EGF, which facilitate MSC migration towards tumour sites (Figure 7.1) (Nakamizo *et al.*, 2005; Motaln *et al.*, 2010). The MSCs are able to differentiate into stromal fibroblasts, which also interact with and influence the tumour cells through paracrine signals and various soluble factors (Erez *et al.*, 2010). The crosstalk between bone marrow-derived MSCs and breast CSCs leads to an increase in the stem cell population ($ALDH1^+$) in SUM159 cells, involving the IL-6 and CXCL7 loop, generated as a result of the interaction between two cell types (Korkaya *et al.*, 2011b). Co-culture of tumour cells with MSCs also increases the percentage of tumour cells expressing the breast CSC markers $CD24^-CD44^+$. The expansion of the CSC population can be observed by the addition of MSC-derived conditioned medium, without any cell–cell contact, which shows that effect is due to the soluble factors (IL-6, CXCL7) released by MSCs in the conditioned medium.

CSCs contribute to tumour cell invasion and metastasis via the cascade of autocrine and paracrine signals from the tumour microenvironment. Well-known stem cell regulatory pathways are frequently dysregulated in

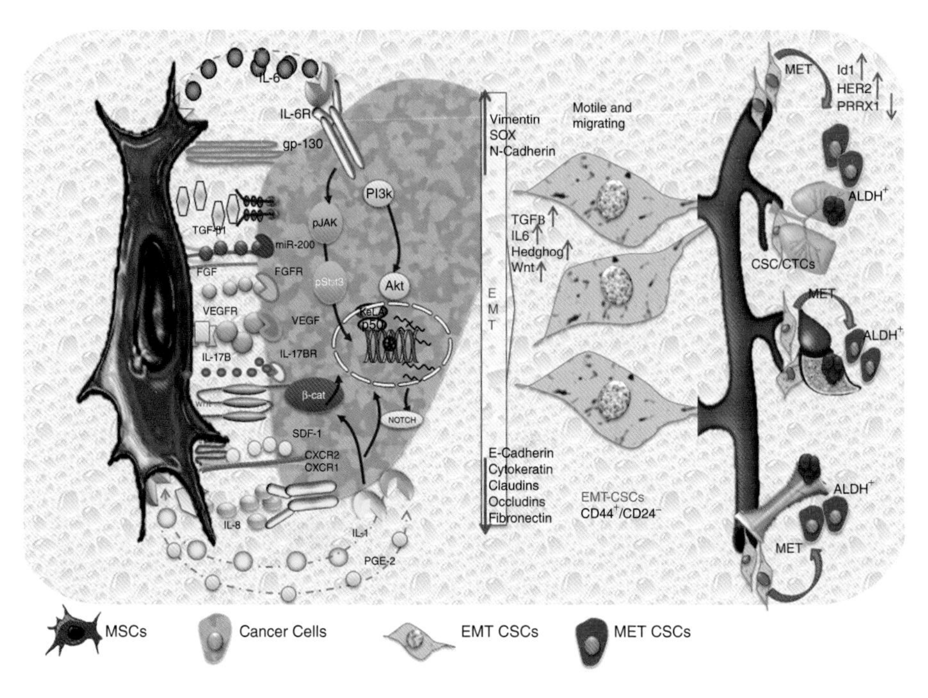

Figure 7.1 Crosstalk between MSCs and breast cancer cells causes migration of EMT CSCs and subsequent colonization at distant metastatic sites by MET CSCs.

tumour cells, including the Notch, Hh, Wnt, PI3K, NF-κB and Jak/STAT pathways (Korkaya *et al.*, 2011a). In addition to mutational alterations, the tumour microenvironment also plays a vital role in the epigenetic activation of these developmental pathways during breast carcinogenesis. The tumour microenvironment contains MSCs and other cells that generate extrinsic factors such as Sonic hedgehog (Shh), Wnt, bone morphogenic proteins (BMPs), fibroblast growth factors (FGFs) and Notch, regulating the stem cell fate (Ivanova *et al.*, 2002). When the equilibrium between these cellular signalling pathways is disrupted, dysregulated self-renewal may play a role in tumour development. It has been shown that tumour cells, as well as multiple cellular elements in the microenvironment, co-evolve during the process of carcinogenesis. Bidirectional paracrine signals coordinately regulate tumourigenic cell populations, including CSCs (Iliopoulos *et al.*, 2009). Karnoub *et al.* (2007) previously demonstrated that paracrine signals produced by stromal cells play important roles in promoting breast cancer metastasis. MSCs within the tumour-associated stroma are critical determinants of cancer cell behaviour and phenotype. MSCs and derived cell types create a CSC niche to enable tumour progression via release of PGE_2 and cytokines. Weinberg and colleagues reported that carcinoma cell-derived IL-1 induces prostaglandin E_2 (PGE_2) secretion by MSCs. The resulting PGE_2 operates in an autocrine manner, cooperating with paracrine IL-1 signalling to induce expression of cytokines by the MSCs. The PGE_2 and the cytokines then proceed to act in a paracrine fashion on the carcinoma cells to induce activation of β-catenin signalling and formation of CSCs (Li *et al.*, 2012). CXCR4 and CXCL12 play an important role in the entry of breast cancer cells into the bone marrow, as well as in attaining breast cancer cell quiescence (Moharita *et al.*, 2006). Inhibition of CXCR4, one of the signalling receptors for CXCL12, effectively inhibits both primary tumour growth and metastasis (Duda *et al.*, 2011).

Gene expression profiling has uncovered the transcription factor SOX4, which shows upregulated activity during TGF-β-induced EMT in normal and cancerous breast epithelial cells. SOX4 activation is important in EMT-induced tumour growth (Tiwari *et al.*, 2013) and metastasis in both *in vitro* and *in vivo* models. MSCs are triggered by the expression of different micro-RNAs (miRs), including miR-335, known to regulate cellular interactions between the MSCs and breast tumours. The variable expression of miR-335 influences the activation of MSCs and the metastasis of breast cancer cells (Hass and Otte, 2012); miR-335 exerts its effect on more than 62 genes, including transcription factor SOX4 and the extracellular matrix component tenascin C, leading to enhancement of metastasis. Several studies have shown that IL-6, a proinflammatory cytokine, directly regulates the self-renewal of breast cancer cells through the activation of stat3, in a process mediated by the IL-6 receptor/GP130 complex (Sansone *et al.*, 2007; Korkaya *et al.*, 2012). During inflammation, IL-6-mediated Stat3 signalling selectively induces a protumourigenic microenvironment (Yu *et al.*, 2009). Stat3 activation in turn leads to transcriptional activation of NF-κB in inflammatory cells, which secrete additional IL-6 and IL-8, acting on tumour cells. Thus, these cytokines generate a positive feedback loop between stromal cells and tumour cells, which further stimulates the CSC components by accelerating metastasis and therapeutic resistance. In addition, IL-6 has been shown to be a key component of a positive feedback loop involved in the expansion of breast CSCs through the bone marrow-derived MSC microenvironment (Korkaya *et al.*, 2011b).

Notch signalling in bone marrow has been reported to act to maintain a pool of mesenchymal progenitors (Hilton *et al.*, 2008). A recent study suggests that IL-6-meditated Jagged1-Notch promotes breast cancer bone metastasis, demonstrating that Jagged1 promotes tumour growth by stimulating release of IL-6 from osteoblasts and directly activating osteoclast differentiation (Sethi *et al.*, 2011). The IL-8 secreted by MSCs activates multiple intracellular signalling pathways by binding its receptors, CXCR1 and CXCR2 (Korkaya *et al.*, 2011a). Reports show that the aggressive behaviour and poor prognosis in patients with cancer are frequently associated with elevated serum IL-8 levels (Benoy *et al.*, 2004; Yao *et al.*, 2007). Increased expression of the IL-8 receptor, CXCR1, on breast CSCs has previously been reported (Ginestier *et al.*, 2010). Recombinant IL-8 increases breast CSC self-renewal and tumour growth while blockade of this receptor in mouse xenografts reduces tumor growth and metastasis. Utilization of the "repertaxin" a small-molecule inhibitor of CXCR1 significantly reduces the breast CSC population, leading to decreased tumourigenicity and metastasis. MSCs are also known to influence breast cancer cells via the activation of the NF-κB pathway. A number of cytokines, including IL-6 and IL-8, secreted by MSCs are regulated by NF-κB and generate a positive feedback loop to maintain a chronic inflammatory state in tumour cells (Iliopoulos *et al.*, 2010). Interestingly, this loop involves the mircro-RNA let7 as well as Lin28, a factor involved in breast CSC self-renewal and expansion. NF-κB activity in breast cancer metastasis is likely induced by infiltrating stromal cells (Tan *et al.*, 2011). NF-κB has also been implicated in the regulation of mouse mammary stem cells during pregnancy. Elevated levels of progesterone during pregnancy induce RANK ligand (RANKL) in differentiated breast epithelial cells. RANKL in turn stimulates breast CSC self-renewal via activation of NF-κβ in these cells (Asselin-Labat *et al.*, 2010; Joshi *et al.*, 2010). The increased incidence of aggressive breast cancers associated with pregnancy (Peck *et al.*, 2002) may result from activation of similar pathways in breast CSCs (Asselin-Labat *et al.*, 2010; Joshi *et al.*, 2010). TGF-β potently induces expansion of breast CSCs by inducing transcriptional changes mediated by several key transcription factors. Several studies have demonstrated that paracrine TGF-β1 secreted by MSCs regulates the establishment of EMT in breast cancer cells by targeting the ZEB/miR-200 regulatory loop (Xu *et al.*, 2012). A dramatic increase in the expression of EMT-specific genes in breast cancer cell lines has been observed following exposure to MSC-secreted factors, including TGF-β1 and TNF-α (Asiedu *et al.*, 2011). These results suggest that MSCs may promote breast cancer metastasis by stimulating EMT-mediated expansion of CSCs, leading to tumour recurrence and drug resistance.

7.9 Conclusions

Emerging concepts in breast tumour invasion, metastasis and therapeutic resistance emphasize the role of CSCs and their regulation by contextual signals within the tumour microenvironment. In this context, a feedback signalling loop through stromal MSCs is vital to the induction of EMT and CSC phenotypes, which are essential in invasion and metastasis. MSCs are an important source of cytokines and chemokines and their role in the plasticity of EMT/MET may depend upon tumour cell type and the presence of cell-specific receptors. The identification of EMT and MET CSC markers should facilitate future studies into the role of MSCs in regulating these cells. Protumourigenic and prometastatic functions of stromal MSCs have now been recognized in breast and other solid tumours, and it is important to understand and target this interactive signalling cascade in order to control CSC expansion and metastasis. As MSCs have the tendency to home to tumours, these cells and associated signalling networks represent attractive therapeutic targets. Although the signalling events triggered by the surrounding environments still poorly understood, targeting of abnormal tumour-promoting paracrine signals between CSCs and MSCs may be a feasible approach to the control of CSC expansion. Therefore, therapeutic strategies interfering with these pathways may provide a means of targeting CSCs, since these cells contribute to metastasis, drug resistance and poor clinical outcomes in advanced breast cancer patients.

References

Aboseif, S. 1999. Mesenchymal reprogramming of adult human epithelial differentiation. Differentiation 65(2), 113–118.

Albini, A., Sporn, M.B. 2007. The tumour microenvironment as a target for chemoprevention. Nat. Rev. Cancer 7(2), 139–47.

Al-Hajj, M., Wicha, M.S., Benito-Hernandez, A., Morrison, S.J., Clarke, M.F. 2003. Prospective identification of tumorigenic breast cancer cells. Proc. Nat. Acad. Sci. USA 100(7), 3983–3988.

Arendt, L.M., Rudnick, J.A., Keller, P.J., Kuperwasser, C. 2010. Stroma in breast development and disease. Semin. Cell Development. Biol. 21(1), 11–18.

Asiedu, M.K., Ingle, J.N., Behrens, M.D., Radisky, D.C., Knutson, K.L. 2011. TGFbeta/TNF(alpha)-mediated epithelial-mesenchymal transition generates breast cancer stem cells with a claudin-low phenotype. Cancer Res. 71(13), 4707–4719.

Asselin-Labat, M.L., Vaillant, F., Sheridan, J.M., Pal, B., Wu, D., Simpson, E.R., *et al.*, 2010. Control of mammary stem cell function by steroid hormone signalling. Nature 465(7299), 798–802.

Bapat, S.A., Malie, A.M., Koppikar, C.B., Kurrey, N.K. 2005. Stem and progenitor-like cells contribute to the aggressive behavior of human epithelial ovarian cancer. Cancer Res. 65(8), 3025–3029.

Baylin, S.B., Ohm, J.E. 2006. Epigenetic gene silencing in cancer – a mechanism for early oncogenic pathway addiction? Nat. Rev. Cancer 6(2), 107–116.

Benoy, I.H., Salgado, R., Van Dam, P., Geboers, K., Van Marck, E., Scharpé, S., *et al.*, 2004. Increased serum interleukin-8 in patients with early and metastatic breast cancer correlates with early dissemination and survival. Clin. Cancer Res. 10(21), 7157–7162.

Beyer Nardi, N., da Silva Meirelles, L. 2006. Mesenchymal stem cells: isolation, in vitro expansion and characterization. Handb. Exp. Pharmacol. (174), 249–282.

Bhowmick, N.A., Chytil, A., Plieth, D., Gorska, A.E., Dumont, N., Shappell, S., *et al.*, 2004. TGF-beta signaling in fibroblasts modulates the oncogenic potential of adjacent epithelia. Science 303(5659), 848–851.

Bian, Z.Y., Li, G., Gan, Y.K., Hao, Y.Q., Xu, W.T., Tang, T.T. 2009. Increased number of mesenchymal stem cell-like cells in peripheral blood of patients with bone sarcomas. Arch. Med. Res. 40(3), 163–168.

Brabletz, T., Jung, A., Reu, S., Porzner, M., Hlubek, F., Kunz-Schughart, L.A., *et al.*, 2001. Variable beta-catenin expression in colorectal cancers indicates tumor progression driven by the tumor environment. Proc. Nat. Acad. Sci. USA 98(18), 10 356–10 361.

Brabletz, T. 2012. To differentiate or not – routes towards metastasis. Nat. Rev. Cancer 12(6), 425–436.

Burness, M.L., Sipkins, D.A. 2010. The stem cell niche in health and malignancy. Semin. Cancer Biol. 20(2), 107–115.

Calabrese, C., Poppleton, H., Kocak, M., Hogg, T.L., Fuller, C., Hamner, B., *et al.*, 2007. A perivascular niche for brain tumor stem cells. Cancer Cell 11(1), 69–82.

Caplan, A.I. 1991. Mesenchymal stem cells. J. Orthop. Res. 9(5), 641–650.

Carreras, A., Almendros, I., Acerbi, I., Montserrat, J.M., Navajas, D., Farré, R. 2009. Obstructive apneas induce early release of mesenchymal stem cells into circulating blood. Sleep 32(1), 117–119.

Celia-Terrassa, T., Meca-Cortés, O., Mateo, F., de Paz, A.M., Rubio, N., Arnal-Estapé, A., *et al.*, 2012. Epithelial-mesenchymal transition can suppress major attributes of human epithelial tumor-initiating cells. J. Clin. Invest. 122(5), 1849–1868.

Chaffer, C.L., Weinberg, R.A. 2011. A perspective on cancer cell metastasis. Science 331(6024), 1559–1564.

Chaffer, C.L., Brennan, J.P., Slavin, J.L., Blick, T., Thompson, E.W., Williams, E.D. 2006. Mesenchymal-to-epithelial transition facilitates bladder cancer metastasis: role of fibroblast growth factor receptor-2. Cancer Res. 66(23), 11 271–11 288.

Chamberlain, G., Fox, J., Ashton, B., Middleton, J. 2007. Concise review: mesenchymal stem cells: their phenotype, differentiation capacity, immunological features, and potential for homing. Stem Cells 25(11), 2739–2749.

Chao, K.C., Yang, H.T., Chen, M.W. 2012. Human umbilical cord mesenchymal stem cells suppress breast cancer tumourigenesis through direct cell-cell contact and internalization. J. Cell Mol. Med. 16(8), 1803–1815.

Chapel, A., Bertho, J.M., Bensidhoum, M., Fouillard, L., Young, R.G., Frick, J., *et al.*, 2003. Mesenchymal stem cells home to injured tissues when co-infused with hematopoietic cells to treat a radiation-induced multi-organ failure syndrome. J. Gene. Med. 5(12), 1028–1038.

Charafe-Jauffret, E., Monville, F., Ginestier, C., Dontu, G., Birnbaum, D., Wicha, M.S. 2008. Cancer stem cells in breast: current opinion and future challenges. Pathobiology 75(2), 75–84.

Chaturvedi, P., Gilkes, D.M., Wong, C.C., Kshitiz, Lou, W., Zhang, H., *et al.*, 2013. Hypoxia-inducible factor-dependent breast cancer-mesenchymal stem cell bidirectional signaling promotes metastasis. J. Clin. Invest. 123(1), 189–205.

Clarke, M.F., Dick, J.E., Dirks, P.B., Eaves, C.J., Jamieson, C.H., Jones, D.L., *et al.*, 2006. Cancer stem cells – perspectives on current status and future directions: AACR Workshop on cancer stem cells. Cancer Res. 66(19), 9339–9344.

Collins, A.T., Berry, P.A., Hyde, C., Stower, M.J., Maitland, N.J. 2005. Prospective identification of tumorigenic prostate cancer stem cells. Cancer Res. 65(23), 10 946–10 951.

Corre, J., Labat, E., Espagnolle, N., Hébraud, B., Avet-Loiseau, H., Roussel, M., *et al.*, 2012. Bioactivity and prognostic significance of growth differentiation factor GDF15 secreted by bone marrow mesenchymal stem cells in multiple myeloma. Cancer Res. 72(6), 1395–1406.

Croker, A.K., Allan, A.L. 2008. Cancer stem cells: implications for the progression and treatment of metastatic disease. J. Cell Mol. Med. 12(2), 374–390.

Deans, R.J., Moseley, A.B. 2000. Mesenchymal stem cells: biology and potential clinical uses. Exp. Hematol. 28(8), 875–884.

De Luca, A., Gallo, M., Aldinucci, D., Ribatti, D., Lamura, L., D'Alessio, A., *et al.*, 2011. Role of the EGFR ligand/receptor system in the secretion of angiogenic factors in mesenchymal stem cells. J. Cell Physiol. 226(8), 2131–2138.

De Ugarte, D.A., Morizono, K., Elbarbary, A., Alfonso, Z., Zuk, P.A., Zhu, M., *et al.*, 2003. Comparison of multi-lineage cells from human adipose tissue and bone marrow. Cells Tissues Organs 174(3), 101–109.

Dominici, M., Le Blanc, K., Mueller, I., Slaper-Cortenbach, I., Marini, F., Krause, D., *et al.*, 2006. Minimal criteria for defining multipotent mesenchymal stromal cells. The International Society for Cellular Therapy position statement. Cytotherapy 8(4), 315–317.

Douville, J., Beaulieu, R., Balicki, D. 2009. ALDH1 as a functional marker of cancer stem and progenitor cells. Stem Cell Dev. 18(1), 17–25.

Duda, D.G., Kozin, S.V., Kirkpatrick, N.D., Xu, L., Kukumura, D., Jain, R.K. 2011. CXCL12 (SDF1alpha)-CXCR4/CXCR7 pathway inhibition: an emerging sensitizer for anticancer therapies? Clin. Cancer Res. 17(8), 2074–2080.

Dvorak, H.F. 1986. Tumors: wounds that do not heal. Similarities between tumor stroma generation and wound healing. N. Engl. J. Med. 315(26), 1650–1659.

Dwyer, R.M., Potter-Beirne, S.M., Harrington, K.A., Lowery, A.J., Hennessy, E., Murphy, J.M., *et al.*, 2007. Monocyte chemotactic protein-1 secreted by primary breast tumors stimulates migration of mesenchymal stem cells. Clin. Cancer Res. 13(17), 5020–5027.

El-Haibi, C.P., Karnoub, A.E. 2010. Mesenchymal stem cells in the pathogenesis and therapy of breast cancer. J. Mammary Gland Biol. Neoplasia 15(4), 399–409.

El-Haibi, C.P., Bell, G.W., Zhang, J., Collmann, A.Y., Wood, D., Scherber, C.M., *et al.*, 2012. Critical role for lysyl oxidase in mesenchymal stem cell-driven breast cancer malignancy. Proc. Nat. Acad. Sci. USA 109(43), 17 460–17 465.

Erez, N., Truitt, M., Olson, P., Arron, S.T., Hanahan, D. 2010. Cancer-associated fibroblasts are activated in incipient neoplasia to orchestrate tumor-promoting inflammation in an NF-kappaB-dependent manner. Cancer Cell 17(2), 135–147.

Erices, A., Conget, P., Minguell, J.J. 2000. Mesenchymal progenitor cells in human umbilical cord blood. Br. J. Haematol. 109(1), 235–242.

Eslaminejad, M.B., Vahabi, S., Shariati, M., Nazarian, H. 2010. *In vitro* growth and characterization of stem cells from human dental pulp of deciduous versus permanent teeth. J. Dent. (Tehran) 7(4), 185–195.

Fang, D., Nguyen, T.K., Leishear, K., Finko, R., Kulp, A.N., Hotz, S., *et al.*, 2005. A tumorigenic subpopulation with stem cell properties in melanomas. Cancer Res. 65(20), 9328–9337.

Fierro, F.A., Sierralta, W.D., Epuñan, M.J., Minguell, J.J. 2004. Marrow-derived mesenchymal stem cells: role in epithelial tumor cell determination. Clin. Exp. Metastasis 21(4), 313–319.

Friedenstein, A.J., Chailakhjan, R.K., Lalykina, K.S. 1970. The development of fibroblast colonies in monolayer cultures of guinea-pig bone marrow and spleen cells. Cell Tissue Kinet. 3(4), 393–403.

Frisch, S.M., Francis, H. 1994. Disruption of epithelial cell-matrix interactions induces apoptosis. J. Cell Biol. 124(4), 619–626.

Fukuda, K., Saikawa, Y., Ohashi, M., Kumagai, K., Kitajima, M., Okano, H., *et al.*, 2009. Tumor initiating potential of side population cells in human gastric cancer. Int. J. Oncol. 34(5), 1201–1207.

Ginestier, C., Hur, M.H., Charafe-Jauffret, E., Monville, F., Dutcher, J., Brown, M., *et al.*, 2007. ALDH1 is a marker of normal and malignant human mammary stem cells and a predictor of poor clinical outcome. Cell Stem Cell 1(5), 555–567.

Ginestier, C., Liu, S., Diebel, M.E., Korkaya, H., Luo, M., Brown, M., *et al.*, 2010. CXCR1 blockade selectively targets human breast cancer stem cells in vitro and in xenografts. J. Clin. Invest. 120(2), 485–497.

Goldstein, R.H., Reagan, M.R., Anderson, K., Kaplan, D.L., Rosenblatt, M. 2010. Human bone marrow-derived MSCs can home to orthotopic breast cancer tumors and promote bone metastasis. Cancer Res. 70(24), 10 044–10 050.

Hass, R., Otte, A. 2012. Mesenchymal stem cells as all-round supporters in a normal and neoplastic microenvironment. Cell Commun. Signal 10(1), 26.

Hemmati, H.D., Nakano, I., Lazareff, J.A., Masterman-Smith, M., Geschwind, D.H., Bronner-Fraser, M., Kornblum, H.I. 2003. Cancerous stem cells can arise from pediatric brain tumors. Proc. Nat. Acad. Sci. USA 100(25), 15 178–15 183.

Hermann, P.C., Huber, S.L., Herrler, T., Aicher, A., Ellwart, J.W., Guba, M., *et al.*, 2007. Distinct populations of cancer stem cells determine tumor growth and metastatic activity in human pancreatic cancer. Cell Stem Cell 1(3), 313–323.

Hilton, M.J., Tu, X., Wu, X., Bai, S., Zhao, H., Kobayashi, T., *et al.*, 2008. Notch signaling maintains bone marrow mesenchymal progenitors by suppressing osteoblast differentiation. Nat. Med. 14(3), 306–314.

Ho, M.M., Ng, A.V., Lam, S., Hung, J.Y. 2007. Side population in human lung cancer cell lines and tumors is enriched with stem-like cancer cells. Cancer Res. 67(10), 4827–4833.

Hombauer, H., Minguell, J.J. 2000. Selective interactions between epithelial tumour cells and bone marrow mesenchymal stem cells. Br. J. Cancer 82(7), 1290–1296.

Honczarenko, M., Le, Y., Swierkowski, M., Ghiran, I., Glodek, A.M., Silberstein, L.E. 2006. Human bone marrow stromal cells express a distinct set of biologically functional chemokine receptors. Stem Cells 24(4), 1030–1041.

Hu, M., Polyak, K. 2008. Molecular characterisation of the tumour microenvironment in breast cancer. Eur. J. Cancer 44(18), 2760–2765.

Iliopoulos, D., Hirsch, H.A., Struhl, K. 2009. An epigenetic switch involving NF-kappaB, Lin28, Let-7 MicroRNA, and IL6 links inflammation to cell transformation. Cell 139(4), 693–706.

Iliopoulos, D., Jaeger, S.A., Hirsch, H.A., Bulvk, M.L., Struhl, K. 2010. STAT3 activation of miR-21 and miR-181b-1 via PTEN and CYLD are part of the epigenetic switch linking inflammation to cancer. Mol. Cell 39(4), 493–506.

Imitola, J., Raddassi, K., Park, K.L., Mueller, F.J., Nieto, M., Teng, Y.D., *et al.*, 2004. Directed migration of neural stem cells to sites of CNS injury by the stromal cell-derived factor 1alpha/CXC chemokine receptor 4 pathway. Proc. Nat. Acad. Sci. USA 101(52), 18 117–18 122.

Ivanova, N.B., Dimos, J.T., Schaniel, C., Hackney, J.A., Moore, K.A., Lemischka, I.R. 2002. A stem cell molecular signature. Science 298(5593), 601–604.

Jaiswal, N., Haynesworth, S.E., Caplan, A.I., Bruder, S.P. 1997. Osteogenic differentiation of purified, culture-expanded human mesenchymal stem cells *in vitro*. J. Cell Biochem. 64(2), 295–312.

Jones, D.L., Wagers, A.J. 2008. No place like home: anatomy and function of the stem cell niche. Nat. Rev. Mol. Cell Biol. 9(1), 11–21.

Joshi, P.A., Jackson, H.W., Beristain, A.G., Di Grappa, M.A., Mote, P.A., Clarke, C.L., *et al.*, 2010. Progesterone induces adult mammary stem cell expansion. Nature 465(7299), 803–807.

Karnoub, A.E., Dash, A.B., Vo, A.P., Sullivan, A., Brooks, M.W., Bell, G.W., *et al.*, 2007. Mesenchymal stem cells within tumour stroma promote breast cancer metastasis. Nature 449(7162), 557–563.

Khakoo, A.Y., Pati, S., Anderson, S.A., Reid, W., Elshal, M.F., Rovira, I.I., *et al.*, 2006. Human mesenchymal stem cells exert potent antitumorigenic effects in a model of Kaposi's sarcoma. J. Exp. Med. 203(5), 1235–1247.

Kidd, S., Spaeth, E., Dembinski, J.L., Dietrich, M., Watson, K., Klopp, A., *et al.*, 2009. Direct evidence of mesenchymal stem cell tropism for tumor and wounding microenvironments using *in vivo* bioluminescent imaging. Stem Cells 27(10), 2614–2623.

Klopp, A.H., Lacerda, L., Gupta, A., Debeb, B.G., Solley, T., Li, L., *et al.*, 2010. Mesenchymal stem cells promote mammosphere formation and decrease E-cadherin in normal and malignant breast cells. PLoS One 5(8), e12180.

Klymkowsky, M.W., Savagner, P. 2009. Epithelial-mesenchymal transition: a cancer researcher's conceptual friend and foe. Am. J. Pathol. 174(5), 1588–1593.

Korkaya, H., Liu, S., Wicha, M.S. 2011a. Breast cancer stem cells, cytokine networks, and the tumor microenvironment. J. Clin. Invest. 121(10), 3804–3809.

Korkaya, H., Liu, S., Wicha, M.S. 2011b. Regulation of cancer stem cells by cytokine networks: attacking cancer's inflammatory roots. Clin. Cancer Res. 17(19), 6125–6129.

Korkaya, H., Kim, G.I., Davis, A., Malik, F., Henry, N.L., Ithimakin, S., *et al.*, 2012. Activation of an IL6 inflammatory loop mediates trastuzumab resistance in HER2+ breast cancer by expanding the cancer stem cell population. Mol. Cell 47(4), 570–584.

Lapidot, T., Sirard, C., Vormoor, J., Murdoch, B., Hoang, T., Caceres-Cortes, J., *et al.*, 1994. A cell initiating human acute myeloid leukaemia after transplantation into SCID mice. Nature 367(6464), 645–648.

Lee, K., Gjorevski, N., Boghaert, E., Radisky, D.C., Nelson, C.M. 2011. Snail1, Snail2, and E47 promote mammary epithelial branching morphogenesis. EMBO J. 30(13), 2662–2674.

Li, H.J., Reinhardt, F., Herschman, H.R., Weinberg, R.A. 2012. Cancer-stimulated mesenchymal stem cells create a carcinoma stem cell niche via prostaglandin E2 signaling. Cancer Discov. 2(9), 840–855.

Littlepage, L.E., Egeblad, M., Werb, Z. 2005. Coevolution of cancer and stromal cellular responses. Cancer Cell 7(6), 499–500.

Liu, S., Ginestier, C., Ou, S.J., Clouthier, S.G., Patel, S.H., Monville, F., *et al.*, 2011. Breast cancer stem cells are regulated by mesenchymal stem cells through cytokine networks. Cancer Res. 71(2), 614–624.

Liu, S., Cong, Y., Wang, D., Sun, Y., Deng, L., Liu, Y., *et al.*, 2014. Breast cancer stem cells transition between epithelial and mesenchymal states reflective of their normal counterparts. Stem Cell Rep. 2(1), 78–91.

Loebinger, M.R., Kyrtatos, P.G., Turmaine, M., Price, A.N., Pankhurst, Q., Lythgoe, M.F., Janes, S.M. 2009. Magnetic resonance imaging of mesenchymal stem cells homing to pulmonary metastases using biocompatible magnetic nanoparticles. Cancer Res. 69(23), 8862–8867.

Ma, X.J., Dahiya, S., Richardson, E., Erlander, M., Sgori, D.C. 2009. Gene expression profiling of the tumor microenvironment during breast cancer progression. Breast Cancer Res. 11(1), R7.

Majumdar, M.K., Thiede, M.A., Mosca, J.D., Moorman, M., Gerson, S.L. 1998. Phenotypic and functional comparison of cultures of marrow-derived mesenchymal stem cells (MSCs) and stromal cells. J. Cell Physiol. 176(1), 57–66.

Majumdar, M.K., Keane-Moore, M., Buyaner, D., Hardy, W.B., Moorman, M.A., McIntosh, K.R., Mosca, J.D. 2003. Characterization and functionality of cell surface molecules on human mesenchymal stem cells. J. Biomed. Sci. 10(2), 228–241.

Mani, S.A., Guo, W., Liao, M.J., Eaton, E.N., Ayyanan, A., Zhou, A.Y., *et al.*, 2008. The epithelial-mesenchymal transition generates cells with properties of stem cells. Cell 133(4), 704–715.

Mansilla, E., Marín, G.H., Drago, H., Sturla, F., Salas, E., Gardiner, C., *et al.*, 2006. Bloodstream cells phenotypically identical to human mesenchymal bone marrow stem cells circulate in large amounts under the influence of acute large skin damage: new evidence for their use in regenerative medicine. Transplant Proc. 38(3), 967–969.

Martin, F.T., Dwyer, R.M., Kelly, J., Khan, S., Murphy, J.M., Curran, C., *et al.*, 2010. Potential role of mesenchymal stem cells (MSCs) in the breast tumour microenvironment: stimulation of epithelial to mesenchymal transition (EMT). Breast Cancer Res. Treat. 124(2), 317–326.

Menon, L.G., Picinich, S., Koneru, R., Gao, H., Lin, S.Y., Koneru, M., *et al.*, 2007. Differential gene expression associated with migration of mesenchymal stem cells to conditioned medium from tumor cells or bone marrow cells. Stem Cells 25(2), 520–528.

Moharita, A.L., Taborga, M., Corcoran, K.E., Bryan, M., Patel, P.S., Rameshwar, P. 2006. SDF-1alpha regulation in breast cancer cells contacting bone marrow stroma is critical for normal hematopoiesis. Blood 108(10), 3245–3252.

Morel, A.P., Lièvre, M., Thomas, C., Hinkal, G., Ansieau, S., Puisieux, A. 2008. Generation of breast cancer stem cells through epithelial-mesenchymal transition. PLoS One 3(8), e2888.

Motaln, H., Schichor, C., Lah, T.T. 2010. Human mesenchymal stem cells and their use in cell-based therapies. Cancer 116(11), 2519–2530.

Muraglia, A., Cancedda, R., Quarto, R. 2000. Clonal mesenchymal progenitors from human bone marrow differentiate *in vitro* according to a hierarchical model. J. Cell Sci. 113(Pt 7), 1161–1166.

Nakamizo, A., Marini, F., Amano, T., Khan, A., Studney, M., Gumin, J., *et al.*, 2005. Human bone marrow-derived mesenchymal stem cells in the treatment of gliomas. Cancer Res. 65(8), 3307–3318.

Noel, D., Djouad, F., Jorgense, C. 2002. Regenerative medicine through mesenchymal stem cells for bone and cartilage repair. Curr. Opin. Investig. Drugs 3(7), 1000–1004.

O'Brien, C.A., Pollett, A., Gallinger, S., Dick, J.E. 2007. A human colon cancer cell capable of initiating tumour growth in immunodeficient mice. Nature 445(7123), 106–110.

O'Brien, C.A., Kreso, A., Jamieson, C.H. 2010. Cancer stem cells and self-renewal. Clin. Cancer Res. 16(12), 3113–3120.

Ocana, O.H., Córcoles, R., Fabra, A., Moreno-Bueno, G., Acloque, H., Vega, S., *et al.*, 2012. Metastatic colonization requires the repression of the epithelial-mesenchymal transition inducer Prrx1. Cancer Cell 22(6), 709–724.

Paget, S. 1989. The distribution of secondary growths in cancer of the breast. 1889. Cancer Metastasis Rev. 8(2), 98–101.

Patel, S.A., Meyer, J.R., Greco, S.J., Corcoran, K.E., Bryan, M., Rameshwar, P. 2010. Mesenchymal stem cells protect breast cancer cells through regulatory T cells: role of mesenchymal stem cell-derived TGF-beta. J. Immunol. 184(10), 5885–5894.

Peck, J.D., Hulka, B.S., Poole, C., Savitz, D.A., Baird, D., Richardson, B.E. 2002. Steroid hormone levels during pregnancy and incidence of maternal breast cancer. Cancer Epidemiol. Biomarkers Prev. 11(4), 361–368.

Pittenger, M.F., Mackay, A.M., Beck, S.C., Jaiswal, R.K., Douglas, R., Mosca, J.D., *et al.*, 1999. Multilineage potential of adult human mesenchymal stem cells. Science 284(5411), 143–147.

Pittenger, M., Vanguri, P., Simonetti, D., Young, R. 2002. Adult mesenchymal stem cells: potential for muscle and tendon regeneration and use in gene therapy. J. Musculoskelet. Neuronal Interact. 2(4), 309–320.

Polyak, K., Haviv, I., Campbell, I.G. 2009. Co-evolution of tumor cells and their microenvironment. Trends Genet. 25(1), 30–38.

Ponte, A.L., Marais, E., Gallay, N., Langonné, A., Delorme, B., Hérault, O., *et al.*, 2007. The *in vitro* migration capacity of human bone marrow mesenchymal stem cells: comparison of chemokine and growth factor chemotactic activities. Stem Cells 25(7), 1737–1745.

Roobrouck, V.D., Ulloa-Montoya, F., Verfaillie, C.M. 2008. Self-renewal and differentiation capacity of young and aged stem cells. Exp. Cell Res. 314(9), 1937–1944.

Sansone, P., Storci, G., Tavolari, S., Guarnieri, T., Giovannini, C., Taffurelli, M., *et al.*, 2007. IL-6 triggers malignant features in mammospheres from human ductal breast carcinoma and normal mammary gland. J. Clin. Invest. 117(12), 3988–4002.

Sasportas, L.S., Kasmieh, R., Wakimoto, H., Hingtgen, S., van de Water, J.A., Mohapatra, G., *et al.*, 2009. Assessment of therapeutic efficacy and fate of engineered human mesenchymal stem cells for cancer therapy. Proc. Nat. Acad. Sci. USA 106(12), 4822–4827.

Schmidt, N.O., Przylecki, W., Yang, W., Ziu, M., Teng, Y., Kim, S.U., *et al.*, 2005. Brain tumor tropism of transplanted human neural stem cells is induced by vascular endothelial growth factor. Neoplasia 7(6), 623–629.

Sethi, N., Dai, X., Winter, C.G., Kang, Y. 2011. Tumor-derived JAGGED1 promotes osteolytic bone metastasis of breast cancer by engaging notch signaling in bone cells. Cancer Cell 19(2), 192–205.

Shin, S.Y., Rath, O., Zebisch, A., Choo, S.M., Kolch, W., Cho, K.H. 2010. Functional roles of multiple feedback loops in extracellular signal-regulated kinase and Wnt signaling pathways that regulate epithelial-mesenchymal transition. Cancer Res. 70(17), 6715–6724.

Shipitsin, M., Polyak, K. 2008. The cancer stem cell hypothesis: in search of definitions, markers, and relevance. Lab. Invest. 88(5), 459–463.

Sipkins, D.A., Wei, X., Wu, J.W., Runnels, J.M., Côté, D., Means, T.K., *et al.*, 2005. *In vivo* imaging of specialized bone marrow endothelial microdomains for tumour engraftment. Nature 435(7044), 969–973.

Son, B.R., Marquez-Curtiz, L.A., Kucia, M., Wysoczynski, M., Turner, A.R., Ratajczak, J., *et al.*, 2006. Migration of bone marrow and cord blood mesenchymal stem cells in vitro is regulated by stromal-derived factor-1-CXCR4 and hepatocyte growth factor-c-met axes and involves matrix metalloproteinases. Stem Cells 24(5), 1254–1264.

Sonabend, A.M., Ulasov, I.V., Tyler, M.A., Rivera, A.A., Mathis, J.M., Lesniak, M.S. 2008. Mesenchymal stem cells effectively deliver an oncolytic adenovirus to intracranial glioma. Stem Cells 26(3), 831–841.

Sordi, V., Malosio, M.L., Marchesi, F., Mercalli, A., Melzi, R., Giordano, T., *et al.*, 2005. Bone marrow mesenchymal stem cells express a restricted set of functionally active chemokine receptors capable of promoting migration to pancreatic islets. Blood 106(2), 419–427.

Stankic, M., Pavlovic, S., Chin, Y., Brogi, E., Padua, D., Norton, L., *et al.*, 2013. TGF-beta-Id1 signaling opposes Twist1 and promotes metastatic colonization via a mesenchymal-to-epithelial transition. Cell Rep. 5(5), 1228–1242.

Studeny, M., Marini, F.C., Champlin, R.E., Zompetta, C., Fidler, I.J., Andreeff, M. 2002. Bone marrow-derived mesenchymal stem cells as vehicles for interferon-beta delivery into tumors. Cancer Res. 62(13), 3603–3608.

Sugaya, K. 2003. Potential use of stem cells in neuroreplacement therapies for neurodegenerative diseases. Int. Rev. Cytol. 228, 1–30.

Takebe, N., Warren, R.Q., Ivy, S.P. 2011. Breast cancer growth and metastasis: interplay between cancer stem cells, embryonic signaling pathways and epithelial-to-mesenchymal transition. Breast Cancer Res. 13(3), 211.

Tan, W., Zhang, W., Strasner, A., Grivennikov, S., Cheng, J.Q., Hoffman, R.M., Karin, M. 2011. Tumour-infiltrating regulatory T cells stimulate mammary cancer metastasis through RANKL-RANK signalling. Nature 470(7335), 548–553.

Thiery, J.P., Sleeman, J.P. 2006. Complex networks orchestrate epithelial-mesenchymal transitions. Nat. Rev. Mol. Cell Biol. 7(2), 131–142.

Tiwari, N., Tiwari, V.K., Waldmeier, L., Balwierz, P.J., Arnold, P., Pachkov, M., *et al.*, 2013. Sox4 is a master regulator of epithelial-mesenchymal transition by controlling Ezh2 expression and epigenetic reprogramming. Cancer Cell 23(6), 768–783.

Trumpp, A., Wiestler, O.D. 2008. Mechanisms of disease: cancer stem cells – targeting the evil twin. Nat. Clin. Pract. Oncol. 5(6), 337–347.

Valdes, F., Alvarez, A.M., Locascio, A., Vega, S., Herrera, B., Fernández, M., *et al.*, 2002. The epithelial mesenchymal transition confers resistance to the apoptotic effects of transforming growth factor Beta in fetal rat hepatocytes. Mol. Cancer Res. 1(1), 68–78.

Vermeulen, L., Tadaro, M., de Sousa Mello, F., Sprick, M.R., Kemper, K., Perez Alea, M., *et al.*, 2008. Single-cell cloning of colon cancer stem cells reveals a multi-lineage differentiation capacity. Proc. Nat. Acad. Sci. USA 105(36), 13 427–13 432.

Wang, M.L., Pan, C.M., Chiou, S.H., Chen, W.H., Chang, H.Y., Lee, O.K., *et al.*, 2012. Oncostatin m modulates the mesenchymal-epithelial transition of lung adenocarcinoma cells by a mesenchymal stem cell-mediated paracrine effect. Cancer Res. 72(22), 6051–6064.

Waterman, R.S., Tomchuck, S.L., Henkle, S.L., Betancourt, A.M. 2010. A new mesenchymal stem cell (MSC) paradigm: polarization into a pro-inflammatory MSC1 or an immunosuppressive MSC2 phenotype. PLoS One 5(4), e10088.

Waterman, R.S., Henkle, S.L., Betancourt, A.M. 2012. Mesenchymal stem cell 1 (MSC1)-based therapy attenuates tumor growth whereas MSC2-treatment promotes tumor growth and metastasis. PLoS One 7(9), e45590.

Weaver, V.M., Fischer, A.H., Peterson, O.W., Bissell, M.J. 1996. The importance of the microenvironment in breast cancer progression: recapitulation of mammary tumorigenesis using a unique human mammary epithelial cell model and a three-dimensional culture assay. Biochem. Cell Biol. 74(6), 833–851.

Wellner, U., Schubert, J., Burk, U.C., Schmalhofer, O., Zhu, F., Sonntag, A., *et al.*, 2009. The EMT-activator ZEB1 promotes tumorigenicity by repressing stemness-inhibiting microRNAs. Nat. Cell Biol. 11(12), 1487–1495.

Xia, X., Chen, W., Ma, T., Xu, G., Liu, H., Liang, C., *et al.*, 2012. Mesenchymal stem cells administered after liver transplantation prevent acute graft-versus-host disease in rats. Liver Transpl. 18(6), 696–706.

Xu, Q., Wang, L., Li, H., Han, Q., Li, J., Qu, X., *et al.*, 2012. Mesenchymal stem cells play a potential role in regulating the establishment and maintenance of epithelial-mesenchymal transition in MCF7 human breast cancer cells by paracrine and induced autocrine TGF-beta. Int. J. Oncol. 41(3), 959–968.

Yamashita, N., Tokunaga, E., Kitao, H., Hisamatsu, Y., Taketani, K., Akiyoshi, S., *et al.*, 2013. Vimentin as a poor prognostic factor for triple-negative breast cancer. J. Cancer Res. Clin. Oncol. 139(5), 739–746.

Yan, X.L., Fu, C.J., Chen, L., Qin, J.H., Zeng, Q., Yuan, H.F., *et al.*, 2012. Mesenchymal stem cells from primary breast cancer tissue promote cancer proliferation and enhance mammosphere formation partially via EGF/EGFR/Akt pathway. Breast Cancer Res. Treat. 132(1), 153–164.

Yang, B., Wu, X., Mao, Y., Bao, W., Gao, L., Zhou, P., *et al.*, 2009. Dual-targeted antitumor effects against brainstem glioma by intravenous delivery of tumor necrosis factor-related, apoptosis-inducing, ligand-engineered human mesenchymal stem cells. Neurosurgery 65(3), 610–624; disc. 624.

Yao, C., Lin, Y., Chua, M.S., Ye, C.S., Bi, J., Li, W., *et al.*, 2007. Interleukin-8 modulates growth and invasiveness of estrogen receptor-negative breast cancer cells. Int. J. Cancer 121(9), 1949–1957.

Yoo, Y.A., Kang, M.H., Lee, H.J., Kim, B.H., Park, J.K., Kim, H.K., *et al.*, 2011. Sonic hedgehog pathway promotes metastasis and lymphangiogenesis via activation of Akt, EMT, and MMP-9 pathway in gastric cancer. Cancer Res. 71(22), 7061–7070.

Young, H.E., Steele, T.A., Bray, R.A., Hudson, J., Floyd, J.A., Hawkins, K., *et al.*, 2001. Human reserve pluripotent mesenchymal stem cells are present in the connective tissues of skeletal muscle and dermis derived from fetal, adult, and geriatric donors. Anat. Rec. 264(1), 51–62.

Yu, H., Pardoll, D., Jove, R. 2009. STATs in cancer inflammation and immunity: a leading role for STAT3. Nat. Rev. Cancer 9(11), 798–809.

Zhang, Z., Wang, X., Wang, S. 2008. Isolation and characterization of mesenchymal stem cells derived from bone marrow of patients with Parkinson's disease. In Vitro Cell Dev. Biol. Anim. 44(5–6), 169–177.

Zhao, M., Sachs, P.C., Wang, X., Dumur, C.I., Idowa, M.O., Robila, V., *et al.*, 2012. Mesenchymal stem cells in mammary adipose tissue stimulate progression of breast cancer resembling the basal-type. Cancer Biol. Ther. 13(9), 782–792.

8 Isolation and Identification of Neural Cancer Stem/Progenitor Cells

David Bakhshinyan[1,2,*], Maleeha A. Qazi[1,2,*], Neha Garg[1,3], Chitra Venugopal[1,3], Nicole McFarlane[1,3] and Sheila K. Singh[1,2,3]

[1]*McMaster Stem Cell and Cancer Research Institute, McMaster University, Hamilton, ON, Canada*
[2]*Department of Biochemistry and Biomedical Sciences, Faculty of Health Sciences, McMaster University, Hamilton, ON, Canada*
[3]*Department of Surgery, Faculty of Health Sciences, McMaster University, Hamilton, ON, Canada*

8.1 Definition of a neural stem cell: functional and marker-based

The discovery of neural stem cells (NSCs) nearly 25 years ago provided the first concrete evidence of the regenerative potential of the nervous system. The initial discovery of a multipotent NSC that can both repeatedly phenocopy itself and generate more restricted neural progenitors (Anderson, 1989; Temple, 1989; Cattaneo and McKay, 1990; Kilpatrick and Bartlett, 1993; Davis and Temple, 1994; Gage and Temple, 2013) laid the groundwork for the unearthing of methods by which to isolate, propagate and differentiate NSCs from the adult central nervous system (CNS) (Reynolds and Weiss, 1992; Snyder *et al.*, 1992; Kilpatrick and Bartlett, 1993). *In vitro* culturing of multipotent NSCs in serum-free conditions allowed epidermal growth

* These authors have contributed equally in creation of this chapter.

Principles of Stem Cell Biology and Cancer: Future Applications and Therapeutics, First Edition.
Edited by Tarik Regad, Thomas J. Sayers and Robert C. Rees.

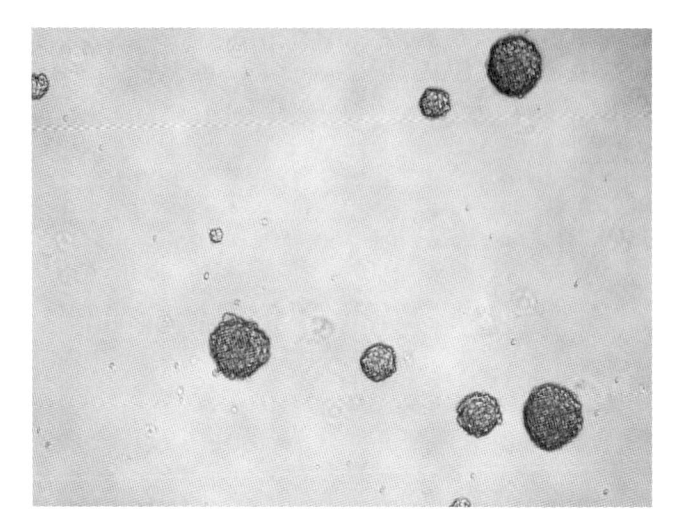

Figure 8.1 Formation of neurospheres from primary human GBM cells. Cells from a primary GBM tumour were grown in NSC media containing growth factors such as EGF, FGF and LIF in low-binding culture plates. After 4–7 days, tumour spheres formed in the culture, indicating the presence of brain tumour stem cells.

factor (EGF)-responsive cells to propagate as floating spheres, later termed 'neurospheres' (Figure 8.1) (Reynolds and Weiss, 1992), which consist of a heterogeneous population of different cell types at various stages of neural lineage commitment.

With the initial findings of NSCs in mouse embryonic/adult brain, attempts to expand human NSCs (hNSCs) to long-term growth were not feasible until 1998, when three research groups developed methods for culturing multipotent human neural progenitor cells (hNPCs) *in vitro* (Svendsen *et al.*, 1998; Carpenter *et al.*, 1999; Vescovi *et al.*, 1999). Human NPCs/NSCs were established from human embryonic forebrain and cultured in the presence of human basic fibroblast growth factor (hbFGF), human leukaemia inhibitory factor (hLIF) and human epidermal growth factor (hEGF), using culture methods similar to those developed for mouse NSCs (Kitchens *et al.*, 1994; Snyder *et al.*, 1992). Over the years, NSCs have been isolated from various areas of the adult brain, including both the CNS (Maslov *et al.*, 2004) and the peripheral nervous system (PNS) (Kruger *et al.*, 2002).

The fundamental properties of NSCs include the capacity to self-renew and to differentiate into three major cell types of the CNS (neurons, astrocytes and oligodendrocytes) without altering their basal functional properties upon *in vitro* passaging. This has been demonstrated *in vivo* using transplantation experiments in animal models of brain lesions or neurodegeneration (Kelly *et al.*, 2004; Lindvall *et al.*, 2004), revealing the amalgamation of NSCs into the host brain and the amelioration of functional defects. NSCs are capable of serial neurosphere formation while maintaining multipotentiality at the clonal level (Louis *et al.*, 2008; Ma *et al.*, 2009). The past decade has elicited the characterization of NSCs by the presence of marker expressions such as CD133, CD15, Sox2, Oct-4, Musashi-1/2 and aldehyde dehydrogenase (ALDH) (Maslov *et al.*, 2004; Corti *et al.*, 2006; Coskun *et al.*, 2008; Kim *et al.*, 2009; Lam *et al.*, 2009; Sun *et al.*, 2009) and by the absence of differentiated markers such as CD24, O4 (oligodendrocytes), MAP2 (neurons), NeuN (neurons), glial fibrillary acidic protein (GFAP) and S100β (Codega *et al.*, 2014). In spite of these developments in the field of the isolation and characterization of NSCs, there still remains no concrete marker or combination of markers available by which to distinguish an NSC from NPCs *in vitro*. Nonetheless, given the improvement in culture conditions and the capacity for simultaneous multiple-marker visualization provided by flow cytometry, we are currently refining our conceptualization of NSCs, both *in vitro* and *in vivo*.

8.2 Cancer stem cell hypothesis

Until the 1970s, the clonal evolution model was used to describe the mechanism of tumour initiation and its progression. Through this model, it was suggested that each cell within a tumour had the potential to acquire stepwise genetic and epigenetic changes that conferred growth advantage and led to tumour progression (Nowell, 1976). However, later experiments demonstrated that only a small subpopulation of bulk tumour cells had the capacity for self-renewal, differentiation and regeneration of a histologically similar tumour when transplanted into immunodeficient mice. In order to account for the functional heterogeneity found in the cells of tumours, the cancer stem cell (CSC) hypothesis was proposed (Reya *et al.*, 2001; Beck *et al.*, 2011; Lee *et al.*, 2011; Jiang *et al.*, 2012; Nguyen *et al.*, 2012). The CSC hypothesis suggests that a small population of tumour cells have the ability to self-renew and to maintain tumour growth by generating more differentiated tumour cells, which are characterized by limited self-renewal capacity and possess a more definitive lineage potential (Heppner and Miller, 1983; Reya *et al.*, 2001). The description of CSCs is solely based on the functional ability of the cell and does not identify its cellular origin, which can be either a normal stem cell that lost the capacity to control its proliferation or a progenitor or more differentiated cell with a newly acquired self-renewal ability. The first conclusive evidence supporting the CSC hypothesis came from the work of Bonnet and Dick (1997), who identified a small population of $CD34^+/CD38^-$ acute myeloid leukaemia (AML) cells that were capable of initiating histologically similar tumours in nonobese diabetic severe combined-immunodeficient (NOD/SCID) mice.

Another important aspect of the CSC hypothesis is the assumption that CSCs have the capacity to evade therapy and are causative of tumour relapse and/or metastasis, as has been shown in several types of solid tumours (Huang *et al.*, 2009; Dimou *et al.*, 2012; Carrasco *et al.*, 2014; Narita *et al.*, 2014; Qian *et al.*, 2014; Todaro *et al.*, 2014). In more recent experiments, the gap between the clonal evolution model and the CSC hypothesis has been narrowed, with three independent studies demonstrating the ability of non-CSC populations in tumours to acquire the capacity to self-renew (Chaffer *et al.*, 2011, 2013; Iliopoulos *et al.*, 2011). Ongoing studies of CSCs have yielded promising treatment opportunities that can target both the bulk tumour and the CSCs (O'Connor *et al.*, 2014). Further refinement of molecular and genetic markers that can accurately distinguish CSC subpopulations from the rest of the cells in a tumour will provide novel therapeutic targets and therapies to allow for better overall survival and prolongation of recurrence-free survival for patients diagnosed with cancer.

8.3 Identification of neural CSCs

8.3.1 Sphere-forming assay, limiting dilution assay and differentiation assay

As adult stem cells are defined by their ability to self-renew and differentiate into multiple cell types, so are CSCs able to self-renew and to give rise to more differentiated cells, which make up the bulk of the tumour. One assay commonly used to investigate the presence of NSCs in the bulk cellular population is the neurosphere-forming assay. After being cultured in NSC media, primary neurospheres are dissociated into single cells and plated into single cell wells. Only a few cells will be able to give rise to a secondary neurosphere, indicating that only a limited population of cells have the capacity to proliferate and self-renew (Tropepe *et al.*, 1999).

The limiting dilution assay is an *in vitro* technique used to measure NSC frequency in the sample. Cells are plated into a 96-well plate in varying densities and allowed to grow for 3–7 days, after which a fraction of wells not containing neurospheres are plotted against the number of cells plated per well. The number of cells required to form one neurosphere is inferred from the point at which 37% of wells do not have neurospheres (Figure 8.2) (Bellows and Aubin, 1989; Tropepe *et al.*, 1999).

Secondary neurospheres are then transferred into media containing growth factors promoting differentiation (Tropepe *et al.*, 1999). The cell culture can further be analysed for the presence of differentiated cells by staining with the neural lineage markers MAP2 (neurons), GFAP (astrocyte) and O4 (oligodendrocytes) to confirm the presence of multipotent cells within secondary neurospheres.

8.3.2 Proliferation assay

The proliferative capacity of NSCs is measured by the number of viable, dividing cells in the culture. A relatively simple method for measuring cellular viability utilizes the fluorescent redox indicator dye resazurin. The oxidized form of this dye is dark blue, but once the dye enters the cell it is converted into a red-fluorescent form (resofurin) detected by spectrometry (Kucharzyk

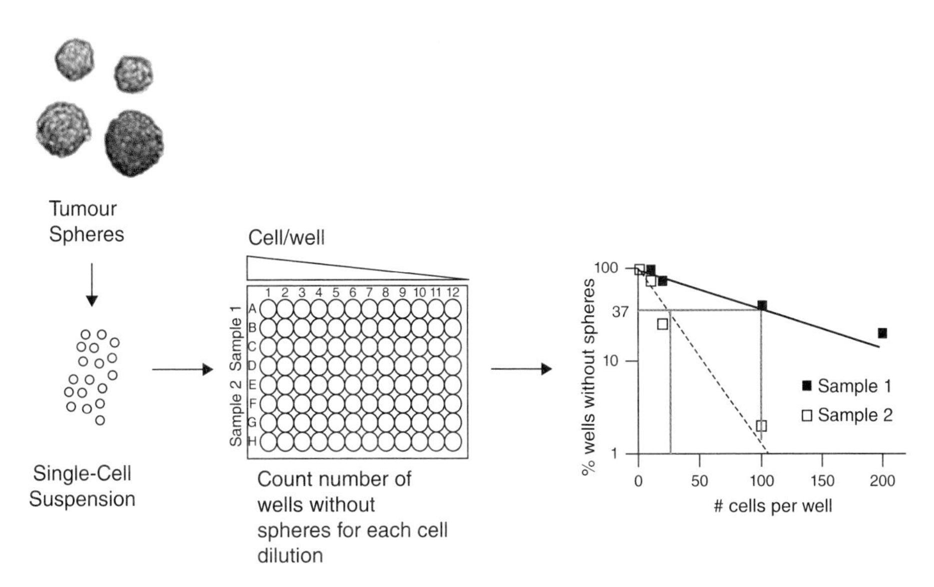

Figure 8.2 Limiting dilution assay to measure NSC frequency in brain tumour samples. Tumour spheres are dissociated into single-cell suspension. Cells are plated into a 96-well plate at concentrations ranging from 200 cells/well to 1 cell/well. After 3–7 days of incubation, the number of wells that do not have any spheres for each cell concentration is counted and plotted as a percentage in a semilogarithmic graph as a function of cell concentration. A line of best fit is plotted through the data points. In this figure, we compare two hypothetical cell populations with different NSC frequencies. Where the lines of best fit intersect the *x*-axis at 37% of empty wells, or 0.37 if using a fraction (red lines), indicates the frequency of NSCs. Hence, in this figure, sample 1 has an NSC frequency of 1 in 100 cells, while sample 2 has an NSC frequency of 1 in 25 cells.

et al., 2010; Vojnits *et al.*, 2012; Walzl *et al.*, 2014). Therefore, a comparison of cell proliferation levels between cell cultures can be made by measuring the amount of red fluorescence emitted from the sample, as more proliferative cells will reduce the resazurin into resofurin at a faster rate (Figure 8.3) (Walzl *et al.*, 2014).

8.4 Identification of BTICs

In the early 2000s, Singh and colleagues reported a small population of $CD133^{+}$ cells termed 'neural CSCs' or 'brain tumour initiating cells' (BTICs) in primary human glioblastoma (GBM) and medulloblastoma (MB) that possessed increased self-renewal, proliferation and differentiation through both *in vitro* (Singh *et al.*, 2003) and *in vivo* (Singh *et al.*, 2004) experiments. The initial *in vitro* studies utilized flow cytometry to sort brain tumour cultures based on CD133 expression, resulting in a separation of CD133 positive and negative populations. The neurosphere assays revealed that only the $CD133^{+}$ cells were capable of growing as nonadherent tumour spheres and

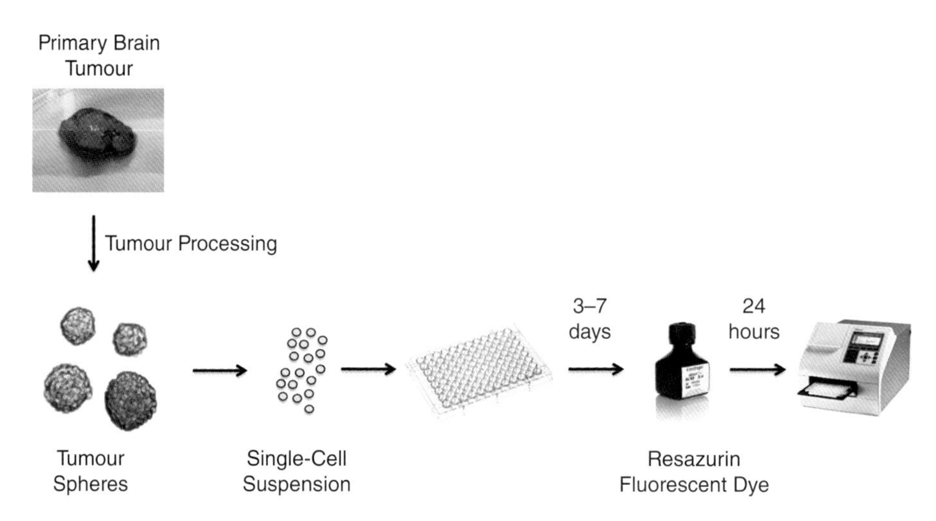

Figure 8.3 Determination of cell proliferation using a resazurin-based assay. Brain tumour cells are grown in NSC media until they form tumour spheres. To assess the cell viability and proliferation, tumour spheres are dissociated into a single-cell suspension and plated in a 96-well plate in a seeding density of 1000–5000 cells/well. Cells are allowed to grow for 5–7 days, after which a resazurin dye is used to measure the fluorescence (at 590 nm), which is a relative measure of cell viability.

were able to maintain their proliferative capacity, indicating stem cell activity. Additionally, when the $CD133^+$ cells were placed in differentiation-inducing conditions, they recapitulated the original tumour type (Singh *et al.*, 2003). Further *in vivo* experiments involving xenotransplantation of $CD133^+$ cells derived from human brain tumours into immunodeficient mice demonstrated that even as few as 100 $CD133^+$ cells could engraft and form a tumour that was histologically and phenotypically identical to the original patient tumour. In comparison, no tumour formation was observed when mice were injected with 100 000 $CD133^-$ cells (Singh *et al.*, 2004). This discovery provided the first conclusive evidence supporting the CSC hypothesis in human brain tumours. In addition to CD133, further research has identified other cell-surface proteins as markers for stem-like cells within brain tumours.

8.5 Current CSC markers

8.5.1 CD133

CD133, also known as prominin-1 (*PROM1*), is a five-transmembrane cell-surface glycoprotein, originally discovered as a murine NSC marker. CD133 has been implicated in several CSC models, as well as in normal neural development (Weigmann *et al.*, 1997; Corbeil *et al.*, 2001; Capela and Temple, 2002; McGrail *et al.*, 2010; Han and Papermaster, 2011; Jaszai *et al.*, 2011). Since its identification as a BTIC marker, CD133 expression has been linked to isolation and identification of human brain-tumour CSCs (Singh *et al.*, 2003). Despite its association with CSCs and its putative functional contribution to driving 'stemness', the actual biological function of CD133 remains unknown.

8.5.2 CD15

In recent years, several studies have sought to identify additional markers for BTICs. One such marker is CD15, also known as stage-specific embryonic antigen 1 (SSEA-1) or Lewis X (LeX). CD15 is expressed in NSCs, NPCs and embryonic stem cells (Capela and Temple, 2002). Like $CD133^+$ cells, cells expressing CD15 possess the capacity to form neurospheres when grown in serum-free conditions, although $CD15^-$ cells do not self-renew to the same extent (Capela and Temple, 2002, 2006). Further evidence of CD15 contributing to stem-like properties comes from an experiment in which $CD15^+$ cells were cultured under conditions that promote differentiation. It was found that only $CD15^+$ cells were capable of giving rise to neurons, astrocytes and oligodendrocytes (Imura *et al.*, 2006). In addition, studies using a mouse model of MB revealed that $CD15^+$ cells had a higher capacity for tumourigenesis than cells lacking CD15 expression (Son *et al.*, 2009).

8.5.3 ALDH

ALDH is an enzyme that reduces the effect of reactive oxygen species (ROS) and has been shown to contribute to cell proliferation and drug resistance. ALDH activity has become an alternative marker of tumour BTICs (Ginestier *et al.*, 2007; Douville *et al.*, 2009; Huang *et al.*, 2009). High ALDH activity has been linked to the identification of mesenchymal and epithelial cells, as well as NSCs. Furthermore, neural CSCs also have an increased ALDH activity, which can potentially contribute to their chemotherapy and radiation resistance (Corti *et al.*, 2006; Gentry *et al.*, 2007). Rasper *et al.* (2010) have shown that even a single $ALDH1^+$ cell derived from human GBM can form a neurosphere and that high ALDH1 activity contributes to the undifferentiated, stem-like state.

8.5.4 Internal markers

In addition to surface markers like CD133 and CD15, a number of intracellular proteins associated with regulation of self-renewal are considered to be markers of neural CSCs, and their expression is strongly associated with increased proliferative capacity and tumourigenic potential. Bmi1, a member of the Polycomb group proteins, is involved in self-renewal of both normal and malignant stem cell populations via repression of the INK4a/ARF locus (Uchida *et al.*, 2000; Vesuna *et al.*, 2009; Merlos-Suarez *et al.*, 2011). Twist1, an upstream regulator of Bmi1, plays an important role in epithelial–mesenchymal transition (EMT) in both normal development and metastatic cancers (Bruggeman *et al.*, 2005, 2007). FoxG1, a downstream target of Bmi1, is often upregulated in MB, resulting in maintenance of NSC multipotency and self-renewal (van der Kooy and Weiss, 2000; Wechsler-Reya and Scott, 2001). Oct4, a transcription factor that is required for reprogramming of cell fates, is essential to maintenance of self-renewal in several malignant stem cell populations (Zhu and Parada, 2002; Rich and Eyler, 2008; Zhu *et al.*, 2009). Nanog is a factor that is required for the reactivation of stemness in multiple CSC types (Wechsler-Reya and Scott, 2001; Clark *et al.*, 2007; Eyler *et al.*, 2008; Rich and Eyler, 2008; Li *et al.*, 2009b).

8.6 Isolation of neural CSCs

Neural CSCs can be isolated and purified through fluorescence- or magnetic-activated cell sorting (FACS or MACS) using BTIC surface markers. The antibodies directed against the BTIC surface markers are labelled with either fluorophores or magnetic nanoparticles. Although both methods will effectively sort tumour populations into BTIC and non-BTIC fractions, there are unique advantages and disadvantages associated with each. FACS allows for a higher purity of the fractionated cell population, as each antibody-labelled cell is individually analysed for fluorescence and sorted according to predetermined parameters, including cell size, granularity and viability (Figure 8.4). However, shear forces caused by pressurized fluids may reduce the post-sort viability of cells (Fong *et al.*, 2009). Cells sorted using

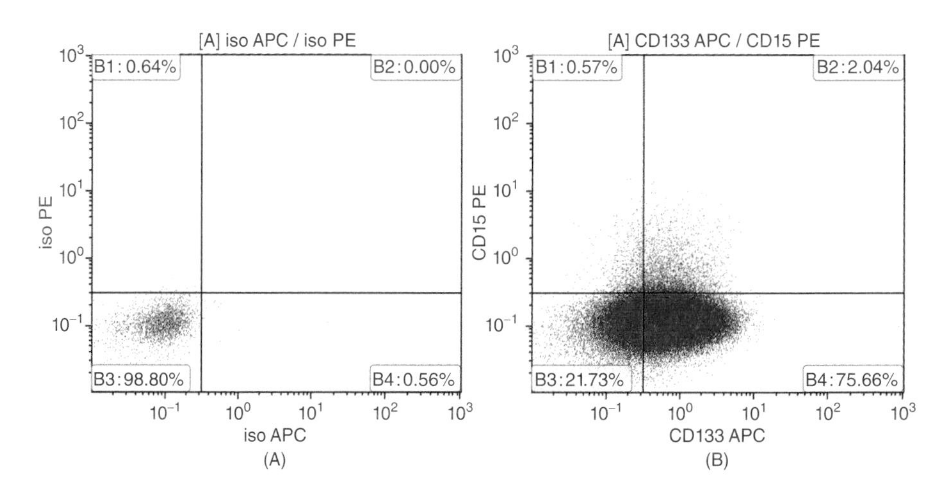

Figure 8.4 Flow analysis of stem cell markers in primary brain tumours. Tumour spheres are dissociated into single-cell suspension and stained with the surface markers CD133-APC, CD15-PE and matched isotype controls. (A) Flow analysis of tumour cells incubated with isotype control shows that there is no nonspecific staining. (B) In the representative sample, the cell population was 75.66% $CD133^+$ and only 0.57% $CD15^+$; 2.04% of cells were positive for both CD133 and CD15.

MACS are not individually analysed for marker expression, resulting in lower purities, but they are also not subjected to strong fluid pressures, resulting in a higher cell viability among sorted populations. Moreover, only FACS can analyse the expression of multiple cell surface markers in a single cell, which can be instrumental in further enrichment of functional BTICs.

Although cell surface markers offer a viable option for the isolation and purification of BTICs from a tumour cell population, there are caveats associated with their use. The cell-surface expression of various BTIC markers is in constant flux and evolves rapidly in response to various environmental factors. For example, cell-surface expression of CD133 can be altered by growing cells in hypoxic conditions (Soeda *et al.*, 2009), targeting glycosylated epitopes (Bidlingmaier *et al.*, 2008) and inducing mitochondrial dysfunction in long-term culture conditions (Griguer *et al.*, 2008). Additionally, the use of trypsin for tissue digestion and dissociation of neurospheres into single cells for FACS may also affect cell surface-marker and receptor expression on BTICs (Fukuchi *et al.*, 2004; Schwab *et al.*, 2008). Therefore, cell sorting based on surface-marker expression offers a snapshot of the biological state of the BTIC, rather than elucidating the truly dynamic nature of a tumour's cell population.

8.7 Role of the microenvironment in tumour progression

Adult NSCs are plastic in their fate programming in culture conditions and *in vivo*, giving rise to the possibility that brain cancers originate from a deregulation of adult NSCs (Vescovi and Snyder, 1999). Another strong indication that endogenous adult NSCs give rise to cancer is their long-term self-renewal capacity *in vivo* (Singh and Dirks, 2007). The distinguishing properties of BTICs versus NSCs are the former's growth factor-independent proliferation, blocked differentiation and genetic and epigenetic aberrations (Zhang *et al.*, 2006; Singh and Dirks, 2007; Dirks, 2008). BTICs also benefit from the vascular niche structure of normal adult NSCs, which provides uncontrolled growth and metastasis (Calabrese *et al.*, 2007). Many groups have shown the importance of purified populations of $CD133^+$ BTICs in accelerating tumour formation, especially in MB and GBM, after transplantation experiments in mice (Singh *et al.*, 2004; Yuan *et al.*, 2004). In glioma stem cells (GSCs), intrinsic factors such as growth factor-independent proliferation and extrinsic factors such as surrounding niches, extracellular matrix (ECM), hypoxia and acidic stress play important roles in the outcome of tumour formation (Heddleston *et al.*, 2011). The enrichment of GSCs in specific niches around tumour vessels, with areas of restricted oxygen (necrosis), supports the importance of the hypoxic niche for nourishment of GSCs (Li *et al.*, 2009a). The GSC requirement for hypoxia inducible factor (HIF)-1α and -2α to promote the stem-like state further clarifies the role of the microenvironment (Heddleston *et al.*, 2009). ECM components associated with the perivascular niche, such as laminin, interact with integrins on GSCs to help drive growth and disease progression (Lathia *et al.*, 2010). Interactions between CSCs and their cellular microenvironment, including parenchymal cells, vascular cells, microglia and peripheral immune cells in the brain, also contribute to tumour progression (Charles *et al.*, 2012). Microglial cells are resident macrophages in the brain that are also associated with tumours; they are often enriched in surgically resected glioma tissue (Graeber *et al.*, 2002; Ghosh and Chaudhuri, 2010). Astrocytes, considered to provide structural support in the brain, are endowed with the capacity to produce neurotrophic factors and astrocyte elevated gene 1 (AEG-1), which function in tumour cell invasion and metastatic progression to regulate metalloproteinases (Hoelzinger *et al.*, 2007; Liu *et al.*, 2010). The conversion of surrounding non-neoplastic astrocytes into a reactive phenotype (Le *et al.*, 2003), along with the confounding role of microglial cells in tumour invasion and growth (Bettinger *et al.*, 2002), adds further complexity to tumour growth and progression. Endothelial cells, which are the source of oxygen and nutrients for tumour cells, also secrete factors for stem cell maintenance (Folkins *et al.*, 2007). Nitric oxide (NO) is the most significant vascular niche-derived factor; it enhances the self-renewal characteristics of CSCs in culture through suppression of NO synthase enzyme, resulting in the prolonged survival of tumour-bearing mice (Charles *et al.*, 2010). For the genesis of new blood vessels, tumour cells exploit neovascularization and angiogenesis by utilizing bone marrow-derived endothelial progenitor cells (Jain *et al.*, 2007). During angiogenesis, activation of endothelial cells by expression of proangiogenic factors and their migration towards tumour cells are augmented by degradation of ECM (Tate and Aghi, 2009), thereby implicating the critical role of these cells in brain cancer progression (Rafii *et al.*, 2002; De Palma and Naldini, 2006).

Collectively, these data implicate the microenvironment in regulating the plasticity of CSCs through modulation of their growth and tumour propagation. The future lies in designing refined models to elucidate the cells of origin of brain tumours, and in finding novel and specific targets for these tumour-initiating cells.

8.8 Epigenetic mechanisms of BTIC regulation

Epigenetics is defined as a heritable change in gene expression that is not complemented by changes in the DNA sequence, and includes modifications such as DNA methylation, histone modifications and RNA-based mechanisms to regulate normal cellular processes (Bird, 2007). Epigenetic imprints are time-honoured early in development, and deregulation of these mechanisms, along with genetic alterations, is one of the major concerns in a variety of cancers, including brain tumours (Brena and Costello, 2007; Esteller, 2008).

Three highly interconnected pathways of epigenetic regulations are: (i) DNA methylation, which consists of the covalent addition of a methyl group (–CH3) at the 5′ position of the pyrimidine ring of cytosine residues and is mediated by DNA methyltransferases (DNMTs) (Miranda and Jones, 2007); (ii) histone modifications, consisting of alterations in histone amino-terminal tails by such mechanisms such as acetylation, methylation, phosphorylation, ubiquination, sumoylation, ADP ribosylation, deimination and proline isomerization (Kouzarides, 2007); and (iii) RNA-based mechanisms, including non-protein-coding RNAs of various lengths, such as micro-RNAs (miRs) and short interfering RNAs (siRs) (Conaco *et al.*, 2006).

The most studied mechanisms of epigenetic change in stem cells are linked with histone modifications and can be pinpointed to methylation and acetylation using the enzymatic activities of histone methyltransferases (HMTs), histone acetyltransferases (HATs), and histone deacetylases (HDACs) (Gonzales-Roybal and Lim, 2013). Histone methylation, involving the addition of a methyl group on a lysine (K) or arginine (R) residue, has different implications in transcription: H3K4, H3K36 and H3K79 are associated with activation of gene expression, while H3K9, H3K27 and H4K20 are associated with its repression (Kouzarides, 2007). Recently, two groups have characterized mutations in somatic histone H3 and chromatin remodelling genes in GBM (Schwartzentruber *et al.*, 2012; Wu *et al.*, 2012). The methylation status of the promoter of a gene called O6-methylguanine-DNA methyltransferase (MGMT), whose normal function is to encode DNA repair to protect cells from carcinogens and, in the case of cancer cells, from chemotherapeutic alkylating agents, is one of the most telling prognostic factors for GBM patient survival (Nagarajan and Costello, 2009). Hypermethylation of MGMT is associated with significantly longer survival in GBM patients treated with the standard therapy of radiation and alkylating agents (Everhard *et al.*, 2006; Hegi *et al.*, 2005). The Polycomb group proteins are well-recognized epigenetic silencers and histone modifiers (Binello and Germano, 2011; Nagarajan and Costello, 2009). Histone-lysine N-methyltransferase (EZH2), a component of Polycomb repressive complex 2 (PRC2), is involved epigenetically in gene silencing by trimethylation and histone modification, and is also highly expressed in GSCs (Suva *et al.*, 2009).

During neural development, temporal and spatial distribution of signalling factors coordinates self-renewal and differentiation in NSCs, and any genetic or epigenetic changes disturbing this balance may lead to the development of tumours. The fate determination of NSCs is controlled by a variety of transcription factors, such as neuron-restrictive silencing factor (NRSF or REST) (Ballas and Mandel, 2005; Faria *et al.*, 2011) and neurogenic basic helix-loop-helix (bHLH), in association with several co-repressors, including (co)repressor for element-1-silencing transcription factor (CoREST), BRCA2-associated-factor 35 (BRAF35), HDACs and methyl-CpG binding protein 2 (Hamby *et al.*, 2008). At the time of neurogenesis in cerebellar NSCs/NPCs, overexpression of REST along with c-myc causes tumours epitomizing MB (Su *et al.*, 2006). In NSCs, DNA is hypomethylated, and an increase in DNA methylation utilizing DNMT1, DNMT3b and DNMT3a occurs with lineage commitment and maturation (Mohn *et al.*, 2008). In MB, miRs add further complexity to epigenetic mechanisms. Some miRs are involved in pathogenesis, such as miRNA-199b-5p, which is downregulated only through epigenetic inactivation, leading to expression of the Hairy and Enhancer of Split-1 gene (HES1) (Garzia *et al.*, 2009). Others, such as miRNA-17/92, have expression in both NSCs and MB, although they are under the control of different transcription factors (Nanog for NSCs (Garg *et al.*, 2013) and N-myc (Northcott *et al.*, 2009) for MB). In MB, promotor hypermethylation of putative tumour suppressor genes (Lindsey *et al.*, 2005), chromatin modifications in histone lysine methylation at H3K9, amplification of the histone lysine demethylases JMJD2C and JMJD2B, deletion of histone lysine methyltransferases (EHMT1 and SMYD4), deletion in genes of the Polycomb group (L3MBTL2 and L3MBTL3) (Northcott *et al.*, 2009) and activation of HDAC5 and HDAC9 (Milde *et al.*, 2010) have all been elucidated in the last decade using a variety of high-throughput techniques.

Therefore, new knowledge of epigenetic modifications can pave the way for a combinatorial approach of standard chemotherapy plus epigenetic therapeutic targets for the treatment of brain tumour patients.

8.9 The therapeutic utility of identifying BTICs

The discovery that a small population of cancer cells within a brain tumour drives tumourigenicity has important implications for our understanding of the tumourigenic process. If BTICs are truly capable of initiating and maintaining tumour growth, then current molecular and pathological analyses of bulk tumour populations may not be sufficient to determine the key molecular aberrations that initiate tumour growth. For example, both MB and GBM have been reclassified into four distinct subgroups derived from gene expression profiles of bulk tumour masses (Northcott *et al.*, 2011; Verhaak *et al.*, 2010). Although subclassification of the two brain tumours is a significant step forward in understanding inter-tumour heterogeneity, the molecular mechanisms governing BTIC populations within these subclasses remains poorly understood. Approaches towards establishing differential gene expression profiles between BTIC and non-BTIC cell populations in brain tumours are essential to understanding the biology of the cells that actually drive tumour formation. Once an understanding of the signalling pathways exclusively governing BTICs has been obtained, novel therapies can be developed that target BTIC cell populations, rather than just the bulk tumour. In fact, current research has started to focus on therapeutic targeting of the BTICs in GBM (Hothi *et al.*, 2012; Triscott *et al.*, 2012). Additionally, since BTICs can be differentiated into cells that express markers for mature neurons, glial cells and oligodendrocytes, therapies may be developed that differentiate BTICs, reducing their self-renewal capacity and hence their tumourigenic potential. In this regard, BMPs have been shown to reduce tumour growth by differentiating $CD133^+$ BTICs and reducing their self-renewal capacity (Piccirillo *et al.*, 2006). Furthermore, due to their self-renewal capacity, BTICs might also be the source of brain tumour dissemination and the establishment of CNS metastasis. Hence, understanding the specific molecular biology of BTICs in brain tumourigenesis is essential to developing therapies that may have the potential to eradicate the disease and prolong patient survival.

Although the conventional treatment modalities of chemotherapy and radiotherapy give short-term survival advantages to patients, they fail to protect against tumour relapse through disease progression to local brain metastases or leptomeningeal infiltration of the brain or spinal cord. The CSC hypothesis contributes to the understanding of this observation, as BTICs are chemo- and radioresistant (Bao *et al.*, 2006; Nakai *et al.*, 2009; Tamura *et al.*, 2010). Rich and colleagues reported that $CD133^+$ glioma cells are more resistant to radiation than $CD133^-$ cells from the same tumour sample (Bao *et al.*, 2006). Glioma cells expressing CD133 preferentially activate DNA damage checkpoints in response to radiation and are more effective at repairing radiation-induced DNA damage. Similarly, BTICs in glioma are also chemoresistant: they present elevated levels of drug transporter proteins, providing resistance against temozolomide, the conventional chemotherapy for treatment of GBM (Hirschmann-Jax *et al.*, 2004; Liu *et al.*, 2006; Nakai *et al.*, 2009). Therapeutic barriers posed by conventional therapy may be overcome with further understanding of the role of BTICs, not only in tumour initiation but also in the promotion of resistance to current treatment modalities. With understanding of the signalling pathways linked to the unique tumourigenic potential of BTICs, novel therapies that selectively target BTICs will achieve greater success in providing long-term patient survival.

Although the CSC hypothesis offers a working framework within which to understand the cellular hierarchy in brain tumours, the possible evolution of BTICs over tumour progression and in response to treatment complicates the study of brain tumourigenesis. Studies conducted in the field of leukaemic stem cells highlight the possibility of CSC evolution over the course of a patient's disease progression (Miyamoto *et al.*, 2000). However, similar studies have not been conducted for brain tumours. The cellular and functional hierarchy established in one type of brain tumour might not be applicable to another, as the underlying genetic pathways driving the tumourigenic process may differ. More importantly, the premise that a tumour-initiating population may perhaps be different than the tumour-maintaining and metastasis-initiating one requires further investigation. Understanding the heterogeneity that might exist within the functional BTICs from a brain tumour is of great importance, as different subsets of BTICs may perform different roles in the progression of the disease. For example, although $CD133^+$ glioma cells are radioresistant, no studies have been conducted to prospectively link the CD133+ BTIC population with tumour recurrence. Furthermore, the treatments offered to brain tumour patients may themselves be a source of further diversification of the BTIC population, leading to the generation of therapy-resistant subclones.

In addition to therapeutically targeting BTICs, epigenetics and the tumour microenvironment also offer avenues of novel therapy development for brain tumours. Epigenetics plays a major role in regulating CSCs in brain tumours, and HDAC inhibitors have shown promise in inhibiting brain tumour cell survival in both MB and GBM (Svechnikova *et al.*, 2008; Hacker *et al.*, 2011; Tang *et al.*, 2012). In MB, the cytotoxic effect of HDAC inhibitors was mediated through sensitization of MB cells to chemotherapy (Hacker *et al.*, 2011). In GBM, HDAC inhibitors not only induced apoptosis, but also induced differentiation, as determined through expression of GFAP (Svechnikova *et al.*, 2008). Similarly, another HDAC inhibitor was shown to synergistically work with the multikinase inhibitor sorafenib to induce cell death in both MB and GBM (Tang *et al.*, 2012). Within the tumour microenvironment, matrix metalloproteinases (MMPs) have been investigated as potential brain tumour therapeutic targets (Lakka *et al.*, 2004; Rao *et al.*, 2007). Development of small molecules that could inhibit the activity of MMPs in brain tumours would lead to reduced tumour invasion, growth and angiogenesis. Hence, in addition to targeting the brain tumour stem cell population, gaining an understanding of the epigenetic signalling pathways and microenvironmental signals in brain tumours would offer a wider array of targets for the development of therapies.

The CSC hypothesis has opened many avenues for further investigation into the tumourigenic process. By understanding the molecular biology of BTICs and focusing our therapeutic modalities on the cells that initiate the tumour, we could work towards prolonged patient survival and eventually a cure. However, although BTICs were initially considered to be a single entity within the cellular hierarchy, the heterogeneity exhibited by brain tumours may actually be a derivative of heterogeneity within BTICs, with multiple, genetically distinct CSC clones. Understanding the complexity of brain tumours and working towards developing better model systems with which to interrogate clonal heterogeneity and capture the dynamics of tumour progression over time would allow for the development of more targeted therapies, thereby promoting increased survival rates in future brain tumour patients.

References

Anderson, D.J. 1989. The neural crest cell lineage problem: neuropoiesis? Neuron 3, 1–12.

Ballas, N., Mandel, G. 2005. The many faces of REST oversee epigenetic programming of neuronal genes. Curr. Opin. Neurobiol. 15, 500–506.

Bao, S., Wu, Q., McLendon, R.E., Hao, Y., Shi, Q., Hjelmeland, A.B., *et al.*, 2006. Glioma stem cells promote radioresistance by preferential activation of the DNA damage response. Nature 444, 756–760.

Beck, B., Driessens, G., Goossens, S., Youssef, K.K., Kuchnio, A., Caauwe, A., *et al.*, 2011. A vascular niche and a VEGF-Nrp1 loop regulate the initiation and stemness of skin tumors. Nature 478, 399–403.

Bellows, C.G., Aubin, J.E. 1989. Determination of numbers of osteoprogenitors present in isolated fetal rat calvaria cells *in vitro*. Dev. Biol. 133, 8–13.

Bettinger, I., Thanos, S., Paulus, W. 2002. Microglia promote glioma migration. Acta Neuropathologica 103, 351–355.

Bidlingmaier, S., Zhu, X., Liu, B. 2008. The utility and limitations of glycosylated human CD133 epitopes in defining cancer stem cells. J. Mol. Med. 86, 1025–1032.

Binello, E., Germano, I.M. 2011. Targeting glioma stem cells: a novel framework for brain tumors. Cancer Sci. 102, 1958–1966.

Bird, A. 2007. Perceptions of epigenetics. Nature 447, 396–398.

Bonnet, D., Dick, J.E. 1997. Human acute myeloid leukemia is organized as a hierarchy that originates from a primitive hematopoietic cell. Nature Med. 3, 730–737.

Brena, R.M., Costello, J.F. 2007. Genome-epigenome interactions in cancer. Human Mol. Genet. 16(Spec. 1), R96–R105.

Bruggeman, S.W., Valk-Lingbeek, M.E., van der Stoop, P.P., Jacobs, J.J., Kieboom, K., Tanger, E., *et al.*, 2005. Ink4a and Arf differentially affect cell proliferation and neural stem cell self-renewal in Bmi1-deficient mice. Genes Dev. 19, 1438–1443.

Bruggeman, S.W., Hulsman, D., Tanger, E., Buckle, T., Blom, M., Zevenhoven, J., *et al.*, 2007. Bmi1 controls tumor development in an Ink4a/Arf-independent manner in a mouse model for glioma. Cancer Cell 12, 328–341.

Calabrese, C., Poppleton, H., Kocak, M., Hogg, T.L., Fuller, C., Hamner, B., *et al.*, 2007. A perivascular niche for brain tumor stem cells. Cancer Cell 11, 69–82.

Capela, A., Temple, S. 2002. LeX/ssea-1 is expressed by adult mouse CNS stem cells, identifying them as nonependymal. Neuron 35, 865–875.

Capela, A., Temple, S. 2006. LeX is expressed by principle progenitor cells in the embryonic nervous system, is secreted into their environment and binds Wnt-1. Dev. Biol. 291, 300–313.

Carpenter, M.K., Cui, X., Hu, Z.Y., Jackson, J., Sherman, S., Seiger, A., Wahlberg, L.U. 1999. *In vitro* expansion of a multipotent population of human neural progenitor cells. Exp. Neurol. 158, 265–278.

Carrasco, E., Alvarez, P.J., Prados, J., Melguizo, C., Rama, A.R., Aranega, A., Rodriguez-Serrano, F. 2014. Cancer stem cells and their implication in breast cancer. Eur. J. Clin. Invest 44(7), 678–687.

Cattaneo, E., McKay, R. 1990. Proliferation and differentiation of neuronal stem cells regulated by nerve growth factor. Nature 347, 762–765.

Chaffer, C.L., Brueckmann, I., Scheel, C., Kaestli, A.J., Wiggins, P.A., Rodrigues, L.O., *et al.*, 2011. Normal and neoplastic nonstem cells can spontaneously convert to a stem-like state. Proc. Nat. Acad. Sci. USA 108, 7950–7955.

Chaffer, C.L., Marjanovic, N.D., Lee, T., Bell, G., Kleer, C.G., Reinhardt, F., *et al.* 2013. Poised chromatin at the ZEB1 promoter enables breast cancer cell plasticity and enhances tumorigenicity. Cell 154, 61–74.

Charles, N., Ozawa, T., Squatrito, M., Bleau, A.M., Brennan, C.W., Hambardzumyan, D., Holland, E.C. 2010. Perivascular nitric oxide activates notch signaling and promotes stem-like character in PDGF-induced glioma cells. Cell Stem Cell 6, 141–152.

Charles, N.A., Holland, E.C., Gilbertson, R., Glass, R., Kettenmann, H. 2012. The brain tumor microenvironment. Glia 60, 502–514.

Clark, P.A., Treisman, D.M., Ebben, J., Kuo, J.S. 2007. Developmental signaling pathways in brain tumor-derived stem-like cells. Dev. Dynam. 236, 3297–3308.

Codega, P., Silva-Vargas, V., Paul, A., Maldonado-Soto, A.R., Deleo, A.M., Pastrana, E., Doetsch, F. 2014. Prospective identification and purification of quiescent adult neural stem cells from their in vivo niche. Neuron 82, 545–559.

Conaco, C., Otto, S., Han, J.J., Mandel, G. 2006. Reciprocal actions of REST and a microRNA promote neuronal identity. Proc. Nat. Acad. Sci. USA 103, 2422–2427.

Corbeil, D., Fargeas, C.A., Huttner, W.B. 2001. Rat prominin, like its mouse and human orthologues, is a pentaspan membrane glycoprotein. Biochem. Biophys. Res. Comm. 285, 939–944.

Corti, S., Locatelli, F., Papadimitriou, D., Donadoni, C., Salani, S., Del Bo, R., *et al.*, 2006. Identification of a primitive brain-derived neural stem cell population based on aldehyde dehydrogenase activity. Stem Cells 24, 975–985.

Coskun, V., Wu, H., Blanchi, B., Tsao, S., Kim, K., Zhao, J., *et al.*, 2008. CD133+ neural stem cells in the ependyma of mammalian postnatal forebrain. Proc. Nat. Acad. Sci. USA 105, 1026–1031.

Davis, A.A., Temple, S. 1994. A self-renewing multipotential stem cell in embryonic rat cerebral cortex. Nature 372, 263–266.

De Palma, M., Naldini, L. 2006. Role of haematopoietic cells and endothelial progenitors in tumor angiogenesis. Biochim. Biophys. Acta 1766, 159–166.

Dimou, A., Neumeister, V., Agarwal, S., Anagnostou, V., Syrigos, K., Rimm, D.L. 2012. Measurement of aldehyde dehydrogenase 1 expression defines a group with better prognosis in patients with non-small cell lung cancer. Am. J. Pathol. 181, 1436–1442.

Dirks, P.B. 2008. Brain tumor stem cells: bringing order to the chaos of brain cancer. J. Clin. Oncol. 26, 2916–2924.

Douville, J., Beaulieu, R., Balicki, D. 2009. ALDH1 as a functional marker of cancer stem and progenitor cells. Stem Cells Dev. 18, 17–25.

Esteller, M. 2008. Epigenetics in cancer. N. Engl. J. Med. 358, 1148–1159.

Everhard, S., Kaloshi, G., Criniere, E., Benouaich-Amiel, A., Lejeune, J., Marie, Y., *et al.*, 2006. MGMT methylation: a marker of response to temozolomide in low-grade gliomas. Ann. Neurol. 60, 740–743.

Eyler, C.E., Foo, W.C., LaFiura, K.M., McLendon, R.E., Hjelmeland, A.B., Rich, J.N. 2008. Brain cancer stem cells display preferential sensitivity to Akt inhibition. Stem Cells 26, 3027–3036.

Faria, C.M., Rutka, J.T., Smith, C., Kongkham, P. 2011. Epigenetic mechanisms regulating neural development and pediatric brain tumor formation. J. Neurosurg. Ped. 8, 119–132.

Folkins, C., Man, S., Xu, P., Shaked, Y., Hicklin, D.J., Kerbel, R.S. 2007. Anticancer therapies combining antiangiogenic and tumor cell cytotoxic effects reduce the tumor stem-like cell fraction in glioma xenograft tumors. Cancer Res. 67, 3560–3564.

Fong, C.Y., Peh, G.S., Gauthaman, K., Bongso, A. 2009. Separation of SSEA-4 and TRA-1-60 labelled undifferentiated human embryonic stem cells from a heterogeneous cell population using magnetic-activated cell sorting MACS) and fluorescence-activated cell sorting FACS. Stem Cell Rev. 5, 72–80.

Fukuchi, Y., Nakajima, H., Sugiyama, D., Hirose, I., Kitamura, T., Tsuji, K. 2004. Human placenta-derived cells have mesenchymal stem/progenitor cell potential. Stem Cells 22, 649–658.

Gage, F.H., Temple, S. 2013. Neural stem cells: generating and regenerating the brain. Neuron 80, 588–601.

Garg, N., Po, A., Miele, E., Campese, A.F., Begalli, F., Silvano, M., *et al.*, 2013. microRNA-17-92 cluster is a direct Nanog target and controls neural stem cell through Trp53inp1. EMBO J. 32, 2819–2832.

Garzia, L., Andolfo, I., Cusanelli, E., Marino, N., Petrosino, G., De Martino, D., *et al.*, 2009. MicroRNA-199b-5p impairs cancer stem cells through negative regulation of HES1 in medulloblastoma. PloS One 4, e4998.

Gentry, T., Foster, S., Winstead, L., Deibert, E., Fiordalisi, M., Balber, A. 2007. Simultaneous isolation of human BM hematopoietic, endothelial and mesenchymal progenitor cells by flow sorting based on aldehyde dehydrogenase activity: implications for cell therapy. Cytotherapy 9, 259–274.

Ghosh, A., Chaudhuri, S. 2010. Microglial action in glioma: a boon turns bane. Immunol. Lett. 131, 3–9.

Ginestier, C., Hur, M.H., Charafe-Jauffret, E., Monville, F., Dutcher, J., Brown, M., *et al.*, 2007. ALDH1 is a marker of normal and malignant human mammary stem cells and a predictor of poor clinical outcome. Cell Stem Cell 1, 555–567.

Gonzales-Roybal, G., Lim, D.A. 2013. Chromatin-based epigenetics of adult subventricular zone neural stem cells. Front. Genet. 4, 194.

Graeber, M.B., Scheithauer, B.W., Kreutzberg, G.W. 2002. Microglia in brain tumors. Glia 40, 252–259.

Griguer, C.E., Oliva, C.R., Gobin, E., Marcorelles, P., Benos, D.J., Lancaster, J.R. Jr, Gillespie, G.Y. 2008. CD133 is a marker of bioenergetic stress in human glioma. PloS One 3, e3655.

Hacker, S., Karl, S., Mader, I., Cristofanon, S., Schweitzer, T., Krauss, J., *et al.* 2011. Histone deacetylase inhibitors prime medulloblastoma cells for chemotherapy-induced apoptosis by enhancing p53-dependent Bax activation. Oncogene 30, 2275–2281.

Hamby, M.E., Coskun, V., Sun, Y.E. 2008. Transcriptional regulation of neuronal differentiation: the epigenetic layer of complexity. Biochim. Biophys. Acta 1779, 432–437.

Han, Z., Papermaster, D.S. 2011. Identification of three prominin homologs and characterization of their messenger RNA expression in *Xenopus laevis* tissues. Mol. Vis. 17, 1381–1396.

Heddleston, J.M., Li, Z., McLendon, R.E., Hjelmeland, A.B., Rich, J.N. 2009. The hypoxic microenvironment maintains glioblastoma stem cells and promotes reprogramming towards a cancer stem cell phenotype. Cell Cycle 8, 3274–3284.

Heddleston, J.M., Hitomi, M., Venere, M., Flavahan, W.A., Yang, K., Kim, Y., *et al.* 2011. Glioma stem cell maintenance: the role of the microenvironment. Curr. Pharma. Des. 17, 2386–2401.

Hegi, M.E., Diserens, A.C., Gorlia, T., Hamou, M.F., de Tribolet, N., Weller, M., *et al.*, 2005. MGMT gene silencing and benefit from temozolomide in glioblastoma. N. Engl. J. Med. 352, 997–1003.

Heppner, G.H., Miller, B.E. 1983. Tumor heterogeneity: biological implications and therapeutic consequences. Cancer Metas. Rev. 2, 5–23.

Hirschmann-Jax, C., Foster, A.E., Wulf, G.G., Nuchtern, J.G., Jax, T.W., Gobel, U., *et al.*, 2004. A distinct 'side population' of cells with high drug efflux capacity in human tumor cells. Proc. Nat. Acad. Sci. USA 101, 14 228–14 233.

Hoelzinger, D.B., Demuth, T., Berens, M.E. 2007. Autocrine factors that sustain glioma invasion and paracrine biology in the brain microenvironment. J. Nat. Cancer Inst. 99, 1583–1593.

Hothi, P., Martins, T.J., Chen, L., Deleyrolle, L., Yoon, J.G., Reynolds, B., Foltz, G. 2012. High-throughput chemical screens identify disulfiram as an inhibitor of human glioblastoma stem cells. Oncotarget 3, 1124–1136.

Huang, E.H., Hynes, M.J., Zhang, T., Ginestier, C., Dontu, G., Appelman, H., *et al.*, 2009. Aldehyde dehydrogenase 1 is a marker for normal and malignant human colonic stem cells (SC) and tracks SC overpopulation during colon tumorigenesis. Cancer Res. 69, 3382–3389.

Iliopoulos, D., Hirsch, H.A., Wang, G., Struhl, K. 2011. Inducible formation of breast cancer stem cells and their dynamic equilibrium with non-stem cancer cells via IL6 secretion. Proc. Nat. Acad. Sci. USA 108, 1397–1402.

Imura, T., Nakano, I., Kornblum, H.I., Sofroniew, M.V. 2006. Phenotypic and functional heterogeneity of GFAP-expressing cells *in vitro*: differential expression of LeX/CD15 by GFAP-expressing multipotent neural stem cells and non-neurogenic astrocytes. Glia 53, 277–293.

Jain, R.K., di Tomaso, E., Duda, D.G., Loeffler, J.S., Sorensen, A.G., Batchelor, T.T. 2007. Angiogenesis in brain tumors. Nat. Rev. Neurosci. 8, 610–622.

Jaszai, J., Fargeas, C.A., Graupner, S., Tanaka, E.M., Brand, M., Huttner, W.B., Corbeil, D. 2011. Distinct and conserved prominin-1/CD133-positive retinal cell populations identified across species. PloS One 6, e17590.

Jiang, W., Peng, J., Zhang, Y., Cho, W.C., Jin, K. 2012. The implications of cancer stem cells for cancer therapy. Int. J. Mol. Sci. 13, 16636–16657.

Kelly, S., Bliss, T.M., Shah, A.K., Sun, G.H., Ma, M., Foo, W.C., *et al.*, 2004. Transplanted human fetal neural stem cells survive, migrate, differentiate in ischemic rat cerebral cortex. Proc. Nat. Acad. Sci. USA 101, 11 839–11 844.

Kilpatrick, T.J., Bartlett, P.F. 1993. Cloning and growth of multipotential neural precursors: requirements for proliferation and differentiation. Neuron 10, 255–265.

Kim, J.B., Sebastiano, V., Wu, G., Arauzo-Bravo, M.J., Sasse, P., Gentile, L., *et al.*, 2009. Oct4-induced pluripotency in adult neural stem cells. Cell 136, 411–419.

Kitchens, D.L., Snyder, E.Y., Gottlieb, D.I. 1994. FGF and EGF are mitogens for immortalized neural progenitors. J. Neurobiol. 25, 797–807.

Kouzarides, T. 2007. Chromatin modifications and their function. Cell 128, 693–705.

Kruger, G.M., Mosher, J.T., Bixby, S., Joseph, N., Iwashita, T., Morrison, S.J. 2002. Neural crest stem cells persist in the adult gut but undergo changes in self-renewal, neuronal subtype potential, factor responsiveness. Neuron 35, 657–669.

Kucharzyk, K.H., Crawford, R.L., Paszczynski, A.J., Hess, T.F. 2010. A method for assaying perchlorate concentration in microbial cultures using the fluorescent dye resazurin. J. Microbiol. Meth. 81, 26–32.

Lakka, S.S., Gondi, C.S., Yanamandra, N., Olivero, W.C., Dinh, D.H., Gujrati, M., Rao, J.S. 2004. Inhibition of cathepsin B and MMP-9 gene expression in glioblastoma cell line via RNA interference reduces tumor cell invasion, tumor growth and angiogenesis. Oncogene 23, 4681–4689.

Lam, C.S., Marz, M., Strahle, U. 2009. gfap and nestin reporter lines reveal characteristics of neural progenitors in the adult zebrafish brain. Dev. Dynam. 238, 475–486.

Lathia, J.D., Gallagher, J., Heddleston, J.M., Wang, J., Eyler, C.E., Macswords, J., *et al.*, 2010. Integrin alpha 6 regulates glioblastoma stem cells. Cell Stem Cell 6, 421–432.

Le, D.M., Besson, A., Fogg, D.K., Choi, K.S., Waisman, D.M., Goodyer, C.G., *et al.*, 2003. Exploitation of astrocytes by glioma cells to facilitate invasiveness: a mechanism involving matrix metalloproteinase-2 and the urokinase-type plasminogen activator-plasmin cascade. J. Neurosci. 23, 4034–4043.

Lee, H.E., Kim, J.H., Kim, Y.J., Choi, S.Y., Kim, S.W., Kang, E., *et al.*, 2011. An increase in cancer stem cell population after primary systemic therapy is a poor prognostic factor in breast cancer. Br. J. Cancer 104, 1730–1738.

Li, Z., Bao, S., Wu, Q., Wang, H., Eyler, C., Sathornsumetee, S., *et al.*, 2009a. Hypoxia-inducible factors regulate tumorigenic capacity of glioma stem cells. Cancer Cell 15, 501–513.

Li, Z., Wang, H., Eyler, C.E., Hjelmeland, A.B., Rich, J.N. 2009b. Turning cancer stem cells inside out: an exploration of glioma stem cell signaling pathways. J. Biol. Chem. 284, 16 705–16 709.

Lindsey, J.C., Anderton, J.A., Lusher, M.E., Clifford, S.C. 2005. Epigenetic events in medulloblastoma development. Neurosurg. Focus 19, E10.

Lindvall, O., Kokaia, Z., Martinez-Serrano, A. 2004. Stem cell therapy for human neurodegenerative disorders-how to make it work. Nat. Med. 10, S42–S50.

Liu, G., Yuan, X., Zeng, Z., Tunici, P., Ng, H., Abdulkadir, I.R., *et al.*, 2006. Analysis of gene expression and chemoresistance of CD133+ cancer stem cells in glioblastoma. Mol. Cancer 5, 67.

Liu, L., Wu, J., Ying, Z., Chen, B., Han, A., Liang, Y., *et al.*, 2010. Astrocyte elevated gene-1 upregulates matrix metalloproteinase-9 and induces human glioma invasion. Cancer Res. 70, 3750–3759.

Louis, S.A., Rietze, R.L., Deleyrolle, L., Wagey, R.E., Thomas, T.E., Eaves, A.C., Reynolds, B.A. 2008. Enumeration of neural stem and progenitor cells in the neural colony-forming cell assay. Stem Cells 26, 988–996.

Ma, D.K., Bonaguidi, M.A., Ming, G.L., Song, H. 2009. Adult neural stem cells in the mammalian central nervous system. Cell Res. 19, 672–682.

Maslov, A.Y., Barone, T.A., Plunkett, R.J., Pruitt, S.C. 2004. Neural stem cell detection, characterization, age-related changes in the subventricular zone of mice. J. Neurosci. 24, 1726–1733.

McGrail, M., Batz, L., Noack, K., Pandey, S., Huang, Y., Gu, X., Essner, J.J. 2010. Expression of the zebrafish CD133/prominin1 genes in cellular proliferation zones in the embryonic central nervous system and sensory organs. Dev. Dynam. 239, 1849–1857.

Merlos-Suarez, A., Barriga, F.M., Jung, P., Iglesias, M., Cespedes, M.V., Rossell, D., *et al.*, 2011. The intestinal stem cell signature identifies colorectal cancer stem cells and predicts disease relapse. Cell Stem Cell 8, 511–524.

Milde, T., Oehme, I., Korshunov, A., Kopp-Schneider, A., Remke, M., Northcott, P., *et al.*, 2010. HDAC5 and HDAC9 in medulloblastoma: novel markers for risk stratification and role in tumor cell growth. Clin. Cancer Res. 16, 3240–3252.

Miranda, T.B., Jones, P.A. 2007. DNA methylation: the nuts and bolts of repression. J. Cell. Physiol. 213, 384–390.

Miyamoto, T., Weissman, I.L., Akashi, K. 2000. AML1/ETO-expressing nonleukemic stem cells in acute myelogenous leukemia with 8;21 chromosomal translocation. Proc. Nat. Acad. Sci. USA 97, 7521–7526.

Mohn, F., Weber, M., Rebhan, M., Roloff, T.C., Richter, J., Stadler, M.B., *et al.*, 2008. Lineage-specific polycomb targets and de novo DNA methylation define restriction and potential of neuronal progenitors. Mol. Cell 30, 755–766.

Nagarajan, R.P., Costello, J.F. 2009. Epigenetic mechanisms in glioblastoma multiforme. Sem. Cancer Biol. 19, 188–197.

Nakai, E., Park, K., Yawata, T., Chihara, T., Kumazawa, A., Nakabayashi, H., Shimizu, K. 2009. Enhanced MDR1 expression and chemoresistance of cancer stem cells derived from glioblastoma. Cancer Invest. 27, 901–908.

Narita, K., Matsuda, Y., Seike, M., Naito, Z., Gemma, A., Ishiwata, T. 2014. Nestin regulates proliferation, migration, invasion and stemness of lung adenocarcinoma. Int. J. Oncol. 44, 1118–1130.

Nguyen, L.V., Vanner, R., Dirks, P., Eaves, C.J. 2012. Cancer stem cells: an evolving concept. Nat. Rev. Cancer 12, 133–143.

Northcott, P.A., Nakahara, Y., Wu, X., Feuk, L., Ellison, D.W., Croul, S., *et al.*, 2009. Multiple recurrent genetic events converge on control of histone lysine methylation in medulloblastoma. Nat. Genet. 41, 465–472.

Northcott, P.A., Korshunov, A., Witt, H., Hielscher, T., Eberhart, C.G., Mack, S., *et al.*, 2011. Medulloblastoma comprises four distinct molecular variants. J. Clin. Oncol. 29, 1408–1414.

Nowell, P.C. 1976. The clonal evolution of tumor cell populations. Science 194, 23–28.

O'Connor, M.L., Xiang, D., Shigdar, S., Macdonald, J., Li, Y., Wang, T., *et al.*, 2014. Cancer stem cells: a contentious hypothesis now moving forward. Cancer Lett. 344, 180–187.

Piccirillo, S.G., Reynolds, B.A., Zanetti, N., Lamorte, G., Binda, E., Broggi, G., *et al.*, 2006. Bone morphogenetic proteins inhibit the tumorigenic potential of human brain tumor-initiating cells. Nature 444, 761–765.

Qian, X., Wagner, S., Ma, C., Coordes, A., Gekeler, J., Klussmann, J.P., *et al.*, 2014. Prognostic significance of ALDH1A1-positive cancer stem cells in patients with locally advanced, metastasized head and neck squamous cell carcinoma. J. Cancer Res. Clin. Oncol. 140(7), 1151–1158.

Rafii, S., Heissig, B., Hattori, K. 2002. Efficient mobilization and recruitment of marrow-derived endothelial and hematopoietic stem cells by adenoviral vectors expressing angiogenic factors. Gene Ther. 9, 631–641.

Rao, J.S., Bhoopathi, P., Chetty, C., Gujrati, M., Lakka, S.S. 2007. MMP-9 short interfering RNA induced senescence resulting in inhibition of medulloblastoma growth via p16(INK4a) and mitogen-activated protein kinase pathway. Cancer Res. 67, 4956–4964.

Rasper, M., Schafer, A., Piontek, G., Teufel, J., Brockhoff, G., Ringel, F., *et al.*, 2010. Aldehyde dehydrogenase 1 positive glioblastoma cells show brain tumor stem cell capacity. Neuro-Oncol. 12, 1024–1033.

Reya, T., Morrison, S.J., Clarke, M.F., Weissman, I.L. 2001. Stem cells, cancer, cancer stem cells. Nature 414, 105–111.

Reynolds, B.A., Weiss, S. 1992. Generation of neurons and astrocytes from isolated cells of the adult mammalian central nervous system. Science 255, 1707–1710.

Rich, J.N., Eyler, C.E. 2008. Cancer stem cells in brain tumor biology. Cold Spring Harb. Symp. Quant. Biol. 73, 411–420.

Schwab, K.E., Hutchinson, P., Gargett, C.E. 2008. Identification of surface markers for prospective isolation of human endometrial stromal colony-forming cells. Human Repro. 23, 934–943.

Schwartzentruber, J., Korshunov, A., Liu, X.Y., Jones, D.T., Pfaff, E., Jacob, K., *et al.*, 2012. Driver mutations in histone H3.3 and chromatin remodelling genes in paediatric glioblastoma. Nature 482, 226–231.

Singh, S., Dirks, P.B. 2007. Brain tumor stem cells: identification and concepts. Neurosurg. Clin. N. Am. 18, 31–38, viii.

Singh, S.K., Clarke, I.D., Terasaki, M., Bonn, V.E., Hawkins, C., Squire, J., Dirks, P.B. 2003. Identification of a cancer stem cell in human brain tumors. Cancer Res. 63, 5821–5828.

Singh, S.K., Hawkins, C., Clarke, I.D., Squire, J.A., Bayani, J., Hide, T., *et al.*, 2004. Identification of human brain tumor initiating cells. Nature 432, 396–401.

Snyder, E.Y., Deitcher, D.L., Walsh, C., Arnold-Aldea, S., Hartwieg, E.A., Cepko, C.L. 1992. Multipotent neural cell lines can engraft and participate in development of mouse cerebellum. Cell 68, 33–51.

Soeda, A., Park, M., Lee, D., Mintz, A., Androutsellis-Theotokis, A., McKay, R.D., *et al.*, 2009. Hypoxia promotes expansion of the CD133-positive glioma stem cells through activation of HIF-1alpha. Oncogene 28, 3949–3959.

Son, M.J., Woolard, K., Nam, D.H., Lee, J., Fine, H.A. 2009. SSEA-1 is an enrichment marker for tumor-initiating cells in human glioblastoma. Cell Stem Cell 4, 440–452.

Su, X., Gopalakrishnan, V., Stearns, D., Aldape, K., Lang, F.F., Fuller, G., *et al.*, 2006. Abnormal expression of REST/NRSF and Myc in neural stem/progenitor cells causes cerebellar tumors by blocking neuronal differentiation. Mol. Cell. Biol. 26, 1666–1678.

Sun, Y., Kong, W., Falk, A., Hu, J., Zhou, L., Pollard, S., Smith, A. 2009. CD133 (Prominin) negative human neural stem cells are clonogenic and tripotent. PloS One 4, e5498.

Suva, M.L., Riggi, N., Janiszewska, M., Radovanovic, I., Provero, P., Stehle, J.C., *et al.*, 2009. EZH2 is essential for glioblastoma cancer stem cell maintenance. Cancer Res. 69, 9211–9218.

Svechnikova, I., Almqvist, P.M., Ekstrom, T.J. 2008. HDAC inhibitors effectively induce cell type-specific differentiation in human glioblastoma cell lines of different origin. Int. J. Oncol. 32, 821–827.

Svendsen, C.N., ter Borg, M.G., Armstrong, R.J., Rosser, A.E., Chandran, S., Ostenfeld, T., Caldwell, M.A. 1998. A new method for the rapid and long term growth of human neural precursor cells. J. Neurosci. Meth. 85, 141–152.

Tamura, K., Aoyagi, M., Wakimoto, H., Ando, N., Nariai, T., Yamamoto, M., Ohno, K. 2010. Accumulation of CD133-positive glioma cells after high-dose irradiation by Gamma Knife surgery plus external beam radiation. J. Neurosurg. 113, 310–318.

Tang, Y., Yacoub, A., Hamed, H.A., Poklepovic, A., Tye, G., Grant, S., Dent, P. 2012. Sorafenib and HDAC inhibitors synergize to kill CNS tumor cells. Cancer Biol. Ther. 13, 567–574.

Tate, M.C., Aghi, M.K. 2009. Biology of angiogenesis and invasion in glioma. Neurotherapeutics 6, 447–457.

Temple, S. 1989. Division and differentiation of isolated CNS blast cells in microculture. Nature 340, 471–473.

Todaro, M., Gaggianesi, M., Catalano, V., Benfante, A., Iovino, F., Biffoni, M., *et al.*, 2014. CD44v6 is a marker of constitutive and reprogrammed cancer stem cells driving colon cancer metastasis. Cell Stem Cell 14, 342–356.

Triscott, J., Lee, C., Hu, K., Fotovati, A., Berns, R., Pambid, M., *et al.*, 2012. Disulfiram, a drug widely used to control alcoholism, suppresses the self-renewal of glioblastoma and over-rides resistance to temozolomide. Oncotarget 3, 1112–1123.

Tropepe, V., Sibilia, M., Ciruna, B.G., Rossant, J., Wagner, E.F., van der Kooy, D. 1999. Distinct neural stem cells proliferate in response to EGF and FGF in the developing mouse telencephalon. Dev. Biol. 208, 166–188.

Uchida, N., Buck, D.W., He, D., Reitsma, M.J., Masek, M., Phan, T.V., *et al.*, 2000. Direct isolation of human central nervous system stem cells. Proc. Nat. Acad. Sci. USA 97, 14 720–14 725.

van der Kooy, D., Weiss, S. 2000. Why stem cells? Science 287, 1439–1441.

Verhaak, R.G., Hoadley, K.A., Purdom, E., Wang, V., Qi, Y., Wilkerson, M.D., *et al.*, 2010. Integrated genomic analysis identifies clinically relevant subtypes of glioblastoma characterized by abnormalities in PDGFRA, IDH1, EGFR, NF1. Cancer Cell 17, 98–110.

Vescovi, A.L., Snyder, E.Y. 1999. Establishment and properties of neural stem cell clones: plasticity *in vitro* and *in vivo*. Brain Pathol. 9, 569–598.

Vescovi, A.L., Parati, E.A., Gritti, A., Poulin, P., Ferrario, M., Wanke, E., *et al.*, 1999. Isolation and cloning of multipotential stem cells from the embryonic human CNS and establishment of transplantable human neural stem cell lines by epigenetic stimulation. Exp. Neurol. 156, 71–83.

Vesuna, F., Lisok, A., Kimble, B., Raman, V. 2009. Twist modulates breast cancer stem cells by transcriptional regulation of CD24 expression. Neoplasia 11, 1318–1328.

Vojnits, K., Ensenat-Waser, R., Gaspar, J.A., Meganathan, K., Jagtap, S., Hescheler, J., *et al.*, 2012. A tanscriptomics study to elucidate the toxicological mechanism of methylmercury chloride in a human stem cell based *in vitro* test. Curr. Med. Chem. 19, 6224–6232.

Walzl, A., Unger, C., Kramer, N., Unterleuthner, D., Scherzer, M., Hengstschlager, M., *et al.* 2014. The resazurin reduction assay can distinguish cytotoxic from cytostatic compounds in spheroid screening assays. J. Biomol. Screen. 19(7), 1047–1059.

Wechsler-Reya, R., Scott, M.P. 2001. The developmental biology of brain tumors. Ann. Rev. Neurosci. 24, 385–428.

Weigmann, A., Corbeil, D., Hellwig, A., Huttner, W.B. 1997. Prominin, a novel microvilli-specific polytopic membrane protein of the apical surface of epithelial cells, is targeted to plasmalemmal protrusions of non-epithelial cells. Proc. Nat. Acad. Sci. USA 94, 12 425–12 430.

Wu, G., Broniscer, A., McEachron, T.A., Lu, C., Paugh, B.S., Becksfort, J., *et al.*, 2012. Somatic histone H3 alterations in pediatric diffuse intrinsic pontine gliomas and non-brainstem glioblastomas. Nat. Genet. 44, 251–253.

Yuan, X., Curtin, J., Xiong, Y., Liu, G., Waschsmann-Hogiu, S., Farkas, D.L., *et al.*, 2004. Isolation of cancer stem cells from adult glioblastoma multiforme. Oncogene 23, 9392–9400.

Zhang, Q.B., Ji, X.Y., Huang, Q., Dong, J., Zhu, Y.D., Lan, Q. 2006. Differentiation profile of brain tumor stem cells: a comparative study with neural stem cells. Cell Res. 16, 909–915.

Zhu, Y., Parada, L.F. 2002. The molecular and genetic basis of neurological tumors. Nat. Rev. Cancer 2, 616–626.

Zhu, H., Acquaviva, J., Ramachandran, P., Boskovitz, A., Woolfenden, S., Pfannl, R., *et al.*, 2009. Oncogenic EGFR signaling cooperates with loss of tumor suppressor gene functions in gliomagenesis. Proc. Nat. Acad. Sci. USA 106, 2712–2716.

9 Colon Stem Cells in Colorectal Cancer

Varun V. Prabhu, Wafik S. El-Deiry and Niklas Finnberg
Laboratory of Translational Oncology and Experimental Cancer Therapeutics, Department of Medical Oncology and Molecular Therapeutics Program, Fox Chase Cancer Center, Philadelphia, PA, USA

9.1 Background

The gastrointestinal tract (GIT) contains cells that rapidly divide and turn over, which contribute to diverse functions and are supported by intestinal stem cells (ISCs) located at the crypt bottom. It is possible that the presence of a rapidly proliferating epithelium in the gut and frequently dividing ISCs might increase the propensity for malignant transformation in the organ. Globally, colorectal cancer (CRC) is one of the most common forms of cancer and the second leading cause of cancer mortality in Europe and the United States (Siegel *et al.*, 2014). The clonal selection model that was formed in 1975 has been the main paradigm by which to describe the emergence of CRC; further characterization of the genetic mechanism that drives colorectal carcinogenesis and progression was made by Bert Vogelstein during the late 1980s and early 1990s (Fearon and Vogelstein, 1990; Puglisi *et al.*, 2013). CRC develops from epithelial cells covering the GIT, as they undergo sequential mutations of the genes that control proliferation, self-renewal and evasion of cell death and the immune system. This model remains the hallmark for understanding the pathogenesis of the disease today. However, it is clear that CRC is a complex disease with several molecular subtypes, and the mechanisms that underlie its intratumoural and intertumoural cellular heterogeneity and which may cause treatment failure remain to be elucidated. There is increasing evidence to suggest that human cancer can be considered a stem cell disease and that not all cancer cells may be equally malignant. There is also

Principles of Stem Cell Biology and Cancer: Future Applications and Therapeutics, First Edition.
Edited by Tarik Regad, Thomas J. Sayers and Robert C. Rees.

growing support for a cancer cellular hierarchical model that assumes that the cancer cells that form a tumour are hierarchically ordered and that only a rare undifferentiated cell present at the apex of this hierarchy has the unique biological properties necessary for tumour initiation, maintenance and metastasis (Puglisi *et al.*, 2013). Provided the characteristics of such cells exist and can be validated, they are considered to be colorectal cancer stem cells (CRCSCs). CRCSCs are believed to contribute to tumour heterogeneity and repopulate the tumour (with a treatment-refractory population of tumour cells) following perturbed tumour homeostasis as a result of, say, cancer therapy. This type of hierarchical model suggests that there are cells within a tumour that harbour different tumourigenic potentials. Certain cells may lose their tumourigenic potential, while others may retain it; such tumour cell populations have been identified in solid and haematological malignancies (Puglisi *et al.*, 2013). Thus, there appear to be several conceptual similarities between normal nonmalignant ISCs and CRCSCs. The concept is based upon a view of carcinogenesis as an aberrant form of organogenesis. Furthermore, the long life and capacity for self-renewal of CRCSCs may hint towards a commonality with the stem cells of normal tissues. Thus, it becomes tempting to draw a linear connection between ISCs in the normal gut and CRCSCs in the tumour. It would appear to be thermodynamically favourable if genetic and/or epigenetic alterations were to contribute to the transformation of an ISC into a CRCSC, whose progeny would constitute the majority of different tumour cells making up the bulk of the cancer. Thus, an ISC could be the progenitor to the CRCSC population in cancers of the gut. However, such an assumption is fraught with pitfalls and is currently not sufficiently supported by evidence, and it also does not exclude the possibility that a CRCSC might arise from a reprogrammed differentiated cell. It has been proposed that a high level of plasticity exists between ISCs and early progenitor cells (transit-amplifying cells, TACs) in the normal small intestine of the mouse (Barker *et al.*, 2012). Thus, it appears that progenitors may redifferentiate to ISCs and therefore that no obligate irreversible boundary exists between progenitor cells and stem cells in the gut. It is tempting to speculate that redifferentiation might occur in a similar manner in CRCs or might even be less stringently controlled in such tissues, considering that the dependency on extracellular growth signals is often lost in advanced malignancies. The genetic lesions that allow cancer cells to proliferate autonomously may allow such cells to wean themselves off the stem cell niches that are crucial for the maintenance of a population of ISCs in the normal gut. Thus, the identity of the normal tissue-originating cell of CRCSCs remains unclear in CRC. The stem cell theory may also have important translational implications, since further understanding of the behaviour of ISCs following injury to the gut may help develop novel means of preventing toxicities following CRC therapy. Great expectations are placed on both preventive and regenerative medicine. On the other side of the spectrum, CRCSCs may make an important impact on translational medicine, as treatment failure is likely related to failure to eradicate tumour cells, which are able to repopulate the tumour host. Thus, how CRCSCs give rise to different treatment-resistant lineages and how the dynamics of different tumour cell populations are affected by a given type of therapy are important areas of study for future research efforts.

9.1.1 ISCs in the maintenance of epithelial homeostasis

Recent experimental evidence from genetically engineered mouse models (GEMs) shows that the epithelial homeostasis in the gut is preserved by ISCs that resides near the crypt base (Clevers, 2013). ISCs are able to self-renew, while at the same time generating TACs (Figure 9.1), which frequently undergo cell division at a lower segment of the intestinal crypt. Following cell division, TACs differentiate into mature distinct epithelial cell lineages as they migrate upward to the top of the vertical axis and shed into the lumen. By contrast, secretory Paneth cells present in the small intestine migrate down towards the bottom of the crypt. The intestinal epithelium has the ability to rapidly repair and compensate for cell loss as part of an injury-induced repair response that prevents loss of structural and functional integrity. This is due to a process called 'epithelial restitution', followed by epithelial cell proliferation and differentiation. The restitution process involves atrophic regions in the gastrointestinal epithelium being covered by neighbouring epithelial cells (Iizuka and Konno, 2011). Epithelial cells surrounding the injury undergo cell migration to cover the naked patch in the epithelial barrier, facilitate cell–cell interaction and reinstate a functional epithelial barrier in the gut. This process typically takes minutes to hours to complete and limits the loss of fluids and electrolytes, all while preserving the submucosal compartments from direct exposure to potential antigens present in the intestinal lumen. Epithelial proliferation and differentiation follow the restitution process to compensate for the loss of cellularity following injury and to restore the function of the gut. It is generally agreed that this process is ISC-driven and involves signalling to stimulate the stem cell pool through the Notch and Wnt pathways in the intestine. The restitution process is mainly dependent on compensatory expansion of the ISCs and epithelial progenitor cells of the gut. Subsequently, ISC cell therapy could potentially have therapeutic value for gastrointestinal disorders, such as inflammatory bowel disease (IBD) and colitis, that have been intimately linked to CRC.

9.1.2 Biomarkers of ISCs

Adult stem cells are characterized by their tissue longevity (stem cells persist for the lifetime of their owner) and multipotency (the ability to produce all cell types of the tissue to which they belong) (Barker *et al.*, 2012). Early studies by Cheng and Leblond (1974) revealed the presence of a slender cell wedged in between Paneth cells in the small intestine, which they called crypt base columnar (CBC) cells (Figure 9.1). By labelling phagocytosing CBC

The crypt bottom of the small intestine

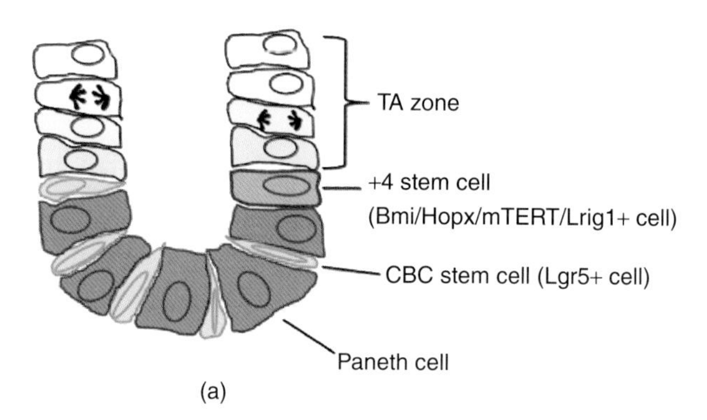

(a)

The crypt bottom of the descending colon

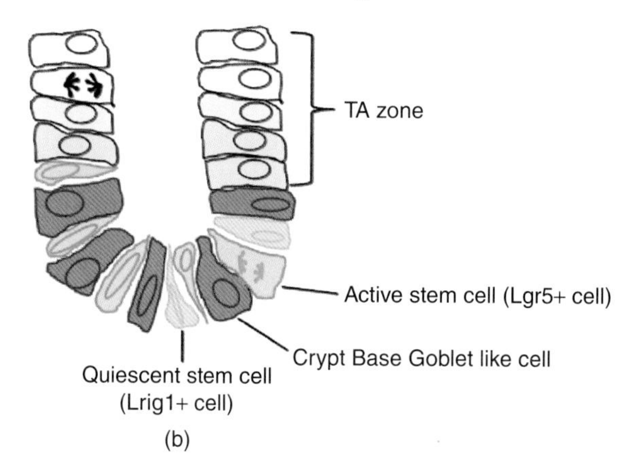

(b)

Figure 9.1 Schematic of the architecture of the crypt bottoms of the small intestine and colon. (A) Cycling CBC stem cells are located interspersed between Paneth cells in the small intestine, which provide the stem cell niche in this segment of the gut. CBC cells express high levels of Lgr5 ($Lgr5^{Hi}$) and function as ancestral cells by generating progenitor cells that form the transit-amplifying zone. CBC functions as a stem cell for all epithelial lineages in the small intestine under conditions of unstressed homeostasis. Stem cells with a lower rate of cell proliferation can be found around the +4 position in the small intestine, and such cells may express or possess elevated activity of Bmi1/Hopx/mTERT/$Lrig1^{+}$. Stem cells at the +4 position become actively dividing following conditions of disrupted epithelial homeostasis after tissue injury. (B) In the descending colon, cycling and noncycling stem cells are intermingled between crypt base goblet cells, which are believed to provide the stem cell niche. Frequently cycling stem cells express Lgr5, whereas quiescent stem cells are positive for Lrig1. Cell division generates an anisotropic movement of progeny, which gradually lose contact with the stem cell niche and successively becomes differentiated lineages.

cells with tritiated thymidine, Cheng and Leblond were able to demonstrate radioactive labelling of more differentiated cells of the crypt. This and other experiments later led Bjerknes and Cheng (1999, 2002) to claim that CBC cells are self-renewing and multipotent ISCs and reside within the stem cell zone of the intestinal crypt. Such findings have paved the way for the development of methodologies that allow for the specific analysis of the cell fates of CBC cells and other cells in the intestine.

More recent ISC markers have been used to identify such cells in GEMs, primarily in the small intestine but also in the colon. The discovery of such markers through the use of GEMs (typically expressing either fluorescent reporter proteins or LacZ controlled by the promoters of putative ISC-specific genes) has boosted our understanding of ISCs in terms of identifying their spatial location and the role they might play in tissue homeostasis of the gut. In the past, knowledge of ISCs was largely based on combination radiation-exposure experiments and DNA label retention to elucidate their location in the organ. More recent experiments have revealed some novel observations about ISCs under normal homeostasis: they do not divide asymmetrically, they are not quiescent and they can self-sustain their own cellular niche within the epithelial cell population of the small intestine and colon. In some cases, it is possible to isolate intestinal cells expressing stem cell markers and expand them to organoids *in vitro* using specific culture conditions. The growth pattern of such organoids resembles the tissue architecture of the small-intestinal epithelium and indicates that ISCs may be maintained in the absence of other nonepithelial cells that were previously considered to be critical for the expansion of the intestinal epithelium. These findings have functional relevance to human epithelial tissues. Human colon cells can grow as spheroids *in vitro*. The colonocytes grow especially well when they express high levels of EPHB2, a receptor tyrosine kinase for the colonies (Jung *et al.*, 2011). This type of culture condition may serve as a tool for the study of mechanisms that control the differentiation and proliferation of ISCs.

9.1.2.1 DNA LRCs Quiescent cells in tissues characterized by their propensity to retain DNA labels such as tritiated thymidine or BrdU over an extended period of time following labelling have classically been considered indicative of the presence of stem cells in the gut. Such cells tend to remain quiescent and may undergo asymmetrical cell divisions to produce progeny. They may also function as an 'injury reserve' that helps maintain homeostasis in the gut following conditions of stress. Subsequently, label-retaining cells (LRCs) may mark quiescent ISCs, which primarily expand the stem cell compartment when necessary. Indeed, small numbers of LRCs have been observed for around 4 weeks after established homeostasis and can re-enter the cell cycle following stress (Potten *et al.*, 1978, 2002). More recent studies have employed transgenic mice expressing H2B-YFP under the Cyp1a1 promoter (Buczacki *et al.*, 2013). This transgenic model has an advantage over nucleotide labelling, which marks noncycling quiescent cells incompletely. An expression pulse of the yellow fluorescent H2B-YFP fusion protein, which binds to DNA, is induced through a bolus injection of β-naphthoflavone (βNF), and nuclear fluorescence is lost with subsequent cell divisions. The Cyp1a1 promoter shows expression in all cells of the crypt–villus axis, with the exception of mature Paneth cells, and H2B-YFP expression is restricted to cells at the crypt base 7 days post-injection with βNF. At 3 weeks following βNF, approximately two YFP-LRCs are observed per crypt, predominantly at the +3 position (Buczacki *et al.*, 2013). By contrast, no LRCs are observed in the colon. Expression profiling of flow-sorted YFP-LRCs shows that these cells are distinct from Paneth cells and from cycling cells from the lower crypt. However, transcripts enriched in Paneth cells, enteroendocrine cells and stem cell markers (Lgr5, Lrig1, CD133, CD44 and Peg3) are enriched in this cell population. Further experiments have concluded that the YFP-LRCs may be precursor cells that are committed to mature into differentiated secretory cells or into the Paneth and enteroendocrine lineages (Buczacki *et al.*, 2013). Following intestinal injury, these cells appear capable of extensive proliferation and give rise to progeny from all epithelial lineages. Subsequently, it has been shown that YFP-LRCs can function as stem cells under conditions that are stressful to the small intestine, and function as a clonogenic reserve. These data demonstrate the plasticity of slowly cycling cell lineages in the GIT and may further aid in identifying CRC cells that cause treatment resistance.

9.1.2.2 CD133/prominin-1 CD133 (known in rodents as prominin-1, Prom1) is a 120 kDa transmembrane cell-surface protein, which it has been suggested is expressed exclusively on CRCSCs from several malignancies (Irollo and Pirozzi, 2013). Indeed, CD133 has been used as a surface marker for the identification and isolation of tumour-initiating ($CD133^+$) cells from CRCs. Furthermore, CD133 has been suggested to characterize normal stem cells and cancer stem cells (CSCs) in human tissues, including colonic mucosa. $CD133^+$ cells have been shown to have self-maintaining properties and can differentiate and re-establish tumour heterogeneity upon serial transplantation *in vivo*. The molecular mechanisms and signalling pathways that control CD133 expression remain largely uncharacterized, and the molecular function of CD133 in stem cell biology remains unclear. The use of CD133 as a biomarker remains controversial with respect to both ISCs and CRCSCs. Studies of CD133 using transgenic mice have produced differing results with respect to CD133 as a selective ISC marker in the mouse colon and intestine (Snippert *et al.*, 2009; Zhu *et al.*, 2009). CD133 expression has been observed in both ISCs and TACs in the small intestine. Other studies have found an increasingly widespread expression of CD133 throughout the colonic mucosa. Thus, it seems unlikely that CD133 is a selective ISC marker; rather, it may harbour a broad expression, which might also include progenitor cells, at least in the mouse gut.

9.1.2.3 TERT Telomerase (TERT) helps with the maintenance of the telomeric ends of chromosomes and has been shown to label long-term LRCs within small-intestinal crypts (De Mey and Freund, 2013). Adult stem cells may express high levels of TERT, and rare cells may express green fluorescent protein (GFP) under the mouse telomerase (mTERT) promoter in the unstressed mouse small intestine, which has been suggested to be distinct from Lgr5+ cells (Montgomery *et al.*, 2011). These quiescent mTERT+ cells occur with a frequency of less than 1 in 150 small-intestinal crypt cells in the unstressed mouse small intestine but become actively cycling following radiation-induced damage. Lineage-tracing experiments suggest that these mTERT+ cells can differentiate into all epithelial intestinal cell types. Given the rarity of mTERT+ cells, it may seem unlikely that such a stem cell would contribute to epithelial homeostasis in the unstressed gut, or even under stressful conditions, considering the microstructure of the crypt. Furthermore, some evidence suggests that elevated mTERT activity is displayed in the cycling small-intestinal crypt, as compared to the standard reported frequency of mTERT+ cells (Schepers *et al.*, 2011). Future experiments will characterize the relevance of TERT-expressing cells in controlling the homeostasis of the gut.

9.1.2.4 Bmi1 Bmi1 constitutes a component of the Polycomb transcriptional repressor complex that functions as a regulator of self-renewal and plays an essential role in maintaining chromatin silencing of neural and haematopoietic progenitor cells (De Mey and Freund, 2013). Bmi1+ cells have been visualized in the small intestine of mice, where they most frequently occupy the +4 position, using a LacZ reporter gene. Bmi1^{+} cells cycle at a lower frequency than CBC cells or Lgr5+ cells and are represented at a much lower frequency in the duodenum and jejunum. By contrast to Lgr5+ cells, Bmi1-expressing cells are not observed in distal parts of the small intestine or colon. It is somewhat unclear if this is related to silencing of the transgene or indicative of segment-specific expression of Bmi1 in ISCs (Sangiorgi and Capecchi, 2008). Indeed, targeted diphtheria toxin-dependent killing of Bmi1+ cells caused crypt death in the small intestine. A follow-up study showed that Bmi1+ cells represent a distinct radiation-resistant population of ISCs from that of Lgr5+ cells in the small intestine (Yan *et al.*, 2012). Furthermore, it has been suggested that Bmi1+ cells function as a stem cell reserve, since following selective diphtheria toxin ablation of Lgr5+ CBCs, Bmi1+ cells become proliferative and rescue crypt homeostasis by providing progeny that replenish Lgr5+ CBCs (Tian *et al.*, 2011). This indicates that, upon injury, Bmi1+ cells may function as stem cells, replenishing CBCs and maintaining tissue homeostasis. Thus, it has been proposed that Lgr5+ cells represent adult stem cells that maintain tissue homeostasis, whereas Bmi1+ cells serve to maintain epithelial compensatory proliferation and regeneration following injury. From this, it can be concluded that the small intestine should harbour two distinct ISC populations that address two operationally distinct conditions.

However, the specificity of enforced Bmi1+ expression to the +4 position has been brought into question, and Bmi1 mRNA can be detected in Lgr5+ cells (Barker *et al.*, 2012). It has been suggested that Bmi1 expression is not specific to the +4 position but can be detected throughout the proliferative epithelium and crypt base. Furthermore, Bmi1-based lineage tracking can be initiated throughout the small-intestinal epithelium, including Lgr5+ CBCs. Thus, it is not currently clear whether Bmi1 selectively marks ISCs or how Bmi1 expression might relate to other ISC markers. On the other hand, chemical targeting of Bmi1 has been suggested to target cancer-initiating cells and serve as a new therapeutic approach for the treatment of CRC (Kreso *et al.*, 2014).

9.1.2.5 Hopx The atypical homeobox-only protein (Hopx) has been reported to play a critical role in the differentiation of several cell lineages, including lung alveolar cells, keratinocytes, T cells and trophoblasts (Yamashita *et al.*, 2013). Although not exclusively expressed at this cell position, Hopx has been proposed as a marker for +4 cells, since a LacZ knock-in allele shows the strongest expression here (Takeda *et al.*, 2011). Indeed, the +4 position is also the most frequent site for lineage-tracing events using a Hopx-CreER transgenic allele. Hopx+ cells may give rise to all intestinal lineages, including Lgr5+ CBCs. It has also been shown that Lgr5+ cells may produce Hopx+ cells at the +4 position. A high level of Hopx expression has been reported in Lgr5+ cells (Barker *et al.*, 2012). These findings further support a model in which transdifferentiation between fast- and slow-cycling ISC lineages occurs even under unstressed conditions. Thus, two stem cell populations may populate the crypts of the GIT. In primary CRC, Hopx has been reported to be methylated and dramatically downregulated, suggesting that Hopx expression may be epigenetically repressed in cancers (Yamashita *et al.*, 2013). Indeed, the methylation status of the Hopx gene has been linked to the differentiation status of CRC. However, it is currently unclear whether Hopx is a marker for CRCSCs.

9.1.2.6 Lgr5 The Wnt target gene, G protein-coupled receptor Lgr5, which acts as a receptor for a small family of Wnt pathways called R-spondins, has been shown to be a specific marker for ISCs in the mouse small intestine and colon (Barker *et al.*, 2012). The expression of Lgr5 is controlled by Wnt signals and Lgr5 is an optional component of the Wnt receptor complex. The generation of mice expressing EGFP under the Lgr5 promoter has helped identify living ISCs in their guts, as has a tool for *in vivo* lineage tracing (Barker *et al.*, 2007). Cells expressing EGFP under the Lgr5 promoter represent a long-lived multipotent stem cell lineage. Lgr5+ cells have been shown to be capable of self-renewal and differentiation into multiple epithelial lineages *in vitro* and *in vivo*. A single Lgr5+ cell was capable of generating crypt-like organoid structures *in vitro*, containing cells with epithelial morphology that expressed markers consistent with multilineage differentiation. Each intestinal crypt contains around 15 Lgr5+ cells, with a slender wedge-like morphology, which are in contact with Paneth cells located at the crypt bottom and divide once every day. Paneth cells provide an extracellular milieu that maintains the stem cell niche, at least *in vitro*, where it has been demonstrated that Paneth cells provide EGF, Wnt and Notch signals to Lgr5+ cells (Sato *et al.*, 2011). In addition, Lgr5+ cells residing at the bases of colonic crypts have also been identified as adult stem cells. In the crypts of the ascending colon, $CD24^+$ or deep secretory cells (DSCs) residing between Lgr5+ cells may represent the Paneth cell equivalents (De Mey and Freund, 2013). CRC is more frequent in the descending colon of humans and is thus of particular interest. In mice, this segment lacks DSCs but contains mature Muc2-secreting cells, referred to as goblet-like cells (GLCs), that express CD24 (De Mey and Freund, 2013). In this colonic segment, high Lgr5 expression is found in columnar 'vacuoled cells', which can be detected next to GLCs, and it has been shown that Lgr5+ cells indeed function as stem cells in the descending colon of mice. There is evidence that Lgr5+ cells in early intestinal adenoma function as tumour stem cells, since deletion of Apc can lead within days to their transformation and fuels the growth of microadenomas (Barker *et al.*, 2009). These transformed Lgr5+ cells represent 5–10% of the cell population in the formed adenomas and retain their hierarchical position even when CRE-recombinase causes concomitant APC deletion and expression of mutant KRAS in Lgr5+ cells (Schepers *et al.*, 2012). Monoclonal antibodies towards Lgr5 have been used to identify a population that behaves like CRCSCs *in vitro* from primary colorectal tumour cells (Kemper *et al.*, 2012). The relevance for the presence of different levels of Lgr5-expressing cells in CRC with respect to clinical parameters remains to be addressed, but data suggest that Lgr5 may be a selective marker for CRCSCs. Recent data employing single-stem-cell *in vivo* live imaging indicate that the short-term dynamics of Lgr5+ cells in the upper part of the stem cell niche (border cells) are occasionally passively displaced into the transit-amplifying zone after the division of proximate cells, which suggests that maintenance of stem cell properties may be uncoupled from division (Ritsma *et al.*, 2014). Lgr5+ cells at the crypt base (central cells) experience a survival advantage compared to border cells and are three times more likely to fully colonize a crypt at steady state. However, all Lgr5+ cells can move in the stem cell niche and are subsequently able to function as a long-term equipotent stem cell pool (Figure 9.1).

9.1.2.7 Lrig1 The ErbB family consists of four receptor tyrosine kinases (EGFR and Her2–4). Maintenance of epithelial homeostasis in the intestinal epithelium is dependent on ErbB signalling, since the ErbB family is expressed in the stem cell niche (Sato *et al.*, 2011). Lrig1 is a transmembrane negative-feedback regulator of ErbB receptor signalling and is increasingly expressed in ISCs. Lrig1 is expressed as frequently as Lgr5, although with limited overlap in the colon (Figure 9.1). Lrig1 expression is considered to mark a quiescent ISC during normal tissue homeostasis in the GIT, but the number of Lrig1+ cells can be expanded following irradiation.

During normal homeostasis, 20% of the Lrig1+ cells are LRCs and 25% express Ki67, whereas after irradiation nonproliferating cells enter the cell cycle. Deletion of the tumour suppressor Apc in Lrig1+ cells leads to the formation of adenomas. Lrig1 may also function as a tumour suppressor, since loss of Lrig1 generates adenomas in the duodenum (Powell *et al.*, 2012). It is unclear to what extent Lgr5 and Lrig1 describe truly distinct cell lineages. Both Lrig1+ and Lgr5+ cells isolated from the colon express CD133, mTERT and Bmi1, and it has been suggested that Lrig1 does not selectively mark a quiescent ISC in the intestine (Barker *et al.*, 2012; Wong *et al.*, 2012). Moreover, a high level of Lrig1-expression is frequently observed in Lgr5+ cells (Munoz *et al.*, 2012; Wong *et al.*, 2012). Thus, it is unclear whether Lrig1 marks quiescent ISCs that are distinct from Lgr5+ cells or if it is a general regulator of the stem cell niche.

9.2 CRCSCs

Recent evidence suggests that various types of cancer, such as leukaemia and solid tumours, are composed of a rare subset of cells with stem cell-like properties, unlike bulk tumour cells, that are alone capable of initiating and sustaining tumours (Zeki *et al.*, 2011). CRCSCs were initially identified as $CD133^+$ in two separate studies in 2007. Only $CD133^+$ cells were capable of initiating tumours and undergoing *in vitro* and *in vivo* passage: $CD133^-$ cells did not form xenograft tumours. The CSC hypothesis suggests that, like adult colorectal tissues, colorectal tumours are hierarchically organized, with cells at various stages of differentiation, and rely on stem cell-like cells for long-term maintenance (O'Brien *et al.*, 2007; Ricci-Vitiani *et al.*, 2007).

9.2.1 CRCSC markers

Since its initial discovery, there has been some debate about the use of CD133 as an ISC and CRCSC marker. Using lineage-tracing analysis, a study has shown that CD133/Prom1+ cells represent mouse ISCs that are capable of initiating tumours upon acquisition of β-catenin mutations (Zhu *et al.*, 2009). However, another study shows that CD133 expression is not restricted to ISCs alone and that differentiated mouse and human normal and tumour epithelial cells also express it. Both $CD133^+$ and $CD133^-$ populations from metastatic human colon tumours are capable of tumour initiation and long-term tumour maintenance. The functional importance of CD133 in tumour growth and CRCSC biology remains unknown (Shmelkov *et al.*, 2008).

Along with CD133, several other CRCSC markers have been identified and characterized (Figure 9.2). Lgr5, the crypt stem cell marker-positive cells have indeed been shown to initiate mouse intestinal adenomas upon acquiring APC deletion and to be responsible for adenoma growth. Lgr5+ cells in adenomas can give rise to Lgr5+ and Lgr5− cells, indicating a stem cell hierarchy

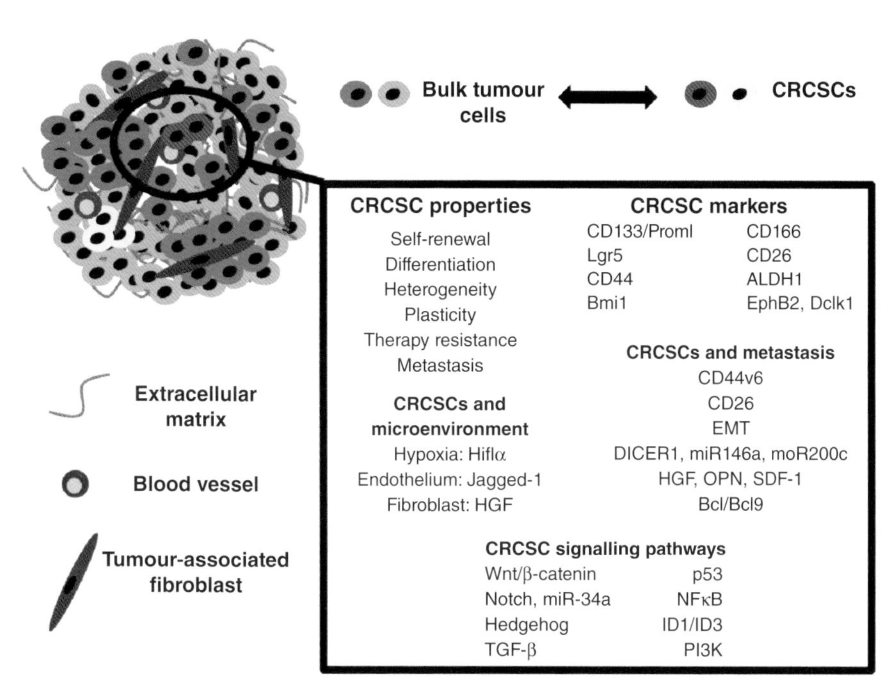

Figure 9.2 Overview of CRCSC biology. Colorectal tumours include a variety of cells at various stages of differentiation. CRCSCs are a rare population of stem cell-like cells within the tumour with enhanced self-renewal properties that drive long-term tumour growth, chemotherapy resistance, relapse and metastasis. Several factors govern the self-renewal and differentiation of CRCSCs, including signalling pathways and the tumour microenvironment. CRCSCs can give rise to multiple lineages within the tumour. Differentiated tumour cells can also give rise to CRCSCs via dedifferentiation. The resulting cellular plasticity contributes to tumour heterogeneity and is a major problem in cancer therapy. Various functional clones of CRCSCs can exist within a single tumour, further contributing to tumour heterogeneity. CRCSC markers, regulators and signalling pathways serve as important therapeutic targets. CRCSC-specific agents can be combined with chemotherapy to target both CRCSCs and bulk tumour cells.

within CRC (Schepers *et al.*, 2012). In similar studies, CD44, CD166 (Dalerba *et al.*, 2007), CD26 (Pang *et al.*, 2010), ALDH1 (Huang *et al.*, 2009), EphB2 (Merlos-Suarez *et al.*, 2011), Bmi1 (Kreso *et al.*, 2014), Dclk1 (Nakanishi *et al.*, 2013) and other CRCSC markers have been characterized (Table 9.1). Interestingly, while most commonly used CRCSC markers, such as CD44, CD133, Lgr5, Bmi1 and ALDH1, are also expressed by ISCs, Dclk1 is expressed on CRCSCs and tuft cells but not ISCs.

9.2.2 CRCSC-mediated chemotherapy resistance and CRC recurrence

The clinical relevance of CRCSCs and their markers has been explored in several studies. CRCSCs are of particular interest in the clinic, due to their

Table 9.1 Biomarkers of gastrointestinal stem cells.

Biomarker	Expression pattern/location	Proliferation status	Tissue regeneration	Intestinal hyperplasia	References
Bmi1	Frequent expression at +4 positions in proximal small intestine; minimal expression in CBC; no lineage tracing in colon	Marks relatively quiescent cells	Yes, 20-fold increased proliferation of Bmi1+ cells following radiation	Yes, deletion of APC in Bmi1+ cells causes small-intestinal adenomas	Sangiorgi and Capecchi (2008), Yan *et al.* (2012)
H2B-YFP label-retaining cells (LRCs)	Crypt base, slight predomination at the +3 position in the small intestine; approximately two per whole crypt; not detected in the colon. Gene expression overlaps with Lgr5+ cells; combined secretory and stem cell expression signature	Quiescent. LRCs function as progenitor cells to Paneth and enteroendocrine cells during normal homeostasis	Yes, increased proliferation following injury; LRCs function as stem cells under conditions of injury (hydroxyurea, doxorubicin, radiation)	N/A	Buczacki *et al.* (2013)
Hopx	Populates the entire small intestine; frequent around +4 position	Relatively quiescent. Lgr5+ cells can give rise to Hopx+ cells *in vitro*	Actively proliferating following radiation-induced injury	N/A	Takeda *et al.* (2011)

Lgr5	Expressed in CBC cells of the small intestine and columnar 'vacuolated' cells of the descending colon	Proliferative	Indispensible for regeneration following radiation-induced injury	Yes, deletion of APC in Lgr5+ cells results in persistent intestinal adenomas	Barker *et al.* (2007), Metcalfe *et al.* (2014)
Lrig1	Broadly expressed in small intestine and colon; highest levels found in Lgr5+ cells	Proliferative	Proliferative following injury; radiation resistant	Negatively regulates ErbB signalling and limits expansion of the stem cell niche. Loss of APC in Lrig1 leads to adenoma formation	Powell *et al.* (2012), Wong *et al.* (2012)
TERT	Infrequently expressed, primarily at the +4 position in the small intestine	Quiescent	Actively proliferating following radiation-induced injury	N/A	Montgomery *et al.* (2011)

chemotherapy resistance (Figure 9.2). For example, CRCSC populations of $CD133^+$ and $CD26^+$ cells are resistant to the commonly used CRC chemotherapy agents 5-fluorouracil and oxaliplatin, while $CD26^-$ and $CD133^-$ cells are sensitive to chemotherapy (Todaro *et al.*, 2007; Pang *et al.*, 2010). ISC- and CRCSC-specific gene signatures have been used to identify CRCSCs in patient tumours. Both signatures have been shown to predict poor prognosis and disease relapse (de Sousa *et al.*, 2011b; Merlos-Suarez *et al.*, 2011). Polymorphisms in CRCSC marker genes such as ALDH1, CD44 and Lgr5 have been shown to predict recurrence in CRC patients treated with 5-fluorouracil-based therapy (Gerger *et al.*, 2011).

9.2.3 CRCSCs and metastasis

Many studies have evaluated whether CRCSCs cause metastasis. CD44 has been described as a marker for CRCSCs, but a recent study looked at the role of CD44 variant 6 (CD44v6) in CRC metastasis. Interestingly, CD44v6 was elevated in metastatic lesions as compared to primary tumours, while $CD44^+$ cells were equally present at both sites. Primary orthotopic tumour growth and metastasis were significantly improved when initiated with $CD44v6^+$ CRC-SCs as compared to $CD44v6^-$ cells. Cancer-associated fibroblasts improved the metastatic potential of CRC cells via secreted cytokines such as hepatocyte growth factor (HGF), osteopontin (OPN) and stromal-derived factor 1a (SDF-1). The cytokines could promote CD44v6 expression and metastasis even with $CD44^-$ cells. The $CD44v6^+$ population was regulated by the Wnt and PI3K pathway, and CD44v6 or PI3K inhibition prevented CRC metastasis (Todaro *et al.*, 2014). In a similar study, $CD26^+$ CRCSCs were identified as the metastatic CRCSC population. By contrast to CD44v6, $CD26^+$ cells were present in both primary and metastatic tumours, and $CD26^+$ cells in the primary tumour predicted metastasis in CRC patients (Pang *et al.*, 2010).

The epithelial–mesenchymal transition (EMT) is an important determinant of tumour metastasis and enriches the CRCSC population (Hwang *et al.*, 2014). $CD44^+$ CRCSCs have been shown to acquire an EMT phenotype under nonadherent conditions, and CD44 expression to be critical for CRC lung metastasis (Su *et al.*, 2011). Similarly, metastatic $CD26^+$ cells also acquire an EMT phenotype, and reduction of CD26 expression reverses this phenotype (Pang *et al.*, 2010). BCL9/BCL9L protein is a part of the TCF/LEF protein complex, partially regulates β-catenin-mediated transcription and is upregulated in colon cancer. Bcl9/Bcl9l deletion in the mouse intestinal epithelium results in tumours with active Wnt signalling, reduced EMT and CRCSC features (Deka *et al.*, 2010). Thus, BCL9/BCL9L protein potentially links the Wnt pathway and CRCSCs with EMT and metastasis. Further, micro-RNAs (miRs) such as miR-146a and miR-200c serve as a link between EMT and CRCSCs (see Section 9.2.5.4).

9.2.4 Heterogeneity and plasticity of CRCSCs

With so many markers for CRCSCs, distinct CRCSC populations can arise according to the combination of markers being expressed (Vermeulen *et al.*, 2008; Kreso *et al.*, 2014). Among ISCs, functionally distinct quiescent Bmi1-expressing and rapidly cycling Lgr5– expressing populations have been identified. Bmi1-expressing cells can give rise to Lgr5-expressing cells, indicating a hierarchy within the ISC population (Tian *et al.*, 2011; Yan *et al.*, 2012). However, another study shows bidirectional interconversion between slow- and rapid-cycling ISC populations, with both populations showing properties of self-renewal and multipotency (Takeda *et al.*, 2011). Within the CRCSC population, various combinations of markers, such as CD44, CD133, CD26 and CD166, give rise to populations that differ in their tumour initiation and growth abilities (Dalerba *et al.*, 2007; Pang *et al.*, 2010). Targeting one marker of self-renewal, such as Bmi1, does not change CD133 or CD44 expression (Kreso *et al.*, 2014). Another study has identified three distinct functional subtypes within CRCSCs. The first subtype is involved in primary tumour growth; the second has long-term self-renewal potential and is important for tumour initiation, growth and metastases; and the third evolves only upon serial passaging (Dieter *et al.*, 2011). The resulting marker-based and functional heterogeneity within the CRCSC population has potential implications for therapeutic targeting of CRCSCs.

Differentiated somatic cells can be reprogrammed to form induced pluripotent stem cells using a specific set of oncogenic reprogramming factors, such as Sox2, Oct4, Nanog, Klf4 and c-myc. Loss of the tumour-suppressor p53 further enhances reprogramming frequency. Like induced pluripotent stem cells, differentiated tumour cells can dedifferentiate to form CRCSCs by accumulating genetic hits, such as an increase in reprogramming factors and p53 loss (Prabhu *et al.*, 2012). A recent paper showed the role of the NF-κB pathway in inducing dedifferentiation of villus cells and $Lgr5^-$ cells with dysregulated Wnt signalling to form tumour-initiating cells (Schwitalla *et al.*, 2013). Thus, CRCSCs can potentially arise from differentiated CRC cells after CRCSC-specific therapy (Figure 9.2).

9.2.5 Regulation of CRCSCs

9.2.5.1 Stem cell signalling pathways Dysregulation of the pathways that govern ISCs is an important feature of CRC (Figure 9.2). Mutations in TGF-β and Smad4, critical regulators of stem cell maintenance, are commonly observed in CRC (Joudeh *et al.*, 2013). APC tumour suppressor loss, Wnt pathway activation and β-catenin mutation result in an abnormal expansion of intestinal crypt stem cells and loss of crypt architecture, ultimately resulting in CRC (Zhu *et al.*, 2009). Notch signalling regulates self-renewal and differentiation in the intestinal epithelium and is elevated in CRCSCs. Notch signalling is important for CRCSC maintenance and resistance to apoptosis (Sikandar *et al.*, 2010). Similarly, the Hedgehog (Hh)–Gli signalling pathway that governs differentiation in the normal intestine is also activated in CRC and is important for CRCSC self-renewal (Varnat *et al.*, 2009). Thus, dysregulated stem cell signalling pathways in CRC may serve as important drug targets in CRCSCs.

9.2.5.2 Other CRC signalling pathways Several other signalling pathways involved in CRC also regulate CRCSCs (Figure 9.2). The tumour-suppressor p53 protects normal cells in response to various stresses by inducing cell-cycle arrest, senescence or apoptosis. Recent evidence suggests that an important aspect of the tumour-suppressive role of p53 is its regulation of stem cell differentiation and self-renewal. Loss of p53 correlates with poorly differentiated tumours in CRC. It results in CRCSC enrichment, while p53 pathway restoration depletes CRCSCs (Prabhu *et al.*, 2012).

As mentioned in Section 9.2.5.1, APC loss is an important initiating event in CRC. The mechanisms involved in ISC expansion and hyperproliferation upon APC loss were recently delineated. RAC1 GTPase-mediated NF-κB activation and reactive oxygen species (ROS) production are involved in Lgr5 stem cell expansion and hyperproliferation, leading to adenoma formation (Myant *et al.*, 2013).

CRCSCs have the capacity to differentiate and give rise to multiple different cell types within CRC tumours, including goblet-like, enterocyte-like and neuroendocrine (NE)-like cells. The PI3K pathway plays an important role in CRCSC multilineage differentiation, as PI3K inhibition shifts the process to a more enterocyte-like differentiation (Vermeulen *et al.*, 2008).

The inhibitor-of-DNA-binding-protein (ID) family regulates the self-renewal of embryonic stem cells, as well as adult tissue stem cells, and ID proteins are overexpressed in CRC. ID1 and ID3 regulate CRCSCs via p21. ID1 and ID3 knockdown results in loss of CRCSC-mediated tumour initiation and self-renewal, decreased p21 levels, increased DNA damage accumulation and increased sensitivity to oxaliplatin (O'Brien *et al.*, 2012).

9.2.5.3 Tumour microenvironment ISCs receive signals of self-renewal and differentiation, such as Wnt, Notch and bone morphogenetic protein, from the surrounding stroma and niche cells (Joudeh *et al.*, 2013). Similarly, CRCSCs rely on signals from the surrounding hypoxic tumour microenvironment and cancer-associated endothelial cells and fibroblasts for maintenance of stemness and initiation of differentiation or proliferation. Myofibroblast-secreted HGF is important to enhancing Wnt-β-catenin signalling in CRC cells and maintaining a CRCSC phenotype. Myofibroblast-secreted factors also induce a CRCSC phenotype in differentiated CRC cells, via Wnt-β-catenin signalling (Vermeulen *et al.*, 2010). CRC-associated endothelial cells secrete a soluble form of Jagged-1 that can activate Notch signalling in CRC cells. The soluble Jagged-1 arises via ADAM17 protease, promoting a CRCSC phenotype, chemotherapy resistance and metastasis (Lu *et al.*, 2013). The hypoxic tumour microenvironment can also help maintain the CRCSC phenotype via Hif1α, Bmi1 and Notch1. Interestingly, this effect of hypoxia on CRC cells is reversible (Yeung *et al.*, 2011).

9.2.5.4 Micro-RNA In cancer, miRs are frequently dysregulated. They regulate target gene expression and cell fate by inhibiting mRNA translation or marking mRNAs for degradation. Recent studies have investigated the role of miRs in CRCSC regulation. The p53 target miR-34a prevents CRCSC self-renewal by inhibiting Notch signalling. It regulates Notch1 levels in a bimodal manner, determines whether CRCSCs undergo self-renewal or differentiation and regulates symmetric versus asymmetric CRCSC division (Bu *et al.*, 2013).

The transcription factor Snail, a key EMT mediator, increases miR-146a, which stabilizes Wnt activity and CRCSC symmetrical division, fuelling tumour growth and cetuximab resistance (Hwang *et al.*, 2014). Micro-RNAs are usually transcribed as long precursors and are converted into mature miRs via DICER1 activity. Loss of DICER1 function results in enrichment of CRCSCs, an EMT phenotype and increased metastasis. The effects of impaired DICER1 are mediated by the resulting loss of several miRs, such as miR-34a and miR-200 family members (Iliou *et al.*, 2014). Loss of miR-200c elevates CRCSCs, EMT and metastasis (Lu *et al.*, 2014).

9.2.6 Approaches to CRCSC targeting

Various approaches to CRC therapy are currently approved or under evaluation in the clinic (Joudeh *et al.*, 2013). There is an urgent unmet need to target the CRCSC subpopulation within CRC that is the cause of tumour progression, therapy resistance, recurrence and metastasis. Currently, there is no US Food and Drug Administration (FDA)-approved agent for the targeting of CRCSCs. Targeting CRCSC markers is one approach; for example, a small-molecule inhibitor of Bmi1 prevents CRCSC-mediated long-term tumour growth and self-renewal (Kreso *et al.*, 2014). CRCSCs have also been targeted with a neutralizing antibody against interleukin 4 (IL-4), a cytokine that protects CRCSCs from cell death (Todaro *et al.*, 2007).

Targeting of stem cell signalling pathways, such as Wnt, Notch and Hh, using small molecules and monoclonal antibodies is currently being investigated for clinical use (Joudeh *et al.*, 2013). For example, the Wnt pathway can be blocked by use of Cox-2 inhibitors or monoclonal antibodies targeting c-Met/HGF interaction (de Sousa *et al.*, 2011a), and Hh signalling can be inhibited with the plant alkaloid cyclopamine (Varnat *et al.*, 2009). Notch signalling can be targeted with γ-secretase inhibitors (Sikandar *et al.*, 2010) or antibodies against the Notch pathway component delta-like ligand 4 (DLL4).

Anti-DLL4 antibodies reduce CRCSC frequency in patient-derived tumour xenografts (Hoey *et al.*, 2009).

CRCSCs are a rare population within CRC; as a result, targeting of CRCSCs alone would lead to a delayed but long-term antitumour effect. Considering that non-stem cells within the tumour can also give rise to CRCSCs via dedifferentiation, it is important to target both non-CRCSCs and CRCSCs. One approach is to combine CRCSC-specific agents with chemotherapy that targets bulk tumour cells, in order to achieve a potent and long-term antitumour effect. For example, the chemotherapeutic agent oxaliplatin was combined with $CD133^+$ CRCSC-targeting anti-IL-4 antibody (Todaro *et al.*, 2007) and an anti-DLL4 antibody targeting CRCSCs was combined with the chemotherapeutic agent irinotecan (Hoey *et al.*, 2009). Another approach is to modulate the signalling pathways that drive or inhibit both CRCSCs and bulk tumour cells. For example, both CRCSCs and bulk tumour cells can be targeted via the p53 pathway (Prabhu *et al.*, 2012), the PI3K-Akt pathway (He *et al.*, 2007; Vermeulen *et al.*, 2008) or NF-κB signalling (Myant *et al.*, 2013), which regulate diverse cellular functions, such as cell-cycle arrest, senescence, apoptosis, stem cell self-renewal and differentiation.

9.3 Conclusions

Exciting new findings are emerging in the field of stem cells of normal and cancerous tissues, which may result in opportunities to develop additional strategies for therapeutic intervention in CRC. However, despite all the advances that have recently been made, it still remains somewhat unclear how increased knowledge of CRCSCs can help in developing novel therapy in the short term. If normal and malignant tissues share a common stem cell then it becomes somewhat unclear how such knowledge will benefit the development of novel therapeutic targeting, since this will not allow for selective design of targeted cancer therapy. In this scenario, it will be the acquired properties of stem cells in cancers relative to those of normal tissue stem cells and their progeny that are of most interest from the perspective of a systemic pharmacologic therapeutic approach. This may particularly be the case when one considers the data on plasticity between stem cells and their progeny. Thus, understanding how the presence of CRCSCs in tumour samples relates to prognosis and treatment responses may be the first step in facilitating clinical translation of discoveries in the CRCSC field. New advances have been made that allow for reliable detection of circulating tumour DNA and cells, and such methods could potentially be used as a method of interrogation of the population dynamics in the tumour and metastatic lesions that are being targeted. It may be of particular interest to understand how therapeutic intervention alters the cellular dynamics between CRCSCs and their progeny. One recent paper showed how knowledge about the reversion rate between differentiated radiosensitive glioblastoma (GBM) cells and radiorefractory cancer stem-like cells helped facilitate mathematical modelling of an optimal radiation dose-delivery schedule in mice carrying PDGF-driven GBMs (Leder *et al.*, 2014). Thus, based on the preclinical data indicating that CRCSCs are treatment-refractory, understanding the population dynamics within the tumour may help in the delivery of conventional therapeutics. Further understanding of how the stem cell niche is maintained in CRC could be very helpful. Data from experimental mouse models indicate that the stem cell niche in early intestinal adenomas may be maintained by noncancerous Paneth cells (Barker *et al.*, 2009). Paneth cells have been shown to contribute to the stem cell niche by activating EGFR, Wnt and Notch signalling in ISCs. It is generally believed that cancer cells wean themselves from extracellular signals controlling differentiation and proliferation as the disease progress. The transgenic mouse models that have been used to date have addressed the importance of ISCs in the initiation of adenomas driven by APC deletion primarily in the small intestine. Thus, such models may have limitations in demonstrating the relationship between ISCs and CRCSCs, given that the adenomas studied are nonmalignant lesions that do not metastasize to distant organs. Furthermore, premalignant or malignant lesions with oncogenic mutations of the Wnt pathway are rarely encountered in the small intestine in clinical settings. Thus, a transgenic mouse model that allows for the study of the potential conversion of ISCs to CRCSCs in the setting of metastatic disease is needed. Further development of CSC biomarkers may be important, since many of the ones currently available may not accurately and/or sufficiently selectively mark CRCSCs. Assessment of biomarkers for CRCSCs is complicated, due to the fact that candidate CRCSC populations are assessed not for their propensity to generate different tumour lineages *per se*, but rather for their propensity to facilitate subcutaneous xenotransplantation in an immunodeficient host. Furthermore, the functional relevance for some of the proposed CRCSCs is unknown and requires further evaluation. CRCSC biomarkers should be evaluated carefully. In order to support and validate the discovery of CRCSCs, it will likely be necessary to establish gene-expression signatures for flow-sorted cells that have been sorted for given markers. It seems unlikely that a single molecular marker would sufficiently label a CRCSC, and therefore multiple markers may have to be evaluated and combined for sensitive isolation and detection. In conclusion, we believe that the CRCSC field holds promise for the development of novel means by which to implicate a more personalized cancer therapy that ultimately may benefit CRC patients by reducing toxicity and improving the prognosis of the disease.

Acknowledgements

We apologize to all those colleagues whose work could not be cited due to the limited number of permitted references.

References

Barker, N., van Es, J.H., Kuipers, J., Kujala, P., van den Born, M., Cozijnsen M., *et al.*, 2007. Identification of stem cells in small intestine and colon by marker gene Lgr5. Nature 449, 1003–1007.

Barker, N., Ridgway, R.A., van Es, J.H., van de Wetering, M., Begthel, H., van den Born, M., *et al.*, 2009. Crypt stem cells as the cells-of-origin of intestinal cancer. Nature 457, 608–611.

Barker, N., van Oudenaarden, A., Clevers, H. 2012. Identifying the stem cell of the intestinal crypt: strategies and pitfalls. Cell Stem Cell 11, 452–460.

Bjerknes, M., Cheng, H. 1999. Clonal analysis of mouse intestinal epithelial progenitors. Gastroenterology 116, 7–14.

Bjerknes, M., Cheng, H. 2002. Multipotential stem cells in adult mouse gastric epithelium. Am. J. Physiol. Gastrointest. Liver Physiol. 283, G767–G777.

Bu, P., Chen, K.Y., Chen, J.H., Wang, L., Walters, J., Shin, Y.J., *et al.*, 2013. A microRNA miR-34a-regulated bimodal switch targets Notch in colon cancer stem cells. Cell Stem Cell 12, 602–615.

Buczacki, S.J., Zecchini, H.I., Nicholson, A.M., Russell, R., Vermeulen, L., Kemp, R., Winton, D.J. 2013. Intestinal label-retaining cells are secretory precursors expressing Lgr5. Nature 495, 65–69.

Cheng, H., Leblond, C.P. 1974. Origin, differentiation and renewal of the four main epithelial cell types in the mouse small intestine. V. Unitarian Theory of the origin of the four epithelial cell types. Am. J. Anat. 141, 537–561.

Clevers, H. 2013. The intestinal crypt, a prototype stem cell compartment. Cell 154, 274–284.

Dalerba, P., Dylla, S.J., Park, I.K., Liu, R., Wang, X., Cho, R.W., *et al.*, 2007. Phenotypic characterization of human colorectal cancer stem cells. Proc. Nat. Acad. Sci. USA 104, 10 158–10 163.

De Mey, J.R., Freund, J.N. 2013. Understanding epithelial homeostasis in the intestine: an old battlefield of ideas, recent breakthroughs and remaining controversies. Tissue Barriers 1, e24965.

De Sousa, E.M., Vermeulen, L., Richel, D., Medema, J.P. 2011a. Targeting Wnt signaling in colon cancer stem cells. Clin. Cancer Res. 17, 647–653.

De Sousa, E.M.F., Colak, S., Buikhuisen, J., Koster, J., Cameron, K., de Jong, J.H., *et al.*, 2011b. Methylation of cancer-stem-cell-associated Wnt target genes predicts poor prognosis in colorectal cancer patients. Cell Stem Cell 9, 476–485.

Deka, J., Wiedemann, N., Anderle, P., Murphy-Seiler, F., Bultinck, J., Eyckerman, S., *et al.*, 2010. Bcl9/Bcl9l are critical for Wnt-mediated regulation of stem cell traits in colon epithelium and adenocarcinomas. Cancer Res. 70, 6619–6628.

Dieter, S.M., Ball, C.R., Hoffmann, C.M., Nowrouzi, A., Herbst, F., Zavidij, O., *et al.*, 2011. Distinct types of tumor-initiating cells form human colon cancer tumors and metastases. Cell Stem Cell 9, 357–365.

Fearon, E.R., Vogelstein, B. 1990. A genetic model for colorectal tumorigenesis. Cell 61, 759–767.

Gerger, A., Zhang, W., Yang, D., Bohanes, P., Ning, Y., Winder, T., *et al.*, 2011. Common cancer stem cell gene variants predict colon cancer recurrence. Clin. Cancer Res. 17, 6934–6943.

He, X.C., Yin, T., Grindley, J.C., Tian, Q., Sato, T., Tao, W.A., *et al.*, 2007. PTEN-deficient intestinal stem cells initiate intestinal polyposis. Nat. Genet. 39, 189–198.

Hoey, T., Yen, W.C., Axelrod, F., Basi, J., Donigian, L., Dylla, S., *et al.*, 2009. DLL4 blockade inhibits tumor growth and reduces tumor-initiating cell frequency. Cell Stem Cell 5, 168–177.

Huang, E.H., Hynes, M.J., Zhang, T., Ginestier, C., Dontu, G., Appelman, H., *et al.*, 2009. Aldehyde dehydrogenase 1 is a marker for normal and malignant human colonic stem cells (SC) and tracks SC overpopulation during colon tumorigenesis. Cancer Res. 69, 3382–3389.

Hwang, W.L., Jiang, J.K., Yang, S.H., Huang, T.S., Lan, H.Y., Teng, H.W., *et al.*, 2014. MicroRNA-146a directs the symmetric division of Snail-dominant colorectal cancer stem cells. Nat. Cell Biol. 16, 268–280.

Iizuka, M., Konno, S. 2011. Wound healing of intestinal epithelial cells. World J. Gastroenterol. 17, 2161–2171.

Iliou, M.S., Da Silva-Diz, V., Carmona, F.J., Ramalho-Carvalho, J., Heyn, H., Villanueva, A., *et al.*, 2014. Impaired DICER1 function promotes stemness and metastasis in colon cancer. Oncogene 33, 4003–4015.

Irollo, E., Pirozzi, G. 2013. CD133: to be or not to be, is this the real question? Am. J. Trans. Res. 5, 563–581.

Joudeh, J., Allen, J.E., Das, A., Prabhu, V., Farbaniec, M., Adler, J., El-Deiry, W.S., 2013. Novel antineoplastics targeting genetic changes in colorectal cancer. Adv. Exp. Med. Biol. 779, 1–34.

Jung, P., Sato, T., Merlos-Suárez, A., Barriga, F.M., Iglesisas, M., Rossell, D., *et al.*, 2011. Isolation and in vitro expansion of human colonic stem cells. Nat. Med. 17, 1225–1227.

Kemper, K., Prasetyanti, P.R., De Lau, W., Rodermond, H., Clevers, H., Medema, J.P. 2012. Monoclonal antibodies against Lgr5 identify human colorectal cancer stem cells. Stem Cells 30, 2378–2386.

Kreso, A., Van Galen, P., Pedley, N.M., Lima-Fernandes, E., Frelin, C., Davis, T., *et al.*, 2014. Self-renewal as a therapeutic target in human colorectal cancer. Nat. Med. 20, 29–36.

Leder, K., Pitter, K., Laplant, Q., Hambardzumyan, D., Ross, B.D., Chan, T.A., *et al.*, 2014. Mathematical modeling of PDGF-driven glioblastoma reveals optimized radiation dosing schedules. Cell 156, 603–616.

Lu, J., Ye, X., Fan, F., Xia, L., Bhattacharya, R., Bellister, S., *et al.*, 2013. Endothelial cells promote the colorectal cancer stem cell phenotype through a soluble form of Jagged-1. Cancer Cell 23, 171–185.

Lu, Y.X., Yuan, L., Xue, X.L., Zhou, M., Liu, Y., Zhang, C., *et al.*, 2014. Regulation of colorectal carcinoma stemness, growth, and metastasis by an miR-200c-Sox2-negative feedback loop mechanism. Clin. Cancer Res. 20(10), 2631–2642.

Merlos-Suarez, A., Barriga, F.M., Jung, P., Iglesias, M., Céspedes, M.V., Rossell, D., *et al.*, 2011. The intestinal stem cell signature identifies colorectal cancer stem cells and predicts disease relapse. Cell Stem Cell 8, 511–524.

Metcalfe, C., Kljavin, N.M., Ybarra, R., De Sauvage, F.J. 2014. Lgr5+ stem cells are indispensable for radiation-induced intestinal regeneration. Cell Stem Cell 14, 149–159.

Montgomery, R.K., Carlone, D.L., Richmond, C.A., Farilla, L., Krenendonk, M.E.G., Henderson, D.E., *et al.*, 2011. Mouse telomerase reverse transcriptase (mTert) expression marks slowly cycling intestinal stem cells. Proc. Nat. Acad. Sci. USA 108, 179–184.

Munoz, J., Stange, D.E., Schepers, A.G., van de Wetering, M., Koo, B.K., Itzkovitz, S., *et al.*, 2012. The Lgr5 intestinal stem cell signature: robust expression of proposed quiescent '+4' cell markers. EMBO J. 31, 3079–3091.

Myant, K.B., Cammareri, P., McGhee, E.J., Ridgway, R.A., Huels, D.J., Cordero, J.B., *et al.*, 2013. ROS production and NF-kappaB activation triggered by RAC1 facilitate WNT-driven intestinal stem cell proliferation and colorectal cancer initiation. Cell Stem Cell 12, 761–773.

Nakanishi, Y., Seno, H., Fukuoka, A., Ueo, T., Yamaga, Y., Maruno, T., *et al.*, 2013. Dclk1 distinguishes between tumor and normal stem cells in the intestine. Nat. Genet. 45, 98–103.

O'Brien, C.A., Pollett, A., Gallinger, S., Dick, J.E. 2007. A human colon cancer cell capable of initiating tumour growth in immunodeficient mice. Nature 445, 106–110.

O'Brien, C.A., Kreso, A., Ryan, P., Hermans, K.G., Gibson, L., Wang, Y., *et al.*, 2012. ID1 and ID3 regulate the self-renewal capacity of human colon cancer-initiating cells through p21. Cancer Cell 21, 777–792.

Pang, R., Law, W.L., Chu, A.C.Y., Poon, J.T., Lam, C.S.C., Chow, A.K.M., *et al.*, 2010. A subpopulation of CD26+ cancer stem cells with metastatic capacity in human colorectal cancer. Cell Stem Cell 6, 603–615.

Potten, C.S., Al-Barwari, S.E., Searle, J. 1978. Differential radiation response amongst proliferating epithelial cells. Cell Tissue Kinet. 11, 149–160.

Potten, C.S., Owen, G., Booth, D. 2002. Intestinal stem cells protect their genome by selective segregation of template DNA strands. J. Cell Sci. 115, 2381–2388.

Powell, A.E., Wang, Y., Li, Y., Poulin, E.J., Means, A.L., Washington, M.K., *et al.*, 2012. The pan-ErbB negative regulator Lrig1 is an intestinal stem cell marker that functions as a tumor suppressor. Cell 149, 146–158.

Prabhu, V.V., Allen, J.E., Hong, B., Zhang, S., Cheng, H., El-Deiry, W.S. 2012. Therapeutic targeting of the p53 pathway in cancer stem cells. Expert Opin. Ther. Targets 16, 1161–1174.

Puglisi, M.A., Tesori, V., Lattanzi, W., Gasbarrini, G.B., Gasbarrini, A. 2013. Colon cancer stem cells: controversies and perspectives. World J. Gastroenterol. 19, 2997–3006.

Ricci-Vitiani, L., Lombardi, D.G., Pilozzi, E., Biffoni, M., Todaro, M., Peschle, C., De Maria, R. 2007. Identification and expansion of human colon-cancer-initiating cells. Nature 445, 111–115.

Ritsma, L., Ellenbroek, S.I., Zomer, A., Snippert, H.J., de Sauvage, F.J., Simons, B.D., *et al.*, 2014. Intestinal crypt homeostasis revealed at single-stem-cell level by *in vivo* live imaging. Nature 507, 362–365.

Sangiorgi, E., Capecchi, M.R. 2008. Bmi1 is expressed *in vivo* in intestinal stem cells. Nat. Genet. 40, 915–920.

Sato, T., van Es, J.H., Snippert, H.J., *et al.*, 2011. Paneth cells constitute the niche for Lgr5 stem cells in intestinal crypts. Nature 469, 415–418.

Schepers, A.G., Vries, R., van den Born, M., van de Wetering, M., Clevers, H. 2011. Lgr5 intestinal stem cells have high telomerase activity and randomly segregate their chromosomes. EMBO J. 30, 1104–1109.

Schepers, A.G., Snippert, H.J., Stange, D.E., van den Born, M., van Es, J.H., van de Wetering, M., Clevers, H. 2012. Lineage tracing reveals Lgr5+ stem cell activity in mouse intestinal adenomas. Science 337, 730–735.

Schwitalla, S., Fingerle, A.A., Cammareri, P., Nebelsiek, T., Göktuna, S.I., Ziegler, P.K. *et al.*, 2013. Intestinal tumorigenesis initiated by dedifferentiation and acquisition of stem-cell-like properties. Cell 152, 25–38.

Shmelkov, S.V., Butler, J.M., Hooper, A.T., Hormigo, A., Kushner, J., Milde, T., *et al.*, 2008. CD133 expression is not restricted to stem cells, and both CD133+ and CD133- metastatic colon cancer cells initiate tumors. J. Clin. Invest. 118, 2111–2120.

Siegel, R., Desantis, C., Jemal, A. 2014. Colorectal cancer statistics, 2014. CA Cancer J. Clin. 64, 104–117.

Sikandar, S.S., Pate, K.T., Anderson, S., Dizon, D., Edwards, R.A., Waterman, M.L., Lipkin, S.M. 2010. NOTCH signaling is required for formation and self-renewal of tumor-initiating cells and for repression of secretory cell differentiation in colon cancer. Cancer Res. 70, 1469–1478.

Snippert, H.J., van Es, J.H., van den Born, M., Begthel, H., Stange, D.E., Barker, N., Clevers, H. 2009. Prominin-1/CD133 marks stem cells and early progenitors in mouse small intestine. Gastroenterology 136, 2187–2194 e1.

Su, Y.J., Lai, H.M., Chang, Y.W., Chen, G.Y., Lee, J.L. 2011. Direct reprogramming of stem cell properties in colon cancer cells by CD44. EMBO J. 30, 3186–3199.

Takeda, N., Jain, R., LeBoeuf, M.R., Wang, Q., Lu, M.M., Epstein, J.A. 2011. Interconversion between intestinal stem cell populations in distinct niches. Science 334, 1420–1424.

Tian, H., Biehs, B., Warming, S., Leong, K.G., Rangell, L., Klein, O.D., de Sauvage, F.J. 2011. A reserve stem cell population in small intestine renders Lgr5-positive cells dispensable. Nature 478, 255–259.

Todaro, M., Alea, M.P., Di Stefano, A.B., Cammareri, P., Vermeulen, L., Iovino, F., *et al.*, 2007. Colon cancer stem cells dictate tumor growth and resist cell death by production of interleukin-4. Cell Stem Cell 1, 389–402.

Todaro, M., Gaggianesi, M., Catalano, V., Benfante, A., Iovino, F., Biffoni, M., *et al.*, 2014. CD44v6 is a marker of constitutive and reprogrammed cancer stem cells driving colon cancer metastasis. Cell Stem Cell 14, 342–356.

Varnat, F., Duquet, A., Malerba, M., Zbinden, M., Mas, C., Gervaz, P., *et al.*, 2009. Human colon cancer epithelial cells harbour active HEDGEHOG-GLI signalling that is essential for tumour growth, recurrence, metastasis and stem cell survival and expansion. EMBO Mol. Med. 1, 338–351.

Vermeulen, L., Todaro, M., de Sousa Mello, F., Sprick, M.R., Kemper, K., Perez Alea, M., *et al.*, 2008. Single-cell cloning of colon cancer stem cells reveals a multi-lineage differentiation capacity. Proc. Nat. Acad. Sci. USA 105, 13 427–13 432.

Vermeulen, L., de Sousa, E.M.F., van der Heijden, M., Cameron, K., de Jong, J.H., Borovski, T., *et al.*, 2010. Wnt activity defines colon cancer stem cells and is regulated by the microenvironment. Nat. Cell Biol. 12, 468–476.

Wong, V.W., Stange, D.E., Page, M.E., Buczacki, S., Wabik, A., Itami, S., *et al.*, 2012. Lrig1 controls intestinal stem-cell homeostasis by negative regulation of ErbB signalling. Nat. Cell Biol. 14, 401–408.

Yamashita, K., Katoh, H., Watanabe, M. 2013. The homeobox only protein homeobox (HOPX) and colorectal cancer. Int. J. Mol. Sci. 14, 23 231–23 243.

Yan, K.S., Chia, L.A., Li, X., Ootani, A., Su, J., Lee, J.Y., *et al.*, 2012. The intestinal stem cell markers Bmi1 and Lgr5 identify two functionally distinct populations. Proc. Nat. Acad. Sci. USA 109, 466–471.

Yeung, T.M., Gandhi, S.C., Bodmer, W.F. 2011. Hypoxia and lineage specification of cell line-derived colorectal cancer stem cells. Proc. Nat. Acad. Sci. USA 108, 4382–4387.

Zeki, S.S., Graham, T.A., Wright, N.A. 2011. Stem cells and their implications for colorectal cancer. Nat. Rev. Gastroenterol. Hepatol. 8, 90–100.

Zhu, L., Gibson, P., Currle, D.S., Tong, Y., Richardson, R.J., Bayazitov, I.T., *et al.*, 2009. Prominin 1 marks intestinal stem cells that are susceptible to neoplastic transformation. Nature 457, 603–607.

10 Prostate Cancer and Prostate Cancer Stem Cells

Magdalena E. Buczek, Jerome C. Edwards and Tarik Regad
The John van Geest Cancer Research Centre, Nottingham Trent University, Nottingham, UK

10.1 Development and origin of the prostate

The prostate is a male sex accessory organ located directly beneath the bladder and adjacent to the anus (Timms, 2008). The normal function of the prostate is to produce and excrete seminal secretions, which provide a lubricant and a fluid medium for spermatozoa (Thomson and Marker, 2006). The prostate develops from the urogenital sinus (UGS) structure, comprising a mesodermally derived mesenchymal layer surrounding an endodermally derived epithelial layer. The initiation of prostatic morphogenesis is mediated by the outgrowth of solid epithelial buds from the epithelial–urogenital sinus (E-UGS) into the mesenchymal–urogenital sinus (M-UGS) (Timms *et al.*, 1994). The initial development of prostate growth results from induction of paracrine signalling between the two types of layers by mesenchymal and epithelial interactions, a phenomenon also seen in the development of the lungs, the kidneys and the gut (Cunha *et al.*, 2004). The prostatic buds form solid cords of epithelial cells that grow into the M-USG in a spatial pattern, initiating the process of branching. Epithelial and mesenchymal/stromal interactions have been shown to be responsible for this cytodifferentiation (Wang *et al.*, 2001). Following ductal canalization, the epithelium reorganizes into four final differentiated prostate cell types: basal, luminal, neuroendocrine (NE) and transit-amplifying cells (TACs), which are defined by their distinct morphology, their locations and their markers' expression patterns (Prajapati *et al.*, 2013).

Principles of Stem Cell Biology and Cancer: Future Applications and Therapeutics, First Edition.
Edited by Tarik Regad, Thomas J. Sayers and Robert C. Rees.

The luminal region comprises the majority of the prostate cells, forming a layer of baso-apical cells that secrete prostatic proteins such as prostatic-specific antigen (PSA), prostatic acid phosphatase (PAP) and fluids into the ductal lumen. These cells are terminally differentiated, androgen-dependent and characterized by the expression of cytokeratins 8 and 18, CD57 and p27 (a cell-cycle inhibitor), along with high levels of androgen receptors (ARs) (De Marzo *et al.*, 1998; Wang *et al.*, 2001; Long *et al.*, 2005). The basal epithelial cells form a layer of flattened cuboidal cells along the basement membrane and express cytokeratins 5 and 14, as well as p63 (a homologue of the tumour-suppressor gene p53), Bcl-2 (an antiapoptotic factor), CD44 and the hepatocyte growth factor (HGF). Basal cells exist independent of androgen signalling and express undetectable levels of ARs (Bonkhoff and Remberger, 1993; Wang *et al.*, 2001; Long *et al.*, 2005). NE cells are a rare cell population found along the basal and luminal cell layers, characterized by their androgen independence and the expression of chromogranin A, synaptophysin and neurotensin (Abrahamsson, 1999; Abate-Shen and Shen, 2000; Amorino and Parsons, 2004). TACs, an intermediate cell population, express both basal and luminal cell markers (CK5, CK8, CK14, ARs and PSA) (Bonkhoff and Remberger, 1993; Bonkhoff *et al.*, 1994; Xue *et al.*, 1998; Abrahamsson, 1999; Hudson *et al.*, 2000). Finally, the prostate ductal structure is surrounded by a stromal layer, which acts as a peripheral boundary to the prostate gland. The stromal compartment is built from several types of cells, including smooth-muscle cells, fibroblasts and myofibroblasts, and is characterized by the expression of CD34, CD44, CD90, CD117 and vimentin (Takao and Tsujimura, 2008).

10.2 Adult prostate stem cells

Prostate stem cells (PSCs) are a small population of undifferentiated cells possessing unique self-renewal and differentiation capacities (Schalken and van Leenders, 2003). They are thought to generate all the cell types found within the prostate. The evidence for their existence in adult prostate tissue came from studies on mice models displaying a regressive prostate phenotype, resulting from androgen deprivation. Androgen restoration favoured the regenerative ability of the prostate in those mice, reverting it to normal size; this process of serial ablation/restoration could be repeated multiple times on the same tissue (Tsujimura *et al.*, 2002).

Adult PSCs are thought to be of basal origin, as they are able to regenerate and initiate prostate following multiple cycles of androgen regression (DeKlerk and Coffey, 1978; Kyprianou and Isaacs, 1988; Montpetit *et al.*, 1988). They are believed to give rise to highly proliferative TACs, which, in turn, differentiate into NE and terminally differentiated luminal cells (Bonkhoff and Remberger, 1996). This is supported by the fact that 70% of proliferating human epithelial cells reside in the basal compartment, as well as by identification of subsets of basal/luminal and basal/NE intermediate profiles, which further suggests a basal origin (Verhagen *et al.*, 1992; Bonkhoff *et al.*, 1994). Human PSC population represent approximately 1% of the total cancer cell population, presenting high colony-forming efficiency *in vitro*, accompanied by the ability to form prostatic-gland structures *in vivo*. This subset population has been associated with a collection of cell-surface markers, namely CD133 (prominin-1, Prom1), α2β1 integrin and CK6a (Collins *et al.*, 2001; Richardson *et al.*, 2004; Schmelz *et al.*, 2005). Further studies have identified several other PSC markers, including aldehyde dehydrogenase (ALDH), ATP binding cassette transporter family membrane efflux pump (ABCG2), tumour-associated calcium signal transducer 2 (Trop-2), p63 and CD44 (Liu *et al.*, 1997; Bhatt *et al.*, 2003; Lawson *et al.*, 2007; Goldstein *et al.*, 2008; Burger *et al.*, 2009; Yao *et al.*, 2009; Pignon *et al.*, 2013). Basal-specific p63 have been shown to be essential in the development of limbs and epithelial organs, such as mammary and prostatic glands in mice (Mills *et al.*, 1999; Yang *et al.*, 1999; Signoretti *et al.*, 2000). PSCs' balance of quiescence, self-renewal potential and ability to differentiate into a heterogenous, highly specialized population is maintained by specific interactions provided by surrounding cells, along with elements of extracellular matrix (ECM), together comprising a complex tumour microenvironment or niche (Fuchs *et al.*, 2004).

The signalling crosstalk between PSCs and neighbouring cells in the stem cell niche is theorized to be mediated by three main pathways: Wnt, Sonic hedgehog (Shh) and Notch. This is supported by experiments in which inhibition of the Wnt receptor resulted in the disruption and loss of intestinal epithelial cell niches (Kuhnert *et al.*, 2004). *Notch-1* shows a high level of expression in mouse prostate epithelium from its birth and remains highly expressed throughout morphogenesis, leading to its decline in adulthood (Shou *et al.*, 2001). Notch-1 knockout in transgenic mice causes a marked impairment in the ability of stem cells to regenerate tissue and, later, a significant inhibition of cell growth, differentiation and branching processes in both *in vitro* and *in vivo* androgen ablation-restoration experiments (Wang *et al.*, 2004). Shh expression has been shown to be involved in the promotion of early ductal morphogenesis via activation of transcription factor Gli-1 during epithelial evagination of the UGS (Lamm *et al.*, 2002; Freestone *et al.*, 2003) (Figure 10.1). Several transcription factors involved in prostate development appear to play roles in the maintenance of PSC stemness. Members of the *Hox* homeobox gene family are expressed in prostate growth, regulating branching and stages of differentiation – in particular *HOXB13*, which has also been found expressed in adult prostate (Edwards *et al.*, 2005). Similarly, the expression of *Nkx3.1* is maintained throughout life and is thought to be important for tissue homeostasis. *Nkx3.1* plays a role in the protection of prostate epithelium from oxidative DNA damage, and its loss both disturbs prostate differentiation and fosters the mutational inactivation of related genes, such as *PTEN* (Bowen *et al.*, 2000; Kim *et al.*, 2002; Ouyang *et al.*, 2005). *Fox*A1, a member of the Forkhead box gene (*Fox*) family, also plays

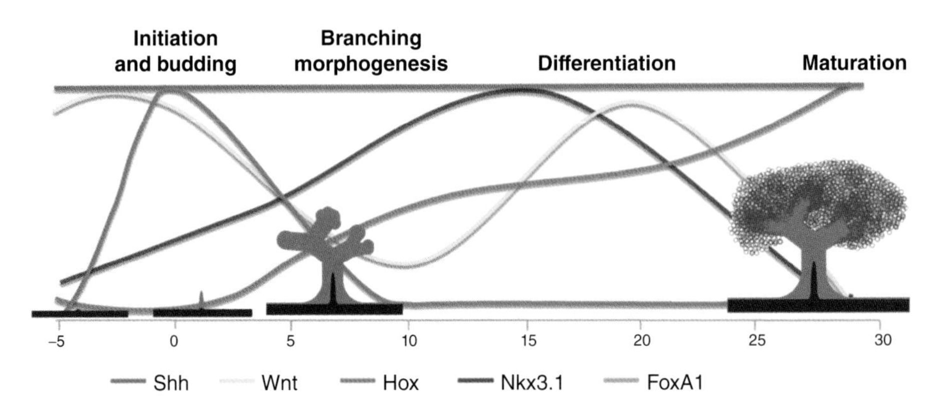

Figure 10.1 Gene expression and branching morphogenesis during rat prostate development within 5 weeks prepartum and 30 weeks postpartum. The key developmental genes identified in prostate development are recorded based on relative expression of each other. The display of the branching pattern is represented in correlation with the temporal gene expression measured by RT-rPCR.

an important role in ductal morphogenesis and epithelial cell maturation in prostate embryogenesis (Prins and Putz, 2008). Its marked expression is also sustained in adult life and has been found to be required for both probasin expression via direct interactions with *FoxA* *cis*-regulatory elements and PSA expression via interactions with ARs on gene promoters (Gao *et al.*, 2010). Sustained expression of a number of these nuclear transcription factors and core signalling pathways mediates early morphogenesis events in the adult prostate (embryonic prostate) (Figure 10.1).

The existence of this distinct PSC population is advantageous for adult prostate maintenance due to its unique self-renewal and differentiation potential, which provide a great regenerative capacity. Conversely, when targeted by potential mutations, PSCs risk spreading abnormally functioning offspring cells and thus increasing the risk of developing life-threatening prostatic disorders such as benign prostate hyperplasia (BPH) and prostate cancer.

10.3 Prostate cancer and prostate cancer stem cells

10.3.1 Common mutations associated with prostate cancer

Prostate cancer is a common malignancy in men, often showing no symptoms of disease and taking a variable clinical pathological course. Most cases of prostate cancer behave in an indolent manner. However, some cases are characterized by a highly aggressive phenotype, as seen in tumours resistant to standard androgen-deprivation therapy (Kirby *et al.*, 2011). In agreement with the stochastic (clonal) evolution model, the level of prostate cancer malignancy appears to correlate with the prevalence of somatic mutations that affect key pathways involved in prostate physiology (e.g. AR signalling) or signalling pathways, associated with cell survival, apoptosis and DNA repair (Grasso *et al.*, 2012). Several studies have given evidence of different mutation patterns characteristic of different stages of prostate cancer progression, some of which are potentially associated with malignant transformation (Taylor *et al.*, 2010; Kumar *et al.*, 2011; Barbieri *et al.*, 2012; Grasso *et al.*, 2012). The current diagnostic techniques for prostate cancer screening involve digital rectal examination (DRE), transrectal ultrasonography (TRUS) and biopsy histopathological, but these procedures are incapable of providing sufficient indication for the appropriate course of therapy (Madu and Lu, 2010). An insight into prostate cancer mutation patterns could help us better assess the implications of the prostate tumourigenesis 'mutation signature' and highlight patterns in aggressiveness and disease progression, providing a molecularly tailored treatment strategy (Kumar *et al.*, 2011). The last 2 decades of cancer research have provided an advancement in genome characterization technologies, bringing most significant prostate cancer-specific aberrations into sharper focus.

Prostate cancer shows a relatively low frequency of point mutations, ranging from one to two alterations per million base pairs, which is much lower than in carcinogen-driven tumours such as lung cancer or melanoma, but a similar rate to that seen in breast, ovarian and renal cancers (Greenman *et al.*, 2007; Pleasance *et al.*, 2010; Berger *et al.*, 2011). Low levels of sequence alterations may explain the lower aggressiveness of prostate cancer compared to other tumours. The prevalence of mutations observed in prostate malignancies is highly dependent on their purity, stage, histological grade (Gleason score) and previous exposure to treatment (Baca and Garraway, 2012). The most common and prostate cancer-specific aberration identified for both primary and advanced tumours is a structural rearrangement in ETS DNA-binding family genes, mostly affecting ERG family members (ETV1, ETV4 and ETV5) (Tomlins *et al.*, 2008). In this mutation, the ETS family-member ERG oncogene is adjoined to a highly active androgen-driven promoter TMPRSS2, leading to its high expression in the prostate epithelium. More recent observations have associated TMPRSS2-ERG fusions with prostate-specific deletion of chromosome 3p14, implicating FOXP1, RYBP and SHQ1 as potential cooperative tumour suppressors (Taylor *et al.*, 2010). Moreover, the overexpression of ERG oncoprotein appears to be involved in a positive regulation of c-Myc and prostaglandin-mediated signalling, potentially contributing to tumour progression (Mohamed *et al.*, 2011; Sreenath *et al.*, 2011). ERG activation has been found in 50–70% of PSA-positive prostate cancer; up to 90% of such activations were found to be TMPRSS2-ERG fusions, and despite variations in ethnic distribution, this mutation is to date the most specific described for prostate cancer (Mosquera *et al.*, 2007; Tomlins *et al.*, 2008).

The androgen-regulated pathway has been shown to be crucial, not only for prostate development, but also for prostate oncogenesis (Prins and Putz, 2008). ARs, through their nuclear translocation and interactions with transcriptional factors, regulate the transcription of several prostate genes involved in cellular proliferation and differentiation. Clinical observations show that up to 90% of prostate cancers are dependent on androgens at initial diagnosis. However, on androgen-deprivation treatment, the majority of them develop castration-resistant prostate cancer (CRPC), associated with transactivation of the AR (Kirby *et al.*, 2011). This is consistent with the exome sequencing of prostate cancer, which shows little or no alteration in the AR locus early in prostate cancer and a dramatic AR amplification and point-mutation rate in up to 50% of CRPC and metastatic specimens tested (Visakorpi *et al.*, 1995; Koivisto *et al.*, 1997; Grasso *et al.*, 2012; Linja and Visakorpi, 2004). Kumar *et al.* (2011) showed a significant novel nonsynonymous variant of AR gain-of-function mutation in CRPC. Malfunctioning AR signalling in primary prostate cancer may be caused by mutations in AR-interacting genes such as NCOR2, NRIPI1, TNK2 and EP300 (Taylor *et al.*, 2010). AR signalling is also regulated by a range of abnormally expressed chromatin/histone remodelers, such as members of the MLL complex (MLL2, MLL, ASH2L), UTX, ASHL1, CHD1 and the direct AR cofactor FOXA1 (Barbieri *et al.*, 2012; Grasso *et al.*, 2012).

Early alterations in prostate cancer oncogenesis predominantly involve signalling that regulates growth, proliferation and normal prostate development. Phosphoinositide-3-kinase (PI3K) is one of the main signalling pathways involved in cellular proliferation that is aberrantly mutated in the early stages of prostate cancer development (Barbieri *et al.*, 2012; Grasso *et al.*, 2012). Overactivation of PI3K signalling is caused by frequent inactivation of the tumour-suppressor gene PTEN, which is unable to counteract PI3K signalling. Loss of heterozygosity at the PTEN locus has been found in up to 70% of primary prostate cancers, whereas 'inactivation' mutations have been found in 5–10%, although with a marked increase of frequency in advanced tumours (Cairns *et al.*, 1997; Gray *et al.*, 1998; McMenamin *et al.*, 1999; Barbieri *et al.*, 2012). Overstimulation of the PI3K pathway is even more enhanced when PTEN mutations are detected alongside amplification of PI3K itself (PiK3CA amplification in 13–29% of primary tumours and 50% of CRPCs). Alterations in INPP4B and PHLPP PAPs were recently implicated in PI3K signalling regulation (Edwards *et al.*, 2003; Gao *et al.*, 2006; Sun *et al.*, 2009 ; Agell *et al.*, 2011). Moreover, loss of PTEN has also been correlated with TP53 loss and overexpression of c-Myc or ERG (King *et al.*, 2009; Wang and Shen, 2011; Kim *et al.*, 2012).

Inactivation of cell-cycle-inhibitory genes appears to disrupt the senescence associated with cancerogenous signalling and possibly the bypass of AR-regulated growth in metastatic and CRPC tumours (Baca and

Garraway, 2012). Accordingly, mutations are also commonly found in key tumour-suppressor genes, such a TP53 and RB1, causing the inactivation of p53 protein, which is responsible for positive regulation of p21 cyclin-dependent kinase cell-cycle inhibitor (Holcomb *et al.*, 2009). Mutations within the tumour-suppressor gene TP53 encourage the activation of cell proliferation and suppression of DNA repair (Dong, 2006; Kumar *et al.*, 2011; Barbieri *et al.*, 2012; Grasso *et al.*, 2012). These mutations occur more frequently in prostate cancer that has undergone pharmacological treatment and radiation, resulting in reduced time to development of distant metastasis (Grignon *et al.*, 1997; Dong, 2006). In fact, prostate cancers metastasizing to bone have been found to have the most frequent *TP53* mutations (Meyers *et al.*, 1998). It is also likely that mutations in *TP53* impair genomic stability, leading to genomic amplification of the *AR* gene during hormone therapy (Dong, 2006). Exome sequencing of prostate cancer has also revealed frequent aberrations in the tumour-suppressor gene *Rb1*, reaching up to 20% in both primary and CRPC samples (Kubota *et al.*, 1995; Barbieri *et al.*, 2012; Grasso *et al.*, 2012). Another mutation associated with prostate cancer implicates another key cell-cycle regulator, *CDKN1B*, which encodes p27, a cyclin-dependent kinase mutated in up to 3% of primary prostate cancer cases, which correlates with poor pathological prognostic outcome (Vis *et al.*, 2000; Dreher *et al.*, 2004; Barbieri *et al.*, 2012). Disruption in *CDKN1B* promotes prostate cancer growth in coordination with the inactivation of *PTEN*, which implies an interaction between p27 and the PI3K/Akt pathway (Di Cristofano *et al.*, 2001). These mutations affect the regulation of cell-cycle arrest, contributing to faulty DNA-repair machinery, and therefore potentially being involved in prostate cancer progression.

The downregulation and aberration of *Nkx3-1* have been associated with prostatic intraepithelial neoplasia (PIN) lesions and early stages of prostate oncogenesis. *Nkx3-1* disruption has been identified in approximately 75% of prostate cancer loss of heterozygosity (Emmert-Buck *et al.*, 1995; Asatiani *et al.*, 2005). Furthermore, in a murine model, the overexpression of Nkx3 results in the suppression of tumour development, strongly implicating Nkx3 as a tumour suppressor gene. Nkx3 has not been found expressed in any other male urogenital system, allowing it to be used as an effective biomarker for carcinogenesis prediction (Bhatia-Gaur *et al.*, 1999). BRCA2 gene mutation is also represented as a hallmark in prostate cancer development. Originally identified as a hallmark of poor prognosis for breast and ovarian cancer, BRCA2 normally mediates DNA repair. Patients with BRCA2 gene mutations show an increasing frequency pattern, with prostate cancer progression into castration-resistant phenotype (Thorne *et al.*, 2011). Finally, it is hypothesized that a high prostate cancer proliferative potential is partly adopted by an embryonic Wnt signalling pathway. Generally silent in differentiated cells, Wnt ligands have been found to be upregulated in prostate cancer, with a marked increase of expression in advanced forms (Kypta and Waxman, 2012). For example, elevated levels of Wnt1, Wnt5a,

Wnt7b and Wnt11 expression have been correlated with prostate cancer aggressiveness (Chen *et al.*, 2004; Li *et al.*, 2008; Uysal-Onganer *et al.*, 2010). In addition, DKK1 expression increases during prostate cancer initiation but decreases during metastasis (Hall *et al.*, 2008). The correlation of Wnt activation and skeletal metastasis may be important for therapy.

It is important to note that heterogeneity of a tumour is not entirely attributable to mutations. Epigenetic changes, interactions with the microenvironment and spatial and temporal differences also alter phenotypic expression, posing a challenge when it comes to determining the origin of the cell (Visvader and Lindeman, 2012).

10.3.2 PCSCs

The cancer stem cell (CSC) hypothesis of tumour development was first postulated in the principal study of Park *et al.* (1971), which showed that a small proportion of cells (0.01–1.00%) in tumour isolates are clonogenic and extensively proliferative *in vitro* and *in vivo*, indicating that these cells might represent tumour stem cells. The CSC hypothesis postulates the existence of a small subset of cancerous cells, originated from stem cells that accumulated mutations and maintained the inherent ability of self-renewal, which, given their pluripotent nature, are accountable for tumour initiation and maintenance of tumour heterogeneity (Shipitsin and Polyak, 2008; Visvader and Lindeman, 2012). Like normal stem cells, CSCs are able to create a hierarchical organization and to reconstitute the bulk of the tumour (Lawson *et al.*, 2005; Magee *et al.*, 2012; Yu *et al.*, 2012). The CSC hypothesis may also explain the frequent ineffectiveness of standard cancer treatments: this small population of cells are thought to be therapy-resistant, often leading to recurrence and giving rise to metastasis (Wang and Shen, 2011). These observations suggest that thorough CSC characterization could lead to a better understanding of the mechanisms that mediate tumour initiation, progression and metastasis. It is important to distinguish CSCs from tumour-initiating cells (TICs). CSC populations, due to their intrinsic biological stem cell properties, are able to reconstitute a tumour in a recipient animal. This tumour can be identical to the parental one from which it was derived, and can be serially xenotransplanted indefinitely. The definition of TICs is wider and refers to the functional concept of a CSC but does not imply hierarchical organization (Wang and Shen, 2011; Yu *et al.*, 2012).

The origin of CSCs has yet to be securely defined, due to the multiple ways in which this biological phenomenon occurs (Yu *et al.*, 2012). Stem cells differentiate and generate a population of TACs or progenitor cells as a natural order of organogenesis. Maturation-arrests theory states that sporadic mutations may occur within the stem cell/progenitor cells before their full differentiation, activating self-renewal signalling and creating cancerous stem cells (Figure 10.2) (Cozzio *et al.*, 2003; Huntly and Gilliland, 2005; Zhao *et al.*, 2007). The stage of differentiation of the tumour is dependent upon the degree

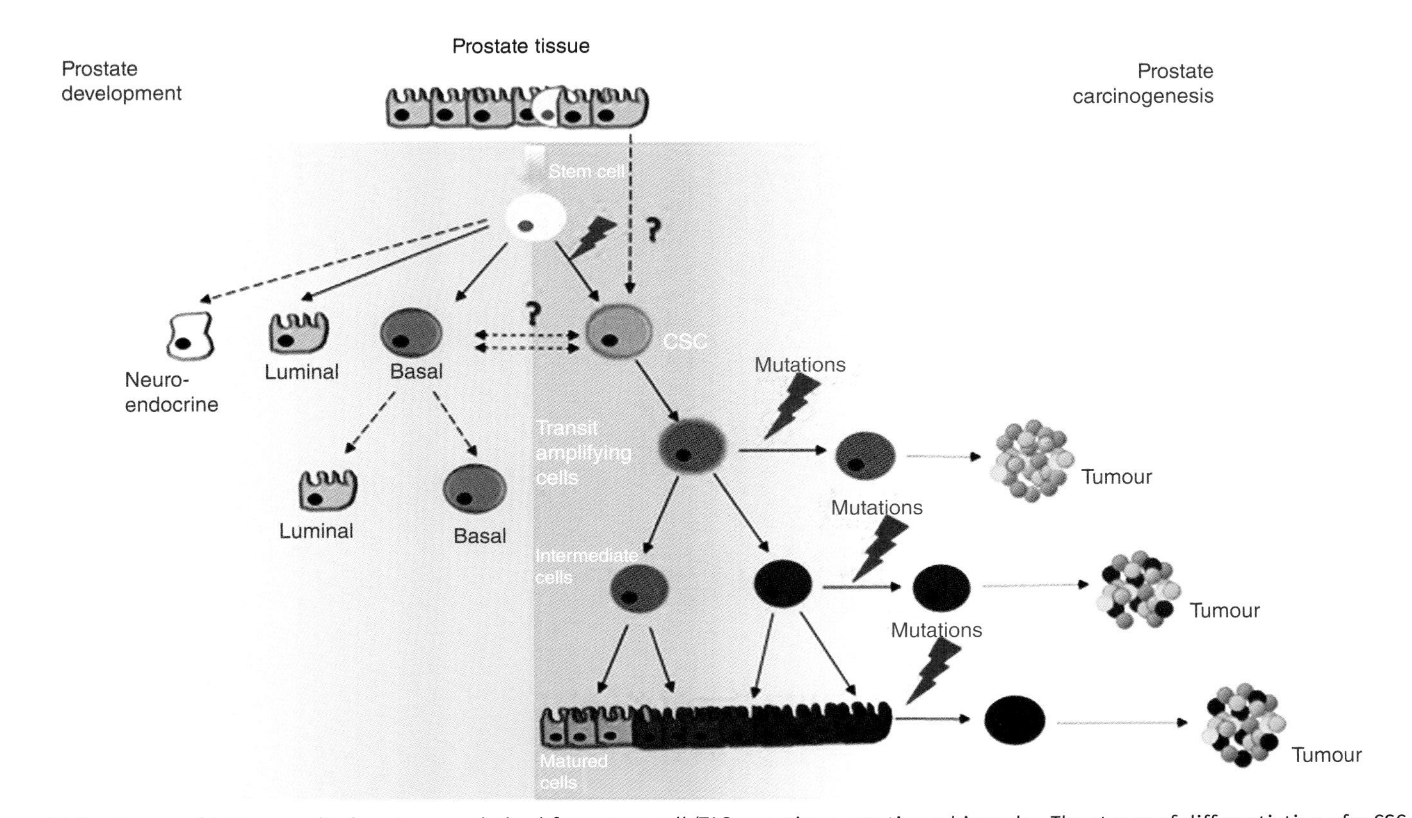

Figure 10.2 Source of heterogeneity in a tumour, derived from stem cell/TAC mutations, creating a hierarchy. The stages of differentiation of a CSC give rise to the genetic variation of the cell of origin, resulting in tumour variance.

of differentiation at the time of maturation arrest (Sell, 2010). The reverse to the maturation arrest theory, the anaplastic theory, claims that CSCs arise due to mutations of fully differentiated mature cells (Sell, 2010). Once a cell has committed and completed its 'cell fate', a mutation may occur, resulting in the alteration of regulatory cell functions. As a consequence, the cell reprogrammes and reacquires its stem-like properties (Yu *et al.*, 2012). Finally, the 19th-century embryonal-rest hypothesis states that residual embryonic or germinal stem cells give rise to CSCs after being left during tissue ontogenesis; this hypothesis become discredited during the 20th century (Sell, 2010).

Given their exceptional potential, the identification of CSCs from different types of cancer has become a major challenge for cancer researchers in the last 2 decades. Current efforts to define CSC populations are based on their functional assessment through the use of xenotransplantation assays of cell populations that have been isolated from primary human cancer tissue. Thus, CSCs are identified as separate populations within the total prostate cancer, based on their ability to initiate tumour formation in animal models following transplantation (Wang and Shen, 2011). Their self-renewal ability is determined by analysis of the tumour formation following serial transplantation. *In vitro* assessments of CSC properties involve the analysis of their ability to form spheres, eject Hoechst dye 33342 and differentiate into nonrenewable heterogeneous cancers (Wicha *et al.*, 2006).

Commonly, CSCs are identified by screening for cancer cell populations expressing stem cell-specific biomarkers or genes and/or by examining the self-renewal and drug resistance properties of a putative population in various *in vitro* functional assays and *in vivo* tumorigenicity models (Wang and Shen, 2011). In a majority of studies, stemness or the ability to self-renew has become attributed to the expression of particular protein markers on the cancer cell surface (Yu *et al.*, 2012). Starting from leukaemic ($CD34^+/CD38^-$), and breast ($CD44^+/CD24^-$) CSCs, CSC populations have been identified in several other cancers, including brain, lung, colon, melanoma, pancreatic, ovarian and prostate (Al-Hajj *et al.*, 2003; Galli *et al.*, 2004; Collins *et al.*, 2005; Kim *et al.*, 2005; Ponti *et al.*, 2005; Ricci-Vitiani *et al.*, 2007; Schatton *et al.*, 2008).

Prostate CSC populations were first identified by fluorescence-activated cell sorting (FACS) from primary human tumours and showed high proliferation potential in clonogenic assays, as well as the ability to give rise to luminal cells. Prostate-cancer $CD44^+$ populations have been shown to have significantly higher proliferation rates, self-renewal abilities and tumorigenic properties than their $CD44^-$ counterparts (Patrawala *et al.*, 2006). Notably, $CD44^+$ cells show an increased expression of several 'stemlike' genes, including *Oct-3/4*, *Bmi*, *β-catenin* and *SMO*, as well as the ability to differentiate into $CD44^-$ and AR^+ cells *in vivo* and *in vitro* (Patrawala *et al.*, 2006). Hurt *et al.* (2008) have also identified a particular LNCaP $CD44^+/CD24^-$ population that is able to form prostatospheres *in vitro* and to reconstitute tumours in nonobese diabetic severe combined-immunodeficient

(NOD/SCID) mice from as few as 100 cells. Other populations with a marker signature of CD44$^+$/CD24$^-$/α2β1^{high} have been identified in sphere-forming assays from the DU145 prostate cancer cell line (Rybak *et al.*, 2011). CD133$^+$ cell populations are also capable of self-renewal and increased proliferation (Collins *et al.*, 2005). Gene expression analysis of CD44$^+$/CD133$^+$/α2β1^{high} originated from DU145 prostate cancer cell lines reveal high levels of expression of *c-myc* and *β-catenin* and a lower expression of *Bax* (Wei *et al.*, 2007). However, CD133 has also been found to be abundant in nonmalignant tissues, which limits its use as a target of cancer therapies (Visvader and Lindeman, 2012). Rajasekhar *et al.* (2011) characterized prostate CSCs by investigating the *in vitro* and *in vivo* expression of TRA-1-60, CD151, CD166 and NF-κB activity using prostate cancer cells derived from human prostate cancer tissue. The study showed the ability of these triple-positive cells to form spheroids and to recapitulate the original parent tumour heterogeneity in serial xenotransplantations, indicating a tumour cell hierarchy in human prostate cancer development. Also, this sphere-forming CSC population did not express the main prostate markers, such as PSA, AR and Nkx3.1, but showed a marked increase in the expression of stem cell-associated markers like Met-receptor kinase, inhibitor of differentiation 1 (Id1), Musashi-1, phospho-histone 3 and Ki67 (Rajasekhar *et al.*, 2011).

Accumulating data reveal significant resemblances in gene expression between CSCs and embryonic stem cells (ESCs). These similarities have been found through gene expression analysis of CSCs from hepatic, colorectal, nasopharyngeal, brain, breast and ovarian cancers (Zbinden *et al.*, 2010; He *et al.*, 2012; Ling and Jolicoeur, 2012; Zhang *et al.*, 2012; Luo *et al.*, 2013). Genes expressed in CSCs include the embryonic stem cell-specific transcriptional factor NANOG, the sex-determining region Y-box 2 (SOX2) and the octamer-binding transcription factor 4 (OCT4, also called POUF1) heterodimers (Wang *et al.*, 2013). NANOG regulates cell proliferation and cell-cycle arrest via interaction with cyclins D1, D2, D3 and cyclin kinases 1 and 6, as well as through the inhibition of p53 signalling, thereby supporting cancer survival (Choi *et al.*, 2012). Overexpression of NANOG is also involved in metastasis through the inhibition of E-cadherin, FOXO1, FOXJ1 and FOXB1, together with Oct4 induction of the transcription factor Slug, which is involved in epithelial–mesenchymal transition (EMT) (Chiou *et al.*, 2010; He *et al.*, 2012). In fact, knockdown of *NANOG* significantly reduces the clonogenic and tumorigenic properties of the DU145 prostate cancer cell line (Jeter *et al.*, 2011). On the other hand, SOX2 promotes the apoptosis-resistant phenotype of the DU145 prostate cell line *in vitro* and *in vivo* using a NOD/SCID xenograft model (Jia *et al.*, 2011). NANOG and OCT4 also enhance the expression of ABCG2, a member of the ATP binding cassette (ABC) subfamily membrane efflux channels that mediate the exclusion of cytotoxic agents (Chiou *et al.*, 2010; Linn *et al.*, 2010; Jeter *et al.*, 2011). These observations support the existence of a rare, drug-resistant population with a significantly high survival potential, and may also explain the frequent ineffectiveness of conventional cancer treatment strategies. In addition, chemotherapy is an effective means of targeting rapidly dividing cell populations but remains inefficacious in targeting putative CSC populations, due to their quiescent and dormant nature (Klonisch *et al.*, 2008).

10.4 Conclusions

Corresponding to the CSC concept, targeting of CSC populations would eradicate the tumourigenic stem cell pool that is at the source of the tumour. However, the heterogeneous and undifferentiated nature of a population of CSCs also poses a challenge in terms of targeting, due to the variation in phenotype and tumourigenic behaviour (Bilusic and Madan, 2011). Defining CSC populations through a set of markers expressed on their surfaces offers a significant opportunity for specific antibody-based drug delivery to the CSC niche. For example, nanoparticles containing paclitaxel coupled with anti-CD133 antibody show marked cytotoxicity *in vitro* and *in vivo*, selectively eliminating $CD133^{+}$ breast CSCs (Swaminathan *et al.*, 2013). Prostate cancer has a relatively slow disease development, which favours immunotherapy applications, as it offers a significant amount of time for the development of a potent immune response. To date, Sipuleucel-T is the only US Food and Drug Administration (FDA)-approved vaccine against prostate-associated peptide for minimally symptomatic metastatic prostate cancer, although PSA-TRICOM and ipilimumab are currently being tested in phase III clinical trial (Kantoff *et al.*, 2010; Buonerba *et al.*, 2011). These therapeutic options provide an improvement in overall survival by hampering tumour growth. However, they fail to cause tumour size to diminish (Madan *et al.*, 2013). Unlike prostate cancer-targeting vaccines, immunization against CSCs gives an opportunity to eradicate the cell population conferring tumour aggressiveness features. It now appears that the design of an effective prostate CSC vaccine is a matter of identifying prostate CSC-specific markers. Another promising approach is to sensitize CSCs to conventional chemotherapy by inhibiting or blocking drug efflux pumps. This has previously been achieved in lung CSCs, through inhibition of the c-Kit receptor, which restored cisplatin sensitivity (Levina *et al.*, 2010). Similarly, inhibition of interleukin 4 (IL-4) abolished chemoresistance to oxaliplatin and 5-fluorouracil in colon CSCs (Todaro *et al.*, 2008). However, temporary inhibition of certain ATP-driven drug efflux channels could lead to signalling rearrangement and acquisition of another drug-specific resistance. Targeting of CSC key signalling cascades might be a potential way of disabling their 'stemness'.

References

Abate-Shen, C., Shen, M.M. 2000. Molecular genetics of prostate cancer. Genes Dev. 14, 2410–2434.

Abrahamsson, P. 1999. Neuroendocrine differentiation in prostatic carcinoma. Prostate 39, 135–148.

Agell, L., Hernández, S., Salido, M., de Muga, S., Juanpere, N., Arumí-Uria, M., *et al.*, 2011. PI3K signaling pathway is activated by PIK3CA mRNA overexpression and copy gain in prostate tumors, but PIK3CA, BRAF, KRAS and AKT1 mutations are infrequent events. Mod. Pathol. 24, 443–452.

Al-Hajj, M., Wicha, M.S., Benito-Hernandez, A., Morrison, S.J., Clarke, M.F. 2003. Prospective identification of tumorigenic breast cancer cells. Proc. Nat. Acad. Sci. USA 100, 3983–3988.

Amorino, G.P., Parsons, S.J. 2004. Neuroendocrine cells in prostate cancer. Crit. Rev. Eukaryot. Gene Expr. 14(4), 287–300.

Asatiani, E., Huang, W.-X., Wang, A., Rodriguez Ortner, E., Cavalli, L.R., Haddad, B.R., Gelmann, E.P. 2005. Deletion, methylation, and expression of the NKX3.1 suppressor gene in primary human prostate cancer. Cancer Res. 65, 1164–1173.

Baca, S.C., Garraway, L.A. 2012. The genomic landscape of prostate cancer. Front. Endocrinol. (Lausanne). 3, 69.

Barbieri, C.E., Baca, S.C., Lawrence, M.S., Demichelis, F., Blattner, M., Theurillat, J., *et al.*, 2012. Exome sequencing identifies recurrent SPOP, FOXA1 and MED12 mutations in prostate cancer. Nat. Genet. 44, 685–689.

Berger, M.F., Lawrence, M.S., Demichelis, F., Drier, Y., Cibulskis, K., Sivachenko, A.Y., *et al.*, 2011. The genomic complexity of primary human prostate cancer. Nature 470, 214–220.

Bhatia-Gaur, R., Donjacour, A.A., Sciavolino, P.J., Kim, M., Desai, N., Young, P., *et al.*, 1999. Roles for Nkx3.1 in prostate development and cancer. Genes Dev. 13, 966–77.

Bhatt, R.I., Brown, M.D., Hart, C.A., Gilmore, P., Ramani, V.A.C., George, N.J., Clarke, N.W. 2003. Novel method for the isolation and characterisation of the putative prostatic stem cell. Cytom. Part A 54, 89–99.

Bilusic, M., Madan, R.A. 2011. Therapeutic cancer vaccines: the latest advancement in targeted therapy. Am. J. Ther. 19(6), e172–e181.

Bonkhoff, H., Remberger, K. 1993. Widespread distribution of nuclear androgen receptors in the basal cell layer of the normal and hyperplastic human prostate. Virchows Arch. A 422, 35–38.

Bonkhoff, H., Remberger, K. 1996. Differentiation pathways and histogenetic aspects of normal and abnormal prostatic growth: a stem cell model. Prostate 28, 98–106.

Bonkhoff, H., Stein, U., Remberger, K. 1994. Multidirectional differentiation in the normal, hyperplastic, and neoplastic human prostate: simultaneous demonstration of cell-specific epithelial markers. Hum. Pathol. 25, 42–46.

Bowen, C., Bubendorf, L., Voeller, H.J., Slack, R., Willi, N., Sauter, G., *et al.*, 2000. Loss of NKX3.1 expression in human prostate cancers correlates with tumor progression. Cancer Res. 60, 6111–6115.

Buonerba, C., Ferro, M., Di Lorenzo, G. 2011. Sipuleucel-T for prostate cancer: the immunotherapy era has commenced. Expert Rev. Anticancer Ther. 11, 25–28.

Burger, P.E., Gupta, R., Xiong, X., Ontiveros, C.S., Salm, S.N., Moscatelli, D., Wilson, E.L. 2009. High aldehyde dehydrogenase activity: a novel functional marker of murine prostate stem/progenitor cells. Stem Cells 27, 2220–2228.

Cairns, P., Okami, K., Halachmi, S., Halachmi, N., Esteller, M., Herman, J.G., *et al.*, 1997. Frequent inactivation of PTEN/MMAC1 in primary prostate cancer. Cancer Res. 57, 4997–5000.

Chen, G., Shukeir, N., Potti, A., Sircar, K., Aprikian, A., Goltzman, D., Rabbani, S.A. 2004. Up-regulation of Wnt-1 and beta-catenin production in patients with advanced metastatic prostate carcinoma: potential pathogenetic and prognostic implications. Cancer 101, 1345–1356.

Chiou, S.-H., Wang, M.-L., Chou, Y.-T., Chen, C.-J., Hong, C.-F., Hsieh, W.-J., *et al.*, 2010. Coexpression of Oct4 and Nanog enhances malignancy in lung adenocarcinoma by inducing cancer stem cell-like properties and epithelial-mesenchymal transdifferentiation. Cancer Res. 70, 10 433–10 444.

Choi, S.C., Choi, J.H., Park, C.Y., Ahn, C.M., Hong, S.J., Lim, D.S. 2012. Nanog regulates molecules involved in stemness and cell cycle-signaling pathway for maintenance of pluripotency of P19 embryonal carcinoma stem cells. J. Cell. Physiol. 227, 3678–3692.

Collins, A.T., Habib, F.K., Maitland, N.J., Neal, D.E. 2001. Identification and isolation of human prostate epithelial stem cells based on α2β1-integrin expression. J. Cell Sci. 114, 3865–3872.

Collins, A.T., Berry, P.A., Hyde, C., Stower, M.J., Maitland, N.J. 2005. Prospective identification of tumorigenic prostate cancer stem cells. Cancer Res. 65, 10 946–10 951.

Cozzio, A., Passegué, E., Ayton, P.M., Karsunky, H., Cleary, M.L., Weissman, I.L. 2003. Similar MLL-associated leukemias arising from self-renewing stem cells and short-lived myeloid progenitors. Genes Dev. 17, 3029–3035.

Cunha, G.R., Ricke, W., Thomson, A., Marker, P.C., Risbridger, G., Hayward, S.W., *et al.*, 2004. Hormonal, cellular, and molecular regulation of normal and neoplastic prostatic development. J. Steroid Biochem. Mol. Biol. 92, 221–236.

De Marzo, A.M., Meeker, A.K., Epstein, J.I., Coffey, D.S. 1998. Prostate stem cell compartments: expression of the cell cycle inhibitor p27Kip1 in normal, hyperplastic, and neoplastic cells. Am. J. Pathol. 153, 911–919.

DeKlerk, D.P., Coffey, D.S. 1978. Quantitative determination of prostatic epithelial and stromal hyperplasia by a new technique. Biomorphometrics. Invest. Urol. 16, 240–245.

Di Cristofano, A., De Acetis, M., Koff, A., Cordon-Cardo, C., Pandolfi, P.P. 2001. Pten and p27KIP1 cooperate in prostate cancer tumor suppression in the mouse. Nat. Genet. 27, 222–224.

Dong, J.-T. 2006. Prevalent mutations in prostate cancer. J. Cell. Biochem. 97, 433–447.

Dreher, T., Zentgraf, H., Abel, U., Kappeler, A., Michel, M.S., Bleyl, U., Grobholz, R. 2004. Reduction of PTEN and p27kip1 expression correlates with tumor grade in prostate cancer. Analysis in radical prostatectomy specimens and needle biopsies. Virchows Arch. 444, 509–517.

Edwards, J., Krishna, N.S., Witton, C.J., Bartlett, J.M.S. 2003. Gene amplifications associated with the development of hormone-resistant prostate cancer. Clin. Cancer Res. 9, 5271–5281.

Edwards, S., Campbell, C., Flohr, P., Shipley, J., Giddings, I., Te-Poele, R., *et al.*, 2005. Expression analysis onto microarrays of randomly selected cDNA clones highlights HOXB13 as a marker of human prostate cancer. Br. J. Cancer 92, 376–381.

Emmert-Buck, M.R., Vocke, C.D., Pozzatti, R.O., Duray, P.H., Jennings, S.B., Florence, C.D., *et al.*, 1995. Allelic loss on chromosome 8p12-21 in microdissected prostatic intraepithelial neoplasia. Cancer Res. 55, 2959–2962.

Freestone, S.H., Marker, P., Grace, O.C., Tomlinson, D.C., Cunha, G.R., Harnden, P., Thomson, A.A. 2003. Sonic hedgehog regulates prostatic growth and epithelial differentiation. Dev. Biol. 264, 352–362.

Fuchs, E., Tumbar, T., Guasch, G. 2004. Socializing with the neighbors: stem cells and their niche. Cell. 116(6), 769–778.

Galli, R., Binda, E., Orfanelli, U., Cipelletti, B., Gritti, A., De Vitis, S., *et al.*, 2004. Isolation and characterization of tumorigenic, stem-like neural precursors from human glioblastoma. Cancer Res. 64, 7011–7021.

Gao, H., Ouyang, X., Banach-Petrosky, W.A., Gerald, W.L., Shen, M.M., Abate-Shen, C. 2006. Combinatorial activities of Akt and B-Raf/Erk signaling in a mouse model of androgen-independent prostate cancer. Proc. Nat. Acad. Sci. USA 103, 14 477–14 482.

Gao, N., Le Lay, J., Qin, W., Doliba, N., Schug, J., Fox, A.J., *et al.*, 2010. Foxa1 and Foxa2 maintain the metabolic and secretory features of the mature beta-cell. Mol. Endocrinol. 24, 1594–1604.

Goldstein, A.S., Lawson, D.A., Cheng, D., Sun, W., Garraway, I.P., Witte, O.N. 2008. Trop2 identifies a subpopulation of murine and human prostate basal cells with stem cell characteristics. Proc. Nat. Acad. Sci. USA 105, 20 882–20 887.

Grasso, C.S., Wu, Y.-M., Robinson, D.R., Cao, X., Dhanasekaran, S.M., Khan, A.P., *et al.*, 2012. The mutational landscape of lethal castration-resistant prostate cancer. Nature 487, 239–243.

Gray, I.C., Stewart, L.M., Phillips, S.M., Hamilton, J.A., Gray, N.E., Watson, G.J., *et al.*, 1998. Mutation and expression analysis of the putative prostate tumour-suppressor gene PTEN. Br. J. Cancer 78, 1296–1300.

Greenman, C., Stephens, P., Smith, R., Dalgliesh, G.L., Hunter, C., Bignell, G., *et al.*, 2007. Patterns of somatic mutation in human cancer genomes. Nature 446, 153–158.

Grignon, D.J., Caplan, R., Sarkar, F.H., Lawton, C.A., Hammond, E.H., Pilepich, M.V., *et al.*, 1997. p53 status and prognosis of locally advanced prostatic adenocarcinoma: a study based on RTOG 8610. J. Natl. Cancer Inst. 89, 158–165.

Hall, C.L., Daignault, S.D., Shah, R.B., Pienta, K.J., Keller, E.T. 2008. Dickkopf-1 expression increases early in prostate cancer development and decreases during progression from primary tumor to metastasis. Prostate 68, 1396–1404.

He, A., Qi, W., Huang, Y., Feng, T., Chen, J., Sun, Y., *et al.*, 2012. CD133 expression predicts lung metastasis and poor prognosis in osteosarcoma patients: a clinical and experimental study. Exp. Ther. Med. 4, 435–441.

Holcomb, I.N., Young, J.M., Coleman, I.M., Salari, K., Grove, D.I., Hsu, L., *et al.*, 2009. Comparative analyses of chromosome alterations in soft-tissue metastases within and across patients with castration-resistant prostate cancer. Cancer Res. 69, 7793–7802.

Hudson, D.L., O'Hare, M., Watt, F.M., Masters, J.R.W. 2000. Proliferative heterogeneity in the human prostate: evidence for epithelial stem cells. Lab. Investig. 80, 1243–1250.

Huntly, B.J.P., Gilliland, D.G. 2005. Leukaemia stem cells and the evolution of cancer-stem-cell research. Nat. Rev. Cancer 5, 311–321.

Hurt, E.M., Kawasaki, B.T., Klarmann, G.J., Thomas, S.B., Farrar, W.L. 2008. CD44+ CD24(−) prostate cells are early cancer progenitor/stem cells that provide a model for patients with poor prognosis. Br. J. Cancer 98, 756–765.

Jeter, C.R., Liu, B., Liu, X., Chen, X., Liu, C., Calhoun-Davis, T., *et al.*, 2011. NANOG promotes cancer stem cell characteristics and prostate cancer resistance to androgen deprivation. Oncogene 30, 3833–3845.

Jia, X., Li, X., Xu, Y., Zhang, S., Mou, W., Liu, Y., *et al.*, 2011. SOX2 promotes tumorigenesis and increases the anti-apoptotic property of human prostate cancer cell. J. Mol. Cell Biol. 3, 230–238.

Kantoff, P.W., Higano, C.S., Shore, N.D., Berger, E.R., Small, E.J., Penson, D.F., *et al.*, 2010. Sipuleucel-T immunotherapy for castration-resistant prostate cancer. N. Engl. J. Med. 363(5), 411–422.

Kim, M.J., Cardiff, R.D., Desai, N., Banach-Petrosky, W.A., Parsons, R., Shen, M.M., Abate-Shen, C. 2002. Cooperativity of Nkx3.1 and Pten loss of function in a mouse model of prostate carcinogenesis. Proc. Nat. Acad. Sci. USA 99, 2884–2889.

Kim, C.F.B., Jackson, E.L., Woolfenden, A.E., Lawrence, S., Babar, I., Vogel, S., *et al.*, 2005. Identification of bronchioalveolar stem cells in normal lung and lung cancer. Cell 121, 823–835.

Kim, J., Roh, M., Doubinskaia, I., Algarroba, G.N., Eltoum, I.-E.A., Abdulkadir, S.A. 2012. A mouse model of heterogeneous, c-MYC-initiated prostate cancer with loss of Pten and p53. Oncogene 31(3), 322–332.

King, J.C., Xu, J., Wongvipat, J., Hieronymus, H., Carver, B.S., Leung, D.H., *et al.*, 2009. Cooperativity of TMPRSS2-ERG with PI3-kinase pathway activation in prostate oncogenesis. Nat. Genet. 41, 524–526.

Kirby, M., Hirst, C., Crawford, E.D. 2011. Characterising the castration-resistant prostate cancer population: a systematic review. Int. J. Clin. Pract. 65, 1180–1192.

Klonisch, T., Wiechec, E., Hombach-Klonisch, S., Ande, S.R., Wesselborg, S., Schulze-Osthoff, K., Los, M. 2008. Cancer stem cell markers in common cancers – therapeutic implications. Trends Mol. Med. 14(10), 450–460.

Koivisto, P., Kononen, J., Palmberg, C., Tammela, T., Hyytinen, E., Isola, J., *et al.*, 1997. Androgen receptor gene amplification: a possible molecular mechanism for androgen deprivation therapy failure in prostate cancer. Cancer Res. 57, 314–319.

Kubota, Y., Shuin, T., Uemura, H., Fujinami, K., Miyamoto, H., Torigoe, S., *et al.*, 1995. Tumor suppressor gene p53 mutations in human prostate cancer. Prostate 27, 18–24.

Kuhnert, F., Davis, C.R., Wang, H.-T., Chu, P., Lee, M., Yuan, J., *et al.*, 2004. Essential requirement for Wnt signaling in proliferation of adult small intestine and colon revealed by adenoviral expression of Dickkopf-1. Proc. Nat. Acad. Sci. USA 101, 266–271.

Kumar, A., White, T.A., MacKenzie, A.P., Clegg, N., Lee, C., Dumpit, R.F., *et al.*, 2011. Exome sequencing identifies a spectrum of mutation frequencies in advanced and lethal prostate cancers. Proc. Nat. Acad. Sci. USA 108, 17 087–17 092.

Kyprianou, N., Isaacs, J. 1988. Identification of a cellular receptor for transforming growth factor-β in rat ventral prostate and its negative regulation by androgens. Endocrinology 123, 2124–2131.

Kypta, R.M., Waxman, J. 2012. Wnt/β-catenin signalling in prostate cancer. Nat. Rev. Urol. 9(8), 418–428.

Lamm, M.L., Catbagan, W.S., Laciak, R.J., Barnett, D.H., Hebner, C.M., Gaffield, W., *et al.*, 2002. Sonic hedgehog activates mesenchymal Gli1 expression during prostate ductal bud formation. Dev. Biol. 249, 349–366.

Lawson, D.A., Xin, L., Lukacs, R., Xu, Q., Cheng, D., Witte, O.N. 2005. Prostate stem cells and prostate cancer. Cold Spring Harb. Symp. Quant. Biol. 70, 187–196.

Lawson, D.A., Xin, L., Lukacs, R.U., Cheng, D., Witte, O.N. 2007. Isolation and functional characterization of murine prostate stem cells. Proc. Nat. Acad. Sci. USA 104, 181–186.

Levina, V., Marrangoni, A., Wang, T., Parikh, S., Su, Y., Herberman, R., *et al.*, 2010. Elimination of human lung cancer stem cells through targeting of the stem cell factor-c-kit autocrine signaling loop. Cancer Res. 70, 338–346.

Li, Z.G., Yang, J., Vazquez, E.S., Rose, D., Vakar-Lopez, F., Mathew, P., *et al.*, 2008. Low-density lipoprotein receptor-related protein 5 (LRP5) mediates the prostate cancer-induced formation of new bone. Oncogene 27, 596–603.

Ling, H., Jolicoeur, P. 2012. Notch-1 signaling promotes the cyclinD1-dependent generation of mammary tumor-initiating cells that can revert to bi-potential progenitors from which they arise. Oncogene 32(29), 3410–3419.

Linja, M.J., Visakorpi, T. 2004. Alterations of androgen receptor in prostate cancer. J. Steroid Biochem. Mol. Biol. 92(4), 255–264.

Linn, D.E., Yang, X., Sun, F., Xie, Y., Chen, H., Jiang, R., *et al.*, 2010. A role for OCT4 in tumor initiation of drug-resistant prostate cancer cells. Genes Cancer 1, 908–916.

Liu, A.Y., True, L.D., LaTray, L., Nelson, P.S., Ellis, W.J., Vessella, R.L., *et al.*, 1997. Cell–cell interaction in prostate gene regulation and cytodifferentiation. Proc. Nat. Acad. Sci. USA 94, 10 705–10 710.

Long, R.M., Morrissey, C., Fitzpatrick, J.M., Watson, R.W.G. 2005. Prostate epithelial cell differentiation and its relevance to the understanding of prostate cancer therapies. Clin. Sci. 108, 1–12.

Luo, G., Long, J., Cui, X., Xiao, Z., Liu, Z., Shi, S., *et al.*, 2013. Highly lymphatic metastatic pancreatic cancer cells possess stem cell-like properties. Int. J. Oncol. 42, 979–984.

Madan, R.A., Gulley, J.L., Kantoff, P.W. 2013. Demystifying immunotherapy in prostate cancer: understanding current and future treatment strategies. Cancer J. 19, 50–58.

Madu, C.O., Lu, Y. 2010. Novel diagnostic biomarkers for prostate cancer. J. Cancer 1, 150–177.

Magee, J.A., Piskounova, E., Morrison, S.J. 2012. Cancer stem cells: impact, heterogeneity, and uncertainty. Cancer Cell 21, 283–296.

McMenamin, M.E., Soung, P., Perera, S., Kaplan, I., Loda, M., Sellers, W.R. 1999. Loss of PTEN expression in paraffin-embedded primary prostate cancer correlates with high Gleason score and advanced stage. Cancer Res. 59, 4291–4296.

Meyers, F.J., Gumerlock, P.H., Sung Gil Chi, Borchers, H., Deitch, A.D., DeVere White, R.W. 1998. Very frequent p53 mutations in metastatic prostate carcinoma and in matched primary tumors. Cancer 83, 2534–2539.

Mills, A.A., Zheng, B., Wang, X.J., Vogel, H., Roop, D.R., Bradley, A. 1999. p63 is a p53 homologue required for limb and epidermal morphogenesis. Nature 398, 708–713.

Mohamed, A.A., Tan, S.-H., Sun, C., Shaheduzzaman, S., Hu, Y., Petrovics, G., *et al.*, 2011. ERG oncogene modulates prostaglandin signaling in prostate cancer cells. Cancer Biol. Ther. 11, 410–417.

Montpetit, M., Abrahams, P., Clark, A.F., Tenniswood, M. 1988. Androgen-independent epithelial cells of the rat ventral prostate. Prostate 12, 13–28.

Mosquera, J.-M., Perner, S., Demichelis, F., Kim, R., Hofer, M.D., Mertz, K.D., *et al.*, 2007. Morphological features of TMPRSS2 – ERG gene fusion prostate cancer. J. Pathol. 212, 91–101.

Ouyang, X., DeWeese, T.L., Nelson, W.G., Abate-Shen, C. 2005. Loss-of-function of Nkx3.1 promotes increased oxidative damage in prostate carcinogenesis. Cancer Res. 65, 6773–6779.

Park, C.H., Bergsagel, D.E., McCulloch, E.A. 1971. Mouse myeloma tumor stem cells: a primary cell culture assay. J. Natl. Cancer Inst. 46, 411–422.

Patrawala, L., Calhoun, T., Schneider-Broussard, R., Li, H., Bhatia, B., Tang, S., *et al.*, 2006. Highly purified CD44+ prostate cancer cells from xenograft human tumors are enriched in tumorigenic and metastatic progenitor cells. Oncogene 25, 1696–1708.

Pignon, J.-C., Grisanzio, C., Geng, Y., Song, J., Shivdasani, R.A., Signoretti, S. 2013. p63-expressing cells are the stem cells of developing prostate, bladder, and colorectal epithelia. Proc. Nat. Acad. Sci. USA 110, 8105–8110.

Pleasance, E.D., Stephens, P.J., O'Meara, S., McBride, D.J., Meynert, A., Jones, D., *et al.*, 2010. A small-cell lung cancer genome with complex signatures of tobacco exposure. Nature 463, 184–190.

Ponti, D., Costa, A., Zaffaroni, N., Pratesi, G., Petrangolini, G., Coradini, D., *et al.*, 2005. Isolation and in vitro propagation of tumorigenic breast cancer cells with stem/progenitor cell properties. Cancer Res. 65, 5506–5511.

Prajapati, A., Gupta, S., Mistry, B., Gupta, S. 2013. Prostate stem cells in the development of benign prostate hyperplasia and prostate cancer: emerging role and concepts. Biomed Res. Int. 2013, 107954.

Prins, G.S., Putz, O. 2008. Molecular signaling pathways that regulate prostate gland development. Differentiation 76, 641–659.

Rajasekhar, V.K., Studer, L., Gerald, W., Socci, N.D., Scher, H.I. 2011. Tumour-initiating stem-like cells in human prostate cancer exhibit increased NF-κB signalling. Nat. Commun. 2, 162.

Ricci-Vitiani, L., Lombardi, D.G., Pilozzi, E., Biffoni, M., Todaro, M., Peschle, C., De Maria, R. 2007. Identification and expansion of human colon-cancer-initiating cells. Nature 445, 111–115.

Richardson, G.D., Robson, C.N., Lang, S.H., Neal, D.E., Maitland, N.J., Collins, A.T. 2004. CD133, a novel marker for human prostatic epithelial stem cells. J. Cell Sci. 117, 3539–3545.

Rybak, A.P., He, L., Kapoor, A., Cutz, J.-C., Tang, D. 2011. Characterization of sphere-propagating cells with stem-like properties from DU145 prostate cancer cells. Biochim. Biophys. Acta 1813, 683–694.

Schalken, J.A., van Leenders, G. 2003. Cellular and molecular biology of the prostate: stem cell biology. Urology 62, 11–20.

Schatton, T., Murphy, G.F., Frank, N.Y., Yamaura, K., Waaga-Gasser, A.M., Gasser, M., *et al.*, 2008. Identification of cells initiating human melanomas. Nature 451, 345–349.

Schmelz, M., Moll, R., Hesse, U., Prasad, A.R., Gandolfi, J.A., Hasan, S.R., *et al.*, 2005. Identification of a stem cell candidate in the normal human prostate gland. Eur. J. Cell Biol. 84, 341–354.

Sell, S. 2010. On the stem cell origin of cancer. Am. J. Pathol. 176, 2584–2494.

Shipitsin, M., Polyak, K. 2008. The cancer stem cell hypothesis: in search of definitions, markers, and relevance. Lab. Invest. 88, 459–463.

Shou, J., Ross, S., Koeppen, H., de Sauvage, F.J., Gao, W.Q. 2001. Dynamics of notch expression during murine prostate development and tumorigenesis. Cancer Res. 61, 7291–7297.

Signoretti, S., Waltregny, D., Dilks, J., Isaac, B., Lin, D., Garraway, L., *et al.*, 2000. p63 is a prostate basal cell marker and is required for prostate development. Am. J. Pathol. 157, 1769–1775.

Sreenath, T.L., Dobi, A., Petrovics, G., Srivastava, S. 2011. Oncogenic activation of ERG: a predominant mechanism in prostate cancer. J. Carcinog. 10, 37.

Sun, X., Huang, J., Homma, T., Kita, D., Klocker, H., Schafer, G., *et al.*, 2009. Genetic alterations in the PI3K pathway in prostate cancer. Anticancer Res. 29, 1739–1743.

Swaminathan, S.K., Roger, E., Toti, U., Niu, L., Ohlfest, J.R., Panyam, J. 2013. CD133-targeted paclitaxel delivery inhibits local tumor recurrence in amousemodel of breast cancer. J. Control. Release 171, 280–287.

Takao, T., Tsujimura, A. 2008. Prostate stem cells: the niche and cell markers. Int. J. Urol. 15, 289–294.

Taylor, B.S., Schultz, N., Hieronymus, H., Gopalan, A., Xiao, Y., Carver, B.S., *et al.*, 2010. Integrative Genomic Profiling of Human Prostate Cancer. Cancer Cell 18, 11–22.

Thomson, A.A., Marker, P.C. 2006. Branching morphogenesis in the prostate gland and seminal vesicles. Differentiation 74, 382–392.

Thorne, H., Willems, A.J., Niedermayr, E., Hoh, I.M.Y., Li, J., Clouston, D., *et al.*, 2011. Decreased prostate cancer-specific survival of men with BRCA2 mutations from multiple breast cancer families. Cancer Prev. Res. (Phila.) 4, 1002–1010.

Timms, B.G. 2008. Prostate development: a historical perspective. Differentiation 76, 565–577.

Timms, B.G., Mohs, T.J., Didio, L.J. 1994. Ductal budding and branching patterns in the developing prostate. J. Urol. 151, 1427–1432.

Todaro, M., Perez Alea, M., Scopelliti, A., Medema, J.P., Stassi, G. 2008. IL-4-mediated drug resistance in colon cancer stem cells. Cell Cycle 7, 309–313.

Tomlins, S.A., Laxman, B., Varambally, S., Cao, X., Yu, J., Helgeson, B.E., *et al.*, 2008. Role of the TMPRSS2-ERG gene fusion in prostate cancer. Neoplasia 10(2), 177–188.

Tsujimura, A., Koikawa, Y., Salm, S., Takao, T., Coetzee, S., Moscatelli, D., *et al.*, 2002. Proximal location of mouse prostate epithelial stem cells: a model of prostatic homeostasis. J. Cell Biol. 157, 1257–1265.

Uysal-Onganer, P., Kawano, Y., Caro, M., Walker, M.M., Diez, S., Darrington, R.S., *et al.*, 2010. Wnt-11 promotes neuroendocrine-like differentiation, survival and migration of prostate cancer cells. Mol. Cancer 9, 55.

Verhagen, A.P., Ramaekers, F.C., Aalders, T.W., Schaafsma, H.E., Debruyne, F.M., Schalken, J.A. 1992. Colocalization of basal and luminal cell-type cytokeratins in human prostate cancer. Cancer Res. 52, 6182–6187.

Vermeulen, L., de Sousa e Melo, F., Richel, D.J., Medema, J.P. 2012. The developing cancer stem-cell model: Clinical challenges and opportunities. Lancet Oncol. 13(2), e83–e89.

Vis, A.N., Noordzij, M.A., Fitoz, K., Wildhagen, M.F., Schröder, F.H., van der Kwast, T.H. 2000. Prognostic value of cell cycle proteins p27(kip1) and MIB-1, and the cell adhesion protein CD44s in surgically treated patients with prostate cancer. J. Urol. 164, 2156–2161.

Visakorpi, T., Kallioniemi, A.H., Syvanen, A.C., Hyytinen, E.R., Karhu, R., Tammela, T., *et al.*, 1995. Genetic changes in primary and recurrent prostate cancer by comparative genomic hybridization. Cancer Res 55, 342–347.

Visvader, J.E., Lindeman, G.J. 2012. Cancer stem cells: current status and evolving complexities. Cell Stem Cell 10, 717–728.

Wang, Z.A., Shen, M.M. 2011. Revisiting the concept of cancer stem cells in prostate cancer. Oncogene 30, 1261–1271.

Wang, Y., Hayward, S., Cao, M., Thayer, K., Cunha, G. 2001. Cell differentiation lineage in the prostate. Differentiation 68, 270–279.

Wang, Y., Chan, S.L., Miele, L., Yao, P.J., Mackes, J., Ingram, D.K., *et al.*, 2004. Involvement of Notch signaling in hippocampal synaptic plasticity. Proc. Nat. Acad. Sci. USA 101, 9458–9462.

Wang, M.-L., Chiou, S.-H., Wu, C.-W. 2013. Targeting cancer stem cells: emerging role of Nanog transcription factor. Onco. Targets. Ther. 6, 1207–1220.

Wei, C., Guomin, W., Yujun, L., Ruizhe, Q. 2007. Cancer stem-like cells in human prostate carcinoma cells DU145: the seeds of the cell line? Cancer Biol Ther. 6(5), 763–768.

Wicha, M.S., Liu, S., Dontu, G. 2006. Cancer stem cells: an old idea – a paradigm shift. Cancer Res. 66, 1883–1890; disc. 1895–1896.

Xue, Y., Smedts, F., Debruyne, F.M.J., de la Rosette, J.J., Schalken, J.A. 1998. Identification of intermediate cell types by keratin expression in the developing human prostate. Prostate 34, 292–301.

Yang, A., Schweitzer, R., Sun, D., Kaghad, M., Walker, N., Bronson, R.T., *et al.*, 1999. p63 is essential for regenerative proliferation in limb, craniofacial and epithelial development. Nature 398, 714–718.

Yao, M., Taylor, R.A., Richards, M.G., Sved, P., Wong, J., Eisinger, D., *et al.*, 2009. Prostate-regenerating capacity of cultured human adult prostate epithelial cells. Cells Tiss. Org. 191, 203–212.

Yu, C., Yao, Z., Jiang, Y., Keller, E.T. 2012. Prostate cancer stem cell biology. Minerva Urol. Nefrol. 64, 19–33.

Zbinden, M., Duquet, A., Lorente-Trigos, A., Ngwabyt, S.-N., Borges, I., Ruiz i Altaba, A. 2010. NANOG regulates glioma stem cells and is essential in vivo acting in a cross-functional network with GLI1 and p53. EMBO J. 29, 2659–2674.

Zhang, W.C., Ng, S.C., Yang, H., Rai, A., Umashankar, S., Ma, S., *et al.*, 2012. Glycine decarboxylase activity drives non-small cell lung cancer tumor-initiating cells and tumorigenesis. Cell 148, 259–272.

Zhao, C., Blum, J., Chen, A., Kwon, H.Y., Jung, S.H., Cook, J.M., *et al.*, 2007. Loss of ??-catenin impairs the renewal of normal and CML stem cells *in vivo*. Cancer Cell 12, 528–541.

11
Stem Cells and Pancreatic Cancer

Audrey M. Hendley[1,2,3] and Jennifer M. Bailey[1,2,3]

[1]*Department of Surgery, Johns Hopkins University School of Medicine, Baltimore, MD, USA*

[2]*The McKusick-Nathans Institute of Genetic Medicine, Johns Hopkins University School of Medicine, Baltimore, MD, USA*

[3]*Department of Internal Medicine, Division of Gastroenterology, Hepatology and Nutrition, The University of Texas Health Science Center at Houston, Houston, TX, USA*

11.1 Introduction

Pancreatic ductal adenocarcinoma (PDAC) is the most common type of pancreatic cancer, accounting for more than 85% of cases. The majority of patients diagnosed with PDAC have limited options for surgical treatment or chemotherapy. PDAC has the lowest 5-year survival rate of any pancreatic cancer, largely because patients are commonly diagnosed with metastatic disease, as the tumours metastasize early and do not respond well to chemotherapy. For patients for whom surgical intervention is an option, the statistics for survival remain grim, as disease recurrence is common, giving them a 2-year survival rate of less than 5%. While these properties are common to many types of cancer, for pancreatic cancer they represent an avenue for potential therapeutic intervention. The discovery of a unique population of pancreatic cancer stem cells (PCSCs) that are primarily responsible for tumour growth and metastases will indelibly aid in the discovery of novel therapies for the treatment of pancreatic cancer.

The cancer stem cell (CSC) hypothesis proposes that a subset of cancer cells have the capacity to give rise to a new tumour containing both CSC populations and differentiated tumour cells that recapitulate the original tumour. Unlike tumour cells from the bulk population, these CSCs, alternatively

Principles of Stem Cell Biology and Cancer: Future Applications and Therapeutics, First Edition.
Edited by Tarik Regad, Thomas J. Sayers and Robert C. Rees.

termed 'tumour-initiating cells' (TICs), have the ability to lie dormant for years, have self-renewal capabilities and are long-lived, giving them a unique ability to cause tumours in immunodeficient mice. These defining characteristics make targeting CSCs incredibly complex, as standard chemotherapeutic strategies target pathways and characteristics unique to the bulk tumour population.

11.2 Pancreas

The pancreas is a digestive organ comprising two major compartments: the exocrine portion and the endocrine compartment. Pancreatic acinar cells, centroacinar cells and ductal cells make up the exocrine component of the pancreas, while alpha cells, beta cells, delta cells, PP cells and epsilon cells located in the islets of Langerhans constitute the endocrine portion (Figure 11.1). Pancreatic acinar cells synthesize digestive enzymes, including proteases like trypsinogen, chymotrypsinogen, elastase and carboxypeptidase, pancreatic lipase, nucleases and amylase, which are secreted into the extensive branched pancreatic ductal tree and eventually flow into the duodenum to aid in digestion. The pancreatic centroacinar and ductal cells produce a bicarbonate-rich secretion when stimulated by the hormone secretin, which mixes with the acinar cell secretions to compose the pancreatic juice. The

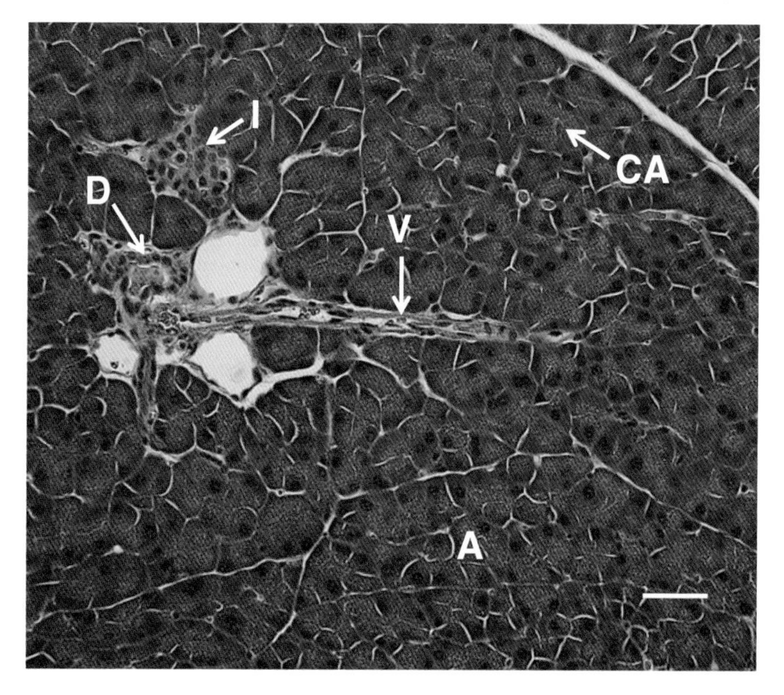

Figure 11.1 Hematoxylin and Eosin staining of the human pancreas. The hormone-producing endocrine gland of the pancreas is composed of the Islets of Langerhans (I). Hormones secreted by islet cells are transported via blood vessels (V). The digestive exocrine compartment is composed of the acinar cells (A), ductal cells (D) and centroacinar cells (CA).

bicarbonate secretion functions to neutralize acidic fluids coming from the stomach prior to entry into the duodenum and small intestine.

11.3 In search of a cell of origin: mouse models of pancreatic cancer

In 2008, the pancreatic cancer team at Johns Hopkins University published a landmark paper in *Science* that reported a comprehensive genetic analysis of 24 patients with advanced PDAC. Using exome sequencing and microarrays, the investigators found 12 core signalling pathways to be altered in the vast majority of pancreatic cancers and discovered mutations in KRAS in 100% of pancreatic cancers. Genetic alterations in the tumour-suppressor genes CDKN2, SMAD4 and TP53 and the protooncogene KRAS were identified as the most common genetic mutations in pancreatic cancer (Jones *et al.*, 2008). The pancreatic cancer research community widely uses mouse models in which the protooncogene KRAS can be activated with Cre/Lox technology. Using this technology, expression of oncogenic KRAS has been targeted to pancreatic progenitor cells in the developing pancreas and to adult mouse pancreatic cell types. Both of these strategies have been used extensively throughout the literature to gain insight into early and advanced stages of pancreatic cancer progression and metastasis (Herreros-Villanueva *et al.*, 2012).

11.3.1 Embryonic activation of oncogenic Kras

The characteristic features of PCSCs, such as self-renewal and pluripotentcy (which are very similar to those of normal tissue stem cells), make them prime candidates to be the cells of origin of PDAC. Numerous investigators have interrogated the tumourigenic capacity of multiple cell types in the pancreas. The advent of Cre/Lox-based techniques has enabled cell-type responses to oncogenic Kras to be determined. Initial experiments using the Pdx1-cre driver to express $Kras^{G12D}$ (referred to as Kras*) in foregut endoderm-derived multipotent pancreatic progenitors induced pancreatic intraepithelial neoplasia (PanIN), which histologically mimicked human PanIN as early as 2.5 months of age. In this model, activation of Kras* occurs through a glycine–aspartic acid substitution in codon 12 of the translated protein, a substitution that is commonly found in human PDAC and results in constitutive downstream signalling of Ras effector pathways. By 9 months of age, the mice displayed acinar–ductal metaplasia (ADM) and the full spectrum of PanIN 1–3, and a small fraction had invasive PDAC (Hingorani *et al.*, 2003). When activating the $Trp53^{R172H/+}$ allele in addition to activation of Kras* using the Pdx1-cre driver, the mice have a median survival of 5 months, develop the full spectrum of preinvasive lesions by 10 weeks of age and develop metastatic PDAC (Hingorani *et al.*, 2005). While highly informative, these mouse models fail to address the question of how adult cell types in the pancreas respond to Kras*.

11.3.2 Adult cell-type activation of oncogenic Kras

Mouse modelling has revealed that the adult acinar cells in the pancreas can undergo ADM and form PanIN as early as 1 month after the expression of Kras* (Habbe *et al.*, 2008; Bailey *et al.*, 2014) (Figure 11.2). In a very elegant manner, Kopp *et al.* (2012) compared the capacity of acinar cells and ductal

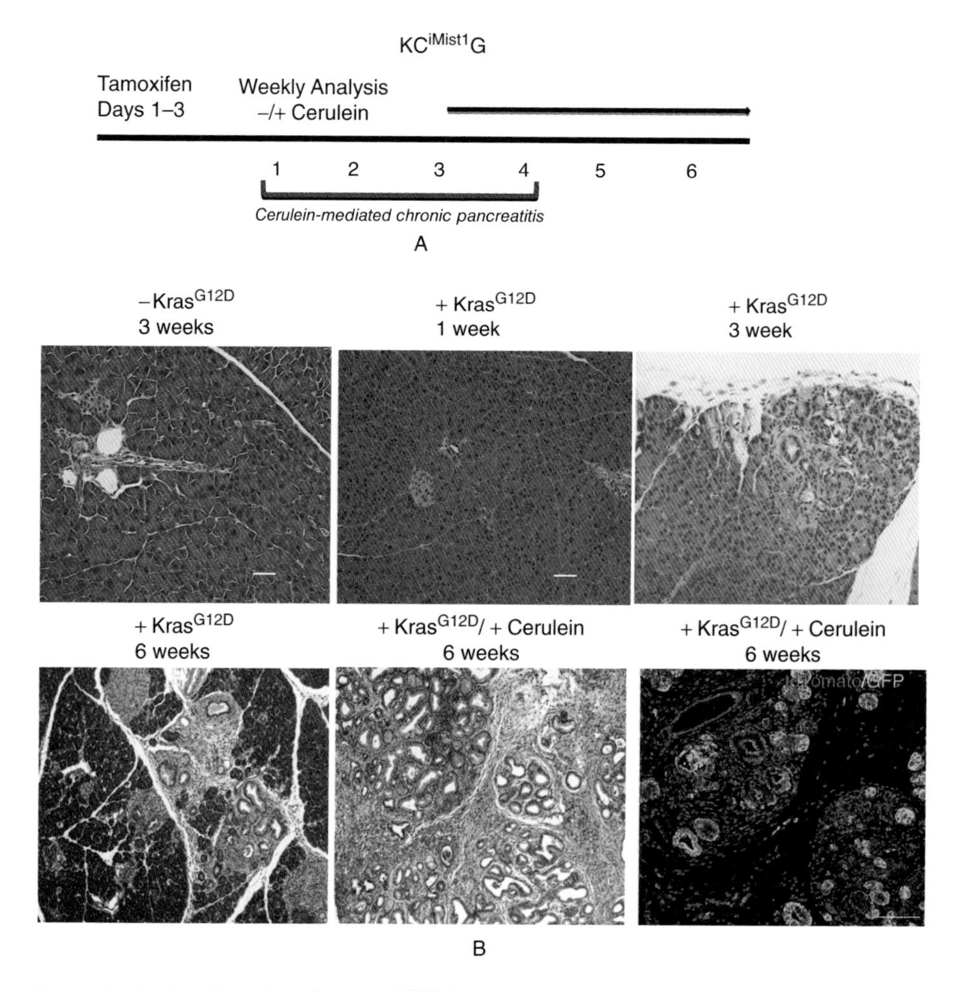

Figure 11.2 PanIN model using the $KC^{iMist1}G$ transgenic mouse model of pancreatic cancer. (A) In order to generate PanIN, a $Mist1Cre^{ERT2}$ (C^{iMist1}) mouse was crossed to a LSL-$Kras^{G12D}$ (K) mouse and an mTmG (G) mouse, the latter of which contains a membrane-targeted GFP that is expressed after tamoxifen is given to it. In a mouse model of cerulein-mediated chronic pancreatitis, cerulein is administered for 3 weeks post-tamoxifen injection. (B) No PanIN was detected in the absence of Kras* expression or in the presence of Kras* expression at 1 week post-tamoxifen injection. ADM and PanIN occupied about <2% of the total pancreatic cross-sectional area 3 weeks after Kras* was expressed. In the presence of cerulein-mediated chronic pancreatitis, PanIN occupied >90% of the pancreatic cross-sectional area 6 weeks after Kras* was expressed. Scale bars are 50 μm.

cells to form PanIN in response to Kras*. The authors used the *Sox9CreER* mouse model to express Kras* in the ductal cells and *Ptf1a*^{CreER} to express Kras* specifically in the acinar cells. Expression of Kras* in the Sox9 compartment failed to induce PanIN as late as 6 months after Kras* was expressed. Despite the resistance of the Sox9 ductal compartment to undergoing metaplasia, one intriguing discovery from this paper was that Sox9 gene expression was induced in acinar cells in response to Kras* and that Sox9 expression was required for the acinar cells to form PanIN, implicating a significant contribution of a 'ductal reprogramming' event in the formation of PanIN by acinar cells (Kopp *et al.*, 2012). In the context of chronic inflammation, the insulin-producing cells of the pancreas have also shown a capacity to undergo malignant transformation in response to Kras* expression (Gidekel Friedlander *et al.*, 2009).

More recently, work by DelGiorno *et al.* (2014) and Bailey *et al.* (2014) implicated a pancreatobiliary origin of pancreatic cancer. Delgiorno *et al.* (2014) expressed Kras* in the Sox17 compartment, which led to the induction of metaplasia and the presence of Dclk1-expressing pancreatic tuft cells. Sox17, in concert with Kras* expression, increased the rate of transformation by acinar cells and indicated a transdifferentiation to a biliary phenotype as an initiating event in pancreatic cancer. Bailey *et al.* (2014) published the presence of tuft cells in ADM and PanIN-containing metaplasia (Figure 11.3A); isolation of these pancreatic Kras*-expressing tuft cells revealed progenitor capabilities, indicating the presence of a 'PanIN' stem cell. These results are informative, as about 70% of pancreatic tumours arise in the head of the pancreas, near the region of the duodenum, indicating this area may have an increased propensity for metaplasia and cancer, especially in response to chronic inflammatory conditions. In the non-neoplastic setting, biliary tree stem cells were recently described as behaving like stem cells in culture conditions (Wang *et al.*, 2013). Furthermore, the maturation lineages of proximal peribiliary glands and distal pancreatic duct glands start near the duodenum and show expression of markers involved in pluripotency, proliferation and self-replication. In the context of recent work published by DelGiorno *et al.* (2014), the function of these cells and the exact effects of inflammation and oncogenic mutations on their neoplastic potential is an area requiring further investigation.

11.4 Current methods of detection

A number of experimental strategies are used to define PCSCs. To date, empirical experiments have involved the isolation of PCSCs based on their cell-surface markers, enzymatic activity or proteasome activity. The cell-surface markers and other strategies used to identify these specific subpopulations are not uniform across tumour types and are the subject of much debate. In these experiments, fluorescence-activated cell sorting (FACS) is

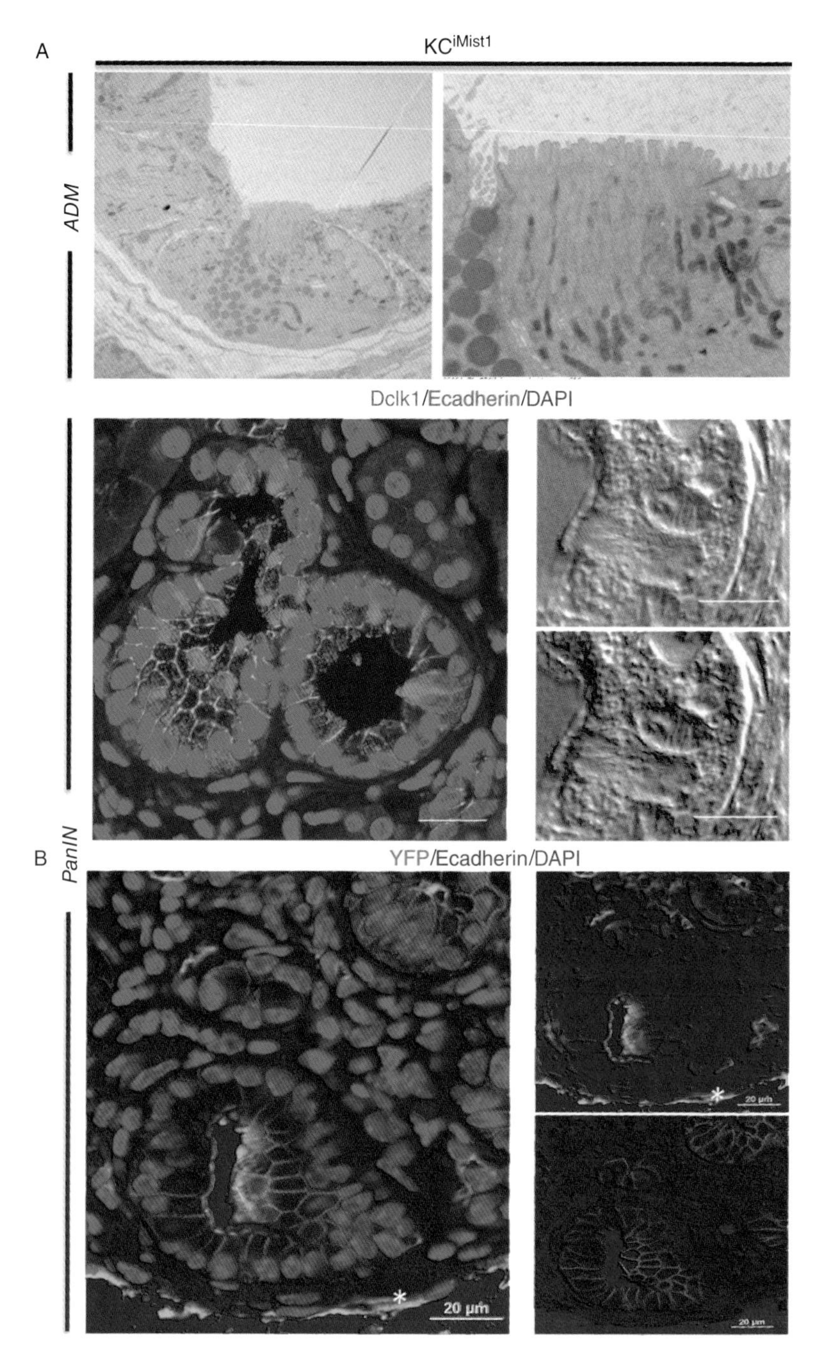

Figure 11.3 Acinar cell differentiation in response to Kras* expression. (A) Transmission electron micrograph showing unique cells in ADM-containing microvilli characteristic of a gastrointestinal tuft cell morphology. (B) Dclk1+ tuft cells are present in ADM and PanIN derived from pancreatic acinar cells expressing Kras*. Lineage tracing using the LSL-YFP allele indicates cells derived from acinar cells. Note the *, indicating cells that have lost the expression of Ecadherin and have an elongated nucleus and cytoplasm. These are delaminated cells derived from acinar cells expressing Kras*.

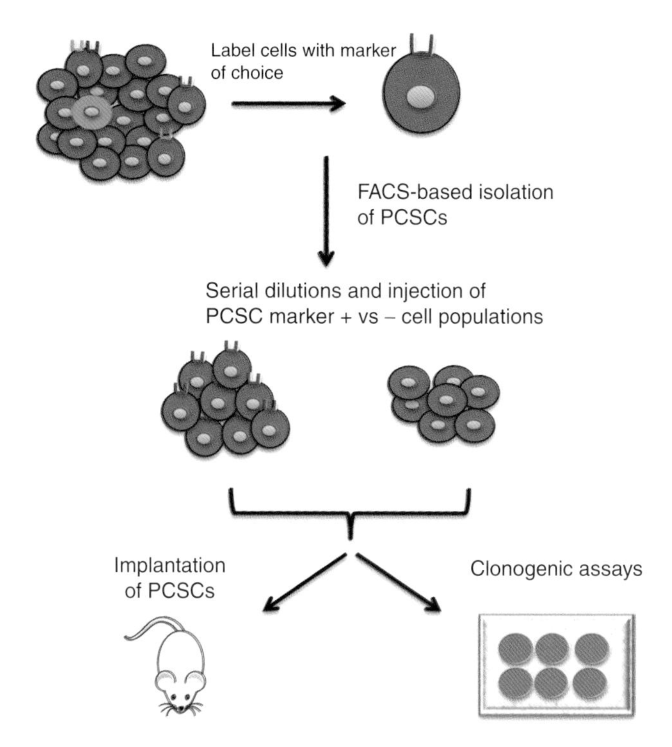

Figure 11.4 Methodology for PCSC detection. Pancreatic cancer cells are labelled with fluorescent markers by one of several methods, including genome manipulation in mice, antibodies that bind specific epitopes or assessment of enzymatic activity using the ALDEFLUOR assay. Labelled cell populations are then isolated by FACS and can be used in limited cell dilutions to implant into mice, in order to assess TIC frequency, or to plate on nonadherent conditions, to assess clonogenic potential.

employed to isolate specific low-abundance cell populations (Figure 11.4). These different populations can then be injected or implanted into mice in limited cell dilutions to determine the frequency of tumour incidence and growth *in vivo*. The identification of the first PCSCs employed this method to isolate relevant cell populations from human xenograft cell lines. The tumours that formed from PCSC populations histologically resembled the parental tumour, indicating that the isolated cells could recapitulate the differentiation status of the parent tumour. Other experimental strategies involve FACS-based isolation of cells, plating of serial dilutions of the different populations and subsequent growth on nonadherent tissue culture dishes. Using this method, scientists can study the growth of tumour spheres or colonies *in vitro*.

As with any biological measurement, there are limitations associated with each assay. The lack of standardization in protocols for sphere-forming assays, tissue digestion and FACS-based cell isolation, xenograft implantation and injection and the antibodies and reagents used present fundamental challenges when comparing observations between laboratories. Implantation of xenografts into mice often involves growing tumours subcutaneously, which is not the normal niche of pancreatic tumours and thus may not fully recapitulate the environment of pancreatic cancer cells. Pancreatic cancer cell lines are not always low passage, which introduces the potential for genetic and epigenetic alterations and phenotypes in these cell lines not present in the original tumour cells they were created from. Despite the known limitations of these methods, several putative PCSC populations have been identified, and much insight has been gained into PCSC biology through their use.

11.5 Markers

11.5.1 CD24, CD44 and ESA

The first population of PCSCs was described by Li *et al.* (2007) In order to discover this population, they used xenografts grown from human pancreatic cancer cells and a panel of cell-surface markers to isolate cells with high levels of CD24, CD44 and ESA on the cell surface using flow cytometry. The investigators sorted cells for single-, double- or triple-positive or negative status to determine the $CD44^+/CD24^+/ESA^+$ cell population, which comprised 0.2–0.8% of cells in 10 tumours studied and was superior in its ability to form tumours in mice. This cell population was highly tumourigenic, with as few as 100 cells able to form tumours in 50% of mice. In stark contrast, only 1 in 12 mice injected with the triple-negative population were able to form tumours with as many as 10 000 cells. Thus, the triple-positive cell population had a 100-fold increased tumour-promoting ability (Li *et al.*, 2007).

Identification of the $CD24^+/CD44^+/ESA^+$ PCSC population was based on unique functional properties of these cells; however, it is not known how or if these markers contribute to PCSC function and/or maintenance. CD24, CD44 and ESA may play a role in cell–cell interactions through cell-adhesion properties. CD24 has been shown to be a ligand for P-selectin, a cell-adhesion molecule present on the surface of activated endothelial cells and platelets that can bind to carcinoma cells (Aigner *et al.*, 1997). Mouse fibroblasts that overexpress CD44 aggregate into cell clusters, but this is inhibited upon administration of a monoclonal antibody that binds to CD44, suggesting that CD44 can mediate cell–cell adhesion (St John *et al.*, 1990; Webb *et al.*, 1990). CD44 also inhibits Hippo signalling (Xu *et al.*, 2010). ESA (or epithelial cell-adhesion molecule, Ep-CAM) is a homophilic cell-adhesion molecule that may mediate cell–cell interactions between epithelial cells in PCSCs (Litvinov *et al.*, 1994). The high expression levels of CD24, CD44 and ESA might also be a byproduct of signalling pathways regulating PCSC features in these cells, so further investigation to manifest the functional relevance of these markers is warranted.

11.5.2 CD133 and CXCR4

CD133 has been identified as a stem cell in both normal and cancerous human tissues, including brain, blood, vasculature and epithelial organs. CD133 expression in pancreatic cancer is an independent prognostic factor and has been significantly associated with metastasis to the lymphatic system (Maeda *et al.*, 2008). Hermann *et al.* (2007) reported the identification of a novel $CD133^+$ PCSC population capable of initiating tumour formation in athymic mice via orthotopic implantation with as little as 500 cells. AC133/CD133/prominin-1 is a transmembrane glycoprotein that localizes specifically to cell-membrane protrusions in various cell types (Shmelkov *et al.*, 2005). The $CD133^+$ population of the L3.6pl pancreatic cancer cell line readily forms pancreatic spheres in culture, while the $CD133^-$ population does not, indicating the clonogenic potential of $CD133^+$ PCSCs. Consistent with the emerging impression that CSCs are often refractory to drug treatment, these cells show a dramatic resistance to gemcitabine treatment, as measured by cell-cycle analysis and cellular apoptosis, when compared to the $CD133^-$ population. Gemcitabine treatment dramatically reduces the size of tumours in mice; however, the $CD133^+$ cell population expands in response to this chemotherapeutic treatment (Hermann *et al.*, 2007). Although this protein has been identified as a cell-surface marker of a number of normal and malignant stem cell populations, its function is not completely understood, and thus its biological contribution to stem cell function is unclear.

Hermann *et al.* (2007) described a unique subpopulation, $CD133^+CXCR4^+$ cells, present at the invasive front in pancreatic tumour histology and not expressed in the bulk tumour. They described an increased migratory ability for $CD133^+CXCR4^+$ cells in cell-migration assays and in mice, as detected in circulating cells. They also showed that pancreatic cancer patients with increased numbers of $CXCR4^+$ tumour cells suffered from a heavier metastatic burden, suggesting a prominent role for CXCR4 and its binding partner SDF-1 in pancreatic cancer metastatic biology (Hermann *et al.*, 2007).

11.5.3 ALDH

Cells with high aldehyde dehydrogenase (ALDH) activity, as detected by the ALDEFLUOR assay, have been identified as a hallmark of CSC populations in a growing number of human tissues, including breast, blood, lung, bladder, ovaries, prostate, colon, thyroid, head and neck, brain, cervix and pancreas. Rasheed *et al.* (2010) demonstrated enhanced tumourigenicity for the $ALDH^+$ PCSC population and showed that expression of ALDH in pancreatic adenocarcinoma correlated with worse overall survival in patients undergoing primary resection. This was the first study suggesting a clinical prognostic significance for ALDH in pancreatic cancer. ALDH expression was examined using immunohistochemistry in resected primary tumours from 269 patients and the results were analysed using a Kaplan–Meier survival analysis. Rasheed *et al.* (2010) also examined the clonogenic potential of the $ALDH^+$ PCSC population, by plating and replating sorted $ALDH^+$ and $ALDH^-$ populations of CAPAN-1 cells, as detected by the ALDEFLUOR assay. Using this assay, the $ALDH^+$ cells formed a striking 11.6-fold more colonies when compared to the $ALDH^-$ cells during the fourth round of replating, suggesting an increased potential for self-renewal for $ALDH^+$ cells. Using cell-invasion assays, migration assays and quantitative RT-PCR, Rasheed *et al.* (2010) also showed that the $ALDH^+$ PCSC population had mesenchymal cell properties and increased invasive and migratory potential when compared to $ALDH^-$ cells, which suggests a role for ALDH in invasion and metastatic disease. Like the $CD133^+$ PCSC population, the $ALDH^+$ PCSC population is resistant to and expands upon treatment with gemcitabine.

Rovira *et al.* (2010) demonstrated that the pancreatic $ALDH^+$ population, which resides in the developing and adult mouse pancreas, has properties consistent with an adult pancreatic progenitor population. These cells rapidly expand in the setting of chronic injury, are enriched for transcripts highly expressed in embryonic multipotent pancreatic progenitor cells and are capable of forming self-renewing ‘pancreatospheres’. ALDH activity may therefore constitute a marker of normal pancreatic cells with progenitor capabilities, in addition to identifying PCSCs.

11.5.4 c-Met

Li *et al.* (2007) reported in an issue of *Gastroenterology* that high expression levels of c-Met ($c\text{-}Met^{high}$) mark a functionally unique and novel population of PCSCs capable of increased self-renewal capacity in *in vitro* sphere assays and possessing a greater capacity for tumour formation in nonobese diabetic severe combined-immunodeficient (NOD/SCID) mice when compared to cells with low-level or no c-Met expression ($c\text{-}Met^{low}$ or $c\text{-}Met^-$). To demonstrate enhanced tumourigenity, the researchers injected single-cell suspensions of $cMet^{high}$ and $cMet^{low}$ (or $cMet^-$) populations from primary human pancreatic cancer xenografts into NOD/SCID mice. When as few as 100 $c\text{-}Met^{high}$ cells were injected into NOD/SCID mice, 7 of 20 animals (35%) formed tumours that recapitulated the human primary pancreatic cancer histologically, while injection of 100 $cMet^{low}$ or $c\text{-}Met^-$ cells into NOD/SCID mice resulted in 0 of 20 animals forming tumours. The authors also described a 15–53% overlap between the previously reported $CD44^+$ PCSC marker and $c\text{-}Met^{high}$ expression. The $c\text{-}Met^{high}$ $CD44^+$ population demonstrated a potent increase in tumourigenity when compared to $c\text{-}Met^{high}$ alone, as evidenced by a 16/18 (89%) rate of tumour formation for $c\text{-}Met^{high}$ $CD44^+$ cells in animals injected with 500 cells, as compared to a 9/20 (45%) rate of tumour formation for $c\text{-}Met^{high}$ alone (Berman *et al.*, 2003).

Like ALDH, c-Met is a marker of normal pancreatic cells with progenitor capabilities that is expressed in the developing and adult pancreas. Suzuki *et al.* (2004) showed that the c-Met$^+$c-Kit$^-$CD45$^-$TER119$^-$ population from the embryonic pancreas was enriched for epithelial-like cells (ECs) in colonies that formed from plating this sorted population at clonal density. They demonstrated by RT-PCR that these ECs were capable of expressing an array of ductal, acinar and islet lineage-specific markers when cultured for several days, suggesting a multipotentcy potential of the c-Met$^+$c-Kit$^-$CD45$^-$TER119$^-$ population (Suzuki *et al.*, 2004).

11.5.5 26S proteosome

Adikrisna *et al.* (2012) reported the identification of a small population of pancreatic cells characterized by low activity of the 26S proteosome with PCSC features. To accomplish this, they expressed a recombinant ornithine decarboxylase with green fluorescent protein (GFP) fused to the degron sequence of the protein from a retroviral vector in human PDAC cell lines, a method that had previously been used to identify CSC populations in breast cancer and glioma (Vlashi *et al.*, 2009). The cell populations with low 26S proteosomal activity accumulated recombinant ornithine decarboxylase and displayed high levels of GFP fluorescence, denoted as Gdeg^{high}. Using the real-time assay for sphere formation published in *Nature Protocols* in 2008, the Gdeg^{high} population formed tumour spheres when plated at 100 cells, but not at the single-cell concentration, and the Gdeg^{low} population did not form spheres in this assay (Chojnacki and Weiss, 2008). Gdeg^{high} formed tumours when as few as 10 cells were injected subcutaneously into NOD/SCID mice, while Gdeg^{low} began to form tumours when 10 000 cells were injected into NOD/SCID mice, suggesting that the Gdeg^{high} population is about 100-fold more tumourigenic than the Gdeg^{low} population. The authors reported that the Gdeg^{high} population represents about 0.5% of pancreatic cells, which is consistent with the low percentage of other identified PCSC populations (Adikrisna *et al.*, 2012). In an effort to identify agents that can inhibit PCSC growth, the authors performed a screen in which quercetin was identified as an antitumour agent that acts both alone and synergistically with gemcitabine to inhibit tumour growth in NOD/SCID mice in both Gdeg^{low} and Gdeg^{high} populations. Quercetin is a flavonoid (a plant pigment derived from flavone) found in fruits, vegetables, leaves and grains. It has previously been shown to exert anti-inflammatory and chemopreventive properties (May *et al.*, 2008) and to display antitumour effects in liver cancer cells (Askari *et al.*, 2012; Chen *et al.*, 2013; Sharmila *et al.*, 2013). Furthermore, quercetin has been shown to be a proapoptotic agent *in vitro* in pancreatic cancer cell lines and in K562 erythroleukemia cells, and *in vivo* in orthotopically implanted pancreatic xenografts, and may offer a unique therapeutic strategy for the treatment of pancreatic cancer (Akan and Garip, 2013; Angst *et al.*, 2013).

11.5.6 Dclk1 and acetylated alpha-tubulin

Dclk1 is a CSC marker in the intestine (May *et al.*, 2008; Nakanishi *et al.*, 2013). Bailey *et al.* (2014) identified a functionally unique population of cells characterized by high-level expression of Dclk1 and acetylated alpha-tubulin that display stem cell properties consistent with pancreatic TICs. They reported a near-complete overlap between high-level expression of Dclk1 and of acetylated alpha-tubulin in neoplastic pancreatic cells, and that a subset of these double-positive cells displayed a gastrointestinal-tuft cell morphology in PanIN (Figure 11.3A). In *in vitro* sphere-forming assays, Dclk1^{HI} and AcTub^{HI} cells from both the KC^{iMist1}G mouse model of pancreatic cancer and human PDAC cell lines exhibited a superior capacity for tumour sphere formation when compared to their controls, which is indicative of an increased clonogenic potential for this population. AcTub^{HI} cells from human xenografts also displayed enhanced tumour-initiating frequency when injected into immunodeficient mice. Taken together, these studies describe the first population of cells with PCSC capabilities that are both functionally and morphologically distinct from other heterogeneous pancreatic tumour cell types (Bailey *et al.*, 2014).

11.6 Signalling pathways mediating PCSCs

11.6.1 Hedgehog pathway

The distinctive functional characteristics of CSC populations that permit some of them to thrive despite drug treatment, in combination with their contribution to tumour maintenance and growth, make CSCs and the pathways responsible for their regulation paramount therapeutic targets. Among the pathways that potentially mediate PCSC function, the Hedgehog (Hh) signalling pathway represents a promising candidate. Hh pathway activation begins with binding of one of three Hh ligands – Sonic hedgehog (SHH), Indian hedgehog (IHH) or Desert hedgehog (DHH) – to the Hh receptor Patched, followed by relief of restraint of Smoothened-mediated Hh target gene regulation through a cascade of downstream signalling interactions involving Gli family proteins (Bale and Yu, 2001). Activation of Hh signalling is an essential component of vertebrate organogenesis, and mutations in genes encoding Patched have been associated with many types of cancers, including basal cell carcinoma, medulloblastoma and rhabdomyosarcoma, and with rare heritable diseases, such as Gorlin's syndrome (Bale and Yu, 2001).

Berman *et al.* (2003) described Hh pathway activation in various endoderm-derived gut tumours, including oesophageal, stomach, biliary tract and pancreatic tumours. In the same issue of *Nature*, Thayer *et al.* (2003) described the formation of lesions that closely resemble human PanIN 1 and PanIN 2, containing similar genetic mutations as pancreatic adenocarcinoma, when SHH is misexpressed in the Pdx1 foregut endoderm giving rise to the developing pancreas. Furthermore, SHH remains widely expressed in various primary pancreatic cancer and metastatic pancreatic cancer cell lines (Thayer *et al.*, 2003). With respect to PCSCs, expression of SHH has been shown to be upregulated 46.3-fold in the $CD44^+CD24^+ESA^+$ PCSC population in pancreatic cancer xenografts when compared to normal pancreatic epithelial cells (Li *et al.*, 2007). Inhibition of Hh signalling with cyclopamine, a naturally occurring substance found in the corn lily *Veratrum californicum*, dramatically reduces invasion and metastasis in pancreatic cancer and reduces the abundance of $ALDH^+$ PCSCs by around threefold in the E3LZ10.7 metastatic pancreatic cancer cell line. Administration of gemcitabine, cyclopamine or a neutralizing antibody for Hh ligands to orthotopically transplated athymic mice results in abrogation of metastasis and decreased overall tumour size (Feldmann *et al.*, 2007; Bailey *et al.*, 2009). Furthermore, Sulforaphane, an active ingredient in cruciferous vegetables such as broccoli, has been shown to inhibit the self-renewal capabilities of PCSCs through inhibition of Hh activation (Rodova *et al.*, 2012). A year later, a study from the same lab demonstrated that sulforaphane reduced the tumour growth of orthotopically implanted primary human PCSCs isolated based on a $CD133^+CD44^+CD24^+ESA^+$ phenotype in NOD/SCID mice and decreased mRNA levels of Oct4 and Nanog in these mice, which suggests that sulforaphane can inhibit the transcription of 'stemness' genes in PCSCs (Li *et al.*, 2013). These inhibition studies demonstrate a seminal role for Hh signalling in the regulation of PCSC function.

11.6.2 c-Met signalling pathway

The functional significance of high-level expression of CD44, CD24 and ESA in PCSCs is at present not fully understood, but there is evidence to suggest that a heparan sulfate-modified CD44 (CD44-HS) variant (CD44v3) promotes c-Met signalling through interaction with hepatocyte growth factor/scatter factor (HGF/SF), the ligand for the receptor tyrosine kinase c-Met. CD44-HS has been shown to bind to HGF/SF through an interaction dependent on its heparan sulfate side chain. When compared with CD44s, CD44-HS significantly promotes phosphorylation of c-Met through HGF/SF, phosphorylation of multiple proteins in the c-Met downstream signalling pathway and activation of c-Met target genes, thereby acting as a functional co-receptor for the c-Met signalling pathway (van der Voort *et al.*, 1999). CD44v3 is expressed moderately in a panel of pancreatic cancer cell lines (Ringel *et al.*, 2001). Expression or activation of c-Met and HGF/SF through genetic mutations, autocrine signalling or gene amplification occurs in at least 32 human cancers, including carcinomas, sarcomas, haematopoetic cancers, skin cancers and others (Graveel *et al.*, 2013). Although the protooncogene c-Met and HGF/SF are highly expressed in human pancreatic cancer, the expression of CD44 variants in PCSCs is at present not well characterized (Ebert *et al.*, 1994). The contribution of CD44-HS to PCSC biology remains an important avenue of investigation, as it may promote c-Met activation in PCSCs and contribute to important functional properties associated with this pathway, including increased invasiveness, cell growth, angiogenesis and cell motility.

11.6.3 Epithelial–mesenchymal transition

Cancer cells undergoing epithelial–mesenchymal transition (EMT) gain migratory and invasive properties, and, as such, it is thought that these cells have a greater propensity to facilitate metastasis. Various populations of PCSCs have been shown to have increased migratory and/or invasive capabilities, including $CD133^+$, $CXCR4^+$ and $ALDH^+$ cells. As described by Rasheed *et al.* (2010), $ALDH^+$ cells have mesenchymal cell features, including downregulation of CDH1 and upregulation of SNAI2, which is a gene-expression profile consistent with EMT. Recent work by Rhim *et al.* (2012) described a delamination and EMT phenotype in preneoplastic and preinvasive PanIN. The acinar cells have a delamination phenotype in response to Kras* expression (Figure 11.3B), indicating a very critical window in which to discover therapies to treat circulating tumour cells prior to the onset of metastatic disease. The increased migratory and invasive capabilities of PCSCs make them plausible candidates for seeds for metastasis.

11.7 Therapeutics (future directions)

In accordance with the CSC hypothesis, differentiated bulk tumour cells have an increased proliferative capacity but only the CSC compartment has unlimited self-renewal capabilities, and thus eradicating CSCs will theoretically impede tumour growth and expansion. Increased proliferative abilities in the non-CSC population are not enough to advance long-term tumour growth. As such, PCSC populations are prime therapeutic targets. Understanding the key features of PCSC biology is a necessary step towards the development of novel treatment strategies aimed at eradicating this population from pancreatic tumours.

References

Adikrisna, R., Tanaka, S., Muramatsu, S., Aihara, A., Ban, D., Ochiai, T., *et al.*, 2012. Identification of pancreatic cancer stem cells and selective toxicity of chemotherapeutic agents. Gastroenterology 143, 234–245 e7.

Aigner, S., Sthoeger, Z.M., Fogel, M., Weber, E., Zarn, J., Ruppert, M., *et al.*, 1997. CD24, a mucin-type glycoprotein, is a ligand for P-selectin on human tumor cells. Blood 89, 3385–3395.

Akan, Z., Garip, A.I. 2013. Antioxidants may protect cancer cells from apoptosis signals and enhance cell viability. APJCP 14, 4611–4614.

Angst, E., Park, J.L., Moro, A., Lu, Q.Y., Lu, X., Li, G., *et al.*, 2013. The flavonoid quercetin inhibits pancreatic cancer growth in vitro and in vivo. Pancreas 42, 223–229.

Askari, G., Ghiasvand, R., Feizi, A., Ghanadian, S.M., Karimian, J. 2012. The effect of quercetin supplementation on selected markers of inflammation and oxidative stress. J. Res. Med. Sci. 17, 637–641.

Bailey, J.M., Mohr, A.M., Hollingsworth, M.A. 2009. Sonic hedgehog paracrine signaling regulates metastasis and lymphangiogenesis in pancreatic cancer. Oncogene 28, 3513–3525.

Bailey, J.M., Alsina, J., Rasheed, Z.A., Mcallister, F.M., Fu, Y.Y., Plentz, R., *et al.*, 2014. DCLK1 marks a morphologically distinct subpopulation of cells with stem cell properties in preinvasive pancreatic cancer. Gastroenterology 146, 245–256.

Bale, A.E., Yu, K.P. 2001. The hedgehog pathway and basal cell carcinomas. Hum. Mol. Genet. 10, 757–762.

Berman, D.M., Karhadkar, S.S., Maitra, A., Montes De Oca, R., Gerstenblith, M.R., Briggs, K., *et al.*, 2003. Widespread requirement for Hedgehog ligand stimulation in growth of digestive tract tumours. Nature 425, 846–851.

Chen, Z.J., Dai, Y.Q., Kong, S.S., Song, F.F., Li, L.P., Ye, J.F., *et al.*, 2013. Luteolin is a rare substrate of human catechol-O-methyltransferase favoring a para-methylation. Mol. Nut. Food Res. 57, 877–885.

Chojnacki, A., Weiss, S. 2008. Production of neurons, astrocytes and oligodendrocytes from mammalian CNS stem cells. Nat. Protoc. 3, 935–940.

Delgiorno, K.E., Hall, J.C., Takeuchi, K.K., Pan, F.C., Halbrook, C.J., Washington, M.K., *et al.*, 2014. Identification and manipulation of biliary metaplasia in pancreatic tumors. Gastroenterology 146, 233–244 e5.

Ebert, M., Yokoyama, M., Friess, H., Buchler, M.W., Korc, M. 1994. Coexpression of the c-met proto-oncogene and hepatocyte growth factor in human pancreatic cancer. Cancer Res. 54, 5775–5778.

Feldmann, G., Dhara, S., Fendrich, V., Bedja, D., Beaty, R., Mullendore, M., *et al.*, 2007. Blockade of hedgehog signaling inhibits pancreatic cancer invasion and metastases: a new paradigm for combination therapy in solid cancers. Cancer Res. 67, 2187–2196.

Gidekel Friedlander, S.Y., Chu, G.C., Snyder, E.L., Girnius, N., Dibelius, G., Crowley, D., *et al.*, 2009. Context-dependent transformation of adult pancreatic cells by oncogenic K-Ras. Cancer Cell 16, 379–389.

Graveel, C.R., Tolbert, D., Vande Woude, G.F. 2013. MET: a critical player in tumorigenesis and therapeutic target. Cold Spring Harb. Perspect. Biol. 5(7), pii: a009209.

Habbe, N., Shi, G., Meguid, R.A., Fendrich, V., Esni, F., Chen, H., *et al.*, 2008. Spontaneous induction of murine pancreatic intraepithelial neoplasia (mPanIN) by acinar cell targeting of oncogenic Kras in adult mice. Proc. Nat. Acad. Sci. USA 105, 18 913–18 918.

Hermann, P.C., Huber, S.L., Herrler, T., Aicher, A., Ellwart, J.W., Guba, M., *et al.*, 2007. Distinct populations of cancer stem cells determine tumor growth and metastatic activity in human pancreatic cancer. Cell Stem Cell 1, 313–323.

Herreros-Villanueva, M., Hijona, E., Cosme, A., Bujanda, L. 2012. Mouse models of pancreatic cancer. World J. Gastroenterol. 18, 1286–1294.

Hingorani, S.R., Petricoin, E.F., Maitra, A., Rajapakse, V., King, C., Jacobetz, M.A., *et al.*, 2003. Preinvasive and invasive ductal pancreatic cancer and its early detection in the mouse. Cancer Cell 4, 437-50.

Hingorani, S.R., Wang, L., Multani, A.S., Combs, C., Deramaudt, T.B., Hruban, R.H., *et al.*, 2005. Trp53R172H and KrasG12D cooperate to promote chromosomal instability and widely metastatic pancreatic ductal adenocarcinoma in mice. Cancer Cell 7, 469–483.

Jones, S., Zhang, X., Parsons, D.W., Lin, J.C., Leary, R.J., Angenendt, P., *et al.*, 2008. Core signaling pathways in human pancreatic cancers revealed by global genomic analyses. Science 321, 1801–1806.

Kopp, J.L., Von Figura, G., Mayes, E., Liu, F.F., Dubois, C.L., Morris, J.P.T., *et al.*, 2012. Identification of Sox9-dependent acinar-to-ductal reprogramming as the principal mechanism for initiation of pancreatic ductal adenocarcinoma. Cancer Cell 22, 737–750.

Li, C., Heidt, D.G., Dalerba, P., Burant, C.F., Zhang, L., Adsay, V., *et al.*, 2007. Identification of pancreatic cancer stem cells. Cancer Res. 67, 1030–1037.

Li, S.H., Fu, J., Watkins, D.N., Srivastava, R.K., Shankar, S. 2013. Sulforaphane regulates self-renewal of pancreatic cancer stem cells through the modulation of Sonic hedgehog-GLI pathway. Mol. Cell. Biochem. 373, 217–227.

Litvinov, S.V., Velders, M.P., Bakker, H.A., Fleuren, G.J., Warnaar, S.O. 1994. Ep-CAM: a human epithelial antigen is a homophilic cell-cell adhesion molecule. J. Cell Biol. 125, 437–446.

Maeda, S., Shinchi, H., Kurahara, H., Mataki, Y., Maemura, K., Sato, M., *et al.*, 2008. CD133 expression is correlated with lymph node metastasis and vascular endothelial growth factor-C expression in pancreatic cancer. Br. J. Cancer 98, 1389–1397.

May, R., Riehl, T.E., Hunt, C., Sureban, S.M., Anant, S., Houchen, C.W. 2008. Identification of a novel putative gastrointestinal stem cell and adenoma stem cell marker, doublecortin and CaM kinase-like-1, following radiation injury and in adenomatous polyposis coli/multiple intestinal neoplasia mice. Stem Cells 26, 630–637.

Nakanishi, Y., Seno, H., Fukuoka, A., Ueo, T., Yamaga, Y., Maruno, T., *et al.*, 2013. Dclk1 distinguishes between tumor and normal stem cells in the intestine. Nat. Gen. 45, 98–103.

Rasheed, Z.A., Yang, J., Wang, Q., Kowalski, J., Freed, I., Murter, C., *et al.*, 2010. Prognostic significance of tumorigenic cells with mesenchymal features in pancreatic adenocarcinoma. J. Nat. Cancer Inst. 102, 340–351.

Rhim, A.D., Mirek, E.T., Aiello, N.M., Maitra, A., Bailey, J.M., Mcallister, F., *et al.*, 2012. EMT and dissemination precede pancreatic tumor formation. Cell 148, 349–361.

Ringel, J., Jesnowski, R., Schmidt, C., Kohler, H.J., Rychly, J., Batra, S.K., Lohr, M. 2001. CD44 in normal human pancreas and pancreatic carcinoma cell lines. Teratog. Carcinog. Mutagen 21, 97–106.

Rodova, M., Fu, J., Watkins, D.N., Srivastava, R.K., Shankar, S. 2012. Sonic hedgehog signaling inhibition provides opportunities for targeted therapy by sulforaphane in regulating pancreatic cancer stem cell self-renewal. PloS One 7, e46083.

Rovira, M., Scott, S.G., Liss, A.S., Jensen, J., Thayer, S.P., Leach, S.D. 2010. Isolation and characterization of centroacinar/terminal ductal progenitor cells in adult mouse pancreas. Proc. Nat. Acad. Sci. USA 107, 75–80.

Sharmila, G., Bhat, F.A., Arunkumar, R., Elumalai, P., Raja Singh, P., Senthilkumar, K., Arunakaran, J. 2013. Chemopreventive effect of quercetin, a natural dietary flavonoid on prostate cancer in *in vivo* model. Clin. Nutr. 33(4), 718–726.

Shmelkov, S.V., St Clair, R., Lyden, D., Rafii, S. 2005. AC133/CD133/prominin-1. Int. J. Biochem. Cell Biol. 37, 715–719.

St John, T., Meyer, J., Idzerda, R., Gallatin, W.M. 1990. Expression of CD44 confers a new adhesive phenotype on transfected cells. Cell 60, 45–52.

Suzuki, A., Nakauchi, H., Taniguchi, H. 2004. Prospective isolation of multipotent pancreatic progenitors using flow-cytometric cell sorting. Diabetes 53, 2143–2152.

Thayer, S.P., Di Magliano, M.P., Heiser, P.W., Nielsen, C.M., Roberts, D.J., Lauwers, G.Y., *et al.*, 2003. Hedgehog is an early and late mediator of pancreatic cancer tumorigenesis. Nature 425, 851–856.

van der Voort, R., Taher, T. E., Wielenga, V. J., Spaargaren, M., Prevo, R., Smit, L., *et al.*, 1999. Heparan sulfate-modified CD44 promotes hepatocyte growth factor/scatter factor-induced signal transduction through the receptor tyrosine kinase c-Met. J. Biol. Chem. 274, 6499–6506.

Vlashi, E., Kim, K., Lagadec, C., Donna, L.D., Mcdonald, J.T., Eghbali, M., *et al.*, 2009. *In vivo* imaging, tracking, and targeting of cancer stem cells. J. Nat. Cancer Inst. 101, 350–359.

Wang, Y., Lanzoni, G., Carpino, G., Cui, C.B., Dominguez-Bendala, J., Wauthier, E., *et al.*, 2013. Biliary tree stem cells, precursors to pancreatic committed progenitors: evidence for possible life-long pancreatic organogenesis. Stem Cells 31, 1966–1979.

Webb, D. S., Shimizu, Y., Van Seventer, G.A., Shaw, S., Gerrard, T.L. 1990. LFA-3, CD44, and CD45: physiologic triggers of human monocyte TNF and IL-1 release. Science 249, 1295–1297.

Xu, Y., Stamenkovic, I., Yu, Q. 2010. CD44 attenuates activation of the hippo signaling pathway and is a prime therapeutic target for glioblastoma. Cancer Res. 70, 2455–2464.

12 NANOG in Cancer Development

Bigang Liu and Dean G. Tang

Department of Molecular Carcinogenesis, The University of Texas MD Anderson Cancer Center, Science Park, Smithville, TX, USA

12.1 Introduction

Embryonic stem cells (ESCs) are pluripotent cells derived from the inner cell mass of developing blastocysts and can be maintained indefinitely in culture (Evans and Kaufman, 1981; Martin, 1981). In addition to their unlimited capacity for self-renewal, ESCs are also pluripotent (they can give rise to all cell types in the body). These two biological properties give ESCs enormous promise in the treatment of devastating and currently incurable disorders such as spinal cord injury, type 1 diabetes and neurological diseases. The maintenance of pluripotency and self-renewal in ESCs is functionally regulated by a core network of pluripotency transcription factors, consisting of OCT4, SOX2, NANOG and SALL4, alongside many other factors (Nichols *et al.*, 1998; Boiani and Scholer, 2005). Many of these transcription factors interact with one another to form a self-reinforcing and complicated transcription network that can collaboratively control the differentiation of ESCs (Wu *et al.*, 2006; Lim *et al.*, 2008; Yang *et al.*, 2008). Interestingly, emerging evidence is implicating many pluripotency transcription factors, including OCT4, SOX2, LIN28, KLF4, MYC and NANOG, in the development and progression of cancer. For example, elevated expression of OCT4 is detected in most oesophageal squamous-cell carcinoma clinical samples, and its expression is positively correlated with metastasis (Zhou *et al.*, 2011). Similarly, overexpression of OCT4 in melanoma cells induces their dedifferentiation, confers cancer stem cell (CSC) properties and promotes tumourigenicity (Kumar *et al.*, 2012). SOX2 is highly expressed in breast

Principles of Stem Cell Biology and Cancer: Future Applications and Therapeutics, First Edition.
Edited by Tarik Regad, Thomas J. Sayers and Robert C. Rees.

tumours and is associated with early-stage tumour formation (Leis *et al.*, 2012). In particular, aberrant NANOG expression and/or function has been linked to a variety of cancers. In this review, we mainly focus on NANOG and discuss its functional roles in cancer development, with a particular emphasis on its regulation of CSC properties.

12.2 The function of NANOG in ESCs

NANOG, named from the mythological Celtic land of the ever young, Tir nan Og, is a homeodomain-harbouring transcription factor crucial for the maintenance of the pluripotency and self-renewal of undifferentiated ESCs and embryonal carcinoma cells (ECCs). Human embryonic *Nanog* gene (referred to as *Nanog1*) has been mapped to chromosome 12 at 12p13.31. Interestingly, 10 processed *Nanog* pseudogenes, named *NanogP2–P10* and a tandem duplicate designated *NanogP1*, have been identified in the human genome. Among all Nanog pseudogenes, only *NanogP8*, located on chromosome 15q14, has a complete open reading frame (ORF) (Booth and Holland, 2004) and possesses an Alu element in the 3′-UTR homologous to the one in *Nanog1* gene, demonstrating that *NanogP8* is a retrogene rather than a pseudogene. Syntenic mouse *Nanog* gene is located on chromosome 6. Similarly, two pseudogenes (*NanogPa*, *NanogPb*) and two retrogenes (*NanogPc*, *NanogPd*) have been discovered in the mouse genome (Booth and Holland, 2004; Robertson *et al.*, 2006).

Human NANOG protein has 305 amino acids, which can be divided into five functional subdomains: N-terminal domain (ND, amino acids 1–95), homeodomain (HD, amino acids 96–155), C1-terminal domain (CD1, amino acids 96–197), tryptophan-rich domain (WR, amino acids 198–243) and C2-terminal domain (CD2, amino acids 244–305) (Figure 12.1) (Oh *et al.*, 2005; Pan and Pei, 2005; Do *et al.*, 2007; Chang *et al.*, 2009). Rich in serine, threonine and proline, the ND is tightly regulated through phosphorylation and other post-translational modifications, and is involved in transcription interference. The HD, in the central region, contains a DNA-binding motif and is required for NANOG nuclear localization and transactivation. The WR region mediates NANOG dimerization, which is required for its regulation of

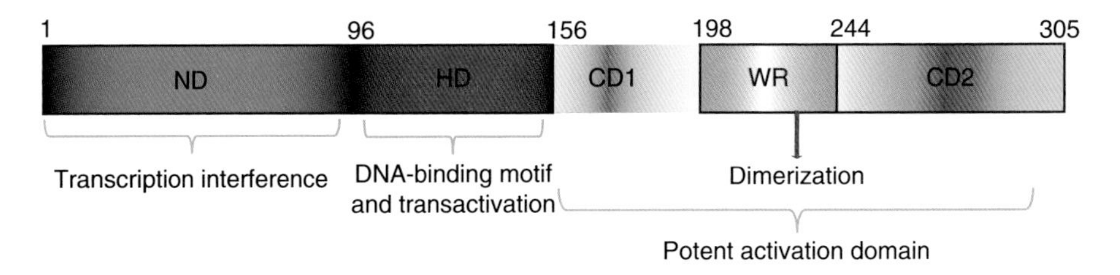

Figure 12.1 Schematic structure of the human Nanog protein. ND, N-terminus domain; HD, homeodomain; CD1 and CD2, C1 and C2-terminal domains; WR, tryptophan-rich domain.

pluripotency. Finally, the CD1 and CD2 regions possess potent transcription activation activity (Mullin *et al.*, 2008; Wang *et al.*, 2008).

Together with OCT4 and SOX2, NANOG plays a key role in maintaining the pluripotency and self-renewal of ESCs and ECCs (Chambers *et al.*, 2003; Mitsui *et al.*, 2003). NANOG expression is highly restricted to the inner cell mass of blastocysts, and Nanog mRNA is detected in pluripotent mouse and human stem cell lines but absent from the differentiated cells (Chambers *et al.*, 2003). The expression level of NANOG is crucial to determining the developmental fate of ESCs. It has been shown that overexpression of NANOG in mouse ESCs (mESCs) overcomes the requirement for leukaemia inhibitory factor (LIF) to maintain its pluripotency (Chambers *et al.*, 2003; Pei, 2009), whereas depletion of NANOG in mESCs results in loss of pluripotency, reduction of cell proliferation and differentiation into extraembryonic endoderm lineage (Darr *et al.*, 2006). Similarly, NANOG overexpression in human ESCs (hESCs) is able to maintain the pluripotency in the absence of feeder cells (Darr *et al.*, 2006) and, conversely, siRNA-mediated downregulation of NANOG in hESCs and human ECCs induces the differentiation of extraembryonic cell lineages and leads to loss of pluripotency (Hyslop *et al.*, 2005). These latter studies suggest that NANOG acts as a gatekeeper of pluripotency in human embryonic development. Moreover, as an important component of the 'Thomson cocktail' (OCT4, SOX2, NANOG and LIN28), NANOG is essential to the establishment of the ground state of pluripotency in promoting somatic cell reprogramming (Yu *et al.*, 2007; Theunissen *et al.*, 2011). Once the pluripotent ground state is established, NANOG is no longer required (Silva *et al.*, 2009).

In summary, NANOG functions as a core intrinsic element of the transcriptional network that sustains the self-renewal of ESCs.

12.3 NANOG and cancer development

NANOG is not expressed in most normal adult tissues. However, there is increasing evidence that the NANOG protein is re-expressed in various human cancers and this reactivation has been frequently reported to promote tumour progression. In fact, NANOG has been proposed, in many studies, as a potential biomarker for the evaluation of cancer prognosis.

12.3.1 Overexpression of NANOG in cancers and positive correlation of its expression levels with poor survival of cancer patients

It was first reported in 2006 that NANOG functions as an oncogenic factor to promote cell proliferation (Zhang *et al.*, 2006). Subsequently, many types of cancer have been found to highly express NANOG. For example, as the most common type of human cancer worldwide, with the highest mortality (Siegel *et al.*, 2014), lung cancer has been reported by several groups to express high levels of NANOG protein. Based on immunohistochemical analysis of 118 clinical samples from lung adenocarcinoma (LAC) patients, Chiou *et al.* (2010) observed that the expression levels of NANOG and OCT4 were higher in LAC samples than in paired normal lung tissues. Through multimarker analyses, they found that patients with lower NANOG expression survived longer than those with higher expression. Importantly, patients triple-positive for NANOG, OCT4 and SLUG had the worst survival outcome, whereas patients triple-negative for these markers exhibited the most favourable survival outcome. The authors further showed that lentivirus-mediated overexpression of NANOG and OCT4 dramatically enhanced the tumourigenesis and metastasis of LAC by inducing epithelial–mesenchymal transition (EMT). In contrast, small hairpin RNA (shRNA)-mediated knockdown of NANOG and OCT4 suppressed the tumourigenicity of LAC cells (Chiou *et al.*, 2010). Another group, by analysing a larger cohort of non-small-cell lung cancer (NSCLC) patients (n = 309), reported that non-NANOG-expressing patients showed much better prognosis and longer survival than high-NANOG-expressing patients (Li *et al.*, 2013). Moreover, they found that elevated nuclear NANOG protein was positively associated with nuclear β-catenin expression and poor prognosis (Li *et al.*, 2013). These and similar studies indicate that NANOG functions as an oncogene in lung cancer.

Since 2005, our group has systematically investigated the expression, origin and functions of NANOG in different tumours, particularly in prostate cancer, the most common type of cancer among men in the United States (Jeter *et al.*, 2009, 2011; Badeaux *et al.*, 2013; Liu *et al.*, 2014). We have demonstrated that various cancer cells produce Nanog mRNA, which is predominantly transcribed from a retrogene, *NanogP8*. We have detected more NANOG-positive cells in a small panel of human primary prostate cancer (HPCa) samples (n = 14) than in matched benign prostate tissues. Importantly, knockdown of Nanog mediated by small interfering RNA (siRNA) and shRNA reduces the tumourigenicity of prostate, breast and colon cancer cells (Jeter *et al.*, 2009). Conversely, using a tetracycline-inducible system, we have demonstrated that Nanog overexpression reprogrammes bulk cancer cells into stem-like cancer cells, enhances the growth of androgen-independent prostate cancer and increases the regeneration of castration-resistant prostate cancer (CRPC) in xenograft models (Jeter *et al.*, 2011). The latter observations imply that elevated NANOG expression might help drive prostate cancer progression towards a hormone-refractory condition. Interestingly, a recent study reported that activation of the 5′ flanking region of *NanogP8* accelerated the tumour formation of prostate cancer cells in xenograft models (Zhang *et al.*, 2014).

As in lung and prostate cancers, several studies also indicate that NANOG protein functions as an oncogenic factor in breast cancer. Nagata *et al.* (2014) observed that patients with a high level of NANOG protein had much lower disease-free and overall survival rates than patients with low levels (n = 100).

They also found that overexpression of NANOG is associated with and responsible for resistance to both hormonal therapy and chemotherapy, thus presaging a worse prognosis (Nagata *et al.*, 2014). Two other groups have also reported that NANOG, together with other ESC transcription factors, is highly expressed in breast cancer samples (Ezeh *et al.*, 2005; Bourguignon *et al.*, 2008). Surprisingly, Lu *et al.* (2013) recently published their results, with slightly different conclusions. By utilizing an inducible Nanog transgenic mouse model, they found that overexpressing Nanog alone in the mammary tissue was not sufficient to induce mammary tumour. However, Nanog could dramatically promote tumourigenesis and accelerate the metastasis of Wnt-1 transgenic mice (Lu *et al.*, 2013). Their results are somewhat similar to our recent transgenic mouse studies, in which we constitutively expressed human prostate cancer-derived NanogP8 cDNA in the cytokeratin 14 cellular compartment (K14-NanogP8). The K14-NanogP8 animals did not develop spontaneous tumours in the skin or other organs (Badeaux *et al.*, 2013). In fact, high levels of NanogP8 expression in one transgenic line disrupted normal tissue and organ development, leading to runt animals and, surprisingly, dampened tumour development in a two-stage carcinogenesis model, as a result of depleting keratinocyte stem cells (Badeaux *et al.*, 2013). Overall, these transgenic studies (Badeaux *et al.*, 2013; Lu *et al.*, 2013) suggest that Nanog plays important functions in regulating stem cell properties and most likely modulates the oncogenic processes.

For instance, NANOG modulates the progression of colorectal cancer (CRC). With immunohistochemical analysis of a large cohort of CRC samples (n = 175), Meng *et al.* (2010) found that most primary CRC samples overexpressed NANOG protein compared with normal tissues, and the expression levels of NANOG were positively associated with poor prognosis, lymph node metastasis and Duke classification of CRC. They further showed that forced expression of NANOG enhanced the proliferation, invasion and migration of CRC cells (Meng *et al.*, 2010). Another group also reported that NANOG was differentially expressed in five out of seven colon cancer cell lines and that shRNA-mediated knockdown of NANOG inhibited the proliferation of cultured colon cancer cells and tumour formation in mouse xenograft models (Ishiguro *et al.*, 2012). Recent results from the Jessup group support and extend these findings, showing that upregulation of NANOG contributes to the progression of CRC (Zhang *et al.*, 2013). This group has observed that NANOG protein is overexpressed in 80% of clinical CRC samples and that downregulation of NANOG by shRNAs significantly suppresses the tumourigenicity and experimental metastasis of CRC cells in nonobese diabetic severe combined-immunodeficient (NOD/SCID) mice. In contrast, overexpression of NANOG accelerates the tumour formation of CRC cells (Zhang *et al.*, 2013). These studies together suggest that NANOG protein expression could be considered as an independent prognostic factor for CRC patients.

Similarly, NANOG has been linked to the development of gastric cancer. After comparing 105 clinical samples of gastric adenocarcinoma with

70 samples of corresponding nondysplastic tissue by immunostaining, Lin *et al.* (2012) reported elevated NANOG protein expression in tumour samples. Subsequent survival analysis indicated that NANOG overexpression positively correlated with advanced clinical stages and reduced 5-year survival of the patients (Lin *et al.*, 2012). Another study, with a small panel of clinical gastric cancer samples (n = 40), suggests that NANOG may play a role in the early developmental stages of gastric cancer but is not related to the prognosis (Zhang *et al.*, 2010).

NANOG re-expression has also been reported in human hepatocellular carcinoma (HCC) and many other cancers. Several groups have detected high NANOG expression in HCC specimens and demonstrated that forced expression of NANOG promotes tumourigenesis and metastasis of HCC cells (Shan *et al.*, 2012; Sun *et al.*, 2013; Wang *et al.*, 2013c). Aberrant upregulation of NANOG has been detected in germ cell tumours (Hart *et al.*, 2005) and other tumours, including brain, cervix, head and neck, kidney, oral, ovarian and pancreatic cancers, and the expression levels of NANOG have been positively linked to poor clinical outcome in patients with these malignancies (Bussolati *et al.*, 2008; Chiou *et al.*, 2008; Ye *et al.*, 2008; Po *et al.*, 2010; Wen *et al.*, 2010; Zbinden *et al.*, 2010; Yu *et al.*, 2011; Lee *et al.*, 2012).

12.3.2 Enhancement of CSC characteristics by NANOG and the role of NANOG as a CSC marker

CSCs are a distinct subpopulation of cancer cells that possess the capacity of unlimited self-renewal and the ability to generate many cell types in the tumour mass. Their defined characteristics include their ability to induce tumour formation *in vivo* and to form spheres *in vitro*, resistance to chemotherapy and radiotherapy and the capacity to differentiate into many cell types present in the parental tumour (Wicha *et al.*, 2006; Rich, 2007; Visvader and Lindeman, 2008).

CSCs were first identified in acute myeloid leukaemia (AML) (Lapidot *et al.*, 1994) and subsequently in various solid tumours, including breast, brain, colon, prostate, lung, ovarian, head and neck (Al-Hajj *et al.*, 2003; Singh *et al.*, 2003; Patrawala *et al.*, 2005; O'Brien *et al.*, 2007; Eramo *et al.*, 2008; Schatton *et al.*, 2008; Zhang *et al.*, 2008; Qin *et al.*, 2012; Han *et al.*, 2014). Our group has been employing several different strategies to enrich putative CSCs. For example, we routinely enrich CSCs by utilizing flow-cytometry sorting based on certain cell-surface markers, among which are CD133, CD44 and ABCG2, commonly used CSC markers in other solid cancers (Al-Hajj *et al.*, 2003; Salven *et al.*, 2003; Bertolini *et al.*, 2009; Jiang *et al.*, 2012; Wakamatsu *et al.*, 2012). We have also enriched potential CSCs based on side-population sorting (Patrawala *et al.*, 2005; Ho *et al.*, 2007) and on the ability to form spheres that can be serially passaged (Galli *et al.*, 2004; Fang *et al.*, 2005). Putative CSCs purified using these approaches need to be validated by serial transplantation, a functional assay that assesses the two essential properties of CSCs: the ability to self-renew and the capacity to recapitulate the original tumour. The rationale of the assay is that an authentic CSC-rich population, when serially transplanted into immunocompromised mice, should exhibit much higher capacity to form tumours that histopathologically recapitulate the original tumour than the control population (Clarke *et al.*, 2006).

CSCs are relatively quiescent and display low proliferative potential. Thus, conventional chemotherapeutic agents that kill rapidly dividing cells will destroy hardly CSCs (Guan *et al.*, 2003) and the surviving CSCs are able to regenerate the tumour at a later stage. As such, CSCs are generally proposed to be the 'seeds' for relapse after treatment and for tumour metastasis (Brabletz *et al.*, 2005; Wang *et al.*, 2013b). Although more and more CSC populations are being identified and characterized, much remains to be learnt about the cellular and molecular mechanisms they employ to modulate their distinct properties. Therefore, a better understanding of the regulatory mechanisms of CSCs is essential for developing new approaches to targeting these cells. Because CSCs and ESCs share many common features, such as the properties of self-renewal and unlimited proliferation, it is natural to speculate that the core pluripotency transcription factors such as NANOG may also be involved in regulation of CSCs.

Recent data suggest that NANOG is actively involved in tumour progression through the regulation of such CSC properties as self-renewal, tumourigenicity, metastasis and drug resistance. Our lab found that the $CD44^{high}$ CSCs from two prostate carcinoma cell lines, Du145 and PC3, express higher levels of NANOG than do $CD44^{low}$ cancer cells. Furthermore, the $CD44^{+}CD133^{+}$ double-positive and $CD133^{+}$ populations purified from human primary prostate carcinoma samples exhibit five times more NanogP8 mRNA than the non-CSCs. Additionally, the level of NanogP8 mRNA in the side-population breast carcinoma MCF7 CSCs is fourfold higher than in non-CSCs (Jeter *et al.*, 2009). Using Nanog promoter as a reporter system, we have shown that NanogP8-GFP^{+} prostate cancer cells exhibit CSC characteristics manifested by increased clonal growth and tumour regenerative capacities (Jeter *et al.*, 2011). Ablation of endogenous Nanog significantly reduces the tumourigenicity of three types of cancer cells (Jeter *et al.*, 2009), whereas overexpression of Nanog in LNCaP cells through an inducible system enhances tumour growth in a castration-resistant environment (Jeter *et al.*, 2011). Overexpression of NANOG also increases the expression of many CSC-associated genes, such as CD133, ABCG2, ALDH1A1 and CD44. Taken together, our results suggest that NANOG-expressing (prostate) cancer cells posses CSC properties. Another group has shown that the $CD117^{+}ABCG2^{+}$ cell population from the prostate cancer cell line 22RV1 displays CSC characteristics by exhibiting multidrug resistance and high tumour-regenerating activity, and that this CSC population expresses high-level NANOG, as well as the other stem cell molecules OCT4, SOX2, Nestin and CD133 (Liu *et al.*, 2010).

NANOG has also been reported to positively modulate CSC properties in lung cancer cells. Work from Chiou *et al.* (2010) indicates that not only are both NANOG and OCT4 highly expressed in $CD133^+$ CSC population, but that enforced expression of NANOG and OCT4 in A549 cells, an LAC cell line, dramatically increases the stem cell-like properties, as evidenced by an enlarged $CD133^+$ cell pool and enhanced sphere formation, drug resistance, tumour initiation, EMT and metastasis. In another study, utilizing CD44 and CD90 as cell-surface markers to enrich CSCs from primary lung carcinoma cells ($n = 15$), the authors found that the $CD44^{high}$ $CD90^+$ CSCs expressed high levels of NANOG protein (Wang *et al.*, 2013a).

Shan *et al.* (2012) described in detail how NANOG might affect the properties of HCC CSCs. Using a Nanog promoter-tracking system, the group isolated a small population of NANOG-expressing cells and showed that they exhibited all the characteristics of CSCs. The NANOG-expressing HCC cells (i) possessed high clonogenicity and tumourigenicity, (ii) displayed a high capacity for tumour invasion and metastasis and were resistant to chemotherapeutic agents and (iii) were capable of differentiating into mature cancer cells. Ablation of Nanog by shRNA in NANOG-expressing HCC cells impaired their self-renewal, and, conversely, forced expression of NANOG in non-NANOG-expressing liver cancer cells confered self-renewal ability. Importantly, the authors provided evidence that NANOG modulates the self-renewal of liver CSCs, potentially by upregulating insulin growth factor receptor 1 (IGF-1R)-signalling pathway (Shan *et al.*, 2012). Consistent with this study, another group showed that overexpression of NANOG in Huh7 HCC cells promoted sphere formation, clonogenicity, invasion and drug resistance *in vitro*, as well as tumour growth and metastasis *in vivo*, and that NANOG overexpression greatly increased the expression levels of CD133 and MMP2 (Sun *et al.*, 2013). The results from these two studies indicate that NANOG may represent a biomarker for HCC CSCs.

Two other studies have causally connected NANOG to brain tumour progression and to regulation of brain tumour stem cells. Zbinden *et al.* (2010) showed that most glioblastoma (GBM) samples expressed NANOG protein and that knockdown of NANOG reduced the number and size of gliomaspheres and inhibited cell proliferation. The expression of NANOG, mainly NANOGP8, was much higher in $CD133^+$ CSCs than in $CD133^-$ cells, and knockdown of NANOG only reduced the proliferation of $CD133^+$ and not of $CD133^-$ cells (Zbinden *et al.*, 2010). These results indicate that NANOG controls the functions of GBM stem cells. By employing a novel red/green assay and orthotopic xenografting, the authors further demonstrated a requirement for NANOG in GBM tumourigenesis (Zbinden *et al.*, 2010). A similar study showed that NANOG was enriched in stem cells from medulloblastoma and modulated the self-renewal of neural stem cells via the Hedgehog (Hh) signalling pathway (Po *et al.*, 2010).

Several reports have also shown that NANOG is upregulated in the CSC populations of gastric, colorectal and head and neck squamous cell carcinomas (Golestaneh *et al.*, 2012; Liu *et al.*, 2013; Zhang *et al.*, 2013). Whether

NANOG activities are indispensible for the modulation of CSC properties in these cancers requires further investigation.

12.3.3 Increase of cell proliferation and metastasis by NANOG and crosstalks with oncogenic pathways

In addition to the positive regulation of CSC properties, NANOG has been reported to modulate other signalling pathways. For instance, it may positively intersect with cell-cycle signalling, as downregulation of NANOG in breast and lung cancer cells suppresses cell proliferation and cell-cycle progression by reducing the expression of cell cycle-related genes, including cyclins D1, D2 and D3, as well as CDK1 and 6 and c-Myc (Choi *et al.*, 2012; Han *et al.*, 2012). NANOG has also been repeatedly reported to promote EMT and cell invasion, and elevated NANOG expression is strongly correlated with metastasis of several malignancies, including breast, colorectal, gastric, liver, lung and ovarian cancer (Chiou *et al.*, 2010; Meng *et al.*, 2010; Lee *et al.*, 2012; Lin *et al.*, 2012; Shan *et al.*, 2012; Nagata *et al.*, 2014). Intriguingly, NANOG has recently been implicated in the signalling pathways of immune detection and clearance, as elevated NANOG appears to aid CSCs in evading immune surveillance and clearance through the Nanog/Tcl1/Akt signalling axis (Noh *et al.*, 2012). Lastly, several studies indicate that NANOG crosstalks with oncogenic pathways, such as orphan nuclear receptor oestrogen-related receptor β (ESRRB), receptor tyrosine kinases, Akt-dependent pathways and focal adhesion kinase (FAK) signalling, to cooperatively regulate tumour progression (Storm *et al.*, 2007; van den Berg *et al.*, 2008; Ho *et al.*, 2012; Hyder *et al.*, 2012; Shan *et al.*, 2012).

12.4 NANOG1 versus NANOGP8 in cancer development

As mentioned in Section 12.2, the human genome contains 10 Nanog pseudogenes. One of these, *NanogP1*, possesses a similar intron–exon structure to and shares a highly conserved promoter region with *Nanog1*. As a result, *NanogP1* is considered a duplicate pseudogene and encodes a truncated Nanog protein, with 232 amino acids (Booth and Holland, 2004). Indeed, two studies demonstrate that NanogP1mRNA is expressed in human pluripotent cells and acute leukaemia cells (Eberle *et al.*, 2010; Palla *et al.*, 2013). Other pseudogenes, namely *NanogP2–P10*, are intronless paralogues of *Nanog1*, and their expressions are controlled by promoters unrelated to the promoter of *Naong1*. *NanogP2*, *P3*, *P4*, *P5*, *P6*, *P7*, *P9* and *P10* are unable to encode a complete NANOG protein and are thus nonfunctional, because their ORFs contain in-frame stop codons, deletions and frame-shifts that introduce premature stop codons. Interestingly, and importantly, another paralogue,

NanogP8, is an intronless retrogene. It contains an intact ORF that exhibits 99.5% similarity with that of *Nanog1* and possesses the capacity to encode a full protein with 305 amino acids (Booth and Holland, 2004). In support, ectopic expression of NANOGP8 in NIH3T3 cells (Zhang *et al.*, 2006), prostate and breast cancer cells (Jeter *et al.*, 2011) and mouse keratinocytes in transgenic animals (Badeaux *et al.*, 2013) confirms that *NanogP8* can generate the functional NANOG protein.

Based on Genbank sequences, the alignment of the reference sequences between *Nanog1* and *NanogP8* reveals that there are only six nucleotides with two potential amino acid differences between the two (Jeter *et al.*, 2009; Ambady *et al.*, 2010), and these two genes seem to encode nearly identical proteins, indistinguishable by Western blot and immunohistochemistry based on current commercial anti-NANOG1 antibodies. Therefore, we are presently unable to discern, at the protein level, whether *Nanog1* or *NanogP8* is responsible for the elevated expression of NANOG protein in various cancers. As such, some reports hypothesize that NANOG1 and NANOGP8 might possess similar biological functions. On the other hand, it is quite possible that NANOG1 and NANOGP8 play divergent functions in tumourigenesis, based on the following observations and reasoning:

1. *Nanog1* and *NanogP8* are located in separate genome loci, with different upstream regulatory (promoter) elements. It is thus reasonable to speculate that the mechanisms by which their expressions in cancer cells are regulated will be distinctly different. Two studies provide some support for this speculation. The 5′ flanking region of *NanogP8*, but not *Nanog1*, is activated in prostate cancer cells to promote the characteristics of CSCs, including clonogenicity and sphere formation *in vitro*, as well as tumour growth *in vivo* (Zhang *et al.*, 2014). Hypoxia significantly upregulates the expression of NanogP8 mRNA, but not Nanog1 mRNA, which in turn enhances the stem-like properties of prostate cancer cells (Ma *et al.*, 2011).
2. The two predicted amino acid differences between NANOG1 and NANOGP8 (at positions 16 and 253) occur in the functional domains of NANOG. The change at residue 16 (Ala to Glu) is located in the transcriptional repression domain, whereas the change at residue 253 (Gln to His) takes place in a potent transcriptional activation domain (Pan and Pei, 2005; Chang *et al.*, 2009; Do *et al.*, 2009; Jeter *et al.*, 2009; Ambady *et al.*, 2010). Consequently, these two amino acid changes might confer novel functions to NANOGP8. This possibility is partially buttressed by our recent finding that overexpression of NANOGP8 elicits stronger and more divergent biological effects on cancer cells than does overexpression of NANOG1 (Jeter *et al.*, 2011). For example, overexpressing NANOGP8, but not NANOG1, in LNCaP cells increases androgen-independent sphere and foci formation, cell migration and castration-resistant tumour growth. Additionally, only enforced expression of NANOGP8 can promote tumour regeneration of Du145 cells.

Finally, NANOGP8 and NANOG1 seem to regulate the expression of different genes when overexpressed (Jeter *et al.*, 2011).

3. Nanog1 is highly expressed in pluripotent cells, whereas Nanog1 mRNA is undetectable in most types of cancer cells examined (Chambers *et al.*, 2003; Palla *et al.*, 2013). In contrast, NanogP8 mRNA is highly expressed in a variety of cancers. By sequencing the RT-PCR amplicons obtained with various primer combinations, we have shown that NanogP8 mRNA is preferentially expressed in multiple cancer cell lines, including breast cancer (MCF7), colon cancer (Colo320) and prostate cancer (LNCaP, Du145 and PC3), as well as in xenograft tumours and primary prostate cancer samples (Jeter *et al.*, 2009, 2011). Two independent research groups recently confirmed that the NanogP8 mRNA is the major form of Nanog transcripts in Du145 and PC3 cells, as well as in normal prostate tissue, prostatic intraepithelial neoplasia and prostate cancer tissues (Ma *et al.*, 2011; Zhang *et al.*, 2014). Similarly, Serrano's group employed a multi-Nanog reverse transcriptase–PCR strategy to investigate NANOG expression in 2 human pluripotent cells and 17 cancer cell lines, including breast, cervical, colon, liver, LAC, melanoma, osteosarcoma, prostate carcinoma and urinary bladder. Their results indicate that most cancer cell lines tested express detectable levels of NANOG protein that was derived from *NanogP8* but not *Nanog1* locus. As expected, the two human pluripotent cells expressed high level of Nanog1 and NanogP1 protein. In addition, the authors provided evidence that NANOGP8 promotes reprogramming as effectively as does NANOG1 (Palla *et al.*, 2013). Notably, by utilizing the change of nucleotides 144(G to A) between Nanog1 cDNA and NanogP8 cDNA (Jeter *et al.*, 2009; Ambady *et al.*, 2010), which specifically creates a new AlwN1 restriction site in NanogP8 cDNA (CGGAGACTG to CAGAGACTG), two research groups combined RT-PCR and AlwN1 digestion to develop a simple method for distinguishing between the *Nanog1* and *NanogP8* transcripts (Ishiguro *et al.*, 2012; Zhang *et al.*, 2013). Their results reveal that clinical CRC samples, as well as several CRC cell lines, express both Nanog1 mRNA and NanogP8 mRNA. They further detected Nanog1/NanogP8 mRNA in 80% CRC metastases (8/10), with 75% of the metastasis samples mainly expressing NanogP8 mRNA. Moreover, downregulation of Nanog1 by shRNA suppresses sphereogenicity and decreases the size of the sidepopulation (SP) in CRC, whereas forced expression of NanogP8 rescues these phenotypes, suggesting that NANOGP8 can replace NANOG1 in the regulation of stemness in CRC.

Collectively, the preceding discussions emphasize that NANOGP8, rather than NANOG1, plays the predominant role in the regulation of tumour development and progression. On the other hand, in order to truly and convincingly decipher the relative functional contributions of NANOG1 versus NANOGP8 in cancer, it is necessary to develop unique antibodies that differentially recognize NANOGP8 versus NANOG1 protein.

12.5 Evidence for the unique biochemical properties of NANOG protein

Given that NANOG has been implicated in a number of cancers, as discussed already, detailed molecular characterization of the NANOG protein in cancer cells will be very important for cancer diagnosis and for developing novel therapeutics targeting this molecule. The cDNA of human Nanog1/NanogP8 contains 915 bp and encodes a protein with 305 amino acids (Booth and Holland, 2004), and the predicated molecular weight of Nanog protein is 35 kD. Surprisingly, much of the published literature, by employing Western blot analysis with a variety of anti-NANOG1 antibodies, reports the PUTATIVE Nanog protein with different molecular weights of 29–80 kD in ESCs, ECCs and somatic cancer cells (Table 12.1) (Mitsui *et al.*, 2003; Hatano *et al.*, 2005; Hyslop *et al.*, 2005; Kim *et al.*, 2005; Hamazaki *et al.*, 2006; Pereira *et al.*, 2006; Wu *et al.*, 2006; Zhang *et al.*, 2006, 2009, 2013; Storm *et al.*, 2007; Bourguignon *et al.*, 2008; Kochupurakkal *et al.*, 2008; Liang *et al.*, 2008; Siu *et al.*, 2008; Torres and Watt, 2008; Chan *et al.*, 2009; Eberle *et al.*, 2010; Kuijk *et al.*, 2010; Kalbermatten *et al.*, 2011). These putative NANOG proteins are generally not further confirmed by more robust techniques such as mass spectrometry, so caution should be taken in interpreting the results over whether the reported species represent the authentic Nanog protein in these cells types.

Very recently, we systematically analysed the molecular behaviours of NANOG1/NANOGP8 proteins, especially endogenous NANOG protein (Liu *et al.*, 2014). By combining WB using multiple anti-NANOG1 antibodies, immunoprecipitation, mass spectrometry and studies using recombinant proteins, we have observed multiple NANOG1 protein species in NTERA-2 ECCs with molecular weights ranging from ~25 to ~100 kD, with the 42 kD band as the major form detected on WB (Table 12.2). Remarkably, the recombinant NANOGP8 proteins, derived from the cDNAs of primary prostate cancer specimens, also migrate at ~28 to ~180 kD and exhibit differential reactivity to different anti-NANOG1 antibodies. Interestingly, recombinant NANOGP8 proteins can spontaneously form high-molecular-weight protein species, with the 42 kD species as the most stable and soluble form of the NanogP8 protein. Unexpectedly, we could hardly detect endogenous NanogP8 protein in most long-term cultured and xenograft-derived somatic cancer cells by immunoprecipitation, implying that the conformation of endogenous NANOGP8 protein in somatic cancer cells might be different from that of the NANOG1 protein in ECCs, or that the endogenous NANOGP8 protein is modified by unknown mechanisms such that it is rapidly turned over. Altogether, our results reveal that the NANOG protein may possess unique and intrinsic biochemical properties that allow it to adopt a suitable conformation in response to various environment stresses, which render it able to migrate on SDS-PAGE with versatile molecular masses (Liu *et al.*, 2014).

Table 12.1 Examples of Nanog antibodies and their recognized protein bands.

Protein band(s)	Tissue or cells	Antibody	Remarks	Reference(s)
29 kD	NTERA-2 cell lysates	**Abnova mouse mAb** (Clone 2C11)	Raised against C-terminal Nanog	Lin *et al.* (2012)
35 kD				
48 kD				
~34 kD	Cancer cell lysates;	**SC goat pAb-N17** (SC-30331; N17)	Affinity-purified pAb against a hNanog N-terminus peptide	Galli *et al.* (2004)
~43 kD	Using rhNanog			A (see below)
≥34 kD	HepG2 & OS732 cells;	**R&D goat pAb** (AF1997)	Affinity-purified pAb against rhNanog aa 153–305 peptide	Boiani, M., Scholer, H.R. (2005)
35 kD	Human ESCs & EC cells;			
36~37 kD	Trophoblastic samples;			Clarke *et al.* (2006)
40~42 kD	Using rhNanog;			Leis *et al.* (2012)
~42 kD	Human colorectal cancer			Golestaneh *et al.* (2012); Lapidot *et al.* (1994)
				B (see below)
~37 kD	Mouse ESCs	**Chemicon Rb pAb** (AB5731)	Affinity purified pAb raised against mNanog N-terminus peptide	C (see below)
17 kD				Liu *et al.* (2013)
72 kD	Mouse ESCs			
A cluster of bands				
>35 kD	Several bands in mouse ESCs	**Home-made Rb pAb**	Raised against mNanog aa 1–95	Meng *et al.* (2010)
~36 kD	293 T cells transfected with hNanog	**Home-made Rb pAb**	Raised against hNanog (aa 168–183)	Ma *et al.* (2011)
~50 kD				
37 kD	Two pull down Nanog protein bands	**Home-made Rb pAb**	FLAG-tagged Nanog stably expressed in mESCs	Martin *et al.* (1981)
~45 kD				
37 kD	Mouse ESCs	**Bethyl Labs Rb pAb** (BL1662)	Affinity purified pAb raised against mNanog	Liu *et al.* (2010)
49 kD				
~38 kD	Human ESCs	**Abcam Rb pAb** (21603)	Using full-length mNanog as immunogen	Liu *et al.* (2013)
~39 kD	Mouse ESCs			
A cluster of bands (unknown MW)				
40 kD	Human ES cells;	**SC Rb pAb-H155** (SC-33759; H155)	Using hNanog aa 151–305 peptide as immunogen	D (see below) 47
~40 kD	Adipose-derived SC cells;			Lee *et al.* (2012)
~40 kD	Lymphoma cell lines			Ho *et al.* (2003)
~40 kD	Mouse ESCs (2~3 bands)	**Home-made Rb pAb**	Raised against mNanog	Ambady *et al.* (2010)

(*continued overleaf*)

Table 12.1 (*continued*)

Protein band(s)	Tissue or cells	Antibody	Remarks	References
~45 kD ~38 kD A cluster of bands (unknown MW)	Human ESCs;	**Abcam Rb pAb** (21624)	Affinity-purified pAb using hNanog aa 29–49 as immunogen	E (see below) Li *et al.* (2013)
~42 kD	Human ESCs	**Kamiya Rb pAb** (PC-102)	Affinity-purified pAb using hNanog aa 29–49 as immunogen	F (see below)
~42 kD ~32 kD ~34 kD	NTERA-2 lysates	**Cell Signaling pAb** (3580)	Affinity-purified Rb pAb against N-terminus of hNanog	G (see below)
~43 kD 2 bands of unspecified M.W	NTERA-2 lysates	**eBioscience mAb** (Cat# 14-5768)	Affinity-purified mAb using full-length hNanog as immunogen	H (see below) Fang *et al.* (2005)
~44 kD	Mouse ESCs and EG cells (>3 bands with the largest at ~44 kD)	**Home-made Rb pAb**	Raised against mNanog	Lu *et al.* (2013)
~46 kD ~42 kD ~38 kD	NTERA-2 cell lysates	**BioLegend mAb** (674001)	Full length human Nanog recombinant protein as immunogen	I (see below)
40 kD 80 kD	Mouse ESCs; Dog testis and pig testis	**Chemicon Rb pAb**	Raised against mNanog	Liu *et al.* (2014)

Abbreviations: EC, embryonal carcinoma; EG, embryonic germ cells; ESCs, embryonic stem cells; hNanog, human Nanog; mAb, monoclonal antibody; mNanog, mouse Nanog protein; pAb, polyclonal antibody; Rb, rabbit; rhNanog, recombinant human Nanog protein; SC, Santa Cruz.
A, http://datasheets.scbt.com/sc-30331.pdf
B, http://www.rndsystems.com/Products/AF1997
C, http://www.emdmillipore.com/US/en/product/Anti-Nanog-Antibody%2C-NT,MM_NF-AB5731#documentation
D, http://datasheets.scbt.com/sc-33759.pdf
E, http://www.abcam.com/nanog-antibody-chip-grade-ab21624.html#description_images_1
F, http://www.kamiyabiomedical.com/pdf/PC-102.pdf
G, http://www.cellsignal.com/products/primary-antibodies/3580?hit=productId&Ntt=3580
H, http://www.ebioscience.com/media/pdf/tds/14/14-5768.pdf
I, http://www.biolegend.com/purified-anti-nanog-antibody-10616.html

Table 12.2 Nanog proteins in NTERA-2 human ECCs. Presenting the results of various experiments using the eight anti-Nanog Abs. The BioLegend Ab is the only one that does not recognize the 42 kD band as the major protein band and does not recognize the 35 kD band at all. Presented molecular masses are all estimated based on their migrations on SDS-PAGE.

Antibody	Western blot	IP/MALDI	Tandem IP/LTQ	IP
Biolegend Rb pAb (632002)	48 kD > 35 kD > 55 kD			
Cell-signalling Rb mAb (5232)	42 kD ≫ 35 kD; 28, 32, 70, **100** kD			42 kD
Cell-signalling Rb pAb (3580)	42 kD ≫ 35 kD ≫ 65 kD			
eBioscience mAb (Cat# 14-5768)	42 kD ≫ 35 kD ≫ 65 kD			
Kamiya Rb pAb (PC-102)	42 kD ≫ 35 kD; 48, 65, **100** kD		~**25**–70 kD	42 kD
R&D goat pAb (AF1997)	42 kD ≫ 35 kD; 28, 65, 70 kD	42, 60, 75, **100** kD	~**25**–70 kD	42 kD
SC Rb pAb-H-155 (SC-33759; H155)	42 kD ≫ 35 kD; 55, 58 kD			42 kD
SC goat pAb-N17 (SC-30331; N17)	42 kD ≫ 35 kD > 55/58 kD			

IP, immunoprecipitation; IP/MALDI, immunoprecipitation combined with MALDI-TOF/TOF mass spectrometry analysis; Tandem IP/LTQ, tandem immunoprecipitation with different antibodies combined with linear ion-trap mass-spectrometry analysis

12.6 Conclusions

As the current cancer therapeutics, including chemotherapy, radiotherapy and surgery, cannot eradicate most cancers, and as surviving or 'leaked' CSCs may serve as the 'seeds' to initiate metastasis and induce recurrence, elimination of CSCs may represent a prerequisite for eradiation of cancer. Therefore, the CSC hypothesis provides a new avenue for cancer treatment and drug development. Theoretically, any factor or signalling pathway that positively regulates the homeostasis of CSCs will be a potential target for novel cancer intervention. Since NANOG1/NANOGP8 is aberrantly expressed and enhances the manifestation of CSC traits in multiple types of cancer, and because NANOG also crosstalks with many well-defined oncogenic pathways, and, most importantly, because normal somatic cells express a very low level of NANOG1/NANOGP8, NANOG and NANOG-related pathways may represent ideal targets for the development of novel cancer treatments with high efficacy and specificity.

Acknowledgments

Work in the authors' lab has been supported, in part, by grants from NIH (R01-CA155693), Department of Defense (W81XWH-13-1-0352 and PC130483) and CPRIT (RP120380).

References

Al-Hajj, M., Wicha, M.S., Benito-Hernandez, A., Morrison, S.J., Clarke, M.F. 2003. Prospective identification of tumorigenic breast cancer cells. Proc. Nat. Acad. Sci. USA 100, 3983–3988.

Ambady, S., Malcuit, C., Kashpur, O., Kole, D., Holmes, W.F., Hedblom, E., *et al.*, 2010. Expression of NANOG and NANOGP8 in a variety of undifferentiated and differentiated human cells. Int. J. Dev. Biol. 54, 1743–1754.

Badeaux, M.A., Jeter, C.R., Gong, S., Liu, B., Suraneni, M.V., Rundhaug, J., *et al.*, 2013. In vivo functional studies of tumor-specific retrogene NanogP8 in transgenic animals. Cell Cycle 12, 2395–2408.

Bertolini, G., Roz, L., Perego, P., Tortoreto, M., Fontanella, E., Gatti, L., *et al.*, 2009. Highly tumorigenic lung cancer CD133+ cells display stem-like features and are spared by cisplatin treatment. Proc. Nat. Acad. Sci. USA 106, 16 281–16 286.

Boiani, M., Scholer, H.R. 2005. Regulatory networks in embryo-derived pluripotent stem cells. Nat. Rev. Mol. Cell Biol. 6, 872–884.

Booth, H.A., Holland, P.W. 2004. Eleven daughters of NANOG. Genomics 84, 229–238.

Bourguignon, L.Y., Peyrollier, K., Xia, W., Gilad, E. 2008. Hyaluronan-CD44 interaction activates stem cell marker Nanog, Stat-3-mediated MDR1 gene expression, and ankyrin-regulated multidrug efflux in breast and ovarian tumor cells. J. Biol. Chem. 283, 17 635–17 651.

Brabletz, T., Jung, A., Spaderna, S., Hlubek, F., Kirchner, T. 2005. Opinion: migrating cancer stem cells – an integrated concept of malignant tumour progression. Nat. Rev. Cancer 5, 744–749.

Bussolati, B., Bruno, S., Grange, C., Ferrando, U., Camussi, G. 2008. Identification of a tumor-initiating stem cell population in human renal carcinomas. FASEB J. 22, 3696–3705.

Chambers, I., Colby, D., Robertson, M., Nichols, J., Lee, S., Tweedie, S., Smith, A. 2003. Functional expression cloning of Nanog, a pluripotency sustaining factor in embryonic stem cells. Cell 113, 643–655.

Chan, K.K., Zhang, J., Chia, N.Y., Chan, Y.S., Sim, H.S., Tan, K.S., *et al.*, 2009. KLF4 and PBX1 directly regulate NANOG expression in human embryonic stem cells. Stem Cells 27, 2114–2125.

Chang, D.F., Tsai, S.C., Wang, X.C., Xia, P., Senadheera, D., Lutzko, C. 2009. Molecular characterization of the human NANOG protein. Stem Cells 27, 812–821.

Chiou, S.H., Yu, C.C., Huang, C.Y., Lin, S.C., Liu, C.J., Tsai, T.H., *et al.*, 2008. Positive correlations of Oct-4 and Nanog in oral cancer stem-like cells and high-grade oral squamous cell carcinoma. Clin. Cancer Res. 14, 4085–4095.

Chiou, S.H., Wang, M.L., Chou, Y.T., Chen, C.J., Hong, C.F., Hsieh, W.J., *et al.*, 2010. Coexpression of Oct4 and Nanog enhances malignancy in lung adenocarcinoma by inducing cancer stem cell-like properties and epithelial-mesenchymal transdifferentiation. Cancer Res. 70, 10 433–10 444.

Choi, S.C., Choi, J.H., Park, C.Y., Ahn, C.M., Hong, S.J., Lim, D.S. 2012. Nanog regulates molecules involved in stemness and cell cycle-signaling pathway for maintenance of pluripotency of P19 embryonal carcinoma stem cells. J. Cell Physiol. 227, 3678–3692.

Clarke, M.F., Dick, J.E., Dirks, P.B., Eaves, C.J., Jamieson, C.H., Jones, D.L., *et al.*, 2006. Cancer stem cells – perspectives on current status and future directions: AACR Workshop on cancer stem cells. Cancer Res. 66, 9339–9344.

Darr, H., Mayshar, Y., Benvenisty, N. 2006. Overexpression of NANOG in human ES cells enables feeder-free growth while inducing primitive ectoderm features. Development 133, 1193–1201.

Do, H.J., Lim, H.Y., Kim, J.H., Song, H., Chung, H.M. 2007. An intact homeobox domain is required for complete nuclear localization of human Nanog. Biochem. Biophys. Res. Commun. 353, 770–775.

Do, H.J., Lee, W.Y., Lim, H.Y., Oh, J.H., Kim, D.K., Kim, J.H., Kim, T. 2009. Two potent transactivation domains in the C-terminal region of human NANOG mediate transcriptional activation in human embryonic carcinoma cells. J. Cell Biochem. 106, 1079–1089.

Eberle, I., Pless, B., Braun, M., Dingermann, T., Marschalek, R. 2010. Transcriptional properties of human NANOG1 and NANOG2 in acute leukemic cells. Nucleic Acids Res. 38, 5384–5395.

Eramo, A., Lotti, F., Sette, G., Pilozzi, E., Biffoni, M., Di Virgilio, A., *et al.*, 2008. Identification and expansion of the tumorigenic lung cancer stem cell population. Cell Death Differ. 15, 504–514.

Evans, M.J., Kaufman, M.H. 1981. Establishment in culture of pluripotential cells from mouse embryos. Nature 292, 154–156.

Ezeh, U.I., Turek, P.J., Reijo, R.A., Clark, A.T. 2005. Human embryonic stem cell genes OCT4, NANOG, STELLAR, and GDF3 are expressed in both seminoma and breast carcinoma. Cancer 104, 2255–2265.

Fang, D., Nguyen, T.K., Leishear, K., Finko, R., Kulp, A.N., Hotz, S., *et al.*, 2005. A tumorigenic subpopulation with stem cell properties in melanomas. Cancer Res. 65, 9328–9337.

Galli, R., Binda, E., Orfanelli, U., Cipelletti, B., Gritti, A., De Vitis, S., *et al.*, 2004. Isolation and characterization of tumorigenic, stem-like neural precursors from human glioblastoma. Cancer Res. 64, 7011–7021.

Golestaneh, A.F., Atashi, A., Langroudi, L., Shafiee, A., Ghaemi, N., Soleimani, M. 2012. miRNAs expressed differently in cancer stem cells and cancer cells of human gastric cancer cell line MKN-45. Cell Biochem. Funct. 30, 411–418.

Guan, Y., Gerhard, B., Hogge, D.E. 2003. Detection, isolation, and stimulation of quiescent primitive leukemic progenitor cells from patients with acute myeloid leukemia (AML). Blood 101, 3142–3149.

Hamazaki, T., Kehoe, S.M., Nakano, T., Terada, N. 2006. The Grb2/Mek pathway represses Nanog in murine embryonic stem cells. Mol. Cell Biol. 26, 7539–7549.

Han, J., Zhang, F., Yu, M., Zhao, P., Ji, W., Zhang, H., *et al.*, 2012. RNA interference-mediated silencing of NANOG reduces cell proliferation and induces G0/G1 cell cycle arrest in breast cancer cells. Cancer Lett. 321, 80–88.

Han, J., Fujisawa, T., Husain, S.R., Puri, R.K. 2014. Identification and characterization of cancer stem cells in human head and neck squamous cell carcinoma. BMC Cancer 14, 173.

Hart, A.H., Hartley, L., Parker, K., Ibrahim, M., Looijenga, L.H., Pauchnik, M., *et al.*, 2005. The pluripotency homeobox gene NANOG is expressed in human germ cell tumors. Cancer 104, 2092–2098.

Hatano, S.Y., Tada, M., Kimura, H., Yamaguchi, S., Kono, T., Nakano, T., *et al.*, 2005. Pluripotential competence of cells associated with Nanog activity. Mech. Dev. 122, 67–79.

Ho, M.M., Ng, A.V., Lam, S., Hung, J.Y. 2007. Side population in human lung cancer cell lines and tumors is enriched with stem-like cancer cells. Cancer Res. 67, 4827–4833.

Ho, B., Olson, G., Figel, S., Gelman, I., Cance, W.G., Golubovskaya, V.M. 2012. Nanog increases focal adhesion kinase (FAK) promoter activity and expression and directly binds to FAK protein to be phosphorylated. J. Biol. Chem. 287, 18 656–18 673.

Hyder, A., Ehnert, S., Hinz, H., Nussler, A.K., Fandrich, F., Ungefroren, H. 2012. EGF and HB-EGF enhance the proliferation of programmable cells of monocytic origin (PCMO) through activation of MEK/ERK signaling and improve differentiation of PCMO-derived hepatocyte-like cells. Cell Commun. Signal 10, 23.

Hyslop, L., Stojkovic, M., Armstrong, L., Walter, T., Stojkovic, P., Przyborski, S., *et al.*, 2005. Downregulation of NANOG induces differentiation of human embryonic stem cells to extraembryonic lineages. Stem Cells 23, 1035–1043.

Ishiguro, T., Sato, A., Ohata, H., Sakai, H., Nakagama, H., Okamoto, K. 2012. Differential expression of nanog1 and nanogp8 in colon cancer cells. Biochem. Biophys. Res. Commun. 418, 199–204.

Jeter, C.R., Badeaux, M., Choy, G., Chandra, D., Patrawala, L., Liu, C., *et al.*, 2009. Functional evidence that the self-renewal gene NANOG regulates human tumor development. Stem Cells 27, 993–1005.

Jeter, C.R., Liu, B., Liu, X., Chen, X., Liu, C., Calhoun-Davis, T., *et al.*, 2011. NANOG promotes cancer stem cell characteristics and prostate cancer resistance to androgen deprivation. Oncogene 30, 3833–3845.

Jiang, Y., He, Y., Li, H., Li, H.N., Zhang, L., Hu, W., *et al.*, 2012. Expressions of putative cancer stem cell markers ABCB1, ABCG2, and CD133 are correlated with the degree of differentiation of gastric cancer. Gastric Cancer 15, 440–450.

Kalbermatten, D.F., Schaakxs, D., Kingham, P.J., Wiberg, M. 2011. Neurotrophic activity of human adipose stem cells isolated from deep and superficial layers of abdominal fat. Cell Tissue Res. 344, 251–260.

Kim, J.S., Kim, J., Kim, B.S., Chung, H.Y., Lee, Y.Y., Park, C.S., *et al.*, 2005. Identification and functional characterization of an alternative splice variant within the fourth exon of human nanog. Exp. Mol. Med. 37, 601–607.

Kochupurakkal, B.S., Sarig, R., Fuchs, O., Piestun, D., Rechavi, G., Givol, D. 2008. Nanog inhibits the switch of myogenic cells towards the osteogenic lineage. Biochem. Biophys. Res. Commun. 365, 846–850.

Kuijk, E.W., De Gier, J., Lopes, S.M., Chambers, I., Van Pelt, A.M., Colenbrander, B., Roelen, B.A. 2010. A distinct expression pattern in mammalian testes indicates a conserved role for NANOG in spermatogenesis. PLoS One 5, e10987.

Kumar, S.M., Liu, S., Lu, H., Zhang, H., Zhang, P.J., Gimotty, P.A., *et al.*, 2012. Acquired cancer stem cell phenotypes through Oct4-mediated dedifferentiation. Oncogene 31, 4898–4911.

Lapidot, T., Sirard, C., Vormoor, J., Murdoch, B., Hoang, T., Caceres-Cortes, J., *et al.*, 1994. A cell initiating human acute myeloid leukaemia after transplantation into SCID mice. Nature 367, 645–648.

Lee, M., Nam, E.J., Kim, S.W., Kim, S., Kim, J.H., Kim, Y.T. 2012. Prognostic impact of the cancer stem cell-related marker NANOG in ovarian serous carcinoma. Int. J. Gynecol. Cancer 22, 1489–1496.

Leis, O., Eguiara, A., Lopez-Arribillaga, E., Alberdi, M.J., Hernandez-Garcia, S., Elorriaga, K., *et al.*, 2012. Sox2 expression in breast tumours and activation in breast cancer stem cells. Oncogene 31, 1354–1365.

Li, X.Q., Yang, X.L., Zhang, G., Wu, S.P., Deng, X.B., Xiao, S.J., *et al.*, 2013. Nuclear beta-catenin accumulation is associated with increased expression of Nanog protein and predicts poor prognosis of non-small cell lung cancer. J. Transl. Med. 11, 114.

Liang, J., Wan, M., Zhang, Y., Gu, P., Xin, H., Jung, S.Y., *et al.*, 2008. Nanog and Oct4 associate with unique transcriptional repression complexes in embryonic stem cells. Nat. Cell Biol. 10, 731–739.

Lim, C.Y., Tam, W.L., Zhang, J., Ang, H.S., Jia, H., Lipovich, L., *et al.*, 2008. Sall4 regulates distinct transcription circuitries in different blastocyst-derived stem cell lineages. Cell Stem Cell 3, 543–554.

Lin, T., Ding, Y.Q., Li, J.M. 2012. Overexpression of Nanog protein is associated with poor prognosis in gastric adenocarcinoma. Med. Oncol. 29, 878–885.

Liu, T., Xu, F., Du, X., Lai, D., Zhao, Y., Huang, Q., *et al.*, 2010. Establishment and characterization of multi-drug resistant, prostate carcinoma-initiating stem-like cells from human prostate cancer cell lines 22RV1. Mol. Cell Biochem. 340, 265–273.

Liu, J., Ma, L., Xu, J., Liu, C., Zhang, J., Chen, R., Zhou, Y. 2013. Spheroid body-forming cells in the human gastric cancer cell line MKN-45 possess cancer stem cell properties. Int. J. Oncol. 42, 453–459.

Liu, B., Badeaux, M.D., Choy, G., Chandra, D., Shen, I., Jeter, C.R., *et al.*, 2014. Nanog1 in NTERA-2 and recombinant NanogP8 from somatic cancer cells adopt multiple protein conformations and migrate at multiple M.W. species. PLoS One 9, e90615.

Lu, X., Mazur, S.J., Lin, T., Appella, E., Xu, Y. 2013. The pluripotency factor nanog promotes breast cancer tumorigenesis and metastasis. Oncogene 33(20), 2655–2664.

Ma, Y., Liang, D., Liu, J., Axcrona, K., Kvalheim, G., Stokke, T., *et al.*, 2011. Prostate cancer cell lines under hypoxia exhibit greater stem-like properties. PLoS One 6, e29170.

Martin, G.R. 1981. Isolation of a pluripotent cell line from early mouse embryos cultured in medium conditioned by teratocarcinoma stem cells. Proc. Nat. Acad. Sci. USA 78, 7634–7638.

Meng, H.M., Zheng, P., Wang, X.Y., Liu, C., Sui, H.M., Wu, S.J., *et al.*, 2010. Overexpression of nanog predicts tumor progression and poor prognosis in colorectal cancer. Cancer Biol. Ther. 9, 295–302.

Mitsui, K., Tokuzawa, Y., Itoh, H., Segawa, K., Murakami, M., Takahashi, K., *et al.*, 2003. The homeoprotein Nanog is required for maintenance of pluripotency in mouse epiblast and ES cells. Cell 113, 631–642.

Mullin, N.P., Yates, A., Rowe, A.J., Nijmeijer, B., Colby, D., Barlow, P.N., *et al.*, 2008. The pluripotency rheostat Nanog functions as a dimer. Biochem. J. 411, 227–231.

Nagata, T., Shimada, Y., Sekine, S., Hori, R., Matsui, K., Okumura, T., *et al.*, 2014. Prognostic significance of NANOG and KLF4 for breast cancer. Breast Cancer 21, 96–101.

Nichols, J., Zevnik, B., Anastassiadis, K., Niwa, H., Klewe-Nebenius, D., Chambers, I., *et al.*, 1998. Formation of pluripotent stem cells in the mammalian embryo depends on the POU transcription factor Oct4. Cell 95, 379–391.

Noh, K.H., Kim, B.W., Song, K.H., Cho, H., Lee, Y.H., Kim, J.H., *et al.*, 2012. Nanog signaling in cancer promotes stem-like phenotype and immune evasion. J. Clin. Invest. 122, 4077–4093.

O'Brien, C.A., Pollett, A., Gallinger, S., Dick, J.E. 2007. A human colon cancer cell capable of initiating tumour growth in immunodeficient mice. Nature 445, 106–110.

Oh, J.H., Do, H.J., Yang, H.M., Moon, S.Y., Cha, K.Y., Chung, H.M., Kim, J.H. 2005. Identification of a putative transactivation domain in human Nanog. Exp. Mol. Med. 37, 250–254.

Palla, A.R., Piazzolla, D., Abad, M., Li, H., Dominguez, O., Schonthaler, H.B., *et al.*, 2013. Reprogramming activity of NANOGP8, a NANOG family member widely expressed in cancer. Oncogene 33(19), 2513–2519.

Pan, G., Pei, D. 2005. The stem cell pluripotency factor NANOG activates transcription with two unusually potent subdomains at its C terminus. J. Biol. Chem. 280, 1401–1407.

Patrawala, L., Calhoun, T., Schneider-Broussard, R., Zhou, J., Claypool, K., Tang, D.G. 2005. Side population is enriched in tumorigenic, stem-like cancer cells, whereas ABCG2+ and ABCG2− cancer cells are similarly tumorigenic. Cancer Res. 65, 6207–6219.

Pei, D. 2009. Regulation of pluripotency and reprogramming by transcription factors. J. Biol. Chem. 284, 3365–3369.

Pereira, L., Yi, F., Merrill, B.J. 2006. Repression of Nanog gene transcription by Tcf3 limits embryonic stem cell self-renewal. Mol. Cell Biol. 26, 7479–7491.

Po, A., Ferretti, E., Miele, E., De Smaele, E., Paganelli, A., Canettieri, G., *et al.*, 2010. Hedgehog controls neural stem cells through p53-independent regulation of Nanog. EMBO J. 29, 2646–2658.

Qin, J., Liu, X., Laffin, B., Chen, X., Choy, G., Jeter, C.R., *et al.*, 2012. The PSA(−/lo) prostate cancer cell population harbors self-renewing long-term tumor-propagating cells that resist castration. Cell Stem Cell 10, 556–569.

Rich, J.N. 2007. Cancer stem cells in radiation resistance. Cancer Res. 67, 8980–8984.

Robertson, M., Stenhouse, F., Colby, D., Marland, J.R., Nichols, J., Tweedie, S., Chambers, I. 2006. Nanog retrotransposed genes with functionally conserved open reading frames. Mamm. Genome 17, 732–743.

Salven, P., Mustjoki, S., Alitalo, R., Alitalo, K., Rafii, S. 2003. VEGFR-3 and CD133 identify a population of CD34+ lymphatic/vascular endothelial precursor cells. Blood 101, 168–172.

Schatton, T., Murphy, G.F., Frank, N.Y., Yamaura, K., Waaga-Gasser, A.M., Gasser, M., *et al.*, 2008. Identification of cells initiating human melanomas. Nature 451, 345–349.

Shan, J., Shen, J., Liu, L., Xia, F., Xu, C., Duan, G., *et al.*, 2012. Nanog regulates self-renewal of cancer stem cells through the insulin-like growth factor pathway in human hepatocellular carcinoma. Hepatology 56, 1004–1014.

Siegel, R., Ma, J., Zou, Z., Jemal, A. 2014. Cancer statistics, 2014. CA Cancer J. Clin. 64, 9–29.

Silva, J., Nichols, J., Theunissen, T.W., Guo, G., Van Oosten, A.L., Barrandon, O., *et al.*, 2009. Nanog is the gateway to the pluripotent ground state. Cell 138, 722–737.

Singh, S.K., Clarke, I.D., Terasaki, M., Bonn, V.E., Hawkins, C., Squire, J., Dirks, P.B. 2003. Identification of a cancer stem cell in human brain tumors. Cancer Res. 63, 5821–5828.

Siu, M.K., Wong, E.S., Chan, H.Y., Ngan, H.Y., Chan, K.Y., Cheung, A.N. 2008. Overexpression of NANOG in gestational trophoblastic diseases: effect on apoptosis, cell invasion, and clinical outcome. Am. J. Pathol. 173, 1165–1172.

Storm, M.P., Bone, H.K., Beck, C.G., Bourillot, P.Y., Schreiber, V., Damiano, T., *et al.*, 2007. Regulation of Nanog expression by phosphoinositide 3-kinase-dependent signaling in murine embryonic stem cells. J. Biol. Chem. 282, 6265–6273.

Sun, C., Sun, L., Jiang, K., Gao, D.M., Kang, X.N., Wang, C., *et al.*, 2013. NANOG promotes liver cancer cell invasion by inducing epithelial-mesenchymal transition through NODAL/SMAD3 signaling pathway. Int. J. Biochem. Cell Biol. 45, 1099–1108.

Theunissen, T.W., Van Oosten, A.L., Castelo-Branco, G., Hall, J., Smith, A., Silva, J.C. 2011. Nanog overcomes reprogramming barriers and induces pluripotency in minimal conditions. Curr. Biol. 21, 65–71.

Torres, J., Watt, F.M. 2008. Nanog maintains pluripotency of mouse embryonic stem cells by inhibiting NFkappaB and cooperating with Stat3. Nat. Cell Biol. 10, 194–201.

van den Berg, D.L., Zhang, W., Yates, A., Engelen, E., Takacs, K., Bezstarosti, K., *et al.*, 2008. Estrogen-related receptor beta interacts with Oct4 to positively regulate Nanog gene expression. Mol. Cell Biol. 28, 5986–5995.

Visvader, J.E., Lindeman, G.J. 2008. Cancer stem cells in solid tumours: accumulating evidence and unresolved questions. Nat. Rev. Cancer 8, 755–768.

Wakamatsu, Y., Sakamoto, N., Oo, H.Z., Naito, Y., Uraoka, N., Anami, K., *et al.*, 2012. Expression of cancer stem cell markers ALDH1, CD44 and CD133 in primary tumor and lymph node metastasis of gastric cancer. Pathol. Int. 62, 112–119.

Wang, J., Levasseur, D.N., Orkin, S.H. 2008. Requirement of Nanog dimerization for stem cell self-renewal and pluripotency. Proc. Nat. Acad. Sci. USA 105, 6326–6331.

Wang, P., Gao, Q., Suo, Z., Munthe, E., Solberg, S., Ma, L., *et al.*, 2013a. Identification and characterization of cells with cancer stem cell properties in human primary lung cancer cell lines. PLoS One 8, e57020.

Wang, X., Zhu, Y., Ma, Y., Wang, J., Zhang, F., Xia, Q., Fu, D. 2013b. The role of cancer stem cells in cancer metastasis: new perspective and progress. Cancer Epidemiol. 37, 60–63.

Wang, X.Q., Ng, R.K., Ming, X., Zhang, W., Chen, L., Chu, A.C., *et al.*, 2013c. Epigenetic regulation of pluripotent genes mediates stem cell features in human hepatocellular carcinoma and cancer cell lines. PLoS One 8, e72435.

Wen, J., Park, J.Y., Park, K.H., Chung, H.W., Bang, S., Park, S.W., Song, S.Y. 2010. Oct4 and Nanog expression is associated with early stages of pancreatic carcinogenesis. Pancreas 39, 622–626.

Wicha, M.S., Liu, S., Dontu, G. 2006. Cancer stem cells: an old idea--a paradigm shift. Cancer Res. 66, 1883–1890; disc. 1895–1896.

Wu, Q., Chen, X., Zhang, J., Loh, Y.H., Low, T.Y., Zhang, W., *et al.*, 2006. Sall4 interacts with Nanog and co-occupies Nanog genomic sites in embryonic stem cells. J. Biol. Chem. 281, 24 090–24 094.

Yang, J., Chai, L., Fowles, T.C., Alipio, Z., Xu, D., Fink, L.M., *et al.*, 2008. Genome-wide analysis reveals Sall4 to be a major regulator of pluripotency in murine-embryonic stem cells. Proc. Nat. Acad. Sci. USA 105, 19 756–19 761.

Ye, F., Zhou, C., Cheng, Q., Shen, J., Chen, H. 2008. Stem-cell-abundant proteins Nanog, Nucleostemin and Musashi1 are highly expressed in malignant cervical epithelial cells. BMC Cancer 8, 108.

Yu, J., Vodyanik, M.A., Smuga-Otto, K., Antosiewicz-Bourget, J., Frane, J.L., Tian, S., *et al.*, 2007. Induced pluripotent stem cell lines derived from human somatic cells. Science 318, 1917–1920.

Yu, C.C., Chen, Y.W., Chiou, G.Y., Tsai, L.L., Huang, P.I., Chang, C.Y., *et al.*, 2011. MicroRNA let-7a represses chemoresistance and tumourigenicity in head and neck cancer via stem-like properties ablation. Oral Oncol. 47, 202–210.

Zbinden, M., Duquet, A., Lorente-Trigos, A., Ngwabyt, S.N., Borges, I., Ruiz I Altaba, A. 2010. NANOG regulates glioma stem cells and is essential in vivo acting in a cross-functional network with GLI1 and p53. EMBO J. 29, 2659–2674.

Zhang, J., Wang, X., Li, M., Han, J., Chen, B., Wang, B., Dai, J. 2006. NANOGP8 is a retrogene expressed in cancers. FEBS J. 273, 1723–1730.

Zhang, S., Balch, C., Chan, M.W., Lai, H.C., Matei, D., Schilder, J.M., *et al.*, 2008. Identification and characterization of ovarian cancer-initiating cells from primary human tumors. Cancer Res. 68, 4311–4320.

Zhang, X., Neganova, I., Przyborski, S., Yang, C., Cooke, M., Atkinson, S.P., *et al.*, 2009. A role for NANOG in G1 to S transition in human embryonic stem cells through direct binding of CDK6 and CDC25A. J. Cell Biol. 184, 67–82.

Zhang, J., Wang, X., Chen, B., Xiao, Z., Li, W., Lu, Y., Dai, J. 2010. The human pluripotency gene NANOG/NANOGP8 is expressed in gastric cancer and associated with tumor development. Oncol. Lett. 1, 457–463.

Zhang, J., Espinoza, L.A., Kinders, R.J., Lawrence, S.M., Pfister, T.D., Zhou, M., *et al.*, 2013. NANOG modulates stemness in human colorectal cancer. Oncogene 32, 4397–4405.

Zhang, K., Fowler, M., Glass, J., Yin, H. 2014. Activated 5′ flanking region of NANOGP8 in a self-renewal environment is associated with increased sphere formation and tumor growth of prostate cancer cells. Prostate 74, 381–394.

Zhou, X., Huang, G.R., Hu, P. 2011. Over-expression of Oct4 in human esophageal squamous cell carcinoma. Mol. Cells 32, 39–45.

13 Liver Cancer Stem Cells and Hepatocarcinogenesis

Hirohisa Okabe[1], Hiromitsu Hayashi[1], Takatsugu Ishimoto[1], Kosuke Mima[1], Shigeki Nakagawa[1], Hideyuki Kuroki[1], Katsunori Imai[1], Hidetoshi Nitta[1], Daisuke Hashimoto[1], Akira Chikamoto[1], Takatoshi Ishiko[1], Toru Beppu[2] and Hideo Baba[1]

[1]*Department of Gastroenterological Surgery, Graduate School of Life Sciences, Kumamoto University, Kumamoto, Japan*
[2]*Department of Multidisciplinary Treatment for Gastroenterological Cancer, Kumamoto University Hospital, Kumamoto, Japan*

13.1 Introduction

Liver cancer stem cells (LCSCs) are a small population of tumour cells that have been of great interest for more than a decade because of their high tumourigenicity and consequent potential as a therapeutic target. LCSCs possess the abilities to self-renew and to differentiate into heterogeneous nontumourigenic cancer cells, which are the major components of tumours. The existence of LCSCs was first proposed in haematopoietic malignancy (Lapidot *et al.*, 1994; Bonnet and Dick, 1997), and later in a variety of solid cancers (Lobo *et al.*, 2007; Visvader and Lindeman, 2008).

Liver cancer is the fifth most prevalent cancer and the third leading cause of cancer-related death worldwide (Jemal *et al.*, 2008; Forner and Bruix, 2012; Forner *et al.*, 2012). Although the hierarchy of cell differentiation in liver development remains debatable, LCSCs are believed to occupy the top position of the cell differentiation hierarchy in liver cancer. Recent whole-genome sequencing project reports assert that genetic abnormalities in malignant tumours are extremely variable, and finding optimal targets for individual

Principles of Stem Cell Biology and Cancer: Future Applications and Therapeutics, First Edition.
Edited by Tarik Regad, Thomas J. Sayers and Robert C. Rees.

patients remains a challenge (Vogelstein *et al.*, 2013). Although molecular information about tumours is still increasing, no matter how heterogeneous they are, tumours presumably at least contain functions specific to the context of their individual organ. Hence, LCSCs likely possess context-dependent properties similar to those of normal liver stem/progenitor cells, although they cannot function as normal liver stem/progenitor cells because of genetic and epigenetic alterations.

When selecting the therapeutic target for LCSCs, it is vital to understand not just their biology but that of normal liver as well. Cell proliferation in the liver is not constitutively active, unlike in the intestine. Normal stem/progenitor cells in the liver are assumed to be dormant under normal conditions. However, most liver cancers arise during the liver regeneration response to continuous injury, suggesting that liver stem/progenitor cells become activated in this situation.

In this chapter, we review recent updates on the biology of LCSCs and normal liver stem/progenitor cells. We also discuss the clinical application of LCSC markers as therapeutic targets, because most LCSC research depends on robust markers for identification.

13.2 Liver progenitor cells and their function

Normal liver stem/progenitor cells are also called oval cells (OCs), although the term 'oval cell' is debatable in terms of the source and the specific markers used for identification, especially in humans (Roskams *et al.*, 1998; Paku *et al.*, 2001; Grompe, 2003). In this chapter, we refer to normal liver stem/progenitor cells around periportal area as OCs. In order to explain the functional aspects of liver progenitor cells, scientists have used animal models that mimic human diseases; the 2-acetylaminofluorene (2-AAF)/partial hepatectomy (PH) model, in which hepatocyte proliferation is blocked by 2-AAF prior to PH, has been extensively used to characterize OCs in rats (Laishes and Rolfe, 1981). This model is based on the theory that liver stem/progenitor cells arise only when the mitogenic capacity of hepatocytes is insufficient for regenerative responses in rats. In mice, two models with diet-induced OCs have been reported: a choline-deficient ethionine-supplemented (CDE) diet (Akhurst *et al.*, 2001) and a 3,5-diethoxycarbonyl-1,4-dihydro-collidine (DDC) diet (Preisegger *et al.*, 1999; Wang *et al.*, 2003) have been reported. Results from recent studies shed light on the lineage-specific functions of OCs through their clonal expansion. In fact, a lineage-tracing mouse model is the preferred method for monitoring the differentiation status of OCs.

13.2.1 Inducible factors of liver progenitor cells

In vitro and *in vivo* studies have identified several inducible factors involved in the activation of OCs. Matthews *et al.* (2004) showed that OCs expressed interleukin 6 (IL-6) receptor and proliferated due to STAT3 activation *in vitro*. Jung *et al.* (2008) used a mouse model fed with high-fat diet plus ethanol and found that the progenitor cells were responsive to Hedgehog (Hh) ligand, which reportedly decreases apoptosis of cholangiocyte (Omenetti *et al.*, 2007), but were unresponsive to TGF-β. Tumour necrosis factor alpha (TNF-α) receptor 1 knockout (KO) mouse fed on CDE diet showed fewer OCs than wild-type mouse (Knight *et al.*, 2000), and TNF-α induced DNA replication of rat liver epithelial cells through NF-κB activation *in vitro* (Kirillova *et al.*, 1999). Intriguingly, TNF-like weak inducer of apoptosis (TWEAK), a member of the TNF family, does not affect mature hepatocytes, but stimulates OC proliferation selectively through its receptor, fibroblast growth factor (FGF)-inducible 14 protein (Fn14). Although Fn14 is expressed only on a few periductal cells around the portal triad in normal liver, increased Fn14 expression was noted in chronic liver diseases in humans, such as nonalcoholic steatohepatitis (NASH), alcoholic cirrhosis and chronic hepatitis C (Jakubowski *et al.*, 2005).

Bisgaard *et al.* (1999)_showed that ductular proliferating cells expressed INF-γ, INF-γRα, INF-γRβ, urokinase receptor (uPAR), intercellular adhesion molecule 1 (ICAM-1) and IL-18 during the liver regenerative process using AAF/PH rat model, implying that INF-γ may prime ductal cells through the activation of signal transducers, which may augment the mitogenic response of the cells to other cytokines and growth factors. Furthermore, it is assumed that uPAR/uPA complex causes proteolytic activation of plasminogen to plasmin on the surface of activated ductular OCs, which in turn leads to fibrinolysis and degradation of extracellular matrix (ECM) components and consequent activation of latent growth factor, such as hepatocyte growth factor (HGF).

Wnt signalling is crucial for intestinal stem cells. OCs in the liver are reportedly facilitated by Wnt/β-catenin signalling. Forced expression of constitutively active β-catenin mutant has been found to promote the expansion of the OC population in the regenerated liver of rat AAF/PH model (Yang *et al.*, 2008a). Apte *et al.* (2008) showed that hepatocyte-specific β-catenin KO mouse treated with DDC diet had dramatically decreased OC population.

Recent studies have suggested that Notch signalling has a pivotal role in biliaray lineage differentiation. Boulter *et al.* (2012) found that liver stem/progenitor cells induced by DDC diet, which is considered to induce biliary lineage cells, preferentially expressed biliary lineage markers such as Notch1, Jagged1, Hes1 and Hey1; this phenotype was predominantly supported by activated hepatic stellate cells. On the other hand, liver stem/progenitor cells induced by the CDE diet, which is considered to induce hepatocyte lineage cells, expressed hepatocyte-specific markers such as Axin2, Numb and Hnf4α; this phenotype was induced by macrophage. Although transdifferentiation of hepatocytes into biliary epithelial cells remains debatable, Yanger *et al.* (2013) generated hepatocyte-specific Notch1-inducible mouse, and demonstrated that Notch1 activation caused reprogramming of YFP-labelled hepatocytes into biliary lineage cells expressing both YFP and SOX9, which consequently formed YFP$^+$ ductules. Another study supported the transdifferentiation of hepatocytes into biliary epithelium (Limaye *et al.*, 2008).

Ishikawa *et al.* (2012) proved the significance of HGF-cMet signalling in the expansion of an OC population. They generated two mouse strains, hepatocyte-specific and stromal cell-specific *c-met* KO mice, both of which showed decreased OC populations in the DDC-treated liver. The significance of mesenchymal–epithelial transition (MET) and epithelial growth factor receptor (EGFR) signalling in the differentiation of OCs was shown using two types of murine liver stem/progenitor cell lines, engineered from either *Egfr* KO mice or *c-met* KO mice. Kitade *et al.* (2013) demonstrated that MET strongly induced hepatocyte differentiation by activating AKT and signal transducers and activator of transcription (STAT3), whereas EGFR selectively induced Notch1 expression, which promoted cholangiocyte differentiation.

13.2.2 Distribution of potential progenitor cells in human disease and their functional aspect

A significant accumulation of OCs has been reported in alcohol-induced fatty liver disease (ALD) and nonalcoholic fatty liver disease (NAFLD) (Lowes *et al.*, 1999; Nobili *et al.*, 2012; Sancho-Bru *et al.*, 2012), owing to the adaptability of progenitor cells to oxidative stress. Oxidative stress probably has a major role in the pathogenesis of both ALD and NAFLD (Tsukamoto and Lu, 2001; Mehta *et al.*, 2002). The replicative activity of mature hepatocytes is known to be inhibited in ALD and NAFLD (Selzner and Clavien, 2000; Yang *et al.*, 2001). Roskams *et al.* (2003) observed that murine models of fatty liver disease exhibited excessive hydrogen peroxide production and OC accumulation.

Focusing on severe liver damage in humans, such as fulminant hepatitis and submassive liver necrosis, Kiss *et al.* (2001) showed that progenitor cells (or ductular cells composing atypical ductules) expressed transforming growth factor (TGF)-α and -β, HGF, c-met, TGF-βRI/II, stem cell factor (SCF) and uPA. The ductular cells were surrounded by smooth-muscle actin (SMA)-positive activated hepatic stellate cells, which probably support them.

These progenitor cells may have different phenotypes based on their aetiologies. Moreover, their functional aspects are also determined by their microenvironment, which is dependent on the nature of the liver disease.

13.2.3 Liver progenitor cell markers

Several methods exist by which to isolate liver progenitor cells from mouse, rat, human foetus, neonatal liver and adult liver. Herrera *et al.* (2006) established a liver progenitor cell line using a novel method based on stringent hepatocyte culture conditions. Intriguingly, their cell line showed hepatocyte-specific markers such as albumin and alpha-fetoprotein (AFP), as well as stromal cell markers such as CD44, CD90, Vimentin and Nestin. In addition, this cell line proved to be functional in a liver transplantation model of fulminant liver failure (Herrera *et al.*, 2013).

Notably, a recent study purified extrahepatic cells from the hepatopancreatic ductal component and exhibited multipotent progenitor phenotypes of the purified cells, which were capable of differentiating into hepatocytes, cholangiocytes and pancreatic islets *in vivo* (Cardinale *et al.*, 2011). Furthermore, SOX9 was expressed throughout the biliary and pancreatic ductal epithelia, as well as in adult intestinal cells, hepatocytes and pancreatic acinar cells derived from Sox9-expressing progenitors (Furuyama *et al.*, 2011). These progenitors in the biliary epithelial cells of the extrahepatic biliary tract may have multilineage potential.

Huch *et al.* (2013) demonstrated that Lgr5+ cells (not seen in normal liver) could expand *in vitro* and differentiate into both biliary and hepatocyte lineage *in vivo* using DDC-treated mouse liver. Since paneth cells support the neighbouring lgr5+ cell in maintaining their stemness in the intestine (Sato *et al.*, 2011), it is interesting to clarify whether cells surrounding Lgr5+ support their stemness in the liver as well. The expression patterns of progenitor cell markers mentioned here are shown in Figure 13.1.

13.3 Cancer stem cells and their functional aspects

To date, several LCSC markers have been reported, most of which are similar to normal liver stem/progenitor cell markers; this suggests that tumour-initiating cells (TICs) have a similar potential to that of liver progenitor cells. The mechanism by which malignant hepatocytes obtain their biliary lineage features is unclear; however, there are some possible explanations: (i) normal progenitor cells become malignant and possess both biliary and hepatocytic features; or (ii) hepatocytes undergo de-differentiation and gain biliary features, and malignant hepatocytes possessing features from both lineages are generated. The latter theory may be partially supported by reports from studies stating that malignant cholangiocyte can arise from artificially disordered hepatocytes (Fan *et al.*, 2012; Sekiya and Suzuki, 2012). These hypotheses are further supported by a study focusing on the de-differentiation of normal hepatocytes into biliary epithelia in a lineage-tracing mouse model (Yanger *et al.*, 2013). In this section, we discuss the functional aspects and application of LCSC markers from a clinical viewpoint (Table 13.1).

13.3.1 Therapeutic application of LCSC markers

Although the functional aspects of cells expressing LCSC markers such as CD90, EpCAM, CD133, Bmi1, SP, ALDH1, CD24, and CD13 remain elusive, their tumourigenic capacity is proven in xenograft models.

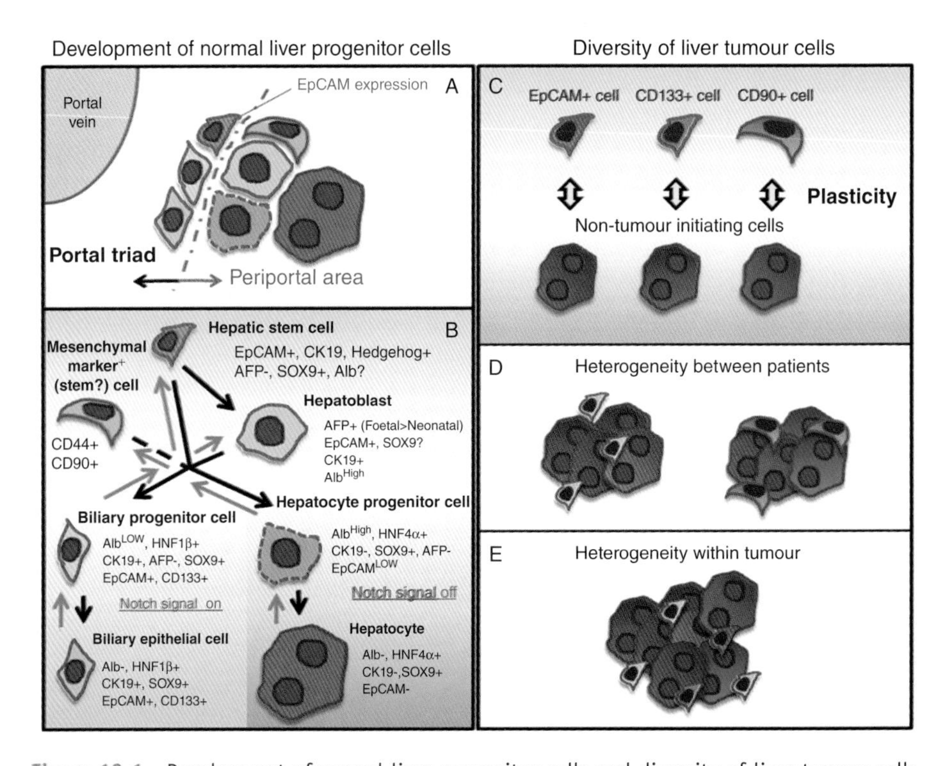

Figure 13.1 Development of normal liver progenitor cells and diversity of liver tumour cells. (A) The Canal of Hering is the site of production of liver stem/progenitor cells. Both biliary and hepatocyte lineage cells around the periportal area can be seen. (B) Hierarchy of liver cell differentiation. The expression patterns are based on different liver developmental stages, including foetal liver, postnatal liver and adult liver. Cells composing ductal plates in foetal liver (or hepatic stem cells) have multipotent capacity. Hepatoblasts in foetal liver express AFP, and another subpopulation probably expresses mesenchymal markers such as CD44 and CD90. Whether this population overlaps with hepatic stem cells or hepatoblasts remains to be elucidated. (C) Liver cancer contains more diverse stem/progenitor cells than normal liver. There are several LCSC markers, some of which are mutually exclusive. In addition, non-tumour-initiating cells may be able to dedifferentiate into LCSCs or undergo plasticity. (D) Physiological differences between patients should be considered when deciding the course of treatment, as all LCSC markers cannot be shared by all patients. (E) Multiple LCSCs have been noted within the same tumour, suggesting that a treatment effective against one type of LCSC may not be suitable for all LCSCs.

13.3.1.1 Epithelial cell-adhesion molecule EpCAM, a 40 kD glycoprotein and haemophilic cell–cell adhesion molecule, is expressed in simple, transitional and pseudostratified epithelia (Winter *et al.*, 2003). In the liver, it is highly expressed in foetal hepatoblasts, OCs and bile duct epithelia. In adults, it is expressed only on the biliary epithelia and not on the hepatocytes, unless injured by chronic mechanical stress (Wang *et al.*, 2010; Sancho-Bru *et al.*, 2012).

Yamashita *et al.* (2008) proposed a classification system defined by EpCAM and AFP based on comprehensive gene-expression analysis of human hepatocellular carcinoma (HCC) samples and found that EpCAM$^+$AFP$^+$

Table 13.1 Markers of liver progenitor cells (LPCs).

Marker	Source	Model and diet	Experiment	References
$CD133^{+}CD45^{-}$	Adult mouse	DDC, CCL4, ANIT	Culture	Rountree *et al.* (2007)
$c\text{-}Kit^{+}CD49^{+}CD29^{+}CD45^{-}TER119^{-}$	Foetal mouse	ED13.5 foetal liver	Culture	Suzuki *et al.* (2000)
$c\text{-}met^{+}CD49f^{+}CD29^{+}CD45^{-}TER119^{-}$	Human foetus		Transplantation into mouse	Dan *et al.* (2006)
$CD13^{+}CD49f^{+}CD133^{+}$	Postnatal mouse		Transplantation	Kamiya *et al.* (2009)
$EpCAM^{+}$	Adult mouse	DDC	Culture	Okabe *et al.* (2009)
$EpCAM^{+}$	Adult mouse	DDC	Transplantation	Kitade *et al.* (2013)
$CD45^{-}CD11b^{-}CD31^{-}MIC1\text{-}1C3^{+}CD133^{+}CD26^{-}$	Adult mouse	DDC	Transplantation	Dorrell *et al.* (2011)
$CD133^{+}CD45^{-}TER119^{-}$	Adult mouse	DDC	Transplantation	Suzuki *et al.* (2008)
$CD117^{+}CD34^{+}Lin^{-}$	Human foetus		Transplantation	Nowak *et al.* (2005)
$Thy1^{+}$	Adult human		Transplantation into mouse	Weiss *et al.* (2008)
$EpCAM^{+}CD49f^{+}$	Adult mouse	From GB	Subcutaneous expansion	Manohar *et al.* (2011)

cells were involved in tumour progression. $EpCAM^{+}$HuH1 cells have higher tumourigenic capacity than $EpCAM^{-}$ cells in nonobese diabetic severe combined-immunodeficient (NOD/SCID) mice. EpCAM expression of HuH1 and HuH7 cells was upregulated by the activation of GSK3β, induced by 6-bromoindirubin-3′-oxime (BIO). β-catenin signalling may be important to maintaining the phenotype of progenitor cells. In addition, blockade of EpCAM increased susceptibility to 5-FU in the HuH1 cell line (Yamashita *et al.*, 2009).

Another notable aspect of this molecule is its application as a robust marker in identifying circulating tumour cells (CTCs) in solid tumours such as HCC (Allard *et al.*, 2004; Nagrath *et al.*, 2007; Pantel *et al.*, 2008). Sun *et al.* (2013) reported that the presence of EpCAM-positive cells in patient blood indicate poor prognosis. In this study, CTCs were present in 66.7% of patients, reinforcing that EpCAM-expressing CTCs are a useful indicator of tumour progression, in addition to common tumour markers such as PIVKA-II and AFP. Interestingly, Yamashita *et al.* (2013) found that EpCAM and CD90 are discrete markers, which implies that EpCAM-based technology may be insufficient for identifying $CD90^{+}$ CTCs, which are a predictor for recurrence of HCC posthepatectomy (Fan *et al.*, 2011).

13.3.1.2 CD90 (Thy-1) CD90 is a 25–37 kDa glycosylphosphatidylinositol-anchored glycoprotein that is expressed on several cell types, including T cells, thymocytes, neurons, endothelial cells and fibroblasts; it is also involved in cell–cell and cell–matrix interactions (Rege and Hagood, 2006; Barker and Hagood, 2009). It has been suggested as a potential LCSC marker of breast cancer and glioblastoma (Liu *et al.*, 2006; Cho *et al.*, 2008), and recent studies have also suggested it as an another putative marker for liver LCSCs. Yang *et al.* (2008b) studied various cell lines and found that $CD90^+$ cells had highly tumourigenic potential *in vivo* and that $CD90^+CD44^+CD45^-$ cells from human blood and primary tumour tissues could generate tumour in immunodeficient SCID/beige mice. Moreover, they could be serially transplanted into secondary and tertiary animals. The authors also suggested that the presence of $CD90^+CD44^+CD45^-$ cells in patients' blood is an indicator of posthepatectomy tumour recurrence (Fan *et al.*, 2011). Although the clinical importance of CD90-expressing cells as LCSCs is well known, the functional role of CD90 needs to be further investigated.

13.3.1.3 SALL4 SALL4, the human homologue of the *drosophila* spalt homeotic gene, is a zinc-finger transcription factor that regulates pluripotency and self-renewal in embryonic stem cells. SALL4 is essential for maintaining the stemness properties of embryonic stem cells through both transcriptional and epigenetic controls, including direct interaction with Oct4 and Nanog (Yang *et al.*, 2010). SALL4 protein is expressed in foetal hepatoblasts and regulates the differentiation of hepatocytes (Oikawa *et al.*, 2009). Mature hepatocytes have been found to lose this expression, whereas human HCC regains it, and upregulated SALL4 expression has been correlated with the molecular characteristics of foetal liver (Yong *et al*, 2013a). SALL4 functions as a transcription repressor by recruiting a histone deacetylase (HDAC) – containing nucleosome remodeling and HDAC (NuRD) complex – and represses the tumour-suppressor phosphatase and tensin homologue (PTEN) to cause activation of the Akt-mediated phosphatidylinositol 3-kinase (PI3K) pathway. Interestingly, the 12-amino-acid peptide, a competitive inhibitor that blocks the interaction between SALL4 and NuRD, which forms a complex with SALL4 as a transcription repressor, blocks the tumour formation in xenograft, suggesting the therapeutic potential of SALL4-positive HCC (Yong *et al*, 2013b). In addition, a recent study proposed that SALL4 may regulate hepatic stem cell markers such as KRT19, EpCAM and CD44 (Zeng *et al.*, 2014).

13.3.1.4 CD133 (prominin-1) CD133, a five-transmembrane-domain cell-surface glycoprotein, has been identified as the primitive stem/progenitor cell marker in haematopoietic cells (Yin *et al.*, 1997). Its role as an LCSC marker has been described in various cancers, including brain, prostate, pancreas and colon (Hemmati *et al.*, 2003; Collins *et al.*, 2005; Li *et al.*, 2007; O'Brien *et al.*, 2007; Ricci-Vitiani *et al.*, 2007). $CD133^+$ cells demonstrated tumourigenic capacity in xenograft mouse models on several HCC cell lines of HuH7, SMMC-7721 and PLC/PRF/5 (Suetsugu *et al.*, 2006; Ma *et al.*, 2007; Yin *et al.*, 2007). On the other hand, as compared to their $CD133^-$ counterpart, $CD133^+$ HuH1 cells did not show any difference in tumourigenecity, although HuH1 cells expressed high levels of CD133 (Yamashita *et al.*, 2009). Also, compared to $CD133^-$ cells, $CD133^+$ cells were found to be more resistant to conventional chemotherapeutic drugs such as doxorubicin and 5-fluorouracil (5-FU), through the activation of Akt/PKB and Bcl-2 survival pathways (Ma *et al.*, 2008b). However, the mechanism by which $CD133^+$ cells activate these cell-survival pathways remains unclear. Results from recent studies have demonstrated that CD133 expression is partially dependent on DNA methylation via TGF-β signalling. You *et al.* (2010) found that TGF-β upregulated CD133 expression though inhibition of DNMT1 and DNMT3 expression, which are pivotal regulators of DNA methylation in solid tumour (Rhee *et al.*, 2002), and subsequently demethylated the CD133 promotor. Kohga *et al.* (2010) suggested that CD133 expression may decrease ADAM9 expression, which functions as a protease-shedding MHC class I-related chain A (MICA) that regulates the sensitivity of tumour cells to natural killer (NK) cells and consequently reduces the susceptibility of CD133-expressing cells to their action.

13.3.1.5 CD44 CD44 is a single-pass type I transmembrane protein that functions as a cellular adhesion molecule for hyaluronic acid, a major component of the ECM. Its function is variable: it depends on the expression of variant isoform (CD44v) (Aruffo *et al.*, 1990; Ponta *et al.*, 2003; Nagano *et al.*, 2013). We found that the dominant isoform of HCC is CD44s (standard isoform), although most gastrointestinal cancers prefer CD44v, which is predominantly composed of CD44v8–10 (Okabe *et al.*, 2013). CD44, which is presumably CD44v-specific for each organ context, has been identified as an LCSC marker in breast, pancreatic and gastric cancers (Li *et al.*, 2007; Wright *et al.*, 2008; Takaishi *et al.*, 2009).

We have reported that CD44s mediates TGF-β induced mechenchymal phenotype (Mima *et al.*, 2012) and that the $CD44^+CD90^+$ cells in the primary tumour also show mesenchymal phenotype (Okabe *et al.*, 2013), while Yang *et al.* (2008b) showed that most $CD90^+$ cells in HCC patients express CD44. However, CD44 upregulation was found to be associated with poor prognosis (Endo and Terada, 2000; Mima *et al.*, 2012), suggesting that although it contributes to tumour progression, it cannot be used to identify LCSCs without the aid of other cell-surface markers. HCC preferably express CD44s, as do several cell types in the stromal component, suggesting that CD44, by itself, is unlikely to be a therapeutic target. Our observations suggest that TGF-β signalling may be a therapeutic target for CD44-expressing mesenchymal HCC cells. However, the mechanism by which TGF-β regulates CD44 expression needs to be further addressed.

13.3.1.6 CD13 (aminopeptidase N) Haraguchi *et al.* (2002) identified CD13, also known as aminopeptidase N, as a novel cell-surface marker for dormant LCSCs. CD13 is a membraneous glycoprotein that plays important roles in colon cancer progression, including in cell proliferation, invasion and angiogenesis. These authors found that CD13 enriched in side population (SP) cells possessed a high tumourigenic potential and a high resistance to 5-FU treatment. $CD13^+$ cells show a quiescent state with G0/G1 phase and can maintain low levels of reactive oxygen species (ROS). Notably, use of CD13 inhibitor with regular chemotherapy diminished CD13-expressing tumour in a xenograft model (Haraguchi *et al.*, 2010). Although CD13 may not be an ideal direct molecular target, because of the surrounding cells in the liver (Christ *et al.*, 2011), the molecular pathway involved in CD13 expression can be a therapeutic target in a unique treatment strategy, as it offers the possibility of targeting dormant LCSCs in HCC.

13.3.1.7 SP (Hoechst33342 dye staining) SP cell-sorting by Hoechst 33342 dye is used to identify LCSCs in HCC. SP cells with low Hoechst staining are considered to be LCSCs, as they express adenosine triphosphate (ATP)-binding cassette (ABC) transporters that efflux Hoechst dye. Chiba *et al.* (2006) isolated SP cells from HuH7 and PLC/PRF/5 cell lines capable of initiating tumour in NOD/SCID mouse. Subsequently, they found that BMI1, a component of polycomb repressive complex 1 (PRC1), was preferentially expressed in SP cells over these cell lines, and that BMI1 knockdown abolished the tumour-initiating ability of SP cells in NOD/SCID mouse (Chiba *et al.*, 2008). BMI1 was demonstrated to be critical for tumour formation in HuH7 and PLC/PRF/5 cell lines. However, the underlying mechanism of the relationship between BMI1 and the function of ABC transporter requires further study.

13.3.1.8 ALDH ALDH is a cytosolic enzyme with more than 17 human isoforms. Its activity is used to identify LCSCs in lung, pancreatic and breast cancers (Ginestier *et al.*, 2007; Jiang *et al.*, 2009; Rasheed *et al.*, 2010). Ma *et al.* (2008a) revealed that $CD133^+/ALDH^+$ cells possess higher tumourigenic potential than their $CD133^+/ALDH^-$ and $CD133^-/ALDH^-$ counterparts, which suggests that ALDII might be used to further enrich the $CD133^+$ subpopulation.

13.3.1.9 CD24 Lee *et al.* (2011a) identified $CD24^+$ HCC cells resistant to cisplatin and capable of self-renewal and tumour initiation. CD24 knockdown successfully decreased the tumourigenicity of HuH7 and PLC/PRF/5, through decreased levels of phosphorylated STAT3 and decreased NANOG activity. The majority of $CD24^+$ cells were also positive for CD133 and EpCAM.

13.3.2 TGF-β signalling in LCSCs

Wu *et al.* (2012) reported the positive effects of TGF-β on liver progenitor cells, finding that rat pluripotent liver progenitor cell (LPC)-like WB-F334 cells treated with TGF-β increased the self-renewal capacity and generated stronger chemoresistance and tumourigenicity in NOD/SCID mice through upregulation of Akt, due to microRNA216a upregulation and subsequent PTEN suppression. The oncogenic role of TGF-β was proved by Morris *et al.* (2012), who showed decreased liver tumour formation in TP53 null mouse after crossbreeding with Tgfr2 KOs lacking TGF-β type 2 receptor. On the other hand, Tang *et al.* (2008) found that LCSCs in mice heterozygous for *elf*, a β-spectrin crucial for the propagation of TGF-β signalling in liver development (Mishra *et al.*, 1999), upregulated IL-6 signalling but inactivated TGF-β signalling in HCC.

13.3.3 Remarkable approaches to TIC-targeted therapy

Based on the results of several recent studies, researchers have proposed an effective treatment strategy involving targeting of LCSCs, albeit one that is limited to preclinical models. Lee *et al.* (2011b) found that Lup-20(29)-en-3β-ol (lupeol), a triterpene found in fruits and vegetables, inhibits the self-renewal ability of hepatic LCSCs in both HCC cell lines and clinical HCC samples *in vitro*. Using a nude mice xenograft model, they showed that lupeol suppressed the *in vivo* tumourigenicity of PLC-8024 and Huh-7 cells. In addition, lupeol sensitized the HCC cells to chemotherapeutic agents through the (PTEN)-Akt-ABCG2 pathway. These results suggest that lupeol may be an effective dietary phytochemical for targeting LCSCs.

Since LCSCs are chemoresistant to conventional chemo- and radiation therapies, Cheung *et al.* (2011) found that granulin–epithelin precursor (GEP) regulated chemoresistance in liver cancer cells (LCCs) through modulation of the expression of ABCB5 drug transporter. Notably, chemoresistant HCC cells that expressed GEP had increased levels of ABCB5, CD133 and EpCAM. Suppression of ABCB5 sensitizes cells to doxorubicin treatment and apoptosis; targeting these proteins in combination with chemotherapy may thus be a new therapeutic strategy.

An alternative approach to the treatment of liver cancer is to induce differentiation of LCSCs. Differentiation therapy can force LCSCs to differentiate and lose their self-renewal capacity. Zhang *et al.* (2013) induced cell differentiation with all-trans retinoic acid (ATRA), a vitamin A metabolite that participates in the regulation of cell proliferation, differentiation and migration. ATRA can induce the reduction of EpCAM expression of HuH7 and CSQT-2 cell lines, and the combinational treatment of ATRA acid and cisplatin can reduce protein kinase B (AKT) phosphorylation, as well as apoptosis of HCC cells *in vitro* and *in vivo*.

13.4 Future perspectives

Varied and robust LCSC markers of liver cancer have been reported in several studies, the results of which have revealed marker-specific LCSC

Table 13.2 Distribution of LCSC marker expressions in HCC cell lines, with genetic background. The expression patterns are derived from a protein-expression analysis, such as fluorescence-activated cell sorting (FACS) or Western blotting. Blank spaces indicate no data are available. Genetic status is provided by cell line, ATCC and JCRB. If the percentage of LCSCs is available, it is given with our unpublished data (†).

		HepG2	PLC/PRF/5	HuH1	HuH7	SKHep1	HLE	Hep3B	HLF	References
Cell type		HB	HCC	HCC	HCC	EA	HCC	HCC	HCC	ATCC, JCRB
p-53		WT	MT	WT	MT	WT	MT	MT	MT	
β-catenin		MT (del)	WT	WT	WT	WT	WT	WT	WT	
CSC marker	EpCAM	High (100%)	High	High (99%)	High (99%)		Low (0.4%)	High (47%)	Low (0.9%)	Yamashita *et al.* (2009), †
	CD90	Low (1%)	Low (0.2%)	Low (0.3%)	Low (0.2%)	High	High (89%)		Low (2%)	Yamashita *et al.* (2009), Yang *et al.* (2008), †
	CD133	Low	Low	High	Middle	None	None	High (90%)	None	Suetsugu *et al.* (2006), †
	CD44	None	None	None	Low	High	High		High	Okabe *et al.* (2013)
	CD13	Low	High (15.4%)	Low	Low (2%)	Low	None		None	Yamashita *et al.* (2009), Haraguchi *et al.* (2010), †
	CD24		High (55%)		High (97.7%)		High (20%)	High (82%)		Lee *et al.* (2011)
	$ALDH^+CD133^+$	Low (1.8%)			Middle (8%)			High (55%)		Ma *et al.* (2008)
	SP	+	+	+	+	Low	Low		Low	Chiba *et al.* (2006), †
	Bmi1	Middle	Middle	Middle	High	Middle	Middle		Middle	Chiba *et al.* (2008), †

characteristics and phenotypes. Since there is much less evidences for the regulation of LCSC markers in tumour microenvironment, we cannot exclude the possibility that non-LCSCs may produce LCSCs or undergo plasticity (Figure 13.1). If there is a robust genomic alteration that is unique to LCSCs, the formation of LCSCs from non-LCSCs is unlikely. However, the results of prior studies provide scant evidence on this aspect.

Based on previous reports and our investigation, Table 13.2 shows expression patterns for LCSC markers in HCC cell lines, as analysed by protein-expression analysis. No LCSC marker is shared among all cell lines. The basic characteristic of a liver cancer is determined by its genetic background, which is extremely variable, as per whole-genome sequencing analysis (Fujimoto *et al.*, 2012); this fact should be considered when designing a treatment strategy. Our understanding of the regulation of LCSC markers is limited, and the available LCSC markers are unlikely to be shared among patients. The diversity of patients, in addition to the diversity of cancer cells within each tumour, should be taken into consideration, in order to stratify the treatment strategy (Figure 13.1). If we can successfully stratify such patients, the unique features of LCSCs versus normal liver stem/progenitor cell will facilitate their identification and subsequent elimination.

References

Akhurst, B., Croager, E.J., Farley-Roche, C.A., Ong, J.K., Dumble, M.L., Knight, B., *et al.*, 2001. A modified choline-deficient, ethionine-supplemented diet protocol effectively induces oval cells in mouse liver. Hepatology 34, 519–522.

Allard, W.J., Matera, J., Miller, M.C., Repollet, M., Connelly, M.C., Rao, C., *et al.*, 2004. Tumour cells circulate in the peripheral blood of all major carcinomas but not in healthy subjects or patients with nonmalignant diseases. Clin. Cancer Res.. 10, 6897–6904.

Apte, U., Thompson, M.D., Cui, S., Liu, B., Cieply, B., Monga, S.P. 2008. Wnt/beta-catenin signaling mediates oval cell response in rodents. Hepatology 47, 288–295.

Aruffo, A., Stamenkovic, I., Melnick, M., Underhill, C.B., Seed, B. 1990. CD44 is the principal cell surface receptor for hyaluronate. Cell 61, 1303–1313.

Barker, T.H., Hagood, J.S. 2009. Getting a grip on Thy-1 signaling. Biochim. Biophys. Acta 1793, 921–923.

Bisgaard, H.C., Muller, S., Nagy, P., Rasmussen, L.J., Thorgeirsson, S.S. 1999. Modulation of the gene network connected to interferon-gamma in liver regeneration from oval cells. Am. J. Pathol. 155, 1075–1085.

Bonnet, D., Dick, J.E. 1997. Human acute myeloid leukemia is organized as a hierarchy that originates from a primitive hematopoietic cell. Nat. Med. 3, 730–737.

Boulter, L., Govaere, O., Bird, T.G., Radulescu, S., Ramachandran, P., Pellicoro, A., *et al.*, 2012. Macrophage-derived Wnt opposes Notch signaling to specify hepatic progenitor cell fate in chronic liver disease. Nat. Med. 18, 572–579.

Cardinale, V., Wang, Y., Carpino, G., Cui, C.B., Gatto, M., Rossi, M., *et al.*, 2011. Multipotent stem/progenitor cells in human biliary tree give rise to hepatocytes, cholangiocytes, and pancreatic islets. Hepatology 54, 2159–2172.

Cheung, S.T., Cheung, P.F., Cheng, C.K., Wong, N.C., Fan, S.T. 2011. Granulin-epithelin precursor and ATP-dependent binding cassette (ABC)B5 regulate liver cancer cell chemoresistance. Gastroenterology 140, 344–355.

Chiba, T., Kita, K., Zheng, Y.W., Yokosuka, O., Saisho, H., Iwama, A., *et al.*, 2006. Side population purified from hepatocellular carcinoma cells harbors cancer stem cell-like properties. Hepatology 44, 240–251.

Chiba, T., Miyagi, S., Saraya, A., Aoki, R., Seki, A., Morita, Y., *et al.*, 2008. The polycomb gene product BMI1 contributes to the maintenance of tumor-initiating side population cells in hepatocellular carcinoma. Cancer Res. 68, 7742–7749.

Cho, R.W., Wang, X., Diehn, M., Shedden, K., Chen, G.Y., Sherlock, G., *et al.*, 2008. Isolation and molecular characterization of cancer stem cells in MMTV-Wnt-1 murine breast tumors. Stem Cells 26, 364–371.

Christ, B., Stock, P., Dollinger, M.M. 2011. CD13, Waving the flag for a novel cancer stem cell target. Hepatology 53, 1388–1390.

Collins, A.T., Berry, P.A., Hyde, C., Stower, M.J., Maitland, N.J. 2005. Prospective identification of tumorigenic prostate cancer stem cells. Cancer Res. 65, 10 946–10 951.

Dan, Y.Y., Riehle, K.J., Lazaro, C., Teoh, N., Haque, J., Campbell, J.S., Fausto, N. 2006. Isolation of multipotent progenitor cells from human fetal liver capable of differentiating into liver and mesenchymal lineages. Proc. Nat. Acad. Sci. USA 103(26), 9912–9917.

Dorrell, C., Erker, L., Schug, J., Kopp, J.L., Canaday, P.S., Fox, A.J., *et al.*, 2011. Prospective isolation of a bipotential clonogenic liver progenitor cell in adult mice. Gene. Dev. 25(11), 1193–1203.

Endo, K., Terada, T. 2000. Protein expression of CD44 (standard and variant isoforms) in hepatocellular carcinoma: relationships with tumor grade, clinicopathologic parameters, p53 expression, and patient survival. J. Hepatol. 32, 78–84.

Fan, S.T., Yang, Z.F., Ho, D.W., Ng, M.N., Yu, W.C., Wong, J. 2011. Prediction of posthepatectomy recurrence of hepatocellular carcinoma by circulating cancer stem cells: a prospective study. Ann. Surg. 254, 569–576.

Fan, B., Malato, Y., Calvisi, D.F., Naqvi, S., Razumilava, N., Ribback, S., *et al.*, 2012. Cholangiocarcinomas can originate from hepatocytes in mice. J. Clin. Invest. 122, 2911–2915.

Forner, A., Bruix, J. 2012. Biomarkers for early diagnosis of hepatocellular carcinoma. Lancet Oncol. 13, 750–751.

Forner, A., Llovet, J.M., Bruix, J. 2012. Hepatocellular carcinoma. Lancet 379, 1245–1255.

Fujimoto, A., Totoki, Y., Abe, T., Boroevich, K.A., Hosoda, F., Nguyen, H.H., *et al.*, 2012. Whole-genome sequencing of liver cancers identifies etiological influences on mutation patterns and recurrent mutations in chromatin regulators. Nat. Genet. 44, 760–764.

Furuyama, K., Kawaguchi, Y., Akiyama, H., Horiguchi, M., Kodama, S., Kuhara, T., *et al.*, 2011. Continuous cell supply from a Sox9-expressing progenitor zone in adult liver, exocrine pancreas and intestine. Nat. Genet. 43, 34–41.

Ginestier, C., Hur, M.H., Charafe-Jauffret, E., Monville, F., Dutcher, J., Brown, M., *et al.*, 2007. ALDH1 is a marker of normal and malignant human mammary stem cells and a predictor of poor clinical outcome. Cell Stem Cell 1, 555–567.

Grompe, M. 2003. The role of bone marrow stem cells in liver regeneration. Semin. Liver Dis. 23, 363–372.

Hashida, H., Takabayashi, A., Kanai, M., Adachi, M., Kondo, K., Kohno, N., *et al.*, 2002. Aminopeptidase N is involved in cell motility and angiogenesis: its clinical significance in human colon cancer. Gastroenterology 122, 376–386.

Haraguchi, N., Ishii, H., Mimori, K., Tanaka, F., Ohkuma, M., Kim, H.M., *et al.*, 2010. CD13 is a therapeutic target in human liver cancer stem cells. J. Clin. Invest. 120, 3326–3339.

Hemmati, H.D., Nakano, I., Lazareff, J.A., Masterman-Smith, M., Geschwind, D.H., Bronner-Fraser, M., *et al.*, 2003. Cancerous stem cells can arise from pediatric brain tumors. Proc. Nat. Acad. Sci. USA 100, 15 178–15 183.

Herrera, M.B., Bruno, S., Buttiglieri, S., Tetta, C., Gatti, S., Deregibus, M.C., *et al.*, 2006. Isolation and characterization of a stem cell population from adult human liver. Stem Cells 24, 2840–2850.

Herrera, M.B., Fonsato, V., Bruno, S., Grange, C., Gilbo, N., Romagnoli, R., *et al.*, 2013. Human liver stem cells improve liver injury in a model of fulminant liver failure. Hepatology 57, 311–319.

Huch, M., Dorrell, C., Boj, S.F., van Es, J.H., Li, V.S., van de Wetering, M., *et al.*, 2013. In vitro expansion of single Lgr5+ liver stem cells induced by Wnt-driven regeneration. Nature 494, 247–250.

Kamiya, A., Kakinuma, S., Yamazaki, Y., Nakauchi, H. 2009. Enrichment and clonal culture of progenitor cells during mouse postnatal liver development in mice. Gastroenterology 137(3), 1114–1126.

Kitade, M., Factor, V.M., Andersen, J.B., Tomokuni, A., Kaji, K., Akita, H., *et al.*, 2013. Specific fate decisions in adult hepatic progenitor cells driven by MET and EGFR signaling. Gene. Dev. 27(15), 1706–1717.

Ishikawa, T., Factor, V.M., Marquardt, J.U., Raggi, C., Seo, D., Kitade, M., *et al.*, 2012. Hepatocyte growth factor/c-met signaling is required for stem-cell-mediated liver regeneration in mice. Hepatology 55, 1215–1226.

Jakubowski, A., Ambrose, C., Parr, M., Lincecum, J.M., Wang, M.Z., Zheng, T.S., *et al.*, 2005. TWEAK induces liver progenitor cell proliferation. J. Clin. Invest. 115, 2330–2340.

Jemal, A., Siegel, R., Ward, E., Hao, Y., Xu, J., Murray, T., *et al.*, 2008. Cancer statistics, 2008. CA Cancer J. 58, 71–96.

Jiang, F., Qiu, Q., Khanna, A., Todd, N.W., Deepak, J., Xing, L., *et al.*, 2009. Aldehyde dehydrogenase 1 is a tumor stem cell-associated marker in lung cancer. Mol. Cancer Res. 7, 330–338.

Jung, Y., Brown, K.D., Witek, R.P., Omenetti, A., Yang, L., Vandongen, M., *et al.*, 2008. Accumulation of hedgehog-responsive progenitors parallels alcoholic liver disease severity in mice and humans. Gastroenterology 134, 1532–1543.

Kirillova, I., Chaisson, M., Fausto, N. 1999. Tumor necrosis factor induces DNA replication in hepatic cells through nuclear factor kappaB activation. Cell Growth Diff. 10, 819–828.

Kiss, A., Schnur, J., Szabo, Z., Nagy, P. 2001. Immunohistochemical analysis of atypical ductular reaction in the human liver, with special emphasis on the presence of growth factors and their receptors. Liver 21, 237–246.

Kitade, M., Factor, V.M., Andersen, J.B., Tomokuni, A., Kaji, K., Akita, H., *et al.*, 2013. Specific fate decisions in adult hepatic progenitor cells driven by MET and EGFR signaling. Genes Dev. 27, 1706–1717.

Knight, B., Yeoh, G.C., Husk, K.L., Ly, T., Abraham, L.J., Yu, C., *et al.*, 2000. Impaired preneoplastic changes and liver tumor formation in tumor necrosis factor receptor type 1 knockout mice. J. Exp. Med. 192, 1809–1818.

Kohga, K., Tatsumi, T., Takehara, T., Tsunematsu, H., Shimizu, S., Yamamoto, M., *et al.*, 2010. Expression of CD133 confers malignant potential by regulating metalloproteinases in human hepatocellular carcinoma. J. Hepatol. 52, 872–879.

Laishes, B.A., Rolfe, P.B. 1981. Search for endogenous liver colony-forming units in F344 rats given a two-thirds hepatectomy during short-term feeding of 2-acetylaminofluorene. Cancer Res. 41, 1731–1741.

Lapidot, T., Sirard, C., Vormoor, J., Murdoch, B., Hoang, T., Caceres-Cortes, J., *et al.*, 1994. A cell initiating human acute myeloid leukaemia after transplantation into SCID mice. Nature 367, 645–648.

Lee, T.K., Castilho, A., Cheung, V.C., Tang, K.H., Ma, S., Ng, I.O. 2011a. CD24(+) liver tumor-initiating cells drive self-renewal and tumor initiation through STAT3-mediated NANOG regulation. Cell Stem Cell 9, 50–63.

Lee, T.K., Castilho, A., Cheung, V.C., Tang, K.H., Ma, S., Ng, I.O. 2011b. Lupeol targets liver tumor-initiating cells through phosphatase and tensin homolog modulation. Hepatology 53, 160–170.

Li, C., Heidt, D.G., Dalerba, P., Burant, C.F., Zhang, L., Adsay, V., *et al.*, 2007. Identification of pancreatic cancer stem cells. Cancer Res. 67, 1030–1037.

Limaye, P.B., Bowen, W.C., Orr, A.V., Luo, J., Tseng, G.C., Michalopoulos, G.K. 2008. Mechanisms of hepatocyte growth factor-mediated and epidermal growth factor-mediated signaling in transdifferentiation of rat hepatocytes to biliary epithelium. Hepatology 47, 1702–1713.

Liu, G., Yuan, X., Zeng, Z., Tunici, P., Ng, H., Abdulkadir, I.R., *et al.*, 2006. Analysis of gene expression and chemoresistance of CD133+ cancer stem cells in glioblastoma. Mol. Cancer 5, 67.

Lobo, N.A., Shimono, Y., Qian, D., Clarke, M.F. 2007. The biology of cancer stem cells. Ann. Rev. Cell Dev. Biol. 23, 675–699.

Lowes, K.N., Brennan, B.A., Yeoh, G.C., Olynyk, J.K. 1999. Oval cell numbers in human chronic liver diseases are directly related to disease severity. Am. J. Pathol. 154, 537–541.

Ma, S., Chan, K.W., Hu, L., Lee, T.K., Wo, J.Y., Ng, I.O., *et al.*, 2007. Identification and characterization of tumorigenic liver cancer stem/progenitor cells. Gastroenterology 132, 2542–2556.

Ma, S., Chan, K.W., Lee, T.K., Tang, K.H., Wo, J.Y., Zheng, B.J., *et al.*, 2008a. Aldehyde dehydrogenase discriminates the CD133 liver cancer stem cell populations. Mol. Cancer Res. 6, 1146–1153.

Ma, S., Lee, T.K., Zheng, B.J., Chan, K.W., Guan, X.Y. 2008b. CD133+ HCC cancer stem cells confer chemoresistance by preferential expression of the Akt/PKB survival pathway. Oncogene 27, 1749–1758.

Manohar, R., Komori, J., Guzik, L., Stolz, D.B., Chandran, U.R., LaFramboise, W.A., Lagasse, E. 2011. Identification and expansion of a unique stem cell population from adult mouse gallbladder. Hepatology 54(5), 1830–1841.

Matthews, V.B., Klinken, E., Yeoh, G.C. 2004. Direct effects of interleukin-6 on liver progenitor oval cells in culture. Wound Rep. Regen. 12, 650–656.

Mehta, K., Van Thiel, D.H., Shah, N., Mobarhan, S. 2002. Nonalcoholic fatty liver disease: pathogenesis and the role of antioxidants. Nutr. Rev. 60, 289–293.

Mima, K., Okabe, H., Ishimoto, T., Hayashi, H., Nakagawa, S., Kuroki, H., *et al.*, 2012. CD44s regulates the TGF-beta-mediated mesenchymal phenotype and is associated with poor prognosis in patients with hepatocellular carcinoma. Cancer Res. 72, 3414–3423.

Mishra, L., Cai, T., Yu, P., Monga, S.P., Mishra, B. 1999. Elf3 encodes a novel 200-kD beta-spectrin: role in liver development. Oncogene 18, 353–364.

Morris, S.M., Baek, J.Y., Koszarek, A., Kanngurn, S., Knoblaugh, S.E., Grady, W.M. 2012. Transforming growth factor-beta signaling promotes hepatocarcinogenesis induced by p53 loss. Hepatology 55, 121–131.

Nagano, O., Okazaki, S., Saya, H. 2013. Redox regulation in stem-like cancer cells by CD44 variant isoforms. Oncogene 32, 5191–5198.

Nagrath, S., Sequist, L.V., Maheswaran, S., Bell, D.W., Irimia, D., Ulkus, L., *et al.*, 2007. Isolation of rare circulating tumour cells in cancer patients by microchip technology. Nature 450, 1235–1239.

Nobili, V., Carpino, G., Alisi, A., Franchitto, A., Alpini, G., De Vito, R., *et al.*, 2012. Hepatic progenitor cells activation, fibrosis, and adipokines production in pediatric nonalcoholic fatty liver disease. Hepatology 56, 2142–2153.

Nowak, G., Ericzon, B.G., Nava, S., Jaksch, M., Westgren, M., Sumitran-Holgersson, S. 2005. Identification of expandable human hepatic progenitors which differentiate into mature hepatic cells *in vivo*. Gut 54(7), 972–979.

O'Brien, C.A., Pollett, A., Gallinger, S., Dick, J.E. 2007. A human colon cancer cell capable of initiating tumour growth in immunodeficient mice. Nature 445, 106–110.

Oikawa, T., Kamiya, A., Kakinuma, S., Zeniya, M., Nishinakamura, R., Tajiri, H., *et al.*, 2009. Sall4 regulates cell fate decision in fetal hepatic stem/progenitor cells. Gastroenterology 136, 1000–1011.

Okabe, M., Tsukahara, Y., Tanaka, M., Suzuki, K., Saito, S., Kamiya, Y., *et al.*, 2009. Potential hepatic stem cells reside in EpCAM+ cells of normal and injured mouse liver. Development 136(11), 1951–1960.

Okabe, H., Ishimoto, T., Mima, K., Nakagawa, S., Hayashi, H., Kuroki, H., *et al.*, 2013. CD44s signals the acquisition of the mesenchymal phenotype required for anchorage-independent cell survival in hepatocellular carcinoma. Br. J. Cancer 110(4), 958–966.

Omenetti, A., Yang, L., Li, Y.X., McCall, S.J., Jung, Y., Sicklick, J.K., *et al.*, 2007. Hedgehog-mediated mesenchymal-epithelial interactions modulate hepatic response to bile duct ligation. Lab. Invest. 87, 499–514.

Paku, S., Schnur, J., Nagy, P., Thorgeirsson, S.S. 2001. Origin and structural evolution of the early proliferating oval cells in rat liver. Am. J. Pathol. 158, 1313–1323.

Pantel, K., Brakenhoff, R.H., Brandt, B. 2008. Detection, clinical relevance and specific biological properties of disseminating tumour cells. Nat. Rev. Cancer 8, 329–340.

Ponta, H., Sherman, L., Herrlich, P.A. 2003. CD44, from adhesion molecules to signalling regulators. Nat. Rev. Mol. Cell Biol. 4, 33–45.

Preisegger, K.H., Factor, V.M., Fuchsbichler, A., Stumptner, C., Denk, H., Thorgeirsson, S.S. 1999. Atypical ductular proliferation and its inhibition by transforming growth factor beta1 in the 3,5-diethoxycarbonyl-1,4-dihydrocollidine mouse model for chronic alcoholic liver disease. Lab. Invest. 79, 103–109.

Rasheed, Z.A., Yang, J., Wang, Q., Kowalski, J., Freed, I., Murter, C., *et al.*, 2010. Prognostic significance of tumorigenic cells with mesenchymal features in pancreatic adenocarcinoma. J. Nat. Cancer Inst. 102, 340–351.

Rege, T.A., Hagood, J.S. 2006. Thy-1 as a regulator of cell-cell and cell-matrix interactions in axon regeneration, apoptosis, adhesion, migration, cancer, and fibrosis. FASEB J. 20, 1045–1054.

Rhee, I., Bachman, K.E., Park, B.H., Jair, K.W., Yen, R.W., Schuebel, K.E., *et al.*, 2002. DNMT1 and DNMT3b cooperate to silence genes in human cancer cells. Nature 416, 552–556.

Ricci-Vitiani, L., Lombardi, D.G., Pilozzi, E., Biffoni, M., Todaro, M., Peschle, C., *et al.*, 2007. Identification and expansion of human colon-cancer-initiating cells. Nature 445, 111–115.

Roskams, T., De Vos, R., Van Eyken, P., Myazaki, H., Van Damme, B., Desmet, V. 1998. Hepatic OV-6 expression in human liver disease and rat experiments: evidence for hepatic progenitor cells in man. J. Hepatol. 29, 455–463.

Roskams, T., Yang, S.Q., Koteish, A., Durnez, A., DeVos, R., Huang, X., *et al.*, 2003. Oxidative stress and oval cell accumulation in mice and humans with alcoholic and nonalcoholic fatty liver disease. Am. J. Pathol. 163, 1301–1311.

Rountree, C.B., Barsky, L., Ge, S., Zhu, J., Senadheera, S., Crooks, G.M. 2007. A CD133-expressing murine liver oval cell population with bilineage potential. Stem Cells 25(10), 2419–2429.

Sancho-Bru, P., Altamirano, J., Rodrigo-Torres, D., Coll, M., Millan, C., Jose Lozano, J., *et al.*, 2012. Liver progenitor cell markers correlate with liver damage and predict short-term mortality in patients with alcoholic hepatitis. Hepatology 55, 1931–1941.

Sato, T., van Es, J.H., Snippert, H.J., Stange, D.E., Vries, R.G., van den Born, M., *et al.*, 2011. Paneth cells constitute the niche for Lgr5 stem cells in intestinal crypts. Nature 469, 415–418.

Sekiya, S., Suzuki, A. 2012. Intrahepatic cholangiocarcinoma can arise from Notch-mediated conversion of hepatocytes. J. Clin. Invest. 122, 3914–3918.

Selzner, M., Clavien, P.A. 2000. Failure of regeneration of the steatotic rat liver: disruption at two different levels in the regeneration pathway. Hepatology 31, 35–42.

Suetsugu, A., Nagaki, M., Aoki, H., Motohashi, T., Kunisada, T., Moriwaki, H. 2006. Characterization of CD133+ hepatocellular carcinoma cells as cancer stem/progenitor cells. Biochem. Biophys. Res. Commun. 351, 820–824.

Sun, Y.F., Xu, Y., Yang, X.R., Guo, W., Zhang, X., Qiu, S.J., *et al.*, 2013. Circulating stem cell-like epithelial cell adhesion molecule-positive tumor cells indicate poor prognosis of hepatocellular carcinoma after curative resection. Hepatology 57, 1458–1468.

Suzuki, A., Zheng, Y.W., Kondo, R., Kusakabe, M., Takada, Y., Fukao, K., *et al.*, 2000. Flow-cytometric separation and enrichment of hepatic progenitor cells in the developing mouse liver. Hepatology 32(6), 1230–1239.

Suzuki, A., Sekiya, S., Onishi, M., Oshima, N., Kiyonari, H., Nakauchi, H., Taniguchi, H. 2008. Flow cytometric isolation and clonal identification of self-renewing bipotent hepatic progenitor cells in adult mouse liver. Hepatology 48(6), 1964–1978.

Takaishi, S., Okumura, T., Tu, S., Wang, S.S., Shibata, W., Vigneshwaran, R., *et al.*, 2009. Identification of gastric cancer stem cells using the cell surface marker CD44. Stem Cells 27, 1006–1020.

Tang, Y., Kitisin, K., Jogunoori, W., Li, C., Deng, C.X., Mueller, S.C., *et al.*, 2008. Progenitor/stem cells give rise to liver cancer due to aberrant TGF-beta and IL-6 signaling. Proc. Nat. Acad. Sci. USA 105, 2445–2450.

Tsukamoto, H., Lu, S.C. 2001. Current concepts in the pathogenesis of alcoholic liver injury. FASEB J. 15, 1335–1349.

Visvader, J.E., Lindeman, G.J. 2008. Cancer stem cells in solid tumours: accumulating evidence and unresolved questions. Nat. Rev. Cancer 8, 755–768.

Vogelstein, B., Papadopoulos, N., Velculescu, V.E., Zhou, S., Diaz, L.A. Jr, Kinzler, K.W. 2013. Cancer genome landscapes. Science 339, 1546–1558.

Wang, X., Foster, M., Al-Dhalimy, M., Lagasse, E., Finegold, M., Grompe, M. 2003. The origin and liver repopulating capacity of murine oval cells. Proc. Nat. Acad. Sci. USA 100(Suppl. 1), 11 881–11 888.

Wang, H., Gao, Y., Jin, X., Xiao, J. 2010. Expression of contactin associated protein-like 2 in a subset of hepatic progenitor cell compartment identified by gene expression profiling in hepatitis B virus-positive cirrhosis. Liver Int. 30, 126–138.

Weiss, T.S., Lichtenauer, M., Kirchner, S., Stock, P., Aurich, H., Christ, B., *et al.*, 2008. Hepatic progenitor cells from adult human livers for cell transplantation. Gut 57(8), 1129–1138.

Winter, M.J., Nagtegaal, I.D., van Krieken, J.H., Litvinov, S.V. 2003. The epithelial cell adhesion molecule (Ep-CAM) as a morphoregulatory molecule is a tool in surgical pathology. Am. J. Pathol. 163, 2139–2148.

Wright, M.H., Calcagno, A.M., Salcido, C.D., Carlson, M.D., Ambudkar, S.V., Varticovski, L. 2008. Brca1 breast tumors contain distinct CD44+/CD24− and CD133+ cells with cancer stem cell characteristics. Breast Cancer Res. 10, R10.

Wu, K., Ding, J., Chen, C., Sun, W., Ning, B.F., Wen, W., *et al.*, 2012. Hepatic transforming growth factor beta gives rise to tumor-initiating cells and promotes liver cancer development. Hepatology 56, 2255–2267.

Yamashita, T., Forgues, M., Wang, W., Kim, J.W., Ye, Q., Jia, H., *et al.*, 2008. EpCAM and alpha-fetoprotein expression defines novel prognostic subtypes of hepatocellular carcinoma. Cancer Res. 68, 1451–1461.

Yamashita, T., Ji, J., Budhu, A., Forgues, M., Yang, W., Wang, H.Y., *et al.*, 2009. EpCAM-positive hepatocellular carcinoma cells are tumor-initiating cells with stem/progenitor cell features. Gastroenterology 136, 1012–1024.

Yamashita, T., Honda, M., Nakamoto, Y., Baba, M., Nio, K., Hara, Y., *et al.*, 2013. Discrete nature of EpCAM+ and CD90+ cancer stem cells in human hepatocellular carcinoma. Hepatology 57, 1484–1497.

Yang, S.Q., Lin, H.Z., Mandal, A.K., Huang, J., Diehl, A.M. 2001. Disrupted signaling and inhibited regeneration in obese mice with fatty livers: implications for nonalcoholic fatty liver disease pathophysiology. Hepatology 34, 694–706.

Yang, W., Yan, H.X., Chen, L., Liu, Q., He, Y.Q., Yu, L.X., *et al.*, 2008a. Wnt/beta-catenin signaling contributes to activation of normal and tumorigenic liver progenitor cells. Cancer Res. 68, 4287–4295.

Yang, Z.F., Ho, D.W., Ng, M.N., Lau, C.K., Yu, W.C., Ngai, P., *et al.*, 2008b. Significance of CD90+ cancer stem cells in human liver cancer. Cancer Cell 13, 153–166.

Yang, J., Gao, C., Chai, L., Ma, Y. 2010. A novel SALL4/OCT4 transcriptional feedback network for pluripotency of embryonic stem cells. PloS One 5, e10766.

Yanger, K., Zong, Y., Maggs, L.R., Shapira, S.N., Maddipati, R., Aiello, N.M., *et al.*, 2013. Robust cellular reprogramming occurs spontaneously during liver regeneration. Genes Dev. 27, 719–724.

Yin, A.H., Miraglia, S., Zanjani, E.D., Almeida-Porada, G., Ogawa, M., Leary, A.G., *et al.*, 1997. AC133, a novel marker for human hematopoietic stem and progenitor cells. Blood 90, 5002–5012.

Yin, S., Li, J., Hu, C., Chen, X., Yao, M., Yan, M., *et al.*, 2007. CD133 positive hepatocellular carcinoma cells possess high capacity for tumorigenicity. Int. J. Cancer 120, 1444–1450.

Yong, K.J., Chai, L., Tenen, D.G. 2013a. Oncofetal gene SALL4 in aggressive hepatocellular carcinoma. N. Engl. J. Med. 369, 1171–1172.

Yong, K.J., Gao, C., Lim, J.S., Yan, B., Yang, H., Dimitrov, T., *et al.*, 2013b. Oncofetal gene SALL4 in aggressive hepatocellular carcinoma. N. Engl. J. Med. 368, 2266–2276.

You, H., Ding, W., Rountree, C.B. 2010. Epigenetic regulation of cancer stem cell marker CD133 by transforming growth factor-beta. Hepatology 51, 1635–1644.

Zeng, S.S., Yamashita, T., Kondo, M., Nio, K., Hayashi, T., Hara, Y., *et al.*, 2014. The transcription factor SALL4 regulates stemness of EpCAM-positive hepatocellular carcinoma. J. Hepatol. 60, 127–134.

Zhang, Y., Guan, D.X., Shi, J., Gao, H., Li, J.J., Zhao, J.S., *et al.*, 2013. All-trans retinoic acid potentiates the chemotherapeutic effect of cisplatin by inducing differentiation of tumor initiating cells in liver cancer. J. Hepatol. 59, 1255–1263.

14 Basic Science of Liver Cancer Stem Cells and Hepatocarcinogenesis

Katherine S. Koch and Hyam L. Leffert
Hepatocyte Growth Control and Stem Cell Laboratory, Department of Pharmacology, School of Medicine, University of California at San Diego, La Jolla, CA, USA

14.1 Introduction

14.1.1 Overview

The basic science of liver cancer stem cells (LCSCs) and hepatocarcinogenesis is best understood by familiarity with the increasingly complex epidemiology of hepatocellular carcinoma (HCC), with the still unknown cell origin(s) of HCC, with the anatomy and function of the liver lobule and with the historical and some novel methods for the study and identification of adult tissue stem cells and LCSCs. These background subjects are briefly discussed in this section.

14.1.2 Epidemiology of HCC

Environmental agents, including carcinogens, diet and genetic predisposition, are associated with toxicity-induced liver cell death and chronic inflammation. Both liver cell death and inflammation can contribute to development of compensatory hepatocyte proliferation and to hepatocarcinogenesis (El-Serag and Mason, 1999; El-Serag and Rudolph, 2007; El-Serag, 2011; American Cancer Society, 2013; World Health Organization, 2014).

HCC will likely cause 745 000 deaths worldwide in 2014 (World Health Organization, 2014). It is the fourth or fifth most common cancer in the world

Principles of Stem Cell Biology and Cancer: Future Applications and Therapeutics, First Edition.
Edited by Tarik Regad, Thomas J. Sayers and Robert C. Rees.

and its incidence rate is increasing (El-Serag, 2011; National Cancer Institute, 2014a, 2014c). In 2014, in the United States alone, approximately 33 190 new cases of HCC will have been diagnosed, with 23 000 subsequent deaths (American Cancer Society, 2014). Major causes of HCC are alcoholism, in the United States and Europe; dietary carcinogens, such as aflatoxin, in Africa and East Asia; hepatitis B virus (HBV), in Africa and Asia; and hepatitis C virus (HCV), in Japan, Europe and the United States (Venook *et al.*, 2010). Chronic HBV and HCV infections are the major factors underlying HCC risk (~10–80%), a risk still higher in co-infected patients (Ikeda *et al.*, 1993; Benvegnù *et al.*, 1994; Chiaramonte *et al.*, 1999; Venook *et al.*, 2010). Cirrhosis of any aetiology, including both alcoholic and biliary cirrhosis, is also a major risk factor, and is present at autopsy in 80–90% of HCC livers (Farinati *et al.*, 1994; Fattovich *et al.*, 2004; Silveira *et al.*, 2008; Hytiroglou *et al.*, 2012; Floreani and Farinati, 2013). Genetic risk factors include α_1-antitrypsin deficiency, haemochromatosis, porphyria cutanea tarda, tyrosinemia and Wilson's disease, nonalcoholic steatohepatitis, and metabolic diseases, such as dyslipidaemia, obesity, type 2 diabetes, glycogen storage disease, as well as autoimmune hepatitis (Bugianesi *et al.*, 2002; Fattovich *et al.*, 2004; Mattner, 2011; Hytiroglou *et al.*, 2012). HCC incidence is also two- to eightfold higher in males than in females (Buch *et al.*, 2008; Howlader *et al.*, 2012). With a new depth and accuracy of cellular identification and structural biopsy interpretation, fuelled also by inbred animal studies and *in vitro* culture investigations, the increasing volume of epidemiological studies provides direction for future research into the heretofore unsolved problems of HCC cell and LCSC origins, and into earlier-detection assays for these cell types.

14.1.3 Problems surrounding HCC cell origins

The cellular origins of HCC are a long-standing problem in liver research (Sell and Leffert, 1982, 2008), compounded by the complex clinical phenotypes of HCCs (van Malenstein *et al.*, 2011). The definitions of cancer stem cells (CSCs) are confusing, since CSCs can both generate and be derived from malignant tumours (Reya *et al.*, 2001; Clarke *et al.*, 2006; Rahman *et al.*, 2011; van Malenstein *et al.*, 2011; Magee *et al.*, 2012; Valent *et al.*, 2012). Generally, the definition of CSCs is a function of their experimental derivation as an isolate from tumour, and as a precursor to a secondary tumour. For the purposes of this article, the definition of a liver cancer precursor cell, or tumour-initiating cell (TIC), will be one preceding the primary tumour and requiring a chronically injured environment for progression to a primary tumour. On the other hand, an LCSC will be a liver cancer precursor, usually isolated from the primary tumour, unless otherwise noted, and fully capable of forming a secondary tumour in an untreated animal. Because the putative parent HCCs are of complex aetiology and phenotype, and of undoubtedly multiple possible origins (Figure 14.1), the biology of LCSCs isolated from them will likely be equally complex.

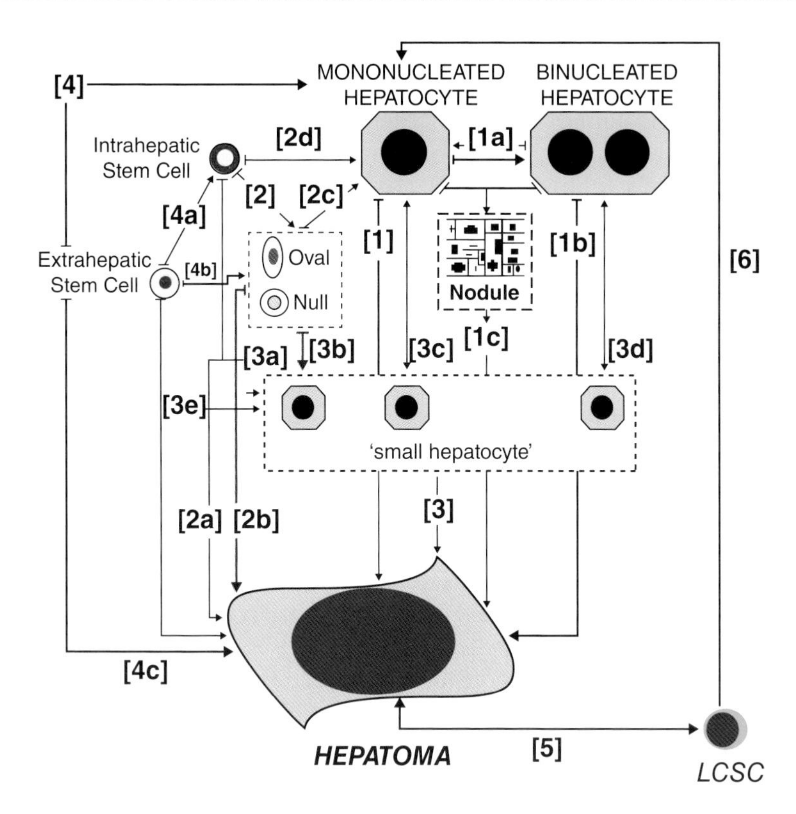

Figure 14.1 Lineage models of candidate HCC progenitor cells (HcPCs): hepatocytes, LSCs such as OCs and/or LCSCs. HCCs might develop from mono- (pathway [1]) or binucleated (pathways [1a] or [1b]) mature hepatocytes; from preneoplastic nodules derived from hepatocytes (pathway [1c]); from intrahepatic multipotent stem cells (MSCs) directly (pathway [2a]); or indirectly (pathway [2]) from bipotential OC or 'null' cell intermediates (pathway [2b]) formed from MSCs or other postulated MSC pathways. Alternatively, HCCs might develop from 'large' mononucleated hepatocytes derived from OCs or null cells (pathway [2c]); directly from intrahepatic stem cells (pathway [2d]); from small hepatocytes (pathway [3]) derived from intrahepatic LSCs like OCs or from OCs derived from injured HSCs (pathway [3a]) (Yang L. *et al.* 2008), AFP^{-} null cells (pathway [3b]) or mono- (pathway [3c]) or binucleated (pathway [3d]) hepatocytes; or from extrahepatic stem cells (pathways [3e], [4] or [4a–4c]). Red arrows indicate likely OC → HCC lineages (pathways [2b] and [3b]). Pathways [5] and [6] reflect putative reversible formation of LCSCs from HCCs, or of HCCs from LCSCs, and concomitant differentiation of LCSCs into mononucleated hepatocytes, respectively (arrows, → or ←, indicate the direction of the cell lineage). Koch and Leffert 2004. Reproduced with permission of Springer Science + Business Media.

Investigations of HCC origins following different injury regimens have focused particularly on hepatocytes and liver oval cells (OCs), small bipotential cells with liver stem cell (LSC)-like properties, both as possible HCC precursors. For instance, after marking proliferating rat hepatocytes with retroviral β-galactosidase (β-gal) reporter-encoding vector transduction following partial hepatectomy, the chemical hepatocarcinogen diethylnitrosamine (DEN) was administered by drinking water, and the resulting hepatic populations and HCCs were examined for β-gal positivity (β-gal$^+$), as well as for the preneoplastic, nodular and HCC marker placental glutathione-S-transferase (GSTp). The percentage of β-gal$^+$ hepatocytes following compensatory growth post partial hepatectomy was the same as that of HCC cells following DEN administration (~18%); clonal growth of nodules and HCCs was indicated by the whole nodule or tumour β-gal positivity; and, whereas all HCCs expressed GSTp, although to different levels, only ~50% of surrounding β-gal$^+$ hepatocytes were GSTp$^+$, suggesting possible hepatocyte HCC precursor subtypes. Disadvantageously uninterpretable in such a marking–mapping system are the unmarked β-gal$^-$ HCC cells (82% β-gal$^-$), originated either from unmarked hepatocyte precursors or from other unmarked precursors, such as OCs (Bralet *et al.*, 2002).

Likewise, in a strategy designed to infect and mark both rat OCs *and* hepatocytes, and a second design to prelabel hepatocytes alone, with a carefully timed β-gal retroviral transduction during an acetylaminofluorene (AAF) hepatocarcinogen regimen known to induce massive OC proliferation and concomitant hepatocyte death, some hepatocytes, preneoplastic foci and nodules, as well as HCCs in *both* transductions, expressed β-gal and GSTp, thus again suggesting that the hepatocyte could be an HCC precursor, and one which might have transited through a less differentiated state (Gournay *et al.*, 2002). The possibility of simultaneous OC infection and proliferation in either of these two systems cannot be rigorously excluded.

Hepatocytes were also implicated as HCC precursors in an Alb-Cre recombinase hepatocyte-targeted *Ikkβ*-deletion mouse system, shown to produce *Ikkβ*-deleted HCC, following DEN administration (Maeda *et al.*, 2005). The patency of this interpretation rests upon the specificity of the albumin expression that drives the Cre recombinase necessary for the *Ikkβ* deletion, and that can be shared by some differentiating OCs.

Subsequently, a system of transplantation has been developed whereby C57BL/6 hepatocytes, isolated from a DEN-treated mouse, are injected instrasplenically into the liver of a recipient mouse (He *et al.*, 2010). These cells will develop HCCs within the recipient liver if it has been subjected to chronic injury and to compensatory proliferation, such as the liver of a MUP/uPA mouse (Weglarz *et al.*, 2000), but will not develop HCCs in a normal liver recipient. The transplantation system has now been used to test both other animal donor cells and other recipient models, as well as to develop a more precise characterization of the HCC progenitor cells (HcPCs) (He *et al.*, 2013). Isolate separations by aggregate formation capacity and by CD44 positivity established the most effective HCC precursors. The small size of HcPCs is similar to that of cells of dysplastic foci that develop many months prior to HCC formation in the DEN-treated mouse or rat. The transcriptomes of HcPCs, OCs and HCC are very similar, but the dysplastic foci following DEN administration are in liver zone 3 near the Cyp2E1 DEN-metabolizing hepatocytes, whereas OCs are thought to be initially located portally in the Canals of Hering. Nevertheless, OCs can proliferate and migrate throughout the liver following certain types of injury (Guest *et al.*, 2010), although they are not noticeable following DEN, as in this model.

In all of these precursor 'fate mapping-like' systems, the results suggest that parenchymal, possibly dedifferentiated, hepatocytes are HCC tumour progenitors. Hepatocytes are the predominant liver epithelial cell (Baratta *et al.*, 2009) and the major sites of metabolic activation of chemical carcinogens and carcinogen toxicity (Guengerich, 2000). While mature hepatocytes are not considered tissue stem cells, they appear to be able to dedifferentiate under certain circumstances. Nevertheless, many other putative intra- and extrahepatic LSCs, including hepatic OCs (Figure 14.1), were not eliminated as HCC precursors in these investigations (see Section 14.2.8). Thus, instead of supporting LSC origins, these observations suggest, but do not prove, an HCC origination model of cumulative genetic, epigenetic alteration and clonal selection from DEN-exposed hepatocytes (Nowell, 1976; Greaves and Maley, 2012; Guichard *et al.*, 2012).

It is still not known whether the cells of origin of HCC can be found within the dysplastic foci observed prior to HCC appearance and following DEN and other carcinogen administration. Whether these dysplastic foci are or are not HCC precursors, the origin of cells in the foci is also not known. Finally, in summary, it is not known whether HCCs originate from OCs, from dedifferentiated hepatocytes or from some other precursor cells.

The development of treatments to target the modulation or elimination of aggressive HCC rests upon the identification and biology of the earliest cells of origin of HCC: hepatocyte, altered hepatocyte, LSC, LCSC or other. Current therapies (National Cancer Institute, 2014b), including treatments of recurrent HCC following tumour resection, might only be providing suitable environments for drug resistance and metastasis, or LCSC proliferation, by focusing treatment on elimination of the wrong cells.

14.1.4 Anatomy and function of the liver lobule

Knowledge of lobular anatomy both in its normal state (Deleve *et al.*, 2011) and in response to different forms of acute or chronic injury provides us with information about the locations of hepatocytes, OCs and other non-parenchymal cells (NPCs); which of these cells proliferate, when, and where they migrate within the lobule; and the possible relevance of this dynamic structure to LSC homeostasis and LCSC formation, latency/proliferation and migration. An adult liver lobule is diagrammed in Figure 14.2.

Mature hepatocytes, with diameters of 20–60 μm, form bilaminar plates of differentiated cells, separated by opposing bile canalicular membranes. These membranes enclose canalicular spaces, into which biliary acids are secreted, for progression into the bile ducts.

Highly oxygenated blood from the hepatic artery comingles with nutrient-rich blood from the portal vein, originating in the intestines, and flows unidirectionally and collectively into and throughout the liver sinusoids.

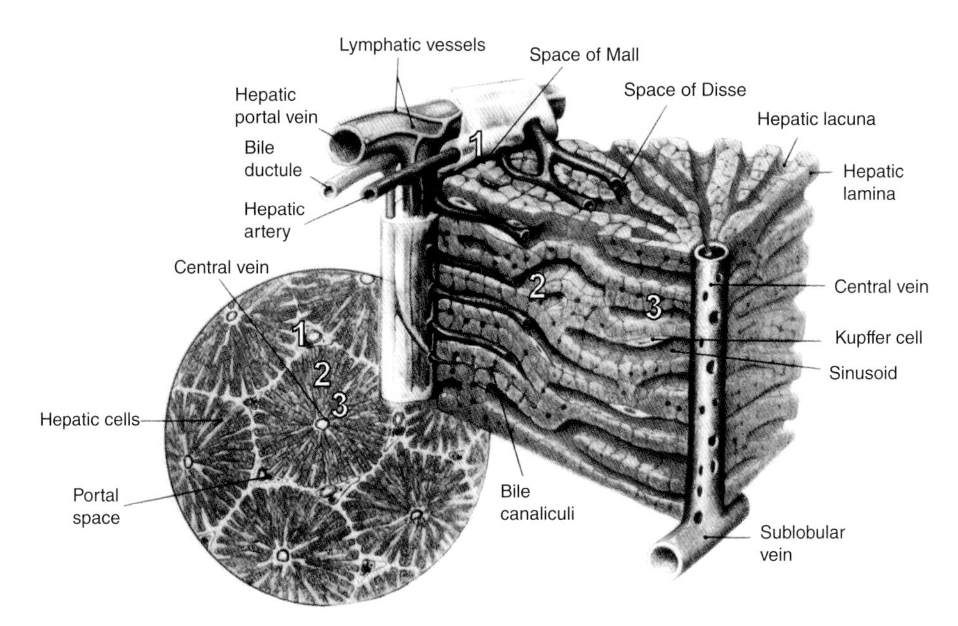

Figure 14.2 Three-dimensional and cross-sectional diagrams of the cellular and structural anatomy of the adult liver lobule. The zones are numbered: 1, Portal triad (Space of Mall); 2, Midzonal - cells between zones 1 and 3; and 3, Centrilobular. Harwin, Lyons and Monsen, 1984. Reproduced with permission of Wolters Kluwer/Lippincott Williams and Wilkins.

This sinusoidal blood flows through a fenestrated endothelial cell layer into the Space of Disse, which contains migratory NPCs such as Kupffer cells, hepatic stellate cells (HSCs) and B and T lymphocytes (Crispe, 2009), and then through an extracellular matrix (ECM) adjacent to hepatocyte basal membranes. The depleted sinusoidal blood finally exits the lobule through the central vein and empties into the inferior vena cava.

Three hepatic zones are positionally defined (Figure 14.2): zone 1, the Portal Triad (Space of Mall), where portal veins, bile ductules and hepatic arterioles are enclosed within a fibrous sheath, proximal to the undifferentiated cells of the Canal of Hering, which separate bile ductules from hepatocyte plates; zone 2, the intermediate region between zones 1 and 3; and, zone 3, the centrilobular region surrounding the central vein. Zones 1–3 are functionally specialized. For example, hepatocytes synthesize and secrete albumin in all zones, 1–3, but alcohol dehydrogenase and hepatocarcinogen metabolizing Cyp450 enzymes are selectively enriched in the hepatocytes of zones 1 and 3, respectively. Bile acids secreted across apical membranes into canalicular spaces flow retrograde to blood, from zone 3 to zone 1; they collect in the bile ductules of the portal triad. Early postnatal life is a time of marked hepatocyte proliferation (from 10% to <1% of cells in S-phase), as lobular structure and zonal formation mature. During this time, α_1-fetoprotein (AFP), a widely used

OC, LSC, TIC and HCC marker, is constitutively expressed in all zones. But as hepatocytes further mature and proliferation subsides to a steady state of ~0.005% of M-phase cells, AFP expression is extinguished directionally, from zone 3 to zone 1, and its blood levels fall to <50 ng/ml (Ramesh *et al.*, 1995).

Hepatocyte receptors are in a steady state with various soluble growth-stimulating and growth-inhibiting endocrine, paracrine, autocrine and juxtacrine factors that emanate from different organs and cells, and with associated factors that bind reversibly to the liver matrix. Depending upon exogenous or endogenous hepatotoxic insults, different stimulatory and inhibitory growth factor steady states shift, as a result of loss of liver mass, stimulating hepatocytes to proliferate in the different zones at different times in order to restore lobular mass. Alternatively, if loss of liver mass is too great for extant hepatocyte recovery, other cells, such as OCs, perhaps other LSCs or even, potentially, LCSCs, proliferate to restore, or disrupt, liver function. The nature of the response is dependent on the type of injury, its chronicity or acuteness and the milieu in which it occurs, such as the age of the animal.

The regenerative response following partial hepatectomy is a basic example of a range of defined injuries and quantifiably measurable responses. For example, regeneration after partial hepatectomy, which affects all liver zones, can be quantified as the time and rate of hepatocyte S- and M-phase entry (deHemptinne and Leffert, 1983). These rates fall with age, in parallel with decreasing telomerase length (Hoare *et al.*, 2010; Schmucker and Sanchez, 2011); regeneration occurs over a longer time frame, but restores the same mass. Quiescent rodent livers (G_0) regenerate in a 30–70% partial hepatectomy dose- and zone-dependent fashion (zone 1 → zone 2 → zone 3); within 3–7 days, hepatocytes divide one to three times to restore liver mass. Zonal effects depend partly upon circulating growth factors, because surgical reversal of bloodflow redirects hepatocyte G_0 → S → G2 → M-phase entry times (zone 3 → zone 2 → zone 1). NPC proliferation is also humorally regulated, and collectively occurs after the start of hepatocyte proliferation. In general, this defined regenerative response provides the background for study of different types of liver injuries, varied by dose, duration, lobular location and recruitment of parenchymal cells and NPCs, both mature and stem cell, to deliver acute to chronic liver insult, resulting in a range of regenerative to fibrotic responses, respectively.

14.1.5 Methods of identification and properties of adult stem cells

In a classical series of experiments, using a spleen colony-forming assay (SCA), a suspension of bone marrow cells (BMCs) from a donor mouse formed discrete colonies, proportional to the injection dose, in the spleen of a lethally irradiated mouse recipient, indicating the existence of bone marrow stem cells (BMSCs) (Siminovitch *et al.*, 1963). This assay provided a way to fate map putative stem cells, revealed key stem cell properties and served as the prototype for the identification and isolation of somatic stem cells, as well as tissue CSCs such as BMCSCs (Spangrude *et al.*, 1988; Reya *et al.*, 2001; Clarke *et al.*, 2006) and CSCs from gliomas (Lathia *et al.*, 2011) and from breast (Russo *et al.*, 2010).

Six principle observations were made: (i) Consistent with clonal origin, the number of intravenously injected BMCs generated linear dose–response curves of macroscopic haematopoietic spleen colonies (Till and McCulloch, 1961). (ii) Radiation log-linear dose–response killing curves of donor BMCs suggested fewer than two hits per colony were necessary for killing (Till and McCulloch, 1961). (iii) Examination of the chromosome spreads of donor BMCs, heavily irradiated shortly after their injection into the recipient mouse in order to generate unique chromosomal markers, showed the majority of cells within the subsequent clones to have the unique colony-specific marker karyotype (Becker *et al.*, 1963). (iv) When lethally irradiated mice were used as recipients of dispersed haematopoietic tissues to determine the number of spleen colony-forming units (CFUs), these colonies were found to be of a variable number per suspension, and each colony was found to contain both undifferentiated and differentiated cells, in variable ratios. Single colonies were therefore isolated, dispersed and used as donor suspensions in a new round of irradiated recipients, at one colony per recipient; the growth of CFUs from single colonies was suggested by statistical analysis to follow a curve most similar to that of a stochastic process of stem cell self-renewal ('birth') or stem cell differentiation ('death'), with the variance between colonies exceeding the mean value (Till *et al.*, 1964). (v) Again using the SCA, physical separation studies distinguished cells with the greatest self-renewing ability as the lowest velocity-sedimenting population, indicating that the probability of self-renewal versus differentiation of a single cell was perhaps not completely random, after all (Worton *et al.*, 1969a). Fractions of bone marrow suspensions separated by both velocity sedimentation and equilibrium density-gradient sedimentation were assayed both for CFU capacity by *in vivo* SCA (CFU-S) and for CFU capacity in an *in vitro* culture assay (CFU-C). Although fractions contained CFU capacity in both CFU assays, there were peaks of CFU *in vitro* capacity that did not overlap with CFU *in vivo* capacity, and vice versa, indicating potential multiple cellular subtypes and putative complex environmental interactions regulating stem cell 'birth' and 'death' – self-renewal and differentiation – respectively (Worton *et al.*, 1969b). (vi) CFU assays showed that a single clone of radiation-karyotypically-marked donor CFU spleen cells was able to repopulate thymus, bone marrow and lymph nodes, as both erythropoietic and granulocytic lineages, an observation consistent with the pluripotency of BMSCs (Wu *et al.*, 1968). Some years later, asymmetric cell division of BMSCs was shown directly by timelapse videos (Wu *et al.*, 2007). All of these studies framed the definition of stem cells, as well as many of the questions still being asked about cell autonomy and cellular interplay with the environment or niche.

Normal stem cells are generally classified as pluripotent, capable of generating the next descendant cell types, or multipotent, generating fewer of the descendant types. Normal stem cells appear to function constitutively in the bone marrow, intestine and skin; their activation is enhanced by toxicity and disease (Potten and Loeffler, 1990; Deasy *et al.*, 2003; Martin-Belmonte and Perez-Moreno, 2012). Stem cell frequencies range from 1×10^{-13} cells/20 g mouse (e.g. the totipotent zygote) to $\leq 1 \times 10^{-4}$ cells/tissue (e.g. bone marrow: Spangrude *et al.*, 1988). Stem cells, by definition, can divide in at least two ways (Cairns, 1975, 2002; Potten *et al.*, 1978; Deasy *et al.*, 2003; Wu *et al.*, 2007): symmetrically (1 stem cell → 2 undifferentiated daughter stem cells) and asymmetrically (1 stem cell → 1 undifferentiated stem cell +1 differentiated daughter). Generally, the stem cell is thought to replenish its pool by self-renewal, but otherwise to be relatively quiescent, while the differentiated daughter cell can expand itself exponentially, according to need or injury.

Hypotheses of the inheritance of parental DNA by nonrandom sister chromatid segregation have been developed to postulate protection from accumulation of mutation and genomic rearrangement within a stem cell population (Cairns, 1975, 2006). The immortal-strand hypothesis is difficult to validate or prove experimentally, and has putatively been shown in some tissues and not in others (Charville and Rando, 2013; Falconer and Lansdorp, 2013; Yadlapalli and Yamashita, 2013; Yennek and Tajbakhsh, 2013), but DNA label-retention analysis, used to take advantage of the functional characteristic of slow turnover by some stem cells, has provided a measure of stem cell location within some organ systems, in spite of the need to activate a quiescent population with low-grade injury in order to label it for visualization. This technique has been used to good effect to label four different stem cell populations in the mouse liver (Kuwahara *et al.*, 2008). By injection of sublethal acetaminophen (APAP) and bromodeoxyuridine (BrdU), followed 14 days later with a large chase dose of APAP, the BrdU is diluted from the differentiated daughter cells, which are presumed to be proliferating rapidly in response to the injury, but is retained in the SC daughter cell, also having incorporated BrdU during the division following the first APAP, but having returned to quiescence by the time of the second injury. The Canals of Hering harboured one population of label-retaining cells (LRCs); OCs have historically been associated with this niche. A second population of LRCs was near the OCs, adjacent to the parenchyma and to the biliary ductules at the Canals of Hering; these were called small peribiliary hepatocytes, and by serial section examination, appear possibly to have been a descendant population of OCs. Both OCs and peribiliary small hepatocytes still retained label at 8 weeks post APAP/BudR. If these *are* found to be descendants of OCs, then an asymmetric division will presumably have produced two stem cells of distinct potencies, an event clearly common throughout normal embryological development but occasioned by toxicity in this case. A third population of LRCs was not within the parenchyma, but instead consisted of occasional, labelled cholangiocytes. The fourth population was small periductular cells with characteristics of 'null' cells, often considered to be a subtype of OCs that does not express AFP or other hepatobiliary markers; 'null cells' historically respond to allyl alcohol toxicity in the rat (Yavorkovsky *et al.*, 1995). There is as yet no definitive lineage connecting these stem cell types to one another, or to hepatocytes, tumour progenitors, TICs, HcPCs, LCSCs, HCCs or any other liver cancer.

14.2 Experimental evidence of LCSCs

14.2.1 Overview

CSC hallmarks of LCSC candidates, reported from investigations of phenotypes shared by various CSCs and nontumour cells (Sell and Leffert, 2008; Magee *et al.*, 2012; Valent *et al.*, 2012), include spheroid formation *in vitro*, side population (SP) formation, the expression of drug resistance and many CSC-associated markers, limiting dilution xenoengraftment; key evidence of asymmetric cell division is limited. There is evidence that OCs express some hallmarks of TICs, LCSCs and HCCs. Future LCSC investigations might include the hepatic properties of polyploidy, carcinogen bystander effects and definitions of niche on formation, on establishment and on migration to a new environment.

14.2.2 LCSCs and spheroids

Many CSCs form spheroids, but whether spheroid formation is a consistent property of an isolate of primary LCSCs is unclear. However, several observations of different aggregated forms have been made in cultured HCC-derived LCSC-like cell (LCSCLC) lines and strains, as well as in HcPC, and even in normal hepatocyte cultures (see Table 14.1).

A study of spheroid formation in cultures of LCSCLCs from $Ikk\beta^{F/F}$ ($Ikk\beta$ normal) mice has been reported (He *et al.*, 2010). Putative LCSCLC strains isolated from DEN-initiated HCCs in $Ikk\beta^{F/F}$ mice were established in culture by supplementation with phenobarbital and epidermal growth factor (EGF). Three of the DEN-induced HCC (dih) strains, dih10, 11 and 12, all expressing AFP and albumin (consistent with formation from AFP^+ HCCs), were selected and found to form spheroids under crowded culture conditions. Because it was previously found that hepatocyte-targeted deletion of $Ikk\beta$ enhanced dih formation in these $Ikk\beta^{\Delta HEP}$ mice compared to their $Ikk\beta^{F/F}$ siblings, an adenovirus, Adv-Cre, infection of the dih10 $Ikk\beta^{F/F}$ strain of LCSCLC was used to delete $Ikk\beta$ from dih10, thereby producing a deficient substrain, $Ikk\beta\Delta$ *dih*10. Under serum-free conditions in Petri plates, the cells of $Ikk\beta\Delta$ *dih*10 formed two and three times more floating spheroids in 1° and 2° spheroid cultures, respectively. Spheroid formation was also enhanced in $Ikk\beta^{F/F}$ *dih* cells transduced with an IκBα super-repressor, again suggesting that IKKβ inhibited spheroid formation. For evaluation of the tumourigenicity of spheroid-forming *dih* cells, $Ikk\beta^{F/F}$ and $Ikk\beta\Delta$ *dih*10 cells

Table 14.1 Spheroid formation in normal, CSC and CSC- and LCSC-like cells.

Cell or tumour type	Cell system(s)	References
Normal rat liver	Hepatocyte cell suspensions	Jeejeebhoy *et al.* (1975)
Paediatric HCC	Single cells from tumour biopsy	Tomuleasa *et al.* (2010)
Mouse HcPCs	Aggregate-forming hepatocytes from DEN-treated BL/6: precursors to HCCs in chronically injured liver	He *et al.* (2013)
$EpCAM^+$ HCC	*In vitro* malignant cell lines	Terris *et al.* (2010)
$CD133^+$ HCC	±BMP4-sensitive *in vitro* malignant cell lines	Zhang *et al.* (2012); Ma (2013)
Prostate	*In vitro* malignant cell lines without or with fibroblasts; intact glands	Wartenberg *et al.* (2001); Khoei *et al.* (2004); Liao *et al.* (2010)
Cervical	*In vitro* malignant cell lines	Olive *et al.* (2004)
Mammary	*In vitro* malignant cell lines	Günther *et al.* (2007); Krohn *et al.* (2009)
Colorectal	*In vitro* malignant cell lines in three-dimensional cultures	Valcárcel *et al.* (2008)
Glioblastoma	Dissociated 1° tumours	Ernst *et al.* (2009)
Melanoma	*In vitro* malignant cell lines	Na *et al.* (2009)
Neck	*In vitro* malignant cell lines	Hirschhaeuser *et al.* (2010)
Mouse DEN-initiated HCC	*In vitro* AFP^+ spheroid-forming cell strains from DEN-initiated HCCs (dih) from $Ikk\beta^{F/F}$ and $Ikk\beta\Delta$ hepatocytes	He *et al.* (2010)

were transplanted subcutaneously into C57BL/6 mice. $Ikk\beta\Delta$ cells grew faster than $Ikk\beta^{F/F}$ cells; after 6 weeks, $Ikk\beta\Delta$ 2° HCC volumes were four times greater than those of $Ikk\beta^{F/F}$ HCCs, and BrdU labelling indices (LIs) were increased, as expected (Koch *et al.*, 2009; He *et al.*, 2010). Likewise, HCCs from $Ikk\beta\Delta$ dih12 cells had greater proliferative indices than those of $Ikk\beta^{F/F}$ dih12 when grown in MUP/uPA mice. These experiments suggest that IKKβ suppresses both spheroid formation and tumourigenicity. Of particular interest is that these spheroid-forming dih cells, isolated from HCCs, are able to generate secondary HCCs after subcutaneous transplantation into C57BL/6 mice, whereas the aggregate-forming HcPC tumour progenitor cells, isolated from DEN-treated liver prior to tumour formation, do not form HCC under this condition (He *et al.*, 2013), distinguishing clearly between tumour progenitor and LCSCLC.

14.2.3 LCSCs and SPs

SP formation has been investigated widely with respect to CSCs and to LCSCs (Chiba *et al.*, 2006). Following staining with Hoechst 33342 dye, cells are separated by fluorescence-activated cell sorting (FACS) into SP and non-SP, based upon active uptake and efflux of the dye by SP via ATP-dependent membrane transporter, ABCG2, and to a lesser extent ABCB1 (Golebiewska *et al.*, 2011; Liu *et al.*, 2011). Separation of SPs is particularly effective when both SP and non-SP fractions are used for comparison to other signature marker distributions, such as the correspondence to shRNA knockdown of cMyc, or spheroid formation, or tumourigenicity (see Holczbauer *et al.*, 2013).

Notably, Hoechst dye can be cytotoxic, and some of the fractional differences in tumour initiation or transplantation, for example, may be a result of this (Sell and Leffert, 2008; Golebiewska *et al.*, 2011). Finally, other ABC transporters, such as ABCB1, also effect uptake and efflux of Hoechst 33342, although with less efficiency than ABCG2. The cellular heterogeneity of transporters within a population, as well as culture and growth state conditions, can lead to variable results, such that the SP determination is not yet an accurate measure of specific transporter activity (Smith *et al.*, 2013).

14.2.4 LCSCs and drug resistance

Enhanced HCC recurrence and reduced patient survival are characterized by overexpression of ATP-dependent membrane ABC transporters, including the multigene multidrug resistance (MDR/P-glycoprotein) family containing ABCB1, which is particularly studied in liver (Kato *et al.*, 2001). These glycoproteins, and other transporter family members, accelerate drug efflux during chemotherapy (Dean *et al.*, 2005; D'Alessandro *et al.*, 2007; Sukowati *et al.*, 2010; Golebiewska *et al.*, 2011; Liu *et al.*, 2011; Oishi and Wang, 2011). HCC procarcinogens like aflatoxin B_1 and AAF upregulate the formation of reactive oxygen species (ROS) and stimulate transporter overexpression (Kuo and Savaraj, 2006). This overexpression has been reported in $CD90^+CD133^+$ LCSCL spheroid-forming 1° HCC cultures (Tomuleasa *et al.*, 2010) and in xenoengraftable $CD44^+CD133^+$ HCC cell lines (Zhu *et al.*, 2010). As previously noted, the expression of some of the transporters to effect rapid efflux of Hoechst dyes is the basis of SP separations, which presently lack specificity of identification of the many members in the large drug-resistance gene family. Although many cells, particularly in the liver, express drug-resistance genes, knowledge of the dynamic spectrum of drug-resistance expression in the putative LCSC population may be critical for potential detection and targeting if LCSCs are found to form a drug-resistant pool of liver cancer precursors which seed recurrent tumours.

14.2.5 LCSCs and LCSC-like markers

Table 14.2 summarizes the properties of some markers considered important in the classification of LCSCs and LCSCLCs (Klonisch *et al.*, 2008; Sell and Leffert, 2008; Liu *et al.*, 2011; Yamashita and Wang, 2013). For example, poor survival and accelerated relapse are reported in human patients with HCCs expressing CD13, CD24, CD44, CD90, CD133, Keratin 19, EpCam, OV6, AFP and *SALL4*. Just as expression in individual HCCs is variable, most

Table 14.2 Markers of putative LCSCs and LCSCLCs.

Marker or marker set	Comments	References
CD13	Enhances growth of human HCC cell lines and inhibits *in vivo* relapse of $CD13^+$ HCCs	Haraguchi *et al.* (2010)
CD24	Mouse TICs form $CD24^+$ HCCs through STAT3-mediated NANOG regulation; sporadic expression in human tissue HCCs	Lee *et al.* (2011)
CD44	$CD44^+CD133^+$ HCC cell lines; both markers confer aggressive xenoengraftment	Zhu *et al.* (2010)
CD90	$CD90^+$ HCC human cell lines and tissue HCCs; $CD90^+CD44^+$ HCC cells generate HCC nodules in liver and produce lung metastases in SCID mice	Yang Z.F. *et al.* (2008)
CD133	Most prevalent HCC CD marker, induced by HBV and HCC infections; HCC expression predicts reduced survival and confers malignant potential by regulating metalloproteinases in HCC cell lines; BMP4 induces differentiation in HCC lines and blocks their tumorigenicity and drug resistance after xenoengraftment; BMP4 can be isolated from normal mouse liver	Ma *et al.* (2007, 2013); Kohga *et al.* (2010); Rountree *et al.* (2011); Sasaki *et al.* (2010); Zhang *et al.* (2012); Ma (2013)
EpCAM	Human $EpCAM^+AFP^+$ HCC tissues display properties of self-renewal and aggressive xenoengraftment	Terris *et al.* (2010)
β-catenin[a]	Human mutation and overexpression, and depletion of hepatocyte β-catenin in adult knockout mice promote LSC and HCC formation; poor survival is also seen in human patients	Wong *et al.* (2001); Wang *et al.* (2011)
Keratin 19	Poor prognosis for patients with keratin 19^+ HCCs; these HCCs also express CD133 and EpCam	Kim *et al.* (2011)
Anti-Ca^{2+} channel MnAb 1B50-1	Targets human TICs from HCC tissues; binds to HCC Ca^{2+} channel $\alpha 2\delta 1$ subunit; HCC lines also express CD13, CD133, AFP and EpCam	Zhao *et al.* (2013)
h-Dlk-1	Associated with foetal liver progenitor cells; expressed in 20% of adult human HCCs, and more in patients younger than 50 with AFP^+ HCC	Yanai *et al.* (2010); Nishina (2012)
CD44, CD133, EpCAM, CD90 and AFP	Unique combinations of CDs and other LCSCLC markers are expressed in HCC tissues from different patients, and in $CD133^+CD90^+$ biopsies	Tomuleasa *et al.* (2010); van Malenstein *et al.* (2011); Wilson *et al.* (2013)

(*continued overleaf*)

Table 14.2 (*continued*)

Marker or marker set	Comments	References
SALL4	Oncofetal protein not found in normal hepatocytes but high in a subset of human xenoengraftable HCCs; confers poor prognosis; *SALL4* elevates PTEN expression; PTEN fusion TAT peptide reduces HCC cell line formation in SCID mice	Yong *at al.* (2013)
$ALDH^{HIGH}$	Co-expression of $CD133^{+}CD90^{+}$ is correlated with human tissue $ALDH^{HIGH}$ LCSCLCs	Lingala *et al.* (2010)
SPs	Expression of CSC and LCSCLC SPs is controversial	Sell and Leffert (2008); Golebiewska *et al.* (2011)

[a] Marker-related effects.
MnAb, monoclonal antibody; TAT, transactivator of transcription protein transduction domain; ALDH, aldehyde dehydrogenase

markers are also expressed variably in isolated putative TICs and LCSCs. One of the longstanding difficulties in identifying and isolating LSCs and LCSCs has been the lack of specific markers by which to distinguish these cells. Identification now rests upon the functional and structural contexts in which these cells are found, in conjunction with expression of a panel of likely markers. The current genome-transcriptome-proteome projects should increase the prospects for specific markers, or perhaps unique subsets of markers.

14.2.6 LCSCs and limiting-dilution xenoengraftment

Co-expression of $CD44^{+}CD133^{+}$ in human HCC cell lines appears to correlate with augmented xenoengraftment in severe combined immunodeficiency disease (SCID) mice (Ma *et al.*, 2007; Zhu *et al.*, 2010; Ma, 2013). Similar observations have been made in HCC lines that overexpress Bm1 (Wang *et al.*, 2008), the expression of which is also associated with reduced patient survival (Chiba *et al.*, 2007). Heightened xenoengraftment of human $CD90^{+}$, $OV6^{+}$ and $SALL4^{+}$ LCSCL tissues in immunocompromised mice has also been reported (Table 14.2) (Yang Z.F. *et al.*, 2008; Yong *et al.*, 2013). Recent attempts to xenoengraft or transplant with low doses of LCSCLCs have also been reported.

In one case, in order to isolate LCSCs from the human HCC line PLC/PRF/5, label-retaining cancer cells (LRCCs) were sorted and collected by FACS for high-Cy5 dUTP (Cy5) labelled DNA, following one labelling cell cycle. After this first sorting into high and lowest Cy5, the high fraction was grown for eight more cycles without Cy5; the high Cy5 ($Cy5^{high}$) and the lowest Cy5 ($Cy5^{low}$) were then again sorted and collected. After this, either 10 $Cy5^{high}$

cells or 10 Cy5^{low} cells were injected subcutaneously into nude/SCID mice and followed for 16 weeks. From the non-LRCC fraction, 2 of 20 mice developed tumours; from the Cy5^{high} fraction, 14 of 20 had tumours (Xin *et al.*, 2012).

In the second report, 100 cells of the CD44-positive fraction of C57BL/6 HcPCs, collected 3–5 months after DEN treatment, were transplanted intrasplenically into MUP/uPA mice. Zero of five and zero of three mice formed tumours after dosing with 100 sorted CD44-negative cells and 100 unsorted cells, respectively, whereas a dose of 100 sorted CD44-positive cells generated tumours in two of four mice. At higher doses, all three groups formed tumours in all mice. These HcPCs required the chronically injured MUP/uPA mouse as recipient, and they would not form subcutaneous tumours in BL/6 mice as the dih cells would do. The dih strains are LCSCLCs, having been isolated from HCCs, whereas the HcPCs are tumour-initiating or tumour progenitor cells, having been isolated before the appearance of a tumour (He *et al.*, 2010, 2013).

Finally, there is one report of xenoengraftment of LCSCLCs isolated from primary human HCC tumour (Muramatsu *et al.*, 2013). These cells were enriched with a nice system to mark asymmetry of division (see Section 14.2.7). Following subcutaneous injection of a range of doses into nonobese diabetic (NOD)/SCID mice, 100 of the self-renewing enriched cells produced HCC in three of six mice, while 100 of the unsorted cells produced tumours in zero of six mice. The low 26S proteasome activity utilized in the asymmetry determination for sorting these LCSCLC populations had previously been found to be distinctive of CSCs of human glioma and breast, and low ROS can be characteristic of haematopoietic stem cells and breast CSCs.

14.2.7 LCSCs and asymmetric cell division

In two of the systems reporting enhanced tumourigenicity in xenoengraftment studies, the asymmetry of cell division was the distinguishing characteristic by which the cells were sorted. Both studies capture live asymmetric division of putative LCSCs: from HCC cell lines in the first case (Hari *et al.*, 2011; Xin *et al.*, 2012) and from cultures established from four primary human resected HCCs, as well as from cell lines, in the second (Muramatsu *et al.*, 2013). In the first case, capture was with confocal microscopy cinematography; in the second, with 21-hour timelapse video microscopy. The first system utilized DNA labelling strategies to examine the roles of nonrandom sister chromatid segregation in asymmetric division and used label retention to examine asymmetry and/or slow cell cycling.

In order to test for nonrandom sister chromatid segregation, human HCC line PLC/PRF/5 cells were grown for one DNA cycle with microporated Cy5-dUTP, then sorted by FACS to enrich for a Cy5^{high} population. These cells were grown through the second cycle with Alexa555-dUTP, and the population was then sorted by FACS into Alexa555^{high} and into Alexa555^{high} + Cy5^{high}. A pair of cells, each with both Alexa555 and Cy5 nuclear labelling, were daughter cells of a symmetrical division, while another pair of cells, one labelled with Alexa555 alone, and the other with both Alexa555 and Cy5, had undergone asymmetric division, presumably by one cell retaining the parental strands. That this asymmetric division had occurred was confirmed by Z-stack registration from confocal microscopy of single cells in cytokinesis, with one dual-labelled nucleus and one singly-labelled nucleus, both still in the same cytoplasm, and both still held together by a bridge of this cytoplasm. To examine label retention, the same PLC/PRF/5 cells were singly labelled for the first cycle with Cy5-dUTP, as previously, and then FACS sorted for collection of $Cy5^{high}$. Following eight more cycles of growth without label, consistently more cells were $Cy5^{high}$ labelled than expected by dilution from symmetric division: 1.54–5.00%, versus 0.39% expected, suggesting that significant label retention had occurred by cell cycle slowing or asymmetric division, possibly by nonrandom chromosome segregation. It was this population of $Cy5^{high}$ cells that was further collected for tumourigenicity studies by xenoengraftment, as described earlier (Xin *et al.*, 2012).

The second study takes advantage of the distinctive metabolic properties of low 26S proteasome and low mitochondrial superoxide, characteristic of CSCs of glioma, breast and haematopoietic systems. The four cultures from freshly isolated human HCC resections were retrovirally transduced with the ZsGreen-fused degron sequence of ODC protein (Gdeg). This is a specific substrate for 26S proteasome degradation, so that expression of the ZsGreen reporter is inversely proportional to 26S proteasome activity. In order to measure ROS, cells can be incubated with MitoSOX Red and can be sorted by FACS for either of these metabolic markers. The percentage of $Gdeg^{high}$ cells in the primary HCC populations was low, at 0.1%. After four passages of growth to expand the transduced human primary HCC cells, they were sorted by FACS into $Gdeg^{high}$ and $Gdeg^{low}$ and grown in separate culture for 21-hour timelapse video microscopy (Figure 14.3). $Gdeg^{high}$ cells divided asymmetrically into a $Gdeg^{high}$ cell and a $Gdeg^{low}$ cell, but $Gdeg^{low}$ cells never divided into $Gdeg^{high}$, only into more $Gdeg^{low}$. As already discussed, the $Gdeg^{high}$ ROS^{low} cells were significantly more tumourigenic in NOD/SCID xenoengraftment than unsorted cells (Muramatsu *et al.*, 2013).

14.2.8 OCs as LSCs

With regard to OCs, the principal question is whether LSC → OC → LCSC transitions occur. Are OCs an LSC, or a type of LSC? Do they become LCSCs? Developmental liver lineage studies indicate that embryonic endodermal and mesodermal interaction precedes the appearance of embryonic liver bud and progenitor cells, and that these in turn precede the formation of embryonic hepatocytes and bile duct epithelia (BDE) (Lemaigre and Zaret, 2004), but the exact lineage of adult LSCs with respect to these embryonic liver progenitor cells is presently unknown with completeness.

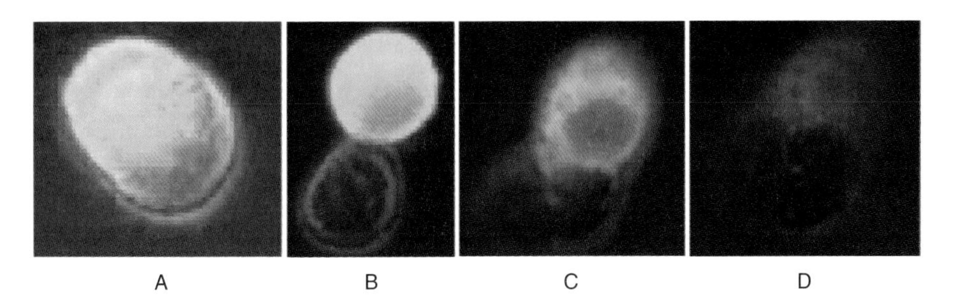

Figure 14.3 Asymmetric cell division of a single Gdeg^{high} HCC cell. Panels (A)-(D), from left to right, are single frames from time lapse videomicroscopy over a period of ~21 hrs. Single cells are ~25 μm in diameter. (A) 0 hour, 1st cytokinesis begins. (B) 7 hour, 1st cell division ends. One green daughter (top); one unstained daughter (bottom). (C) 12 hour, unstained daughter cell divides; progeny remain unstained. Green parental cell does not divide further. (D) 21 hour, unstained daughter cells continue to divide (bottom); single green parental cell does not. See original article and supporting time lapse video; HEP_26345_sm_SuppMov.AVI. Taken from Figure 1B, p. 222, of Muramatsu S. *et al.* (2013) Visualization of stem cell features in human hepatocellular carcinoma reveals in vivo significance of tumor-host interaction and clinical course. *Hepatology*, 58(1), 218–228, © 2013 American Association for the Study of Liver Diseases; Reproduced with kind permission of John Wiley and Sons.

Liver OCs, widely considered to be rare adult LSCs, were first reported in chemical hepatocarcinogenesis studies (Price *et al.*, 1952; Farber, 1956; Abelev, 1971; Shinozuka *et al.*, 1978; reviewed in Sell and Leffert, 1982, 2008; Koch and Leffert, 2004). OCs appear, or increase greatly in number, after some types of injury, but decrease with age and tissue turnover (Guest *et al.*, 2010; Sell, 2010). They are ≤6 μm diameter, whereas binucleated hepatocytes are ~40 × 60 μm in size (see Sell and Leffert, 1982). OCs are rare; given the fact that they are really only detectable following extensive proliferation after injury, and yet, therefore, still assuming that they exist in the normal animal in very small numbers, their numbers have been estimated at 1×10^{-5} cells/0.8 g mouse liver (Braun and Sandgren, 2000). OCs show high nucleus to cytoplasm ratios (see Sell and Leffert, 1982) and express AFP (see Sell, 2010), CK19, CK7 and many other markers, according to growth state, species and specific toxic and carcinogenic conditions.

OCs are thought to be involved in normal hepatocellular repair when hepatocyte proliferation is blocked by carcinogens or other toxic chemicals (Trautwein *et al.*, 1999; Yovchev *et al.*, 2008). Depending upon the toxin or chemical carcinogen, this OC proliferative process can encompass injury to bile ducts, as well as to hepatocytes and NPCs in different lobular zones. Adult OCs are widely said to be bipotential: they appear to be able to form hepatocytes and BDE. When excessive hepatotoxicity occurs, OCs are activated to grow periportally, or adjacent to injury sites, where they are thought to differentiate and compensate for tissue deficits. For example, AFP^{+} OCs are induced periportally in the location of alcohol dehyrogenase (Leffert *et al.*, 1987), after zone 1-specific allyl alcohol injury (Guest *et al.*, 2010), and are ultimately found dispersed but initially centrilobularly, in the location of Cyp2E1 and Cyp2B1/2 carcinogen and drug-metabolizing enzymes (Guengerich, 2000), after zone 3-specific CCl_4 injury and DEN toxicity. It has been suggested for some time, following very careful [3H]thymidine labelling and microscopy studies in mice treated with one dose of Dipin followed by partial hepatectomy, that a small pool of OCs within the Canal of Hering and in contact with immediately adjacent periportal hepatocytes progresses through a series of transitional cells to small hepatocytes, while apparently spreading portally to centrally through the liver acini, to replace damaged and dying hepatocytes. Following common bile duct ligation (BDL), on the other hand, bile duct proliferation occurred in the portal area, which appeared to proceed by way of BDE cells, which were larger and had a lower nucleus to cytoplasm ratio than OCs. In this model, microscopy did not support these cells proceeding to hepatocytes, but only to BDE. It was noted that a progression of OCs could indicate that there were a number of OC subtypes in various stages of growth and differentiation, or that the progression through various OC specifications was sensitive to variable microenvironment (Factor and Radaeva, 1993; Factor *et al.* 1994). It seems possible that these Canal of Hering OCs and portal BDEs represent cells from two of the Kuwahara label-retaining niches (Kuwahara *et al.*, 2008; see Section 14.1.5).

Several recent fate-mapping investigations have suggested independently that some subtypes of mouse OCs, or LSCs, generated following different conditions of adult liver injury, can be traced from cells marked during embryogenesis; or, possibly, following certain types of liver injury, from HSCs, marked during adult quiescence, prior to injury. Short descriptions of these studies follow:

1. The progeny of a *Foxl1*-Cre cross to Rosa26R lacZ mice, subjected at 10–12 weeks of age to BDL, to toxic liver injury with an OC inducer (3,5-diethoxycarbamyl-1-4-dihydrocollidine, DDC) or to a choline deficient/ethionine (CDE) carcinogen diet, and followed for 21 days, exhibited a significant immediate induction of β-gal$^+$ cells in periportal bile ductular reactions, which, in time, extended, by appearance of blue hepatocyte-appearing cells, out towards the parenchyma. By 3 days post BDL, some double-positive, blue β-gal$^+$ CK19$^+$ cells (suggestive of BDE-type differentiation) were seen in the portal bile ductular reactions; by 5 days, some blue β-gal$^+$ hepatocyte-like cells appeared around the portal biliary reactions; and by 14 days, a few double-positive blue β-gal$^+$ HNF-4-α$^+$ cells (suggestive of heptocellular differentiation) were seen as the β-gal$^+$ CK19$^+$ cells began to wane and a population of β-gal$^+$ CK19$^-$HNF-4-α$^-$ cells was peaking. The E12.5 liver-bud primordium did not stain blue in this mouse, suggesting that Foxl1 was not expressed in these embryonic cells; it is likely that the blue β-gal$^+$ cells originated from somewhere other than the hepatic liver bud, or that expression of Foxl1 occurs later in development. There were negligible to no blue cholangiocytes in sham animals. Both DDC injury and CDE carcinogen

diet also induced rapid and large β-gal$^+$ ductular reaction populations, and, within days, subsets of these were β-gal$^+$ CK19$^+$, but followed by many fewer β-gal$^+$ HNF-4-α$^+$ cells in comparison to the BDL animals. In all three injury models, the initially observed blue cells appeared to be in the intraductal cholangiocyte, and in the peribiliary label-retaining niches of Kuwahara (Kuwahara *et al.*, 2008; see Section 14.1.5). Because no triple-positive cells were scored and there was a β-gal$^+$ CK19$^-$HNF-4-α$^-$ population peak, intermediate in time, it seems possible that either the initial Foxl1-positive population is composed of two subtypes or a single population is bipotential (Sackett *et al.*, 2009).

2. At least two *Sox9* promoter-tamoxifen inducible-Cre;Rosa26R mouse model systems were employed to mark Sox9-expressing cells and their descendants, generally with YFP or lacZ and GFP Rosa reporters. Because *Sox9* expression in embryogenesis is exclusive at E15.5 to ductal plate-forming cells, transient tamoxifen injection of the pregnant *Sox9*-Cre;Rosa YFP females at E15.5 caused Cre recombination in the *Sox9*-expressing ductal plate cells, thereby marking them and their descendants with constitutively expressed YFP. Subsequent examination of these livers 4 weeks postnatally indicated YFP$^+$ cholangiocytes of the interlobular bile ducts and intralobular biliary ductules and the Canals of Hering, as well as a small number of zone 1 YFP$^+$ hepatocytes. Following known OC-inducing CDE and DDC diets, given at 4 weeks postnatally, for 21 and 14 days, respectively, to these E15.5 Sox9$^+$ ductal plate YFP-marked animals, these livers exhibited significant YFP$^+$ periportal OC proliferation, suggesting that at least one significant subtype of OC had been derived from the ductal plate cells. Perhaps because these diets do not cause significant liver injury when given briefly, there was no apparent accumulation of new YFP$^+$ hepatocytes greater than the limited population of control YFP$^+$ periportal hepatocyes. There is no evidence as yet for a lineage from the ductal plate *Sox9* OC to the hepatocyte, the test of which will perhaps require a different or more chronic injury model than the protocols used here. Nevertheless, in this system, except for the limited periportal YFP$^+$ hepatocytes, the initial population of hepatocytes did not appear to have originated from the E15.5 Sox9$^+$ ductal plate cells, nor did these marked ductal plate cells or progeny appear to be involved in normal hepatocellular homeostasis up to 4 weeks postnatally (Carpentier *et al.*, 2011; Dorrell *et al.*, 2011). Finally, on the other hand, examination of the liver of a Rosa26R LacZ mouse containing a different *Sox9*-Cre construct, in which the tamoxifen induction-Cre recombinase sequence was placed within the 3′ untranslated region of the native *Sox9* locus, rather than within a bacterial artificial chromosome as in the previously described system, revealed β-gal labelling of a *Sox9*-expressing biliary progenitor pool within 1 day following high-dose tamoxifen injection at 8 weeks postnatally. Within 10–30 days post injection, the hepatocyte population became β-gal labelled from the portal to the centrilobular regions, in

contrast to any progression following from the E15.5 marking in the previous *Sox9* system. Although a subtype of 8-week-postnatal hepatocytes might actually express *Sox9*, sufficiently for Cre-recombination, in which case tamoxifen-induced β-gal labelling of these hepatocytes would be indistinguishable from labelling as a descendant of any other labelling, it might also be possible that the 8-week-postnatal intrahepatic biliary cells were expressing *Sox9*, and had, in fact, become the precursors to hepatocytes. It is not really possible to compare results from tamoxifen induction at E15.5 with those from induction at postnatal 8 weeks. There is, in fact, a suggestion that there is a 'switch' from the early embryonic hepatocyte origin from *Sox9*$^-$ hepatoblasts to postnatal origin from *Sox9*$^+$ intrahepatic biliary cells. It is interesting to note that in this system, the postnatally marked Sox9 biliary cells did not appear to contribute to the hepatocellular proliferative response following DDC injury or partial hepatectomy, but were a precursor to a marked hepatocyte component following a methionine/choline deficient/ethionine (MCDE) diet (Furuyama *et al.*, 2011). Clearly, these animal systems are complex, and promising.

3. The progeny of a glial fibrillary acidic protein (GFAP)-Cre X Rosa GFP mouse cross (GFAP-Cre/GFP) were used to investigate the fate of GFAP-expressing, quiescent hepatic stellate cells (Q-HSC), which downregulate GFAP, synthesize alpha smooth-muscle actin (α-SMA) and begin to proliferate as myofibroblastic hepatic stellate cells (MF-HSCs) following chronic liver injury, such as occurs in rodents with an MCDE diet. In normal, untreated 3–4-month-old GFAP-Cre/GFP mice, there were GFAP$^+$GFP$^+$ sinusoidal Q-HSCs, as well as unexpected GFAP$^+$ GFP$^+$ portal bile ductular cells, and neither GFAP$^+$ nor GFP$^+$ hepatocytes. Following MCDE diet administration, the portal biliary ductular cells remained GFAP$^+$GFP$^+$ and did not proliferate significantly, whereas the Q-HSCs became GFAP$^-$GFP$^+$ and proliferated greatly, particularly within the centrilobular and midzonal sinusoids. Over a 6-week period, involving diet for 3 weeks and return to normal for 3 weeks, a few isolated, and then increasingly large numbers of hepatocytes became GFP$^+$ (GFAP$^-$), from centrilobular to midzonal to portal. Most of the proliferating GFAP$^-$GFP$^+$ HSCs began to express the myofibroblastic α-SMA progressively from central to portal sinusoids, followed by an occasional GFP$^+$α-SMA$^+$ hepatocyte, also expressing progenitor AE1/AE3 reactive cytokeratins, in the same zones; then, as the α-SMA expression declined rapidly in HSCs, and then in hepatocytes, and the GFP$^+$ HSCs disappeared, a large proliferation of albumin-expressing GFP$^+$ hepatocytes, negative for AE1/AE3 cytokeratins, spread periportally, ultimately making up one-third of the hepatocellular population. It would appear that, following MCDE injury in these mice, HSCs might

give rise to mature hepatocytes by undergoing a transition through a hepatic progenitor (Yang L. *et al.*, 2008).

To date, these fate-mapping studies have raised many new questions, but have not yet mapped OCs → HCCs or OCs → LCSCs. Are early alterations proceeding from liver injury found in the proliferating population of responding OCs, or are there changes in other injured (adjacent) cells, such as hepatocytes? And which altered cells lead to HCCs or to LCSCs? It is, perhaps, instructive in this regard that mouse adult hepatic progenitor cells (HPCs), hepatoblasts (HBs) and adult hepatocytes (AHs), all transduced with H-Ras and SV40LT, contained SPs with LCSC properties. All three transduced cells of origin, following single-cell clonal expansion and transplantation of each line, generated HCC, cholangiocarcinoma (CCA) and combined hepatocellular cholangiocarcinoma (CHC), but in consistently different lineage-based ratios, thus, presumably, indicating that single cells from these transduced lines were multipotent but still retained the signature of their lineage of origin. Transcriptome comparison with the cells of origin indicated that the AHs required the greatest dysregulation of gene expression in the acquisition of tumour-initiating capacity or tumour formation, followed by HBs and HPCs requiring less (Holczbauer *et al.*, 2013). It seems quite possible that the OC, its subtypes, and various injury states will be a part of this group of cells which are LCSCs, or are progressing to LCSCs, and to liver cancers.

14.2.9 LCSCs and polyploidy

Postnatal hepatocytes are mononucleated and diploid. With maturation, more than half of hepatocytes become binucleated and polyploid (Gupta, 2000; Baratta *et al.*, 2009; Celton-Morizur *et al.*, 2010). Hepatocytes are 2N, 4N (or 2N + 2N), 8N (or 4N + 4N, or 4N + 2N + 2N, etc.) or 16N (8N + 8N, etc.) per cell, or more. Polyploidy increases with liver age (Enesco and Samborski, 1983) and after 70% partial hepatectomy (Gentric *et al.*, 2012).

It has recently been found that not only is polyploidy generated, usually by failed cytokinesis, and also probably by endomitosis and endoreplication, but it is also reduced by multipolar mitotic spindle formation and/or by chromosome missegregation, to form daughter cells of reduced ploidy or of aneuploidy (Duncan *et al.*, 2010). Of particular interest is that aneuploidy in daughter cells includes abnormal chromosomal numerical value, and can also include genetically monoallelic gene expression as a result of segregation of uniparental chromosome sets. Just as the normally 5–10% of epigenetically monoallelically expressed genes provide opportunities for adaptation with diversity (Gimelbrant *et al.*, 2007), so it is proposed that each aneuploid hepatocyte or clone newly formed by ploidy reduction could represent a unique clonally expandable possibility of viable genetic diversity that, particularly under conditions of xenobiotic or chemical stress common to the liver, represents a heritable genotypic chance of resistance (Duncan *et al.*, 2010). Clearly, resistance is potentially either deleterious or beneficial.

That an extant pool of genotypically slightly diverse cells might well facilitate either resistance to or progression to tumour initiation, or to tumour evolution, is apparent by examination of the evolutionary trees of single-cell DNA sequencing of two human breast cancers and their liver metastases (Navin *et al.*, 2011). The polygenomic tumour contains only three subpopulations, each separated from the others probably by a single major genetic event and clonal expansion of the previous subpopulation, one of which is the metastatic population. There are many 'primary pseudodiploids', which have deviated probably from the earliest subpopulation but have neither expanded nor metastasized. The consequences of the ploidy genotypy, expansion and reduction within the normal or injured hepatocyte population on this process of tumourigenesis, and on the generation of LCSCs (see Zhang *et al.*, 2014, for polyploidy in an ovarian system), await further study.

14.2.10 LCSCs and bystander effects

Bystander effects have been defined as soluble signalling by radioactive metabolites via the microenvironment (Barcellos-Hoff and Brooks, 2001). Similar effects might apply to the induction of HCCs by chemical procarcinogens (Alison, 2005; Fukushima *et al.*, 2005). For instance, $Cyp1A2^-$ OCs, stimulated to proliferate in order to renew a population of hepatocytes lost by hepatocellular death due to carcinogen toxicity (e.g. AAF), might be exposed to the activated carcinogen and to subsequent mutation by diffusion of such activated carcinogen from the extant $Cyp1A2^+$ hepatocyte, acting in a paracrine manner to crossfeed undifferentiated OCs. The chronology of OC differentiation from $Cyp1A2^-$ OC to $ALB^+Cyp1A2^+$ OC may well affect the outcome of such injury. Such bystander effects represent some of the complexities in perturbation, subsequent and secondary genetic or epigenetic OC alteration and respecification of OC or LSC transitions to HCC and/or LCSC.

14.2.11 LCSCs and niche

Four stem cell niches have been identified in liver by the DNA label-retaining cells found within them (Kuwahara *et al.*, 2008; see Section 14.1.5). These cells are all putative stem cells with apparently different sizes and shapes, and each responds with proliferation to different injury models. Their developmental lineages are being slowly deciphered by recent fate-mapping studies, but there are indications that some of these cells may be a step in transition to one of the others, in which case there must be interaction between their respective niches. Many of these cells, OCs in particular, are so rare as to be visible only in proliferation following injury, and often in response to concomitant hepatocellular loss, so that it seems possible that OCs exist only as injured cells. But until proven otherwise, they are considered normal stem cells. There is no definitive evidence that they are precursors to liver cancers, to LCSCs or to tumour progenitor cells, or that their niches are potential sites of origin of LCSCs or of tumour progenitor cells.

The most reliable niche in which to activate tumour progenitor cells is the chronically injured liver, but until these progenitor cells and LCSCs can be identified for visualization before isolation, it is difficult to study them in their microenvironments. On the other hand, significant progress is being made in defining the vascular niche of the liver sinusoidal endothelial cells, and the interplay between regenerative CXCR7 and fibrotic CXCR4 'angiocrine' paracrine pathways, during acute and chronic liver injuries (Ding *et al.*, 2014).

14.3 Conclusions

Although populations of putative LSCs often proliferate massively following delivery of certain types of liver injury, such as carcinogen feeding, the origins and dispositions of these cells are not yet known; they are thought to replenish the hepatocyte population following injury, but there is currently no direct evidence for this. And because these cells appear in large numbers in response to liver injury, which is itself an environment conducive, if not requisite, for liver cancer formation, it is also possible that these proliferating LSCs are tumour progenitors. If so, it is important to determine their relationship to LCSCs, as well as the origins of LCSCs within the tumour which have already progressed to capability of full malignancy, but which, putatively, by definition, cycle within the tumour, or its environment, as a relatively quiescent stem cell parent, capable of giving rise asymmetrically to actively malignant progeny cells. The origins of these cells are also not known. Do they arise before or after overt malignant capability of the tumour? What is the lineal relationship of both of these populations to normal or injured LSCs? A clear understanding of the progression to malignancy of all these cell populations is necessary for the development of early detection and of treatment strategies.

References

Abelev, G.I. 1971. Alpha-fetoprotein in ontogenesis and its association with malignant tumors. Adv. Cancer Res. 14, 295–358.

Alison, M.R. 2005. Liver stem cells: implications for hepatocarcinogenesis. Stem Cell Rev. 1, 253–260.

American Cancer Society. 2013. Liver cancer. Available from: http://www.cancer.org/cancer/livercancer (last accessed 29 November 2014).

American Cancer Society. 2014. Liver cancer overview. Available from: http://www.cancer.org/acs/groups/cid/documents/webcontent/003058-pdf.pdf (last accessed 29 November 2014).

Baratta, J.L., Ngo, A., Lopez, B., Kasabwalla, N., Longmuir, K.J., Robertson, R.T. 2009. Cellular organization of normal mouse liver: a histological, quantitative immunocytochemical, and fine structural analysis. Histochem. Cell Biol. 131, 713–726.

Barcellos-Hoff, M.H., Brooks, A.L. 2001. Extracellular signaling through the microenvironment: a hypothesis relating carcinogenesis, bystander effects, and genomic instability. Radiat. Res. 156, 618–627.

Becker, A.J., McCulloch, E.A., Till, J.E. 1963. Cytological demonstration of the clonal nature of spleen colonies derived from transplanted mouse marrow cells. Nature 197, 452–454.

Benvegnù, L., Fattovich, G., Noventa, F., Tremolada, F., Chemello, L., Cecchetto, A., Alberti, A. 1994. Concurrent hepatitis B and C virus infection and risk of hepatocellular carcinoma in cirrhosis. A prospective study. Cancer 74, 2442–2448.

Bralet, M.P, Pichard, V., Ferry, N. 2002. Demonstration of direct lineage between hepatocytes and hepatocellular carcinoma in diethylnitrosamine-treated rats. Hepatology 36, 623–630.

Braun, K.M., Sandgren, E.P. 2000. Cellular origin of regenerating parenchyma in a mouse model of severe hepatic injury. Am. J. Pathol. 157, 561–569.

Buch, S.C., Kondragunta, V., Branch, R.A., Carr, B.I. 2008. Gender-based outcomes differences in unresectable hepatocellular carcinoma. Hepatol Int. 2, 95–101.

Bugianesi, E., Leone, N., Vanni, E., Marchesini, G., Brunello, F., Carucci, P., *et al.*, 2002. Expanding the natural history of nonalcoholic steatohepatitis: from cryptogenic cirrhosis to hepatocellular carcinoma. Gastroenterology 123, 134–140.

Cairns, J. 1975. Mutation selection and the natural history of cancer. Nature 255, 197–200.

Cairns, J. 2002. Somatic stem cells & the kinetics of mutagenesis & carcinogenesis. Proc. Nat. Acad. Sci. USA 99, 10 567–10 570.

Cairns, J. 2006. Cancer and the immortal strand hypothesis. Genetics 174, 1069–1072.

Carpentier, R., Suñer, R.E., van Hul, N., Kopp, J.L., Beaudry, J.B., Cordi, S., *et al.*, 2011. Embryonic ductal plate cells give rise to cholangiocytes, periportal hepatocytes, and adult liver progenitor cells. Gastroenterology 141(4), 1432–1438, 1438.e1–4.

Celton-Morizur, S., Merlen, G., Couton, D., Desdouets, C. 2010. Polyploidy and liver proliferation: central role of insulin signaling. Cell Cycle 9, 460–466.

Charville, G.W., Rando, T.A. 2013. The mortal strand hypothesis: Non random chromosome inheritance and the biased segregation of damaged DNA. Sem. Cell Devel. Biol. 24, 653–660.

Chiaramonte, M., Stroffolini, T., Vian, A., Stazi, M.A., Floreani, A., Lorenzoni, U., *et al.*, 1999. Rate of incidence of hepatocellular carcinoma in patients with compensated viral cirrhosis. Cancer 85, 2132–2137.

Chiba, T., Kita, K., Zheng, Y.W., Yokosuka, O., Saisho, H., Iwama, A., *et al.*, 2006. Side population purified from hepatocellular carcinoma cells harbors cancer stem cell-like properties. Hepatology 44, 240–251.

Chiba, T., Zheng, Y.W., Kita, K., Yokosuka, O., Saisho, H., Onodera, M., *et al.*, 2007. Enhanced self-renewal capability in hepatic stem/progenitor cells drives cancer initiation. Gastroenterology 133, 937–950.

Clarke, M.F., Dick, J.E., Dirks, P.B., Eaves, C.J., Jamieson, C.H., Jones, D.L., *et al.*, 2006. Cancer stem cells – perspectives on current status and future directions: AACR Workshop on cancer stem cells. Cancer Res. 66, 9339–9344.

Crispe, I.N. 2009. The liver as a lymphoid organ. Ann. Rev. Immunol. 27, 147–163.

D'Alessandro, N., Poma, P., Montalto, G. 2007. Multifactorial nature of hepatocellular carcinoma drug resistance: could plant polyphenols be helpful? World J. Gastroenterol. 13, 2037–2043.

Dean, M., Fojo, T., Bates, S. 2005. Tumor stem cells and drug resistance. Nat. Rev. 5, 275–284.

Deasy BM, Jankowski RJ, Payne TR, Cao, B., Goff, J.P., Greenberger, J.S., Huard, J. 2003. Modeling stem cell population growth: incorporating terms for proliferative heterogeneity. Stem Cells 21, 536–545.

deHemptinne, B., Leffert, H.L. 1983. Selective effects of portal blood diversion and glucagon on rat hepatocyte rates of S-phase entry and deoxyribonucleic acid synthesis. Endocrinology 112, 1224–1232.

DeLeve, L.D., Jaeschke, H., Kalra, V.K., Asahina, K., Brenner, D.A., Tsukamoto, H. 2011. 15th International Symposium on Cells of the Hepatic Sinusoid, 2010. Liver Int. 31, 762–772.

Ding, B.-S., Cao, Z., Lis, R., Nolan, D.J., Guo, P., Simons, M., *et al.*, 2014. Divergent angiocrine signals from vascular niche balance liver regeneration and fibrosis. Nature 505, 97–102.

Dorrell, C., Erker, L., Schug, J., Kopp, J.L., Canaday, P.S., Fox, A.J., *et al.*, 2011. Prospective isolation of a bipotential clonogenic liver progenitor cell in adult mice. Genes Dev. 25, 1193–1203.

Duncan, A.W., Taylor, M.H., Hickey, R.D., Hanlon Newell A.E., Lenzi, M.L., Olson, S.B., *et al.*, 2010. The ploidy conveyer of mature hepatocytes as a source of genetic variation. Nature 467, 707–710.

El-Serag, H.B. 2011. Hepatocellular carcinoma. N. Engl. J. Med. 365, 1118–1127.

El-Serag, H.B., Mason, A.C. 1999. Rising incidence of hepatocellular carcinoma in the United States. N. Engl. J. Med. 340, 745–750.

El-Serag, H.B., Rudolph, K.L. 2007. Hepatocellular carcinoma: epidemiology and molecular carcinogenesis. Gastroenterology 132, 2557–2576.

Enesco, H.E., Samborsky, J. 1983. Liver polyploidy: influence of age and of dietary restriction. Exp. Gerontol. 18, 79–87.

Ernst, A., Hofmann, S., Ahmadi, R., *et al.*, 2009. Genomic and expression profiling of glioblastoma stem cell-like spheroid cultures identifies novel tumor-relevant genes associated with survival. Clin. Cancer Res. 15, 6541–6550.

Factor, V.M., Radaeva, S.A. 1993. Oval cells – hepatocytes relationships in Dipin-induced hepatocarcinogenesis in mice. Exp. Toxicol. Pathol. 45, 239–244.

Factor, V.M., Radaeva, S.A., Thorgeirsson, S.S. 1994. Origin and fate of oval cells in dipin-induced hepatocarcinogenesis in the mouse. Am. J. Pathol. 145, 409–422.

Falconer, E., Lansdorp, P.M. 2013. Strand-seq: a unifying tool for studies of chromosome segregation. Sem. Cell. Devel. Biol. 24, 643–652.

Farber, E. 1956. Similarities in the sequence of early histological changes induced in the liver of the rat by ethionine, 2-acetylamino-fluorene, and 3′-methyl-4-dimethylaminoazobenzene. Cancer Res. 16, 142–148.

Farinati, F., Floreani, A., De Maria, N., Faqiuoli, S., Naccarato, R., Chiaramonte, M. 1994. Hepatocellular carcinoma in primary biliary cirrhosis. J. Hepatol. 21, 315–316.

Fattovich, G., Stroffolini, T., Zagni, I., Donato, F. 2004. Hepatocellular carcinoma in cirrhosis: incidence and risk factors. Gastroenterology 127, S35–S50.

Floreani, A., Farinati, F. 2013. Risk factors associated with hepatocellular carcinoma in primary biliary cirrhosis. Hepatology 58, 1520–1521.

Fukushima, S., Kinoshita, A., Puatanachokchai, R., Kushida, M., Wanibuchi, H., Morimura, K. 2005. Hormesis and dose–response-mediated mechanisms in carcinogenesis: evidence for a threshold in carcinogenicity of non-genotoxic carcinogens. Carcinogenesis 26, 1835–1845.

Furuyama, K., Kawaguchi, Y., Akiyama, H., Horiguchi, M., Kodama, S., Kuhara, T., *et al.*, 2011. Continuous cell supply from a Sox9-expressing progenitor zone in adult liver, exocrine pancreas and intestine. Nat. Genet. 43, 34–41.

Gentric, G., Celton-Morizur, S., Desdouets, C. 2012. Polyploidy and liver proliferation. Clin. Res. Hepatol. Gastroenterol. 36, 29–34.

Gimelbrant, A., Hutchinson, J.N., Thompson, B.R., Chess, A. 2007. Widespread monoallelic expression on human autosomes. Science 318, 1136–1140.

Golebiewska, A., Brons, N.H., Bjerkvig, R., Niclou, S.P. 2011. Critical appraisal of the side population assay in stem cell and cancer stem cell research. Cell Stem Cell 8, 136–147.

Gournay, J., Auvigne, I., Pichard, V., Ligeza, C., Bralet, M.P., Ferry, N. 2002. *In vivo* cell lineage analysis during chemical hepatocarcinogenesis in rats using retroviral-mediated gene transfer: evidence for dedifferentiation of mature hepatocytes. Lab Invest. 82, 781–788.

Greaves, M., Maley, C.C. 2012. Clonal evolution in cancer. Nature 481, 306–313.

Guengerich, F.P. 2000. Metabolism of chemical carcinogens. Carcinogenesis 21, 345–351.

Guest, I., Ilic, Z., Sell, S. 2010. Age dependence of oval cell responses and bile duct carcinomas in male Fischer 344 rats fed a cyclic choline-deficient, ethionine-supplemented diet. Hepatology 52, 1750–1757.

Guichard, C., Amaddeo, G., Imbeaud, S., Ladeiro, Y., Pelletier, L., Maad, I.B., *et al.*, 2012. Integrated analysis of somatic mutations and focal copy number changes identifies key genes and pathways in hepatocellular carcinoma. Nat. Genet. 44, 694–698.

Günther, S., Ruhe, C., Derikito, M.G., Böse, G., Sauer, H., Wartenberg, M. 2007. Polyphenols prevent cell shedding from mouse mammary cancer spheroids and inhibit cancer cell invasion in confrontation cultures derived from embryonic stem cells. Cancer Lett. 250, 25–35.

Gupta, S. 2000. Hepatic polyploidy and liver growth control. Semin. Cancer Biol. 10, 161–171.

Haraguchi, N., Ishii, H., Mimori, K., Tanaka, F., Ohkuma, M., Kim, H.M., *et al.*, 2010. CD13 is a therapeutic target in human liver cancer stem cells. J. Clin. Invest. 120, 3326–3339.

Hari, D., Xin, H.W., Jaiswal, K., Wiegand, G., Kim, B.K., Ambe, C. *et al.*, 2011. Isolation of live label-retaining cells and cells undergoing asymmetric cell division via nonrandom chromosomal cosegregation from human cancers. Stem Cells Dev. 20, 1649–58.

Harwin, F.M., Lyons, L., Monsen, H. 1984. The Liver, a poster. Anatomical Chart Company, Chicago, Illinois. Wolters Kluwer/Lippincott Williams and Wilkins: Publishers.

He, G., Yu, G.Y., Temkin, V., *et al.*, 2010. Hepatocyte IKKbeta/NF-kappaB inhibits tumor promotion and progression by preventing oxidative stress-driven STAT3 activation. Cancer Cell 17, 286–297.

He, G., Dhar, D., Nakagawa, H., Font-Burgada, J., Ogata, H., Jiang, Y., *et al.*, 2013. Identification of liver cancer progenitors whose malignant progression depends on autocrine IL-6 signaling. Cell 155, 384–396.

Hirschhaeuser, F., Walenta, S., Mueller-Klieser, W. 2010. Efficacy of catumaxomab in tumor spheroid killing is mediated by its trifunctional mode of action. Cancer Immunol. Immunother. 59, 1675–1684.

Hoare, M., Das, T., Alexander, G. 2010. Ageing, telomeres, senescence, and liver injury. J. Hepatol. 53, 950–961.

Holczbauer, A., Factor, V.M, Andersen, J.B., Marquardt, J.U., Kleiner, D.E., Raggi, C., *et al.*, 2013. Modeling pathogenesis of primary liver cancer in lineage-specific mouse cell types. Gastroenterology 145, 221–231.

Howlader, N., Noone, A.M., Krapcho, M., Neyman, N., Aminou, R., Altekruse, S.F., *et al.*, 2012. SEER Cancer Statistics Review, 1975–2009 (Vintage 2009 Populations). National Cancer Institute: Bethesda, MD.

Hytiroglou, P., Snover, D.C., Alves, V., Balabaud, C., Bhathal, P.S., Bioulac-Sage, P., *et al.*, 2012. Beyond 'cirrhosis', a proposal from the International Liver Pathology Study Group. Am. J. Clin. Pathol. 137, 5–9.

Ikeda, K., Saitoh, S., Koida, I., Arase, Y., Tsubota, A., Chayama, K. *et al.*, 1993. A multivariate analysis of risk factors for hepatocellular carcinogenesis: a prospective observation of 795 patients with viral and alcoholic cirrhosis. Hepatology 18, 47–53.

Jeejeebhoy, K.N., Ho, J., Greenberg, G.R., Phillips, M.J., Bruce-Robertson, A., Sodtke, U. 1975. Albumin, fibrinogen and transferrin synthesis in isolated rat hepatocyte suspensions. A model for the study of plasma protein synthesis. Biochem. J. 146, 141–155.

Kato, A., Miyazaki, M., Ambiru, S., Yoshitomi, H., Ito, H., Nakagawa, K., *et al.*, 2001. Multidrug resistance gene (MDR-1) expression as a useful prognostic factor in patients with human hepatocellular carcinoma after surgical resection. J. Surg. Oncol. 78, 110–115.

Khoei, S., Goliaei, B., Neshasteh-Riz, A., Deizadji, A. 2004. The role of heat shock protein 70 in the thermoresistance of prostate cancer cell line spheroids. FEBS Lett. 561, 144–148.

Kim, H., Choi, G.H., Na, D.C., Ahn, E.Y., Kim, G.I., Lee, J.E., *et al.*, 2011. Human hepatocellular carcinomas with 'stemness'-related marker expression: keratin 19 expression and a poor prognosis. Hepatology 54, 1707–1717.

Klonisch, T., Wiechec, E., Hombach-Klonisch, S., Ande, S.R., Wesselborg, S., Schulze-Osthoff, K., Los, M., *et al.*, 2008. Cancer stem cell markers in common cancers-therapeutic implications. Trends Mol. Med. 14, 450–460.

Koch, K.S., Leffert, H.L. 2004. Normal liver progenitor cells in culture. In: Sell, S. (ed.) Stem Cells Handbook. Totowa, NJ: Humana Press, pp. 367–384.

Koch K.S., Maeda S., He G., Karim, M., Leffert, H.L. 2009. Targeted deletion of hepatocyte Ikkbeta confers growth advantages. Biochem. Biophys. Res. Commun. 380, 349–354.

Kohga, K., Tatsumi, T., Takehara, T., Tsunematsu, H., Shimizu, S., Yamamoto, M., *et al.*, 2010. Expression of CD133 confers malignant potential by regulating metalloproteinases in human hepatocellular carcinoma. J. Hepatol. 52, 872–879.

Krohn, A., Song, Y.H., Muehlberg, F., Droll, L., Beckmann, C., Alt, E. 2009. CXCR4 receptor positive spheroid forming cells are responsible for tumor invasion *in vitro*. Cancer Lett. 280, 65–71.

Kuo, M.T., Savaraj, N. 2006. Roles of reactive oxygen species in hepatocarcinogenesis and drug resistance of gene expression in liver cancers. Carcinogenesis 45, 701–705.

Kuwahara, R., Kofman, A.V., Landis, C.S., Swenson, E.S., Barendswaard, E., Theise, N.D. 2008. The hepatic stem cell niche: identification by label-retaining cell assay. Hepatology 47, 1994–2002.

Lathia, J.D., Hitomi, M., Gallagher, J., Gadani, S.P., Adkins, J., Vasanji, A., *et al.*, 2011. Distribution of CD133 reveals glioma stem cells self-renew through symmetric and asymmetric cell divisions. Cell Death Dis. 2, e200.

Lee, T.K., Castilho, A., Cheung, V.C., Tang, K.H., Ma, S., Ng, I.O. 2011. CD24^{+} liver tumor-initiating cells drive self-renewal and tumor initiation through STAT3-mediated NANOG regulation. Cell Stem Cell 9, 50–63.

Leffert, H.L., Koch, K.S., Shapiro, P., Skelly, H., Hubert, J., Monken, C., *et al.*, 1987. Primary cultures, monoclonal antibodies and nucleic acid probes as tools for studies of hepatic structure and function. In: Tygstrup, N., Orlandi, F. (eds) Cirrhosis of the Liver: Methodology and Fields of Research. Philadelphia: Elsevier, pp. 121–140.

Lemaigre, F., Zaret, K.S. 2004. Liver development update: new embryo models, cell lineage control, and morphogenesis. Curr. Opin. Genet. Dev. 14, 582–590.

Liao, C.P., Adisetiyo, H., Liang, M., Roy-Burman, P. 2010. Cancer-associated fibroblasts enhance the gland-forming capability of prostate cancer stem cells. Cancer Res. 70, 7294–7303.

Lingala, S., Cui, Y.Y., Chen, X., *et al.*, 2010. Immunohistochemical staining of cancer stem cell markers in hepatocellular carcinoma. Exp. Mol. Pathol. 89, 27–35.

Liu, L.L., Fu, D., Ma, Y., Shen, X.Z. 2011. The power and the promise of liver cancer stem cell markers. Stem Cells Dev. 20, 2023–2030.

Ma, S. 2013. Biology and clinical implications of CD133^{+} liver cancer stem cells. Exp. Cell Res. 319, 126–132.

Ma, S., Chan, K.W., Hu, L., Lee, T.K., Wo, J.Y., Ng, I.O., *et al.*, 2007. Identification and characterization of tumorigenic liver cancer stem/progenitor cells. Gastroenterology 132, 2542–2456.

Ma, Y.C., Yang, J.Y., Yan, L.N. 2013. Relevant markers of cancer stem cells indicate a poor prognosis in hepatocellular carcinoma patients: a meta-analysis. Eur. J. Gastroenterol. Hepatol. 25, 1007–1016.

Maeda, S., Kamata, H., Luo, J.L., Leffert, H., Karim, M. 2005. IKKbeta couples hepatocyte death to cytokine-driven compensatory proliferation that promotes chemical hepatocarcinogenesis. Cell 121, 977–990.

Magee, J.A., Piskounova, E., Morrison, S.J. 2012. Cancer stem cells: impact, heterogeneity, and uncertainty. Cancer Cell 20, 283–296.

Martin-Belmonte, F., Perez-Moreno, M. 2012. Epithelial cell polarity, stem cells, and cancer. Nat. Rev. Cancer 12, 23–38.

Mattner, J. 2011. Genetic susceptibility to autoimmune liver disease. World J. Hepatol. 3, 1–7.

Muramatsu, S., Tanaka, S., Mogushi, K., Adikrisna, R., Aihara, A., Ban, D. *et al.*, 2013. Visualization of stem cell features in human hepatocellular carcinoma reveals in vivo significance of tumor–host interaction and clinical course. Hepatology 58, 218–228.

Na, Y.R., Seok, S.H., Kim, D.J., Han, J.H., Kim, T.H., Jung, H., *et al.*, 2009. Isolation and characterization of spheroid cells from human malignant melanoma cell line WM-266-4. Tumour Biol. 30, 300–309.

National Cancer Institute. 2014a. Adult Primary Liver Cancer Treatment (PDQ®): General Information. Available from: http://www.cancer.gov/cancertopics/pdq/treatment/adult-primary-liver/HealthProfessional/page1 (accessed 08 January 2015).

National Cancer Institute. 2014b. Adult Primary Liver Cancer Treatment (PDQ®): Treatment Option Overview. Available from: http://www.cancer.gov/cancertopics/pdq/treatment/adult-primary-liver/HealthProfessional/page4 (accessed 08 January 2015).

National Cancer Institute. 2014c. Liver (Hepatocellular) Cancer Screening: (PDQ®): Significance. Available from: http://www.cancer.gov/cancertopics/pdq/screening/hepatocellular/HealthProfessional/page2 (accessed 08 January 2015).

Navin, N., Kendall, J., Troge, J., Andrews, S.P., Rodgers, L., McIndoo, J., *et al.*, 2011. Tumour evolution inferred by single-cell sequencing. Nature 472, 90–94.

Nishina H. 2012. hDlk-1: a cell surface marker common to normal hepatic stem/progenitor cells and carcinomas. J. Biochem. 152, 121–123.

Nowell, P.C. 1976. The clonal evolution of tumor cell populations. Science 194, 23–28.

Oishi, N., Wang, X.W. 2011. Novel therapeutic strategies for targeting liver cancer stem cells. Int. J. Biol. Sci. 7, 517–535.

Olive, P.L., Banáth, J.P., Sinnott, L.T. 2004. Phosphorylated histone H2AX in spheroids, tumors, and tissues of mice exposed to etoposide and 3-amino-1,2,4-benzotriazine-1,3-dioxide. Cancer Res. 64, 5363–5369.

Potten, C.S., Loeffler, M. 1990. Stem cells: attributes, cycles, spirals, pitfalls and uncertainties. Lessons for and from the crypt. Development 110, 1001–1020.

Potten, C.S., Hume, W.J., Reid, P., Cairns, J. 1978. The segregation of DNA in epithelial stem cells. Cell 15, 899–906.

Price, J.M., Harman, J.W., Miller, E.C., Miller, J.A. 1952. Progressive microscopic alterations in the livers of rats fed the hepatic carcinogens 3′-methyl-4-dimethyl aminoazo-benzene and 4′-fluoro-4-dimethylaminoazobenzene. Cancer Res. 12, 192–200.

Rahman, M., Deleyrolle, L., Vedam-Mai, V., Azari, H., Abd-El-Barr, M., Reynolds, B.A. 2011. The cancer stem cell hypothesis: failures and pitfalls. Neurosurgery 68, 531–545.

Ramesh, T., Ellis, A.W., Spear, B.T. 1995. Individual mouse α-fetoprotein enhancer elements exhibit different patterns of tissue-specific and hepatic position-dependent activity. Mol. Cell Biol. 15, 4947–4955.

Reya T., Morrison S.J., Clarke M.F., Weissman I.L. 2001. Stem cells, cancer, and cancer stem cells. Nature 414, 105–111.

Rountree C.B., Ding W., Dang H., Vankirk, C., Crooks, G.M. 2011. Isolation of $CD133^+$ liver stem cells for clonal expansion. J. Vis. Exp. 10, 56.

Russo, J., Snider, K., Pereira, J.S., Russo, I.H. 2010. Estrogen induced breast cancer is the result in the disruption of the asymmetric cell division of the stem cell. Horm. Mol. Biol. Clin. Investig. 1, 53–65.

Sackett, S.D., Li, Z., Hurtt, R., Gao, Y., Wells, R.G., Brondell, K., *et al.*, 2009. Foxl1 is a marker of bipotential hepatic progenitor cells in mice. Hepatology 49, 920–929.

Sasaki, A., Kamiyama, T., Yokoo, H., Nakanishi, K., Kubota, K., Haga, H., *et al.*, 2010. Cytoplasmic expression of CD133 is an important risk factor for overall survival in hepatocellular carcinoma. Oncol. Rep. 24, 537–546.

Schmucker, D.L., Sanchez, H. 2011. Liver regeneration and aging: a current perspective. Curr. Gerontol. Geriatr. Res. 2011, 526379.

Sell, S. 2010. On the stem cell origin of cancer. Am. J. Pathol. 176, 2584–2494.

Sell, S., Leffert, H.L. 1982. An evaluation of cellular lineages in the pathogenesis of experimental hepatocellular carcinoma. Hepatology 2, 77–86.

Sell, S., Leffert, H.L. 2008. Liver cancer stem cells. J. Clin. Oncol. 26, 2800–2805.

Shinozuka, H., Lombardi, B., Sell, S., Iammarino, R.M. 1978. Early histological and functional alterations of ethionine liver carcinogenesis in rats fed a choline deficient diet. Cancer Res. 38, 1092–1098.

Silveira, M.G., Suzuki, A., Lindor, K.D. 2008. Surveillance for hepatocellular carcinoma in patients with primary biliary cirrhosis. Hepatology 48, 1149–1157.

Siminovitch L., McCulloch E.A., Till, J.E. 1963. The distribution of colony-forming cells among spleen colonies. J. Cell Physiol. 62, 327–336.

Smith, P.J., Wiltshire, M., Chappell, S.C., *et al.*, 2013. Kinetic analysis of intracellular Hoechst 33342-DNA interactions by flow cytometry: misinterpretation of side population status. Cytometry 83A, 161–169.

Spangrude, G.J, Heimfeld, S., Weissman, I.L. 1988. Purification and characterization of mouse hematopoietic stem cells. Science, 241, 58–62; erratum 244, 1030.

Sukowati, C.H., Rosso, N., Crocè, L.S., Tiribelli, C. 2010. Hepatic cancer stem cells and drug resistance: relevance in targeted therapies for hepatocellular carcinoma. World J. Hepatol. 2, 114–126.

Terris, B., Cavard, C., Perret C. 2010 EpCAM, a new marker for cancer stem cells in hepatocellular carcinoma. J. Hepatol. 52, 280–281.

Till, J.E., McCulloch, E.A. 1961. A direct measurement of the radiation sensitivity of normal mouse bone marrow cells. Radiat Res. 14, 213–222.

Till, J.E., McCulloch, E.A., Siminovitch, L. 1964. A stochastic model of stem cell proliferation, based on the growth of spleen colony-forming cells. Proc. Nat. Acad. Sci. USA 51, 29–36.

Tomuleasa, C., Soritau, O., Rus-Ciuca D., Pop, T., Todea, D., Mosteanu, O., *et al.*, 2010. Isolation and characterization of hepatic cancer cells with stem-like properties from hepatocellular carcinoma. J. Gastrointestin. Liver Dis. 19, 61–67.

Trautwein, C., Will, M., Kubicka, S., Rakemann, T., Flemming, P., Manns, M.P. 1999. 2-acetaminofluorene blocks cell cycle progression after hepatectomy by p21 induction and lack of cyclin E expression. Oncogene 18, 6443–6453.

Valcárcel, M., Arteta, B., Jaureguibeitia, A., Lopategi, A., Martínez, I., Mendoza, L., *et al.*, 2008. Three-dimensional growth as multicellular spheroid activates the proangiogenic phenotype of colorectal carcinoma cells via LFA-1-dependent VEGF: implications on hepatic micrometastasis. J. Trans. Med. 6, 57.

Valent, P., Bonnet, D., De Maria, R., Lapidot, T., Copland, M., Melo, J.V., *et al.*, 2012. Cancer stem cell definitions and terminology: the devil is in the details. Nat. Rev. Cancer 11, 767–775.

van Malenstein, H., van Pelt, J., Verslype, C. 2011. Molecular classification of hepatocellular carcinoma anno 2011. Eur. J. Cancer 47, 1789–1797.

Venook, A.P., Papandreou, C., Furuse, J., de Guevara, L.L. 2010. The incidence and epidemiology of hepatocellular carcinoma: a global and regional perspective. Oncologist 15(Suppl. 4), 5–13.

Wang, H., Pan, K., Zhang, H.K., Weng, D.S., Zhou, J., Li, J.J., *et al.*, 2008. Increased polycomb-group oncogene Bmi-1 expression correlates with poor prognosis in hepatocellular carcinoma. J. Cancer Res. Clin. Oncol. 134, 535–541.

Wang, E.Y., Yeh, S.H., Tsai T.F., Huang, H.P., Jeng, Y.M., Lin, W.H., *et al.*, 2011. Depletion of β-catenin from mature hepatocytes of mice promotes expansion of hepatic progenitor cells and tumor development. Proc. Nat. Acad. Sci. USA 108, 18384–18389.

Wartenberg, M., Dönmez, F., Ling, F.C., Acker, H., Hescheler, J., Sauer, H. 2001. Tumor-induced angiogenesis studied in confrontation cultures of multicellular tumor spheroids and embryoid bodies grown from pluripotent embryonic stem cells. FASEB J. 15, 995–1005.

Weglarz, T.C., Degen, J.L., Sandgren, E.R. 2000. Hepatocyte transplantation into diseased mouse liver. Kinetics of parenchymal repopulation and identification of the proliferative capacity of tetraploid and octaploid hepatocytes. Am. J. Pathol. 157, 1963–1974.

Wilson, G.S., Hu, Z., Duan, W., Tian, A., Wang, X.M., McLeod, D., *et al.*, 2013. Efficacy of using cancer stem cell markers in isolating and characterizing liver cancer stem cells. Stem Cells Dev. 22, 2655–2664.

Wong, C.M., Fan, S.T., Ng, I.O. 2001 Beta-catenin mutation and overexpression in hepatocellular carcinoma: clinicopathologic and prognostic significance. Cancer 92, 136–145.

World Health Organization. 2014. Cancer Fact Sheet No.297. Available from: http://www.who.int/mediacentre/factsheets/fs297/en/ (last accessed 29 November 2014).

Worton, R.G., McCulloch E.A., Till, J.E. 1969a. Physical separation of hemopoietic stem cells differing in their capacity for self-renewal. J. Exp. Med. 130, 91–103.

Worton, R.G., McCulloch, E.A., Till, J.E. 1969b. Physical separation of hemopoietic stem cells from cells forming colonies in culture. J. Cell Physiol. 74, 171–182.

Wu, A.M., Till, J.E., Siminovitch, L., McCulloch, E.A. 1968. Cytological evidence for a relationship between normal hematopoietic colony-forming cells and cells of the lymphoid system. J. Exp. Med. 127, 455–464.

Wu, M., Kwon, H.Y., Rattis, F., Blum, J., Zhao, C., Ashkenazi R., *et al.*, 2007. Imaging hematopoietic precursor division in real time. Cell Stem Cell, 1, 541–454.

Xin, H.-W., Hari, D.M., Mullinax, J.E., Ambe, C.M., Koizumi, T., Ray, S., *et al.*, 2012. Tumor-initiating label-retaining cells in human gastrointestinal cancers undergo asymmetric cell division. Stem Cells 30, 591–598.

Yadlapalli, S., Yamashita, Y.M. 2013. DNA asymmetry in stem cells – immortal or mortal? J. Cell Sci. 126, 1–8.

Yamashita, T., Wang, X.W. 2013. Cancer stem cells in the development of liver cancer. J. Clin. Invest. 123, 1911–1918.

Yanai, H., Nakamura, K., Hijioka S., Kamei, A., Ikari, T., Ishikawa, Y., *et al.*, 2010. Dlk-1, a cell surface antigen on foetal hepatic stem/progenitor cells, is expressed in hepatocellular, colon, pancreas and breast carcinomas at a high frequency. J. Biochem. 148, 85–92.

Yang, L., Jung, Y., Omenetti, A., Witek, R.P., Choi, S., Vandongen, H.M., *et al.*, 2008. Fate mapping evidence that hepatic stellate cells are epithelial progenitors in adult mouse livers. Stem Cells 26, 2104–2113.

Yang, Z.F., Ho, D.W., Ng, M.N., Lau, C.K., Yu, W.C., Ngai, P., *et al.*, 2008. Significance of $CD90^+$ cancer stem cells in human liver cancer. Cancer Cell 13, 153–166.

Yavorkovsky, L., Lai, E., Ilic, Z., Sell, S. 1995. Participation of small intraportal stem cells in the restitution response of the liver to periportal necrosis induced by allyl alcohol. Hepatology 21, 1702–1712.

Yennek, S., Tajbakhsh, S. 2013. DNA asymmetry and cell fate regulation in stem cells. Sem. Cell Devel. Biology 24, 627–642.

Yong, K.J., Gao, C., Lim, J.S., Yan, B., Yang, H., Dimitrov, T., *et al.*, 2013. Oncofetal gene SALL4 in aggressive hepatocellular carcinoma. N. Engl. J. Med. 368, 2266–2276.

Yovchev, M.I., Grozdanov, P.N., Zhou, H., Racherla, H., Guha, C., Dabeva, M.D. 2008. Identification of adult hepatic progenitor cells capable of repopulating injured rat liver. Hepatology 47, 636–647.

Zhang, L., Sun, H., Zhao, F., Lu, P., Ge, C., Li, H., *et al.*, 2012. BMP4 administration induces differentiation of CD133$^+$ hepatic cancer stem cells, blocking their contributions to hepatocellular carcinoma. Cancer Res. 72, 4276–4285.

Zhang, S., Mercado-Uribe, I., Xing, Z., Sun, B., Kuang, J., Liu, J. 2014. Generation of cancer stem-like cells through the formation of polyploid giant cancer cells. Oncogene 33(1), 116–128; erratum 134.

Zhao, W., Wang, L., Han, H., Jin, K., Lin, N., Guo, T., *et al.*, 2013. 1B50-1, a mAb raised against recurrent tumor cells, targets liver tumor-initiating cells by binding to the calcium channel $\alpha2\delta1$ subunit. Cancer Cell 23, 541–556.

Zhu, Z., Hao, X., Yan, M., Yao, M., Ge, C., Gu, J., Li, J. 2010. Cancer stem/progenitor cells are highly enriched in CD133$^+$CD44$^+$ population in hepatocellular carcinoma. Int. J. Cancer 126, 2067–2078.

15 Cancer Stem Cell Biomarkers

Stefano Zapperi[1] and Caterina A.M. La Porta[2]
[1] *CNR-IENI, Milan, Italy*
[2] *Department of Biosciences, University of Milan, Milan, Italy*

15.1 Cancer stem cells

Modern anticancer drug discovery began in the mid-19th century, with cytotoxic chemotherapeutic agents targeting cancer cells with high proliferative rates. It is well known, however, that cancers are composed of heterogeneous cell types (Virchow, 1855; Cogngeim, 1867), suggesting that the nature of the target cells suffering the effects of the oncogenic activity might play an important role in the control of oncogenesis. In contrast with conventional views of cancer, in which all cells are tumourigenic, the cancer stem cell (CSC) hypothesis proposed a decade ago (Reya *et al.*, 2001; Pardal *et al.*, 2003) that tumours contain a subset of cells that both self-renew and give rise to a differentiated progeny. While many aspects of this theory remain speculative and are still evolving, CSCs are viewed as those cells at the apex of the tumour hierarchy, highlighting the role of aberrant differentiation in tumourigenesis. In analogy to normal adult tissue stem cells, CSCs are the driving force of the tumour. In particular, CSCs are usually defined by two minimal properties. The first is the ability to regrow the tumour from which they were isolated or identified, which implies that the tumour-initiating cells can only be defined experimentally *in vivo*. The use of serial transplantation to validate a candidate CSC subpopulation allows the capability to recapitulate the heterogeneity of the primary tumour to be monitored. Both xeno- and syngeneic transplantations might, however, misrepresent the real network of interactions with diverse support, such as fibroblasts, endothelial cells, macrophages, mesenchymal stem cells and many of the cytokines and receptors involved

Principles of Stem Cell Biology and Cancer: Future Applications and Therapeutics, First Edition.
Edited by Tarik Regad, Thomas J. Sayers and Robert C. Rees.

in these interactions (La Porta, 2009). As a consequence of this, contrasting results have appeared in the literature for the frequency of CSCs in xenograft tumours. In a recent paper, our group suggested that the number of CSCs should not be stated in absolute terms, but only relative to the animal model used (La Porta, 2009). The second property necessary to define CSCs is the multipotency of lineage differentiation. In addition to these properties, there are at least two aspects of CSCs that are not completely clear: their origin and their frequency *in vivo*. The use of the correct panel of markers is necessary to demonstrate their origin. The second issue is particularly difficult to address, for two obvious reasons: if CSCs are few, there are problems connected to the sensitivity of the technique and/or the small portion of biopsy analysed. Moreover, the behaviour and frequency of tumour-initiating cells *in vivo* and in animal models could be influenced by various environmental factors. For example, haematopoietic stem cells grow asymmetrically, and the probability of asymmetric cell division depends on the presence of certain cytokines (Brummendorf *et al.*, 1998; Takano *et al.*, 2004; La Porta, 2010). Interestingly, haematopoietic stem cells, along with oligodendrocyte precursors, show an intrinsic property of asymmetric growth and appear to be regulated only in a permissive way by extrinsic factors (Gao and Raff, 1997; Brummendorf *et al.*, 1998). Furthermore, some proteins have been shown to be asymmetrically distributed during haematopoietic stem cell division (Fonseca *et al.*, 2008). A recent review analysed the developmental signalling pathways involved in the self-renewal of haematopoietic stem cells, such as the Notch pathway and Wnt and TGF-β signalling, as well as chemical regulators like retinoic acid (Zon, 2008). It is important to highlight that 'stemness' does not necessarily involve a set of genes common to all stem cells, since the gene expression patterns of stem cells in distinct tissues differ widely (Klein *et al.*, 2007).

Another important aspect is the mechanism by which stem cells decide whether to remain in the niche or to leave. This could be a major factor in the balancing act between stem cell self-renewal and differentiation (Wallenfang and Matunis, 2003). In this connection, melanocyte growth is controlled by keratinocytes and melanoma seems to escape from this control through different mechanisms, including downregulation of receptors (E-cadherin, P-cadherin and desmoglein), upregulation of receptors and signalling molecules important for melanoma cell and melanoma cell–fibroblast interactions (N-cadherin, zanula occludens protein 1) and deregulation of morphogenesis (Notch receptors and their ligands). The investigation of normal melanocyte homeostasis might help us to define how melanoma and, in particular, melanoma CSCs escape the microenvironment created by epidermal keratinocytes and how they develop new cellular partners in fibroblasts and endothelial cells that support their growth and invasion (Haas and Herlyn, 2005).

Finally, the role of senescence in the dynamics of CSCs proliferation has been considered in recent years. Cancer cells are characterized by their persistent proliferation, but just like normal cells (Hayflick and Moorhead, 1961), tumour cells can go senescent, halting their growth (Di Micco *et al.*, 2006; Collado and Serrano, 2010). The molecular basis for the induction of senescence appears to be a combination of several mechanisms, including telomerase shortening, DNA damage and oxidative stress (Collado and Serrano, 2010). It has been suggested that senescence should be present only in preneoplastic cells (Collado and Serrano, 2010), but there is evidence that senescence markers increase during tumour progression (Wasco *et al.*, 2008). Recently, our group investigated this aspect, formulating cancer growth in mathematical terms and obtaining predictions for the evolution of senescence (La Porta *et al.*, 2012). We also performed experiments in human melanoma cells that are compatible with the hierarchical model and showed that senescence is a reversible process, controlled by survivin (La Porta *et al.*, 2012). Our findings show that enhancing senescence is unlikely to provide a useful therapeutic strategy for fighting cancer, unless CSCs are specifically targeted (La Porta *et al.*, 2012). Another important result of our paper is that slightly different assay conditions lead to different CSC fractions, which can sometimes be relatively large (Quintana *et al.*, 2008). There is, in fact, no reason to believe that the CSC population must be small. This idea comes from the analogy with tissue stem cells, which replicate homeostatically, keeping their population constant either by asymmetric division or stochastically (Clayton *et al.*, 2007; Lopez-Garcia *et al.*, 2010), leading to a vanishing concentration of stem cells in the total cell population. CSCs do not replicate homeostatically and therefore their population grows exponentially. Changes in assay conditions can change the duplication rate of CSCs, leading under extreme conditions (e.g. the use of matrigel, mice permissive conditions, etc.) to a relatively large concentration of CSCs, perhaps resolving previous controversies (Quintana *et al.*, 2008, 2010; Boiko *et al.*, 2010).

15.2 Choice of CSC biomarkers

The use of multiple markers to define a CSC subpopulation is, in our view, strongly recommended. A panel of markers defines haematopoietic stem cells as well. The literature is sometimes confusing, since researchers often use potential biomarkers without a specific biological motivation, employing some marker alone in some papers and in a combination in others. In this section, we discuss recent findings for four tumours: breast, colon and brain cancer and melanoma. The picture that comes out is still fuzzy and confused, but we will suggest some ideas to clarify it.

15.2.1 CSCs and breast cancer

Al-Hajj *et al.* (2003) described the presence of a tumourogenic subpopulation in eight out of nine breast cancer patients, based on the cell-surface marker expression of $CD44^{-}CD24^{-/low}$. The authors isolated tumour cells from all nine patients and showed that eight generated new tumours containing additional $CD44^{-}CD24^{-/low}$ after serial passing (Al-Hajj *et al.*, 2003). They also demonstrated that there is a hierarchy of breast cancer cells, with some cells proliferating extensively but the majority, derived from this population, having only limited proliferative potential *in vivo* (Al-Hajj *et al.*, 2003). Sheridan *et al.* (2006) showed that the $CD44^{-}CD24^{-/low}$ subpopulation of breast cancer cells expresses high levels of proinvasive genes and highly invasive properties, but this phenotype was not sufficient to predict capacity for pulmonary metastasis. In fact, they could not find any correlation between $CD44^{-}CD24^{-/low}$ phenotype and the ability to home and proliferate at sites of metastasis (Sheridan *et al.*, 2006). One could speculate that no correlation was found because these biomarkers are not exclusively present in CSC subpopulations. Fillmore and Kuperwasser (2008) analysed eight human breast cancer cell lines for CD44, CD24 and epithelial-specific antigen (ESA) expression and showed that the percentage of $CD44^{-}CD24^{-/low}$ cells did not correlate with tumourogenicity. In contrast, cells characterized by $CD44^{-}CD24^{-}/ESA^{+}$ can self-renew, reconstitute the parental cell line, retain BrdU label and preferentially survive chemotherapy (Fillmore and Kuperwasser, 2008). In a more recent paper, Gupta *et al.* (2011) studied the dynamics of phenotypic proportions in human breast cancer cell lines, using $CD44^{high}/CD24^{-}/EpCAM^{low}$ as a CSC marker. The authors showed that subpopulations of cells purified from the CSC phenotype tend to express it again in time, and speculated that these results provide an indication of phenotypic switching of cancer cells into CSCs (Gupta *et al.*, 2011). A recently published paper showed a similarity between Her2+ intrinsic human breast cancer subtypes and mammary stem cell populations, supporting the origin of CSCs in a stem cell population (Benjamin *et al.*, 2012). Moreover, the authors described a new signature for human CSCs, claiming that new gene sets with prognostic value could also be useful for predicting which patients will respond to certain treatment strategies (Benjamin *et al.*, 2012). In a recent paper, Dickopf1, a protein known to negatively regulate the Wnt pathway, was shown to affect the fate decision of breast CSCs sorted according to $CD44^{+}CD24^{low}$, driving the cells to differentiate (Agur *et al.*, 2011). Interestingly, the results are supported by a mathematical model (Agur *et al.*, 2011).

Aldehyde dehydrogenase (ALDH) was used as a stem cell marker in 33 human breast cell lines (Charafe-Jauffret *et al.*, 2009). ALDH is a detoxifying enzyme that oxidizes intracellular aldehydes and is believed to play a role in the differentiation of stem cells via the metabolism of retinal to retinoic acid (Marcato *et al.*, 2014). Interestingly, ALDH activity can be used to sort a subpopulation of cells that display stem cell properties from normal breast tissue and breast cancer (Marcato *et al.*, 2014). ALDH activity, assessed by ADELFLUOR assay, has been successfully used to isolate CSCs from multiple myeloma and acute leukaemia, as well as from brain tumour (Fang *et al.*, 2005; Klein *et al.*, 2007). However, in melanoma there are contrasting results: ALDH phenotype is not associated with more aggressive subpopulations, arguing against ALDH as a 'universal' marker (Hadnagy *et al.*, 2006). Another interesting pathway that has been extensively studied is the Notch receptor signalling pathway (for a recent review, see Hadnagy *et al.*, 2006). An important point is the toxicity of these potential treatments. While the Notch pathway appears promising, it is active in other tissues, so it might have a great toxicity. Therefore, as suggested by Harrison and colleagues, it seems important to study the complexity of the Notch pathway in order to more successfully target CSCs (Harrison *et al.*, 2010). On the other hand, in a recent study, 275 patients were analysed for $CD44^{+}CD24^{-}$ putative stem cell marker, as well as for other markers (vimentin, ostenectin, connexin 43, ADLH, CK18, GATA3, MUC1), in primary breast cancers of different subtypes and histological stages (D'Amico *et al.*, 2013). This study reveals a high degree of diversity in the expression of several of the selected markers in different tumour subtypes and histologic stages. We would point out that the latter findings could be explained by the observation that none of these markers are really specific for CSCs.

All together, these data show that the therapeutic implications of CSCs in breast cancer are still unclear. The current understanding of the role of CSCs in clinical trials is discussed in a recent review article (Charafe-Jauffret *et al.*, 2009).

15.2.2 CSCs and melanoma

Several published papers, using different putative CSC markers (CD20, CD133, ABCG2, ABCB5, CD271 and CXCR6), show that a CSC subpopulation exists in melanoma (Dou *et al.*, 2007; Monzani *et al.*, 2007; Schatton *et al.*, 2008; Boiko *et al.*, 2010; Taghizadeh *et al.*, 2010; Zhong *et al.*, 2010). However, Quintana *et al.* (2010) argued against the existence of CSCs based on the following observations: a relatively large fraction of melanoma cells (up to ~25%) initiated tumours in severely immunocompromised nonobese diabetic severe combined-immunodeficient (NOD/SCID) IL2Rγnull mice; the fraction of tumour-inducing cells depends upon assay conditions; and several putative CSC markers appear to be reversibly expressed. In conclusion, this paper suggested that the best model by which to confirm the presence of CSCs is a severe immunocompromised mouse. In a follow-up study, the same authors analysed the expression of more than 50 surface markers on melanoma cells derived from several patients (A2B5, cKIT, CD44, CD49B, CD49D, CD49F, CD133, CD166), but focusing on CD133 and CD166 (Quintana *et al.*, 2010). Using these markers, they found no enrichment and a high frequency of tumourigenic cells (Quintana *et al.*, 2010). In a recent paper, however, it was shown that CD133 is highly expressed in melanoma cells and is not a good marker by which to sort CSCs (Monzani *et al.*, 2007). Moreover, Boiko *et al.* (2010), using the same immunocompromised mice, did not confirm Quintana *et al.*'s data (Monzani *et al.*, 2007). Boiko *et al.* (2010) used CD271, nerve growth factor receptor, as a marker by which to identify CSCs. In 2010, our group showed for the first time that a marker connected with a functional property of stem cells, CXCR6 (connected with asymmetric division), is expressed by human melanoma cell lines and biopsies (Taghizadeh *et al.*, 2010). The CXCR6 subpopulation showed a stronger self-renewal capability in a xenograft model (Taghizadeh *et al.*, 2010). We also showed that CXCR6-positive cells were able to grow better when treated with the ligand (CXCL16) (Taghizadeh *et al.*, 2010). A recent paper reports that the growth of B16-F10 melanoma cells in syngeneic mice seems to be maintained by a relatively large proportion (>10%) of tumour cells, through analysis of a side population (Zhong *et al.*, 2010).

Finally, more recently, Kumar *et al.* (2012) showed that Oct4, a highly critical transcription factor in stem cells, can also promote dedifferentiation of melanoma cells to CSC-like cells. These authors also showed that hypoxia can increase the level of oct4 (Kumar *et al.*, 2012) and therefore suggested that, since a transient expression of Oct4 protein is sufficient to induce dedifferentiation of melanoma cells and Oct4 can be regulated by hypoxia, Oct4 might mediate the effect of hypoxia on tumour progression (Kumar *et al.*, 2012). While these results are promising, more studies are needed to elucidate the regulation and function of hypoxia in tumour progression.

15.2.3 CSCs and colon cancer

Colorectal cancer (CRC) is the third most common type of cancer and the second leading cause of tumour-related death in the Western world (Dallas *et al.*, 2009). Despite the well-known genetic mutations that drive the transition from healthy colonic epithelia to dysplastic adenoma and finally to colon adenocarcinoma, current anticancer treatments are often able to eradicate the disease. Indeed, the response rate to current systemic therapies is about 50%, but resistance develops in nearly all patients (Dallas *et al.*, 2009). CSCs are also identified from the expression of one or multiple cell-surface markers associated with cancer stemness, such as CD133 (Haraguchi *et al.*, 2008; Puglisi *et al.*, 2009), CD44 (Haraguchi *et al.*, 2008; Wang *et al.*, 2008), CD166 (Mărgaritescu *et al.*, 2014) or Lgr5 (Dalerba *et al.*, 2007). More functional markers, such as Wnt activity (Vermeulen *et al.*, 2010) and ALDH1 activity (Huang *et al.*, 2009), have been exploited for identification of colon CSCs. However, none of the markers used to isolate stem cells in various cancerous tissues are expressed exclusively by the stem cell fraction. Indeed, most such markers are chosen either because they are expressed in normal stem cells or because they have been found to identify CSCs in other malignancies.

15.2.4 CSCs and brain tumours

Glioblastoma multiforme (GBM) is the most common type of primary malignant brain tumour, accounting for 55% of primary brain tumours (Legler *et al.*, 1999). The prognosis of GBM is very poor; most patients die of tumour recurrence.

Glioma stem cells (GSCs) were among the first CSCs to be described for solid tumours (Legler *et al.*, 1999). The existence of GSCs is now widely accepted, with the most clinically relevant features of CSCs, such as resistance to existing therapies, having been confirmed in GSCs (Singh *et al.*, 2004; Bao *et al.*, 2006; Bleau *et al.*, 2009).

15.3 Imperfect markers

A critical discussion of the recent literature on CSC biomarkers highlights a large amount of controversial information and conflicting results. This may lead to the conclusion that CSCs do not exist in some tumours (Quintana *et al.*, 2008, 2010) or else to the hypothesis that their phenotype is dynamic and stochastically reversible (Gupta *et al.*, 2011). If the CSC phenotype is stochastic and reversible, it becomes complicated to define a precise hierarchy between the cells. In a heterogenous cell population, in which phenotypes can switch stochastically to and from the CSC state, defining CSCs might be difficult. A similar scenario is much like the traditional stochastic model of cancer, in which each cell has the potential to seed a tumour.

We believe that before the CSC hypothesis can be abandoned, we should critically analyse the existing data and assess whether it is possible to reconcile the apparent contradictions. In this respect, an interdisciplinary approach combining traditional cell biology with tools and models from applied mathematics and statistical physics could be of great help. Quantitative methods could contribute to overcome the limitations posed by a purely biological approach, as we discuss in this section, reconsidering some of the evidence usually interpreted against the presence of CSCs.

In analogy with stem cells, it is sometimes argued that CSCs should be very uncommon. Hence, the presence of many putative CSCs is difficult to understand. We have recently studied a mathematical model that can be used to quantitatively reproduce the growth curves for both stem cells and CSCs by varying just one parameter related to stem cell homeostatic conditions (La Porta *et al.*, 2012). In one case, we expect homeostasis, so that stem cells do not proliferate and their fraction in a cell population is vanishingly small. In tumours, however, homeostasis is broken and CSCs proliferate exponentially. Yet, in most cases, the fraction of CSCs in the population is small, since non-CSCs proliferate as well, although for a limited number of divisions. Under some experimental conditions favouring the growth of CSCs (e.g. the use of matrigel, mice permissive conditions; Quintana *et al.*, 2008, 2010), the number of CSCs does not need to be small. There is no reason to believe that the growth rates of CSCs and non-CSCs are independent of the environment, leading to the observed dependence of CSC numbers on assay conditions.

When a cell population has been purified from the CSC phenotype by a suitable marker, we would expect the marker not to be expressed again in the population as time goes on. Evidence of the reversible expression of a marker from purified tumour cell populations has led to the concept of phenotypic switching (Gupta *et al.*, 2011). The same data, however, could be quantitatively explained by assuming that there is no one-to-one correspondence between a given marker expression and CSCs. For instance, some CSCs might prevalently express a marker, but not all CSCs. Conversely, most cancer cell would not express the marker, but some would. The net outcome is that the positive cell subpopulation would be CSC-enriched and therefore much more likely to seed a tumour. On the other hand, the negative subpopulation would still express the marker, and eventually the few CSCs would restore the original phenotypic proportion. The validity of this imperfect-marker scenario can be tested quantitatively by comparing mathematical models with experiments (Zapperi *et al.*, 2012). This has been successfully done for ABCG2 in melanoma: thanks to the model, one can estimate that the ABCG2+ population contains 16% CSCs, and the ABCG2− one only 0.6% (La Porta *et al.*, 2012). This explains the enhanced ability of ABCG2+ cells to seed tumours in xenografts (La Porta *et al.*, 2012). Additional factors that can lead to reversible marker expression are the inevitable errors of the sorting process. These should also be quantified and analysed in terms of mathematical models.

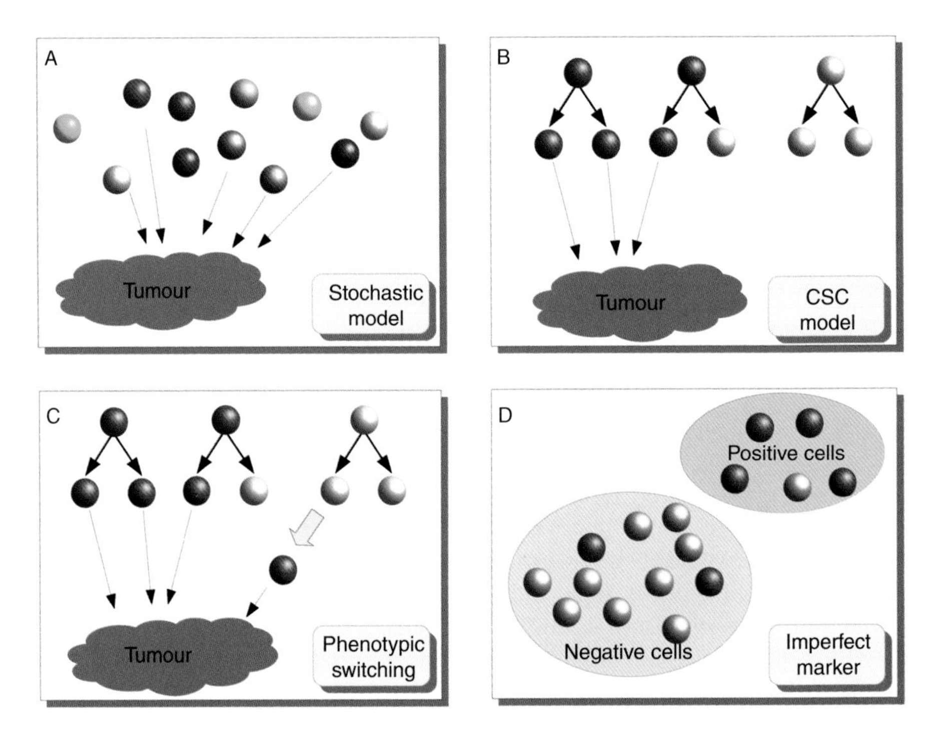

Figure 15.1 Population dynamics models for cancer progression. (A) The stochastic model postulates that cancer cells are heterogeneous but can all potentially seed a tumour. (B) According to the CSC model, cancer cells are organized hierarchically and only CSCs can seed a tumour. (C) If phenotypic switching occurs, non-CSCs can switch back to the CSC state and seed a tumour. (D) The imperfect-marker model recognizes that sorting by a biological marker is not an exact process and that some CSCs can be present in the negative subpopulation.

15.4 Conclusions

In conclusion, we believe that additional effort should be devoted to understanding the functional and biological properties of putative CSCs in cancer. The main issue is to identify biological markers linked to CSC-specific functions and confirm their expression in human bioptic samples. This is also a challenge for the development of diagnostic/prognostic tools by which to follow cancer development. The use of quantitative models can be of great help in orientating research and avoiding pitfalls in the interpretation of the experimental data. An interdisciplinary approach, combining applied mathematics, physics and biology, is needed when traditional biological thinking is unable to solve the problem completely. Figure 15.1 summarizes all the concepts discussed in this chapter.

References

Agur, Z., Kirnasovsky, O.U., Vasserman, G., Tencer-Hershkowicz, L., Kogan, Y., Harrison, H., *et al.*, 2011. Dickkopf1 regulates fate decision and drives breast cancer stem cells to differentiation: an experimentally supported mathematical model. PLoS One 6, e24225.

Al-Hajj, M., Wicha, M.S., Benito-Hernandez, A., Morrison, S.J., Clarke, M.F. 2003. Prospective identification of tumorigenic breast cancer cells. Proc. Nat. Acad. Sci. USA 100, 3983–3988.

Bao, S., Wu, Q., McLendon, R.E., Hao, Y., Shi, Q., Hjelmeland, A.B., *et al.*, 2006. Glioma stem cells promote radioresistance by preferential activation of the DNA damage response. Nature 444, 756–760.

Benjamin, T., Spike, D.D., Engle, J.C., Lin, S.K., Cheung, J.L., Geoffrey, M., Wahl A. 2012. Mammary stem cell population identified and characterized in late embryogenesis reveals similarities to human breast cancer. Cell Stem Cell 10, 183–197.

Bleau, A.M., Hambardzumyan, D., Ozawa, T., Fomchenko, E.I., Huse, J.T., Brennan, C.W., Holland, E.C. 2009. PTEN/PI3K/Akt pathway regulates the side population phenotype and ABCG2 activity in glioma tumor stem-like cells. Cell Stem Cell 4, 226–235.

Boiko, A.D., Razorenova, O.V., van de Rijn, M., Swetter, S.M., Johnson, D.L., Ly, D.P., *et al.*, 2010. Human melanoma initiating cells express neural crest nerve growth factor receptor CD271. Nature 466, 133–137.

Brummendorf, T.H., Dragowska, W., Zijlmans, J., Thornbury, G., Lansdorp, P.M. 1998. Asymmetric divisions sustain long/ term hematopoiesis from single sorted human liver cells. J. Exp. Med. 188, 1117–1124.

Charafe-Jauffret, E., Ginestier, C., Iovino, F., Wicinski, J., Cervera, N., Finetti, P., *et al.*, 2009. Breast cancer cell lines contain functional cancer stem cells with metastatic capacity and a distinct molecular signature. Cancer Res. 69, 1302–1313.

Clayton, E., Doupé, D.P., Klein, A.M., Winton, D.J., Simons, B.D., Jones, P.H. 2007. A single type of progenitor cell maintains normal epidermis. Nature 446(7132), 185–189.

Cogngeim, J. 1867. Ueber entrzundung und eiterung. Path. Anat. Physiol. Med. 40, 1–79.

Collado, M., Serrano, M. 2010. Senescence in tumours: evidence from mice and humans. Nat. Rev. Cancer 10, 51–57.

D'Amico, L., Patanè, S., Grange, C., Bussolati, B., Isella, C., Fontani, L., *et al.*, 2013. Primary breast cancer stem-like cells metastasise to bone, switch phenotype and acquire a bone tropism signature. Br. J. Cancer, 108, 2525–2536.

Dalerba, P., Dylla, S.J., Park, I.K., Liu, R., Wang, X., Cho, R.W., *et al.*, 2007. Phenotypic characterization of human colorectal cancer stem cells. Proc. Nat. Acad. Sci. USA 104(24), 10 158–10 163.

Dallas N.A., Xia L., Fan F., Gray, M.J., Gaur, P., van Buren, G. 2nd, *et al.*, 2009. Chemoresistant colorectal cancer cells, the cancer stem cell phenotype, and increased sensitivity to insulin-like growth factor-I receptor inhibition. Cancer Res. 69(5), 1951–1957.

Di Micco, R., Fumagalli, M., Cicalese, A., Piccinin, S., Gasparini, P., Luise, C., *et al.*, 2006. Oncogene-induced senescence is a DNA damage response triggered by DNA hyper-replication. Nature 444, 638–642.

Dou, J., Pan, M., Wen, P., Li, Y., Tang, Q., Chu, L., *et al.*, 2007. Isolation and identification of cancer stem cell-like cells from murine melanoma cell lines. Cell Mol. Immunol. 4, 467–472.

Du, L., Wang, H., He, L., Zhang, J., Ni, B. Wang, X., *et al.*, 2008. CD44 is of functional importance for colorectal cancer stem cells. Clin. Cancer Res. 14(21), 6751–6760.

Fang, D., Nguyen, T.K., Leishear, K., Finko, R., Kulp, A.N., Hotz, S., *et al.*, 2005. A tumorigenic subpopulation with stem cell properties in melanomas. Cancer Res. 65, 9328–9337.

Fillmore, C.M., Kuperwasser, C. 2008. Human breast cancer cell lines contain stem-like cells that self-renew, give rise to phenotypically diverse progeny and survive chemotherapy. Breast Cancer Res. 10, R25.

Fonseca, A.V., Bauer, N., Corbeil, D. 2008. The stemcellmarker CD133 meets the endosomal compartment – new insights into the cell division of hematopoietic stem cells. Blood Cell Mol. Dis. 41, 194–195.

Gao, F.B., Raff, M. 1997. Cell size control and cell intrinsic maturation program in proliferating oligodendrocyte precursor cells. J. Cell Biol. 138, 1367–1377.

Gupta, P.B., Fillmore, C.M., Jing, G., Shapira, S.D., Tao, K., Kuperwasser C., Lander E.S. 2011. Stochastic state transitions give rise to phenotypic equilibrium in populations of cancer cells. Cell 146, 633–644.

Haas, N.K., Herlyn, M. 2005. Normal human melanocyte homeostasis as a paradigm for understanding melanoma. J. Invest. Dermatol. Symp. Proc. 10, 153–163.

Hadnagy, A., Gaboury, L., Beaulieu, R., Balicki, D. 2006. SP analysis may be used to identify cancer stem cell population. Exp. Cell Res. 312, 3701–3710.

Haraguchi, N., Ohkuma, M., Sakashita, H., Matsuzaki, S., Tanaka, F., Mimori, K., *et al.*, 2008. CD133+CD44+ population efficiently enriches colon cancer initiating cells. Ann. Surg. Oncol. 15(10), 2927–2933.

Harrison, H., Farnie, G., Brennan, K.R., Clarke, R.B. 2010. Breast cancer stem cells: something out of notching? Cancer Res. 15, 8973–8976.

Hayflick, L., Moorhead, P.S. 1961. The serial cultivation of human diploid cell strains. Exp. Cell Res. 25, 585–621.

Huang, E.H., Hynes, M.J., Zhang, T., Ginestier, C., Dontu, G., Appelman, H., *et al.*, 2009. Aldehyde dehydrogenase 1 is a marker for normal and malignant human colonic stem cells (SC) and tracks SC overpopulation during colon tumorigenesis. Cancer Res. 69(8), 3382–3389.

Klein, W.M., Wu, B.P., Zhao, S., Wu, H., Klein-Szanto, A.J., Tahan, S.R. 2007. Increased expression of stem cell markers in malignant melanoma. Mod. Pathol. 20, 102–107.

Kumar, S.M., Liu, S., Lu, H., Zhang, H., Zhang, P.J., Gimotty, P.A. 2012. Acquired cancer stem cell phenotypes through Oct4-mediated dedifferentiation. Oncogene 31(47), 4898–4911.

La Porta, C. 2009. Cancer stem cells: lessons from melanoma. Stem Cell Rev. 5(1), 61–65.

La Porta, C.A.M. 2010. Cancer stem cells: light and shadows. In: Singh, S.R. (ed.) Stem Cell, Regenerative Medicine and Cancer. Hauppauge, NY: Nova Science, pp. 513–525.

La Porta, C.A.M., Zapperi, S., Sethna, J. 2012. Senescence cells in growing tumors: population dynamics in cancer stem cells. PloS Comp. Biol. 8, e1002316.

Legler, J.M., Ries, L.A., Smith, M.A., Warren, J.L., Heineman, E.F., Kaplan, R.S., Linet, M.S. 1999. Cancer surveillance series (corrected): brain and other central nervous system cancers: recent trends in incidence and mortality. J. Nat. Cancer Inst. 91(16), 1382–1390.

Lopez-Garcia, C., Klein, A.M., Simons, B.D., Winton, D.J. 2010. Intestinal stem cell replacement follows a pattern of neutral drift. Science 330, 822–825.

Marcato, P., Dean, C.E., Liu, R., Coyle, K.M., Bydoun, M., Wallace, M., *et al.*, 2014. Aldehyde dehydrogenase 1A3 influences breast cancer progression via differential retinoic acid signaling. Mol. Oncol. 2014, S1574–S7891.

Mărgaritescu, C., Pirici, D., Cherciu, I., Bărbălan, A., Cârtână, T., Săftoiu, A. 2014. CD133/Cd166/Ki-67 triple immunofluorescence assessment for putative cancer stem cells in colon. J. Gastrointestin. Liver Dis. 23, 161–170.

Monzani, E., Facchetti, F., Galmozzi, E., Corsini, E., Benetti, A., Cavazzin, C., *et al.*, 2007. Melanoma contains CD133 and ABCG2 positive cells with enhanced tumorigenic potential. Eur. J. Cancer 43, 935–946.

Pardal, R., Clarke, M.F., Morrison, S.J. 2003. Applying the principle of stem cell biology to cancer. Nat. Rev. Cancer 3, 895–902.

Puglisi, M.A., Sgambato, A., Saulnier, N., Rafanelli, F., Barba, M., Boninsegna, A. *et al.*, 2009. Isolation and characterization of CD 133+ cell population within human primary and metastatic colon cancer. Eur. Rev. Med. Pharm. Sci. 13(1), 55–62.

Quintana, E., Shackleton, M., Sabel, M.S., Fullen, D.R., Johnson, T.M., Morrison, S.J. 2008. Efficient tumour formation by single human melanoma cells. Nature 456, 593–598.

Quintana, E., Shackleton, M., Foster, H.R., Fullen, D.R., Sabel, M.S., Johnson, T.M., Morrison, S.J. 2010. Phenotypic heterogeneity among tumorigenic melanoma cells from patients that is reversible and not hierarchically organized. Cancer Cell 18, 510–523.

Reya, T., Morrison, S.J., Clarke, M.F., Weissman, I.L. 2001. Stem cells, cancer and cancer stem cells. Nature 414, 105–111.

Schatton, T., Murphy, G.F., Frank, N.Y., Yamaura, K., Waaga-Gasser, A.M., Gasser, M., *et al.*, 2008. Identification of cells initiating human melanomas. Nature 451, 345–349.

Sheridan, C., Kishimoto, H., Fuchs, R.K., Mehrotra, S., Bhat-Nakashatri, P., Turner, C.H., *et al.*, 2006. CD44+/Cd24− breast cancer cells exhibit enhanced invasive properties: an early sytep necessary to metastasis. Breast Cancer Res. 8(5), R59.

Singh, S.K., Hawkins, C., Clarke, I.D., Squire, J.A., Bayani, J., Hide, T., *et al.*, 2004. Identification of human brain tumour initiating cells. Nature 432, 396–401.

Taghizadeh, R., Noh, M., Huh, Y.H., Ciusani, E., Sigalotti, L., Maio, M., *et al.*, 2010. CXCR6, a newly defined biomarker of tissue-specific stem cell asymmetric self-renewal, identifies more aggressive human melanoma cancer stem cells. PLoS One 5(12), e15183.

Takano, H., Ema, H., Sudo, K., Nakauchi, H. 2004. Asymmetric division and lineage commitment at the level of hematopoietic stem cells: interference from differentiation in daughter cell and granddaughter cell pairs. J. Exp. Med. 199, 295–302.

Vermeulen, L., de Sousa, F., Melo, E., van der Heijden, M., *et al.*, 2010. Wnt activity defines colon cancer stem cells and is regulated by the microenvironment. Nat. Cell Biol. 12(5), 468–476.

Virchow, R. 1855. Editorial. Virchows Arch. Pathol. Anat. Physiol. Med 1855, 23.

Wallenfang, M.R., Matunis, E. 2003. Developmental biology orienting stem cells. Science 301, 1490–1491.

Wasco, M.J., Pu, R.T., Yu, L., Su, L., Ma, L. 2008. Expression of c-H2AX in melanocytic lesions. Hum. Pathol. 39, 1614–1620.

Zapperi, S., La Porta, C.A.M. 2012. Do cancer cells undergo phenotypic switching? The case for imperfect cancer stem cell markers. Sci. Rep. 2, 441.

Zhong, Y., Guan, K., Zhou, C., Maa, W. Wang, D., Zhang, Y., Zhang, S. 2010. Cancer stem cells sustaining the growth of mouse melanoma are not rare. Cancer Lett. 292, 17–23.

Zon, L. 2008. Intrinsic and extrinsic control of haematopoietic stem cell self renewal. Nature 453, 306–313.

16 Interactomic Analysis of the Stem Cell Marker NANOG in a Prostate Cancer Setting

Kiran Mall and Graham Ball

The John van Geest Cancer Research Centre, Nottingham Trent University, Nottingham, UK

16.1 Introduction

16.1.1 Prostate cancer

Over the last few years, there has been an increase in the number of prostate cancer cases in the male populations of developed countries, as well as in the number of prostate cancer-associated deaths. The correct name for prostate cancer is 'prostate adenocarcinoma' (Schulz *et al.*, 2007). Due to the increased research attention that has been given to prostate cancer over the last few years, the mechanisms involved in its progression and development are being slowly elucidated, but they are still not fully understood (Riegman *et al.*, 1991). Prostate cancer has a highly variable progression; it usually begins with a number of tumours that are seen as 'clinically insignificant', which slowly metastasize over a long period of time to other locations within the body (Schulz *et al.*, 2007). Genetic research is being used to help close many of the current gaps within this knowledge base, which will help to identify and increase the number of known biomarkers, aiding in the early diagnosis, treatment and prognosis of the patient (Riegman *et al.*, 1991).

16.1.2 Stem cells

Stem cells are cells that have the ability to self-renew *in vivo* and to produce differentiated cells of various cell lineages. During the process of cell division,

Principles of Stem Cell Biology and Cancer: Future Applications and Therapeutics, First Edition.
Edited by Tarik Regad, Thomas J. Sayers and Robert C. Rees.

a stem cell produces two daughter cells, one which will gain the ability to self-renew and one which cannot self-renew but has become differentiated. An adult stem cell can be seen as a long-lived target for chance mutations (Strachan and Read, 1999), but differentiated cells, depending on their purpose, generally die within a few days of being produced. A stem cell has the ability to accumulate mutations, which can be passed on to its daughter cells during cell division; the daughter cells also have the ability to accumulate more mutations, which will be passed on in turn when they go through cell division themselves. It is these properties that have led researchers to believe cancerous cells could originate from stem cells that have gained a large number of mutations over their lifetime (Pecorino, 2012). This notion is supported by the realization that stem cells and cancer cells have a number of similarities, including the ability to self-renew. In cancer cells, this ability loses all sense of regulation, making them unable to stop reproducing.

In order for the body to maintain cell balance, its stem cells must work to a precise schedule, keeping a strict balance between the number of cells that can self-renew and the number of cells that can differentiate. If this balance is not maintained then unregulated self-renewal of stem cells can lead to the growth of a tumour, a hallmark of cancer, supporting the hypothesis that tumour cells can arise from stem cells (Reya *et al.*, 2001). A differentiated cell can also undergo a mutation that allows it to change back into a self-renewal cell. Various studies have identified the presence of a small number of cells within tumours that help to maintain the growth and survival of the cancer cells. These cells contain markers in the form of proteins that help to identify them as being different from those cells that would normally be present in the tissue (Chen *et al.*, 2008).

This belief that stem cells might be where cancer cells originate from has led researchers to study them in more detail. Therefore, the mechanisms of self-renewal in stem cells are being elucidated. The Wnt pathway has been found to be one of the major pathways at work in stem cells, suggesting that it is also involved in the process of self-renewal during the development of a cancer (Reya *et al.*, 2003). Wnt1 is a known protooncogene. Mutations within a self-renewal cell can cause the Wnt signalling pathway to become activated, and this mode of activation has been recognized in many different types of cancer, including intestinal and colorectal cancer. The Hedgehog (Hh) pathway has also been found to having a role in self-renewal: it has a large role in embryonic development, tissue self-renewal and carcinogenesis and is essential for the formation of relevant patterns within tissues such as the skin and gut. Like Wnt proteins, Hh proteins (Sonic, Desert and Indian) are secreted intercellularly signalling molecules that act as ligands to trigger a specific signal-transduction pathway (Pecorino, 2012).

16.1.3 Pluripotent stem cells

Stem cells come in the form multipotent cells and pluripotent cells. A pluripotent cell has the ability to produce a single cell in no set pathway that is of a different cell lineage to the original cell. Such cells are present within both developing and adult organisms, aiding in the regulative development of embryos (Selwood and Johnson, 2006). Much research has been carried out to confirm the existence and determine the purpose of pluripotent stem cells. Pluripotent stem cells have been found in solid tumours, such as human teratocarcinomas, through an experiment carried out on differential tissues such as muscle and bone (Sell, 2004). Stem cells within solid tumours express markers that are specific to the organ in which they are found. A study has shown that the majority of human breast cancer samples contain a population of tumourigenic stem cells that express the cell-surface marker $CD44^+CD24^-$/$LowLin^-$ (Dontu *et al.*, 2003). This population was enriched 50–100-fold with cells able to form tumours in mice (Salcido *et al.*, 2010). The results show that the tumours have the same phenotypic heterogeneity as those found in the original tumour population, from both tumourigenic and nontumourigenic cells (Dean *et al.*, 2005).

In another study, the overexpression of the Wnt family of genes showed them to be important regulators of normal cell development, leading to the expansion of the mammary stem cell pool and increasing the susceptibility to cancer of the test subject (Liu *et al.*, 2004). Stem cells with pluripotency and the ability to self-renew have been isolated from human tumours affecting the central nervous system (CNS) (Singh *et al.*, 2004). These cells express CD133, 'a cell surface antigen known originally as a marker of haematopoietic stem cells in other normal tissue' (Richardson *et al.*, 2004). The exact origins of the pluripotent stem cells in a tumour will vary: they may occur as a result of the malignant transformation of a normal stem cell that has accumulated a number of cancerous mutations over its lifespan or from a differentiated cell that has developed the ability to continually self-renew (Cozzio *et al.*, 2003).

A study carried out by Schopperle and DeWolf (2007) showed the identification of podocalyxin, a biomarker for human testis. It was found to be expressed on human embryonic carcinoma cell lines: established malignant pluripotent stem cell lines found in germ cell tumours (Schopperle and DeWolf, 2007). Other research teams have also found that podocalyxin is highly expressed in human embryonic stem cells (hESCs) and can be seen as a marker for pluripotent embryonic stem cells (ESCs) that have not fully undergone the correct method of differentiation (Zeng *et al.*, 2004). Podocalyxin has also been shown to play a role as a pluripotent stem cell marker in a large number of non-testis human cancers, including prostate, breast, liver, brain and blood cancers (Riccioni *et al.*, 2006). Schopperle and DeWolf (2007) also found that podocalyxin forms a complex with glucose transporter 3 (GLUT3). One of the most regularly seen aspects of human malignancy is the upregulation and overexpression of glucose transporters on the surfaces of cancer cells (Macheda *et al.*, 2005). Schopperle and DeWolf (2007) also hypothesized that one of the functions of podocalyxin might be to regulate the expression of these transporter proteins in human cancer and stem cells at their cell surfaces (Wegner *et al.*, 2010). Markert *et al.* (2011) used the transcriptome profiling method of gene set enrichment analysis on prostate cancer cells based on their gene signatures, which 'reflect embryonic stem cells (ESC), induced pluripotent stem cells (iPSC), and polycomb repressive complex-2 phenotypes (PRC2)' (Rhodes and Chinnaiyan, 2005). ESC signatures were found in 13% of 281 prostate-cancer subjects, many with Gleason scores ≥ 8.

16.1.4 Known pluripotent stem cell markers

Transcriptional regulation, epigenetic regulation and miRNAs are the three main aspects of a cell which determine its pluripotency. Genetic studies concentrating on a number of individual genes have enabled the identification of critical pluripotency factors. The two main proteins identified as playing a large role in the early development and continued pluripotency of ESCs are the homeodomain transcription factors: Oct4 and Nanog (Chambers *et al.*, 2003). A transcription factor called Sox2, which has been identified on the HMG-box transcription factor, was fond to heterodimerize with Oct4 for the regulation of several genes in mouse embryonic stem cells (mESCs) (Boyer *et al.*, 2005). Oct4, Nanog and Sox2 all share a large number of the genes that they target. They demonstrate a feedback loop and aid in the control of each other's transcription factors via an autologous regulatory circuit. A number of the targets of Oct4, Sox2 and Nanog encode key transcription factors for the correct differentiation and development of ESCs, as well as being transcriptionally inactive. Another function of Oct4, Sox2 and Nanog is to aid in the regulatory processes involved in the maintenance of pluripotency within ESCs (Loh *et al.*, 2006). Pluripotency factors have also been shown to work alongside undifferentiated ESCs with the help of epigenetic regulators for the activation of genes involved in pluripotency maintenance (Chen and Daley, 2006).

Oct4 and Nanog play a key role in the maintenance of the self-renewal and pluripotency of ESCs (Boiani and Schöler, 2005). Oct4 is a member of the transcription factor family known as Pit-Oct-Unc (POU) and can be found in the pluripotent cells of pregastrulation embryos (Liang *et al.*, 2008). It has been shown to take part in the processes that lead to tumourigenicity and malignancy within lung cancers cells (Chen *et al.*, 2008). Nanog aids in the determination of cellular fate during the development of the embryo. It has been found as a downstream target of Oct4 (Chambers *et al.*, 2003) and its proper function is required in order for Oct4 to remain present in the correct state (Cavaleri and Schöler, 2003). Oct4 and Nanog are two of four factors that solidify the 'reprogramming capability of adult cells into germ-line-competent-induced pluripotent stem cells' (Okita *et al.*, 2007). Analysis of colorectal tumour samples using immunohistochemical methods shows an increased expression of Nanog, which highly correlated with poor prognosis and the presence of metastasis within the lymph node (Meng *et al.*, 2010). These transcription factors have been found in pluripotent ESCs and within the inner cell mass (ICM) of the blastocyst, which is where the ESC is derived from. If any of these factors become disrupted and can no longer function properly then the pluripotency is lost, which causes inappropriate differentiation of ICM and ESCs into trophectoderm and extraembryonic endoderm (Chambers *et al.*, 2003).

Nanog is very diverse in the cells it affects and the mechanisms it uses, showing it to be a homeodomain-containing protein with a fundamental position in the transcriptional network of pluripotentcy (Cole *et al.*, 2008). It has been found to be expressed within 'pluripotent embryo cells, derivative embryonic stem cell and the developing germline of mammals' (Lavial *et al.*, 2007). In human cells, Nanog facilitates the process of molecular reprogramming (Yu *et al.*, 2007) and promotes the transfer of pluripotent attributes following ESC fusion (Silva *et al.*, 2006). Takahashi and Yamanaka (2006) were able to reprogramme somatic cells back into an ESC-like state. This discovery led them to successfully reprogramme mouse embryonic fibroblasts (MEFs) and adult fibroblasts back into pluripotent ESC-like cells using the viral-mediated transduction of the four main transcription factors, Oct4, Sox2, c-Myc and Fbx15. After analysis of their results, the cells which had activated Fbx15 were named 'induced pluripotent stem cells' (iPSCs). These iPSCs showed pluripotent ability, which enabled them to form tetratomas, but the authors could not generate live chimeras (Masui *et al.*, 2007). A study by Chambers *et al.* (2007) suggested that one of the functions of Nanog might be to stabilize the pluripotency of the cell, instead of just maintaining it.

Oct4 can heterodimerize with the HMG-box transcription factor Sox2 in ESCs. Sox2 contributes to pluripotency, at least in part, by regulating Oct4 levels (Masui *et al.*, 2007). Oct4 is rapidly and apparently completely silenced during early cellular differentiation. The key roles played by their unique expression patterns (Hart *et al.*, 2004) make it likely that these regulators are central to the transcriptional regulatory hierarchy that specifies ESC identity. The genes occupied by Oct4, Sox2 and Nanog contribute to pluripotency in hESCs and mESCs (Loh *et al.*, 2006). Location of these genes by genome-wide location factors has yielded three findings: (i) Oct4, Sox2 and Nanog bind together at their own promoters to form an interconnected autoregulatory loop; (ii) the three factors often co-occupy their target genes; and (iii) Oct4, Sox2 and Nanog collectively target two sets of genes: one that is actively expressed and one that is silent in ESCs but remains poised for subsequent expression during cellular differentiation (Boyer *et al.*, 2005).

Nanog, Oct4 and Sox2 work together to maintain pluripotency in cells (Rodda *et al.*, 2005). ESCs show a large number of similarities to cancer cells, including the ability to proliferate at a fast rate with poor differentiation. Nanog has been found to be expressed in various types of human cancers (Glinsky, 2008), which show that it plays a role in tumourigenesis. Zbiden *et al.* (2010) showed that overexpression of Nanog was highly related to poor prognosis, while Chambers *et al.* (2003) reported that its overexpression could increase the mobility and migration of cancer cells in colon cancer. Chiou *et al.* (2008) also showed Nanog to be greatly associated with the progression and poor prognosis of late-stage oral cancers, and Han *et al.* (2012) showed that its inhibition decreased breast cancer growth and induced blocking of the cell cycle.

Nanog has many pseudogenes. Its general form, found in many tumours, including in prostate cancer (Jeter *et al.*, 2009), is Nanog1. Another common form in tumours is NanogP8 (Ambady *et al.*, 2010). In prostate cancer, the early progenitor cells that have shown an association with different surface markers include CD44, CD133 and CXCR4 (Miki *et al.*, 2007). Hypoxia has also been shown to occur in a number of cancers within the inner environment of solid tumours, which is optimal for the existence of undifferentiated tumour cells. If hypoxia is found in prostate cancer, it shows poor prognosis (Mori *et al.*, 2010). Ma *et al.* (2011) found that hypoxia had the ability to upregulate stem cell-like properties of the prostate cancer cells.

It has been suggested that Oct4 plays a pivotal role in the development of the mammalian embryo, as the mRNA of Oct4 is downregulated during differentiation (Boiani and Scholer, 2005) and it is primarily found within blastocysts (Pesce *et al.*, 1998a). Experiments performed on Nanog have shown that, *in vitro*, Nanog mRNA is contained in a large number of pluripotent stem cell lines, including 'embryonic stem, embryonic germ and embryonic carcinoma cells' (Chambers *et al.*, 2007); however, this is not the case in adult cells. During the process of differentiation, pluripotent cells have shown downregulation of Nanog expression (Chiou *et al.*, 2008).

Like Nanog, Oct4 has been seen to be expressed within embryo pluripotent cells. It is downregulated in the three somatic lineages, but only when gametogenesis is taking place as the initiation of male and female meiosis occurs (Pesce *et al.*, 1998b). According to Kim *et al.* (2009), 'Oct4 is re-expressed in unfertilized oocytes after birth and it can be detected until the final stages of oocyte maturation'.

Nanog has also been implicated in other studies in the regulation of the expression of other transcriptional factors, such as POU5F1 and Sox2. It is currently not known what specific mechanisms the cells go through, but it has been hypothesized that one of the mechanisms by which Nanog maintains self-renewal and the undifferentiated state of the cells is through modulation of Oct4 and Sox2 levels. It also controls the molecular fate of the ESC effectors, as shown by the presence of Foxd3 and Setdb1. Foxd3 is present in order to encode a transcriptional repressor of importance in maintaining the ICM of ESC lines (Hanna *et al.*, 2002). According to Dodge *et al.* (2004), 'The Setdb1 gene encodes histone H3Lys9 methyltransferase that is required for survival of mouse ESCs'. Mycn has been reported as one of the key mediators involved in the self-renewal and proliferation of ESCs; both Oct4 and Nanog bind to it in order to allow this process to take place (Cartwright *et al.*, 2005). Two others regulators that have also been identified as targets for the maintenance of pluripotent ESCs are Esrrb and Rif1. Esrrb belongs to the family of hormone receptors found within the nucleus; it is also present in mutant homozygous embryos that show irregular proliferation in trophoblast and a reduction in primordial germ cells (Mitsunaga *et al.*, 2004). Rif1 is an orthologue of a yeast telomeric protein, with a higher expression in mESCs and germ cells (Adams and McLaren, 2004); it has also been found in dysfunctional telomeres and has shown relevance in the cellular response to DNA damage (Xu and Blackburn, 2004). Rif1 has also been found to be a target for Oct4 and Nanog in hESCs; however, as pluripotency studies are still in their infancy, not a lot of information are known about the mechanisms of its regulation (Loh *et al.*, 2006).

16.2 Methodological background

16.2.1 Sample selection

A publically available data set for prostate cancer, containing over 50 000 genes, was selected using the search engine array Express. The data were taken from an investigation carried out by Wang *et al.* (2010) on 'In silico estimates of tissue components in surgical samples based on expression profiling data'. Collectively, 148 samples of RNA were taken from patients with prostate cancer gene-expression profiles using 'stroma, tumour, BPH and atrophic glands'. The patients were then put through Affymetrix U133A arrays, used to hybridize these samples (Wang *et al.*, 2010). Affymetrix U133A arrays use the Gene Chip technology in a single array, which corresponds to '14 500 well-characterized human genes'. They have already helped investigate the progression of a number of diseases within the human body. The four main tissue cells used in Wang *et al.* (2010)'s investigation were tumour cells, stroma cells, benign prostate hyperplasia (BPH) epithelial cells and dilated cystic gland epithelial cells in the prostate. The data were run through various artificial neuronal networks (ANNs) (see Section 16.2.3). Each of the samples was determined with the help of pathologists, who used various histopathological and clinical pathologies to determine the presence of a tumour (this is why the amount of cell types used varies between the samples). Wang *et al.* (2010) used methods outlined in a paper published by Stuart *et al.* (2004) on 'patterns of gene expression in prostate cancer'. This paper explains how linear regression is used to compare prostate tumour samples and determines what the cell types are and how to compare them with nontumour samples. Stuart *et al.* (2004) use the following equation:

Equation one: the following equation states that the average expression level – G_{jk} – of gene j in a sample k is the average of cell type expectations, β_{ij}, weighted by cell type fractions x_{ki}'

$$G_{jk} = \sum_{i} x_{ki}\beta_{ij} + \varepsilon_{jk}$$

While using this equation to compare the expression levels of tumour samples against nontumour samples, Wang *et al.* (2010) then used another method to assess the samples determined by pathologists. The calculations of the standard errors, coefficients and intercepts were all carried out by the two-cell-type model. Eventually, the regression coefficient β is given as the expected cell-type expression level (Stuart *et al.*, 2004).

16.2.2 Data sorting

The data set contained both tumour-positive and tumour-negative data. These data were opened in MS Excel and sorted to determine the presence of Nanog. Once Nanog had been located, the interaction values produced during the original experiment were used to work out a median value for each of the interactions it had with the genes. This median value was used to set an interaction marker. All the interaction values above the median suggested high expression of the gene, while all those below the median suggested a low expression in the presence of Nanog.

16.2.3 Statistical analysis

Before the training programme was started, the data were linearly scaled using a minimum value of 0 and a maximum value of 1. The data were sorted on a scale of 0–1 by using the 'IF' rule in MS Excel. Once the data had been sorted and scaled, the file was saved and re-opened in another system called Statistica, which was use to transpose the data (reduce their size) so that they could be run in the Stepwise ANN. An ANN, consisting of a very simple and highly interconnected procession of neurons, is a computational structure inspired by the biological neuronal system of a mammal (Rahimi-Ajdadi and Abbaspour-Gilandeh, 2011). It uses various mathematical methods to sort through a large number of data, which helps solve various complex problems arising from the large number of relationships present between variables (Rahimi-Ajdadi and Abbaspour-Gilandeh, 2011). The weights placed within such networks are updated using a method known as a back-propagation algorithm, which gives the network its multilayer perception architecture (Lancashire *et al.*, 2008). Generally, the structure of a back-propagation network consists of three layers: an input layer, a hidden layer (or more than one, if desired) and an output layer (Rahimi-Ajdadi and Abbaspour-Gilandeh, 2011). Before the training step can begin, the data must be scaled in a linear fashion (as explained in Section 16.2.4). As the initial weights in our data set were scaled between 0 and 1, the system used two layers of hidden nodes (Lancashire *et al.*, 2008). The structure of a neuron consists of two major weight and transfer functions, which receive the weight and eventually generate an output. The stepwise approach identifies the key protein molecular ions representative of the specific treatment regimens within a large data set. The stepwise approach used during our investigation employed a single input in the model. For each model, random rounds of training, testing and validation via bootstrapped subsets were used in order to provide a measure of confidence in the predications made (Lancashire *et al.*, 2005).

16.2.4 Back propagation

The back-propagation algorithm, first described by Rumelhart *et al.* (1986), is used during clinical research in many medical settings to determine the different interactions between genes and their strengths in prognosis (Cinar *et al.*, 2009). As mentioned in Section 16.2.3, a back-propagation neuronal network is separated into three layers: input layer, hidden layer(s) and output layer. The input layer is used to 'feed' the data (information) into the network. This information is processed by the neurons in the hidden layer, which have no predefined initial values. The results are then fed to the output layer, which processes the information and produces values (Røe *et al.*, 2011).

The back-propagation system involves two phases, known as 'training processes': the feed-forward phase and the back-propagation phase (Cinar *et al.*, 2009). In the feed-forward system, the data are transmitted in a forward direction. The output neurons contain parameters which help to determine and process the data they produce. In the back-propagation system, errors generated during the feed-forward phase are used to change the weights within the neurons, using an algorithm on a learning gradient (Røe *et al.*, 2011). The back-propagation system works by assessing the differences between the data generated by the feed-forward system and set values desired by the output layer and calculating errors. The error values thus produced are run through the back-propagation system in order to calculate links within the input layer (Ronco and Fernandez, 1999).

16.2.5 Cytoscape interaction diagrams

Cytoscape is an open-source Java network visualization and analysis tool, standardized and licensed under the Common Good Public License (CGPL), that provides a large array of useful features (Shannon *et al.*, 2003). It is not specifically designed for use over the Web, except via Java WebStart or as a library for generating static network images for Web display (Lopes *et al.*, 2010). It has four main principles – visualization, compatibility, functionality and strength – each of which aids in the production of accurate interaction diagrams, showing 'hubs' that identify the genes with the highest number of interactions within the data set:

1. Visualization: The system provides two-dimensional representational diagrams, even for a large-scale network analysis containing hundreds of thousands of nodes and edges. It can support a variety of formats, including directed, undirected and weighted graphs. It provides various powerful visual styles, allowing the user to change the properties of the

nodes and edges, and contains a large number of different algorithm layouts, including cyclic and spring-embedded.
2. Compatibility: The system provides various data parsers and filters, which make it compatible with other tools, including a wide range of ANNs.
3. Functionality: The system is highly interactive. The user can zoom in or out to view various parts of the network in more detail. The different properties set for specific nodes or edges can be saved and reloaded exactly how they were. The system can manage different networks very easily, as it has its own built-in network manager. The user can save the various networks at different points within the investigation, allowing them to view and compare changes within the networks and within the data produced during the investigation.
4. Strength: The system's main purpose is to visualize molecular interaction networks and their integration with gene-expression profiles and other datasets.

Cytoscape also allows the user to manipulate the various networks and to compare a large number of different networks. Plug-ins can be created and stored by the user and made available for use by others; this allows specific and specialized analysis of the networks and of various molecular profiles (Pavlopoulos *et al.*, 2008).

16.3 Results and development of networks

16.3.1 Initial marker screen for links to NANOG

A back-propagation algorithm was developed using probes from the data described in Section 16.2, in order to predict the expression level of the NANOG gene. This analysis, which considered 54 677 genes, produced a rank order of probes based on the root-mean-square error of predictions of NANOG (Lancashire *et al.*, 2009). The results showed the interactions of each probe on the array with the target NANOG across the tumour-positive samples, in order to help identify markers which could be used as a surrogate for the presence of any stem cell markers in prostate cancer.

16.3.2 Back-propagation artificial neural network

Once all the data had finished running, the top 100 probes predicting NANOG were selected and put into a network inference algorithm. This approach uses probes to predict probes. So, for each of the top 100 probes, the remaining 99 were used to determine the strength of its interactions. In this way, the interactions between the top 100 probes were identified and stored in an array of 9900 interactions for the top 100 genes. Full details of this algorithm are presented in Lemetre *et al.* (2009). This array of weights was sorted and the top 100 positive and top 100 negative were retained and put into a Cytoscape network (Figure 16.1). The initial Cytoscape network contained

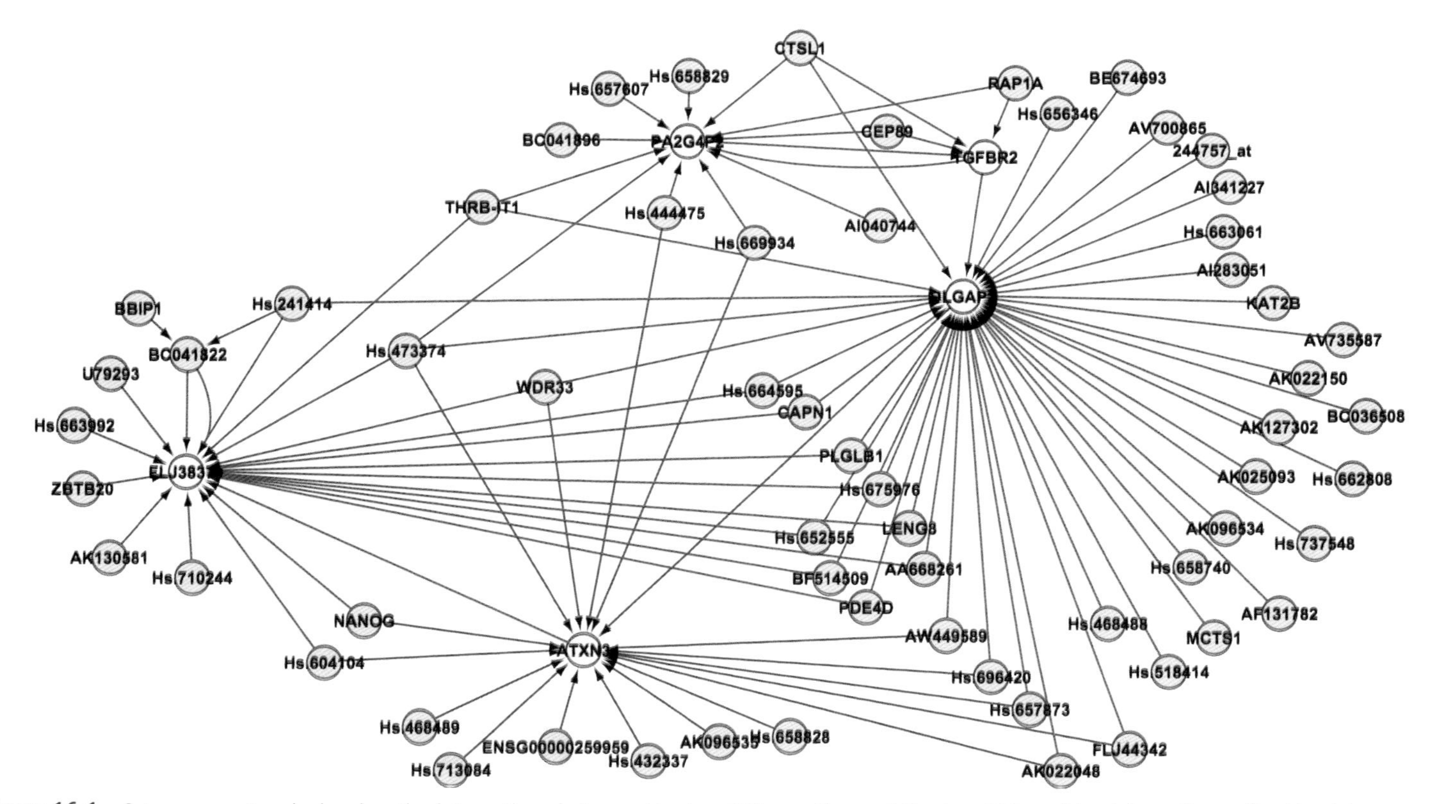

Figure 16.1 Cytoscape network showing the interactions between the top 100 negative and the top 100 positive interactions after the data were run through the back-propagation ANN. The red arrows show a positive interaction and the blue arrows a negative interaction.

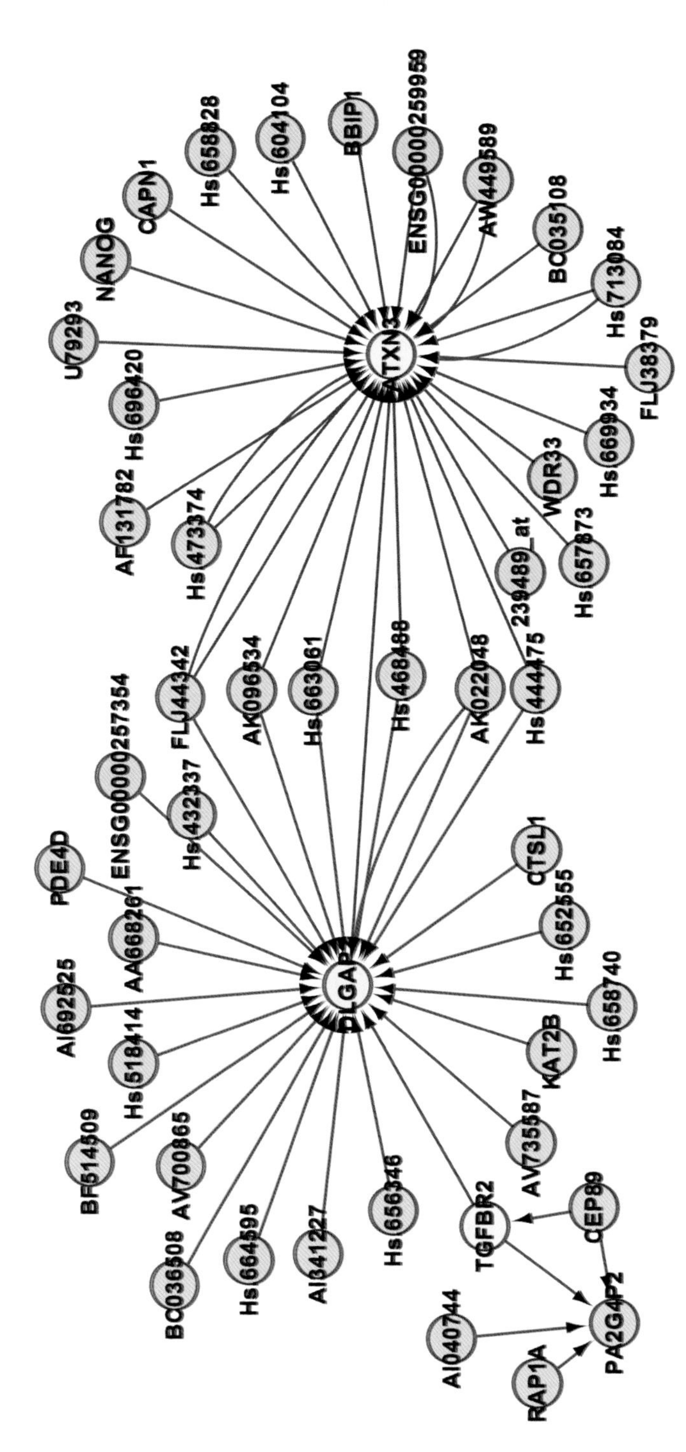

Figure 16.2 Cytoscape network showing the interactions between the top 30 negative and the top 30 positive interactions after the data were run through the back-propagation ANN. The red lines represent a positive interaction and the blue lines a negative interaction.

a large number of genes and seemed to be quiet complicated to understand, so another Cytoscape network with a smaller number of genes was also produced. Thus, the data were further condensed into the top 30 positive and the top 30 negative values, which were put into a second Cytoscape network, highlighting four hubs (Figure 16.2). From these analyses, the probes with the most links to other probes were defined as hubs (Tables 16.1 and 16.2).

Table 16.1 Names, symbols, subcellular locations and chromosomal locations of the hubs highlighted in the Cytoscape network of the top 100 positive and top 100 negative interactions.

Gene	Abbreviation	Name	Chromosomal location	Subcellular location
207334_s_at	TGβR2	Transforming growth factor, beta receptor II	3p22	Cell membrane
216422_at	PA2G4P2	Proliferation-associated 2G4 pseudogene 2	20p12.1	Unknown
242965_at	No information could be found			
235240_at	ATXN3	Ataxin 3	14q21	Nucleus
210227_at	DLGAP2	Discs, large (Drosophila) homologue-associated protein 2	8p23	Cell membrane
1556473_at	FLJ38379	Uncategorized	2q37.3	
215304_at	No information could be found			
230139_at	NFXL1	Nuclear transcription factor, X-box binding-like 1	2p12	
244696_at	AFF3	AF4/FMR2 family, member 3	2q11.2	Nucleus
239606_at	no information could be found			
1562280_at	POU2F1	POU domain, class 2, transcription factor 1	1q24.2	Nucleus
1559490_at	LRCH3	Leucine-rich repeats and calponin homology (CH) domain containing 3	3q29	Secretary glands
242110_at	ARHGAP5	Rho GTPase-activating protein 5	14q12	Cytoplasm

Table 16.2 Detail of the hubs highlighted in the Cytoscape network of the top 30 positive and top 30 negative interactions.

Gene	Abbreviation	Name	Chromosomal location	Subcellular location
207334_s_at	TGβR2	Transforming growth factor, beta receptor II	3p22	Cell membrane
216422_at	PA2G4P2	Proliferation-associated 2G4 pseudogene 2	20p12.1	Unknown
235240_at	ATXN3	Ataxin 3	14q21	Nucleus
210227_at	DLGAP2	Discs, large (Drosophila) homologue-associated protein 2	8p23	Cell membrane

16.3.3 Identified hubs

16.3.3.1 Ataxin-3 Ataxin-3 is a protein gene which causes various neurodegenerative diseases that are members of a family of diseases caused by a CAG repeat that becomes expanded within the polyglutamine domain of this protein (Zoghbhi *et al.*, 2000). The main neurodegenerative diseases caused by this gene include spinocerebellar ataxin type-3 (Kawaguchi *et al.*, 1994), spinocellular ataxia type 1, 2, 3, 6 and 7 (Ross *et al.*, 1999) and Huntington's disease (Dyer and McMurray, 2001). Repression of the process of transcription has been shown in these diseases, suggesting that it is one of their main features. The ataxin-3 molecule is made up of an N-terminal domain (ND), called the Josephin domain (JD), two ubiquitin-interacting motifs (UIMs), a polyglutamine stretch and a small tail, which has variable sequences.

The majority of the studies carried out on this molecule have looked at the transcriptional regulation occurring on it as a result of the poluglutamine disease proteins. Dyer and McMurray (2001) used 'truncated proteins with polyglutamine-rich domains' instead of the full-length proteins, so it is still unclear how the cell undergoes its processes with the full-length protein and how it interacts with the various domains. This finding was highlighted by pathological research carried out on subjects with Huntington's disease, as the results differed from those in previous experiments due to the presence of the full-length protein rather than the truncated one (Dyer and McMurray, 2001), as well as the specificity of transcriptional regulators that have been sequestered by nuclear inclusions of ataxin-3 (Chai *et al.*, 2001).

Ataxin-3 has been described as 'an excellent candidate to study transcriptional regulation' (Li *et al.*, 2002) due to its small size, which allows for the correct function of the other domains present within the molecule and during the investigation of a polyglutamine domain. Three mechanisms have been elucidated for transcriptional repression: sequestration of transcriptional activators/coactivators by inclusion of polyglutamine-containing protein (Steffan *et al.*, 2000), inhibition of the histone acetyltransferase activity of coactivators such as CBP/p300 (Gusterson *et al.*, 2003) and direct co-repressor activity (Wood *et al.*, 2000). During sequestration of transcriptional activators by inclusions formed by pathological polyglutamine proteins, the factors that have shown to be activated are the androgen receptors (ARs), huntingtin, atrophin and ataxin-7 (Chai *et al.*, 2002). The activation of ARs shows that ataxin-3 plays a role within the prostate, and possibly in the mechanisms leading to the progression of prostate cancer.

Sponocerebellar ataxia type 3, also known as Machado-Joseph disease, is a germ-line neurodegenerative disease that has been labelled as autosomal dominant. This disease is caused by the growth of a polyglutamine stretch in the gene (Varshavsky, 1997). Hershko and Ciechanover (1998) found that the gene responsible for the production of STAT-3 also encodes the gene ataxin-3. Ataxin-3 has deubiquitination activity, which can be destroyed by the presence of a mutation in a Cys molecule (C14) found within the catalytic site of the molecule (Hershko and Ciechanover, 1998). The ataxin-3/Josephine family is one of five families classed as deubiquitinating enzymes (Wilkinson *et al.*, 1989). Due to its UIM binding abilities, ataxin-3 (Chai *et al.*, 2004) might have the ability to function as part of the 'protein surveillance pathway', which eventually ends at the proteasome (Moa *et al.*, 2005). Ubiquitination is a reversible modification; the state of substrate unbiquitination is highly dependent on the 'balance between the effects of deubiquitinating enzymes (DUBs)' (Komander *et al.*, 2009). Ataxin-3 carries out the process of deubiquitination by first binding ubiquitins to the polyubiquitin chains and then acting as a component in the eventual delivery of ubiquitinated substrates to the proteasome (Wang *et al.*, 2008). Caspases are involved in many of the different pathways to cancer activation, so the presence of ataxin-3 can aid in the determination of a solid pathway leading to prostate cancer. According to Burnett *et al.* (2003), 'Ataxin 3 has the typical properties of DUB: the enzyme disassembles ubiquitin-lysozyme conjugates, cleaves ubiquitin-7-amido-4-methycoumame (ubiquitin-AMC), and binds to the DUB inhibitor ubiquitin aldehyde (Ubal)'.

Polyglutamine proteins are good substrates for kinases. Examples include polyglutamine ataxin-3, a substrate for casein kinase 2 (Tao *et al.*, 2008); this molecule is phosphorylated by glycogen synthase kinase 3β, which is another polyglutamine substrate (Fei *et al.*, 2007). Ataxin-3 polymerization is increased by the substitution of serine molecule with an alanine molecule, perhaps indicating that the phosphorylation of ataxin-3 becomes aggregated when this occurs. Importantly, the activities of ataxin-3 associated with ubiquitin are required for protection against neurodegeneration. According to Berke *et al.* (2004), 'Ataxin-3 cleavage is blocked by caspase inhibitors, with caspase 1 being one possible cleavage protease and Ataxin-3 fragmentation also correlates with increased Ataxin-3 aggregation'.

Ataxin-3 is an intracellular protein that is expressed in nearly all the tissues of the body (Paulson *et al.*, 1997a). Abnormal folding of the protein, along with aggregation, has been shown to be one of the possible key causative factors in polyglutamine diseases, and various types of interneuron aggregates can occur in the cells of patients affected by polyglutamine diseases (Paulson *et al.*, 1997b).

16.3.3.2 DLGAP2 DLGAP2 is highly expressed within the cells of the brain and testis (Ranta *et al.*, 2000). It can be found on chromosome 8 at the 8p23.3 position, which has been found to be frequently deleted in bladder cancer (Muscheck *et al.*, 2003), suggesting it might be used as a possible tumour suppressor. A study carried out using reverse transcriptions to produce complementary DNA that contained polymorphic sites with RNA isolated from brain and testis in heterozygous human cells (Karolchick, 2003) found that they were expressed in the testis of all samples examined (Luedi *et al.*, 2007). According to Hughes *et al.* (2005), 'It is a basic principle in tumour biology that an alteration in a specific chromosome, which is present in human cancers is functionally important event before their development and progression'. This theory is based on the results of many studies showing that specific chromosomal losses generally become inactivated in the second copy of a tumour suppressor. However, 'chromosomal gains or amplifications lead to the overexpression of oncogenes' (Hughes *et al.*, 2005). An example would be the loss of 17p, which shows the inactivation of TP53, while 'gains or amplifications at 8q24 are often associated with overexpression of oncogenic Myc' (Hughes *et al.*, 2005). Unfortunately, the association between chromosomal changes and the inactivation or oncogenes on individual genes is not always easily demonstrated. Therefore, in prostate cancer, it has not yet been possible to link the common losses at 8p to the inactivation of any particular gene. A vast amount of research has found 'either gain of 8q or loss of 8p, or both, to correlate with increased tumour stage, grade, metastasis, recurrence or death of disease' (Hughes *et al.*, 2005).

Among the large number of prostate cancers, alterations found at chromosome 8p are heterogeneous with losses detected in the 8p12–8p21 and 8p23.1 areas (Gelsi-Boyer *et al.*, 2005). In theory, the deletion of one copy of a gene should decrease the amount of gene expression to half its normal level. However, inactivation of the second copy of a gene in a tumour suppressor has been shown to result from bone mutations, deletions or promoter hypermethylation. According to Hornstein *et al.* (2008), 'In prostate cancer no gene or chromosome 8p is consistently inactivated by any such combination of epigenetic or genetic events. Since deletions in cancer cells usually extend across several loci, it is thus difficult to distinguish which genes down-regulated prostate cancer are decisive for tumour development and progression and which are "bystanders" of 8p.'

A study carried out by Srikanten *et al.* (1999) found that the loss of 6q16.3–q21 was present in prostate tumours. 'The short arm of chromosome 8 was frequently deleted for this group of tumours, specifically 8p23.2. This region has been previously reported as being commonly deleted in prostate cancer' (Cher *et al.*, 1994). Chromosome 8p has been shown to contain a number of tumour-suppressor genes, including DLGAP2 (8p23.2). Latil *et al.* (1997) found through genetic linkage mapping of familial prostate cancer that the chromosome found at 16q harboured deletions in around 50% of prostate tumours at the position of 16q23–qter, suggesting possible genes involved in prostate cancer can be found here (Neville *et al.*, 2002). The DLGAP2 family is one of the least-studied components of the zone within a cell that has the most potential to undergo significant deterioration. After making this discovery, Shertser *et al.* (2012) proposed that mediation and communication with subsynaptic molecules changed the expression of DLGAP, which has been found in both schizophrenia and autism. They showed that abnormal expression of DLGAP2 is associated with maladaptation to trauma (Shertser *et al.*, 2012). The revelation that DLGAP2 has a role in the production of some neurodegenerative diseases shows a link with ataxin-3.

16.3.3.3 TGFβR2 This gene is known as the transforming growth factor β receptor type 2 gene (TGFβR2) and has shown the presence of many frame-shift mutations of the polyA tail (Lu *et al.*, 1998). TGFβ signalling takes place once the two main receptors, TGFβI and TGFβRII, have become fully formed. The signalling of TGFβ is mediated through Smad and no-Smad pathways, so that they can 'regulate transcription, translation, microRNA biogenesis, protein synthase and post-translational modifications' (Pasche *et al.*, 1999). When TGFβ undergoes binding to its receptor complex, this initiates signalling through a family of proteins known as Smad. The presence of mutations in at least two members of this family (Smad 2 and Smad 4) has been associated with the development of a number of tumours, including colorectal, breast and ovarian cancers (Schutte *et al.*, 1996).

The mutator pathway has been described in the progression of tumours and is able to progress at a faster pace through an increased accumulation of mutations. According to Duval and Hamelin (2002), 'In this sense, a significant prevalence of mutations in a given gene has been considered a reliable indicator that such genes are targets rather than passengers. A great number of possible target genes have been proposed.' A large number of research groups consider TGFβR2 to be a target gene in the mutator pathway (Parsons *et al.*, 1995), due to the high frequency of frame shift mutations that are found in this pathway (Rampino *et al.*, 1997). The significance of TGFβ in the progression of a tumour is still unclear, however, and needs to be established. Currently, we know that TGFβ is common in many human cancers, as it is a good inhibitor of cell growth within human epithelial cells. It emits its signal by making contact with two 'transmembrane serine-threonine kinase receptors, TGFβRI and TGFβRII' (Nagaraj and Datta, 2010). The TGFβRII receptor gene contains a polyadenine (A) (Lothe *et al.*, 1993) in its coding region, which has been shown to regularly undergo frame-shift mutations in colorectal cancer, due to the presence of microsatellite insertions (Iacopetta *et al.*, 1998). Although it has many attributes that could aid in tumour progression, if TGFβII were really a great target gene that could be used to determine a contribution to tumour progression then patients with mutations in this gene would show a more advanced tumour with a worse prognosis. During analysis, TGFβR2 mutations in microsatellite insertions causing a positive tumour show a moderate to poor degree of differentiation (Grady *et al.*, 1998).

It has been shown that epigenetic changes are present within nearly all tumours, including DNA methylation processes. Such processes lead to gene expression becoming deregulated and eventual tumour growth (Liu *et al.*, 2009). TGFβ is a pleiotropic cytokine that has been implicated in the regulation of mammalian development, differentiation and homeostasis of all cell types and tissue modifications (Hata and Davis, 2009).

TGFβ binds the TGFβRII that will recruit and begin the transphosphorylation of TGFβRI (Shi and Massagué, 2003). According to Shi and Massagué (2003), 'The activated TGFβRI then phosphorylates Smad 2 and Smad 3 at the C-terminus. Activated 2/3 Smad forms hetero-oligomers with Smad 4 and migrates to the nucleus to regulate transcription'. The Smad complexes interact with countless numbers of transcriptional co-regulators, along with a number of other factors that can mediate gene expression or repression of the target molecule (Vogelmann *et al.*, 2005). Smad 2/3 also interacts with and regulates miRNA processing, and TGFβ signals through a number of non-Smad pathways (Mu *et al.*, 2012). Dysregulation of both Smad and non-Smad pathways is implicated in aberrant TGFβ signalling and its protumourigenic events in advanced caner (Zhang *et al.*, 2011). Matsumura *et al.* (2011) stated that most genes that have undergone methylation in cancer are needed during TGFβ signalling processes, showing they are important to tumour progression. This is consistent with their finding that overexpression of TGFβ shows an association with 'aggressiveness and poor prognosis in prostate cancer' (Zhang *et al.*, 2009).

A study carried out by Sutkowski *et al.* (1992) showed that TGFβ plays the same roles in both normal and cancer prostate cells, as it works to inhibit the process of proliferation and induce apoptosis. According to Guo and Kyprianou (1999), 'Reactivation of TGFβ signalling in androgen-insensitive prostate cancer cells re-establishes TGFβ tumour-suppressive properties'. In prostate cancer cells expressing wild-type AR, dihydrotestosterone antagonizes the proapoptotic effects of TGFβ, but TGFβ still suppresses proliferation in these cells. By contrast, in prostate cells expressing a constitutively activated Smad 4, ARs are translocated to the nucleus alone (Massagué, 1998). In the presence of androgen, the apoptotic effects of TGFβ are enhanced in these cells (Schmid *et al.*, 1993). In aggressive prostate cancer, it is common to find downregulation of TGFβRs, particularly TGFβRII (Shariat *et al.*, 2004).

16.3.3.4 PA2G4P2 PA2G4 is a family of proteins known to regulate the proliferation of other proteins. One member, Ebp1 (BrbB3-binding protein 1) , has been implicated within some cancer cells (Yoo *et al.*, 2000). Expression of Ebp1 within a cell inhibits growth and induces differentiation. This has been seen in both breast and prostate cancer cells by Zhang *et al.* (2003) and in ARs by Zhang *et al.* (2002), which emphasises its role in prostate cancer (Zhang *et al.*, 2005).

Ebp1 is known to transduce signals regulating the growth of cells. It shows a large association with the mechanisms of ErbB3 in the progression of prostate cancer (Koumakpayi *et al.*, 2007). The PA2GP2 proteins show an increased presence within the nucleus and the cytoplasm of various malignant and non-malignant cells (Xia *et al.*, 2001). Lessor *et al.* (2000) suggest that in human breast and prostate cancer, cytoplasmic expression of Ebp1 limits proliferation and differentiation, suggesting it may play an important role in tumourigenesis. Ebp1 shows the ability to repress transcription within the ARs and cyclin D1 by binding directly to the ARs, as well as through its interactions with histone deacetylase HDAC2 and Rb, supporting the hypothesis that it plays a role in the development of prostate cancer (Zhang *et al.*, 2002).

Gannon *et al.* (2008) have shown that tissue samples extracted from an area adjacent to prostate cancer growth have an increased level of Ebp1 compared to normal prostate cancer cells. They hypothesize that this is due to the 'tumour's field effect on the surrounding cells'. There are two isoforms of the cytoplasmic Ebp1 (p42, which inhibits cellular proliferation, and p48, which promotes proliferation and cell survival; Liu *et al.*, 2006), but these have not been investigated; however, they may provide an explanation for the regulation of cellular proliferations within prostate cancer (Gannon *et al.*, 2008). The effects of the overexpression of Ebp1 have also been studied by Zhang *et al.* (2005), in order to help elucidate the process of AR signalling inhibition and AR cell growth. Through expression-profiling methods, they have determined a number of genes which act as targets for the AR receptors, including those that are involved in the independent growth of AR cells during the progression of prostate cancer and those which become downregulated in the presence of Ebp1 (Zhang *et al.*, 2005).

16.3.3.5 FLJ38379 This gene has been highlighted within the Cytoscape network, but, after a large amount of research, all that can be found is that it is an uncharacterized gene. Its chromosomal position shows its presence on chromosome 2 at 2q37.3.

16.4 Discussion

This investigation was started with the aims of determining whether pluripotent stem cells could be used as biomarkers in the initiation and progression of prostate cancer and of determining a specific pluripotent stem cell marker for prostate cancer. Many research groups have hypothesized that a small group of cells known as cancer stem cells have the characteristics required to initiate prostate cancer. This small set of cells has been found in a large number of tumorous growths, but their exact mechanisms of cancer initiation are as yet unknown. They do have the ability to both self-renew and differentiate, which suggests the properties of both stem cells and cancer cells. A data set of both tumour-positive and tumour-negative cells was run though a number of different ANNs, including a back-propagation training algorithm, and the results were condensed into the top 100 positive interactions and the top 100 negative interactions. The data were finally sorted by magnitude, producing 200 positive values, which were used to produce a Cytoscape network highlighting six genes: DLGAP2, TGFβR2, POU2F1, Atatxin-3, PA2G4P2 and FLJ38379.

Prostate cancer is one of the most common cancers in men older than 50. Its prevalence is beginning to increase at a very dangerous rate, especially in developed areas. It can be present in both latent and apparent forms and it is believed that more than 50% of men over the age of 80 years will have the latent version. As the originating cause of prostate cancer is not known, a large amount of research has been carried out to try and determine it. The closest steps we have come to a breakthrough is determining that hormonal, genetic and environmental factors are all causes. It has been observed, however, that prostate cancer does not develop in men who were castrated before puberty, suggesting that some prostate cancer cases are due to androgen abnormalities (Koeneman, 2006).

The growth and formation of the prostate epithelial layer is highly dependent on the presence of androgens and mitogens in the progression from a simple neoplasia to the beginning of prostate cancer development. New research has shown that stem cells may have a role to play in the development of cancer, and teams are discovering the ability of the tissue to differentiate and self-renew in different types of the cell. Stem-like cells have been found in the proximal duct area of the prostate; TGFβ regulates their growth, as does the presence of the Bcl-2 family and stem cell antigen-1 (Szmitko *et al.*, 2003). Urogenital sinus (UGS) mesenchyme cells have the ability to redevelop the prostate, which further suggests the presence of cells (Ozaki and Leonard, 2002). Prostate progenitor cells have been characterized by the expression of p63, CD44 α2β, human telomerase reverse transcriptase and CD133 (Gordon, 2003); these might be kept within the prostate gland until they are needed. The evidence pool for the origination of prostate cancer in stem cells is growing (McNicol and Isral, 2008).

A meta-analysis of DLGAP2, TGFβR2, POU2F1 and PA2G4P2 has increased our knowledge of their roles within normal and healthy cells, and, in some cases, of their roles in tumourigenesis. TGFβR2 has been shown to have a large role to play in the epithelial–mesenchymal transition (EMT) pathway and in aiding in tumourigenesis. As stem cell research is still in its infancy, the specific pathways that EMT cells take to cause tumourigenesis are still not known. Azhar *et al.* (2003) found that TGFβR2 is one of the major regulators of epicardial cells. Three main isoforms of this cytokine (TGFβR1, 2 and 3) are found within the human body, but only two are expressed abundantly during the early stages of epicardial and coronary vascular formation, prompting the question of why TGFβR2 dominates within the cells of the body (except the cardiac cells) (Muraoka-Cook *et al.*, 2005). A heterodimer complex is produced by the binding of the two receptors TGFβR1 and TGFβR2, which bind to the TGFβ2 molecule to allow the correct signals to become transduced within the cell (Robert and Wakefield, 2003). Once this binding has occurred, the TGFβR2 activates the TGFβR1, which leads to the eventual activation of various other signalling molecules found within the cell (Roberts and Sporn, 1990). Studies have also found that genes within the MAPK pathways (e.g. p38) could become activated by TGFβR2, as could those of the JNK and ERK1/2 pathways, to initiate a cellular response with a role to play in invasive cancer cells (Hu *et al.*, 1998). This activation of various pathways suggests that TGFβR2 has a major role to play within tumourigenesis, and these findings could be a stepping stone to the elucidation of its role TGFβR2 in tumourigenesis.

Many cellular processes take part in cancer cell development. EMT has shown the ability to dissolve their intracellular attachments, determine the type of mobility functions they will need and then move from one part of the body to another (De *et al.*, 2004). In order to allow them to settle into their new environment, EMT cells have adapted the ability to reverse their original changes, in a process known as mesenchymal–epithelial transition (MET). This enables them to undergo self-renewal and differentiation without causing any major disruption, meaning they can integrate well into their new environment and begin the production of cancer-causing stem cells without being detected until tumourigenesis is evident (Ridley *et al.*, 2003). Mitogens, TGFβ, Notch and Wnt are all signalling factors that work extracellularly to help the process of EMT. Pechlivanis and Kuhlmann (2006) have suggested that TGFβR2 is a very important EMT inducer, as their studies found that Ras and Raf are two oncogenic stimulants of the EMT process are both independent of the extracellular signalling produced by TGFβR2 (Pechlivanis and Kuhlmann, 2006). The progress of pancreatic carcinoma cells from one location to another during the EMT response is introduced by the loss of their co-receptor beta glycan, which inhibits TGFβ signalling (Wyckoff *et al.*, 2000).

POU2F1 is another name for the transcription factor Oct-1. The POU-domain proteins Sox and Oct have shown in previous experiments to work in conjunction with each other in ESCs. It has been shown that the POU2F1 protein interacts without DNA-binding proteins in order to influence the POU factor-specific interactions, implying that there is a large repertoire that could, if needed, aid in the specificities of POU paralogues (Peri *et al.*, 2003). This repertoire aids in the amount of binding that occurs and helps to differentiate between the different POU domains. POU domains are generally dimers, which allows them to bind to palindromic sequences from various sites (Tantin *et al.*, 2008). Specific biological activities can be elucidated and the access of coactivators determined by the specific conformations present on the POU domain (Kang *et al.*, 2009). The various members of the POU2F family play important roles in the undifferentiation of hESCs: POU2F1 has been found doing this in the foetal liver and spleen, adult colon and islets of Langerhans (Milenkovic *et al.*, 2010).

Gu *et al.* (2007) found that two different prostate cancer cell lineages, HPET and HPET-5, were positive for the expression of the pluripotent stem cell markers Oct4, Nanog and Sox2. Oct4 is a POU transcription factor, which has been found to be expressed in oocytes (Stewart, 2002). Its normal expression aids in the maintenance of cellular pluripoency, but if it becomes upregulated then cellular differentiation results, while if downregulation occurs then the cell loses its ability to dedifferentiate and its pluripotency (Niwa *et al.*, 2000). Gu *et al.* (2007) also found that a transcription factor, Nanog, was present within the intercellular mass (Chambers *et al.*, 2003). They observed that an increase in the expression of Nanog inhibits differentiation, which could disturb the activities of Oct4, causing cellular complications (Mitsui *et al.*, 2003).

Even though these two transcription factors have been shown to play a role in the maintenance of cellular pluriotency, it is not known how they do so (Niwa *et al.*, 2000).

Liu *et al.* (2006) identified another transcription factor, called Sox4. They found its expression to be higher than normal in prostate cancer tissues. They also found that when Sox4 was not expressed in LNCap cells, apoptosis was induced, but when Sox4 expression was upregulated in RWPE-1, the cells became immortalized, which suggests a critical role for Sox4 in the 'growth and survival of prostate cancer cell' (Jeffers *et al.*, 2003). Further analysis of the RWPE-1 pathways highlighted the presence of human papilloma virus (HPV)-causing oncoproteins, which have been found to inactivate the tumour-suppressor attributes of the Rb and p53 tumour-suppressor proteins. Liu *et al.* (2006) hypothesized that the changes in the Rb/p53 pathway caused by the inactivation of their oncoproteins increased the ability of Sox4 to aid in the development of cancer (Schilham *et al.*, 1996). Sox4 has been implicated in a large number of cancers. In order to elucidate the reasons for this, its genomic sequences were analysed using the CONFAC software (Karanam and Moreno, 2004), which showed that hypoxia-inducible factor 1 alpha (HIF-1α) affects the role of Sox4. This suggests that Sox4 is a target of NF-nB, which is required for cellular transformation (Suh and Rabson, 2004).

Another member of the POU transcription family, called Brn-3, is critical in the regulation of cell proliferation and differentiation, and acts to protect the cell from apoptosis (Latchman, 1999). It has a strong effect on the fate of the cell during embryogenesis and aids in the maintenance of mature neuronal cells in adults (McEvilly and Rosenfeld, 1999). Its important roles in the cell might be dangerous if its regulation were lost, as has been seen through its alteration in a number of cancers (Leblond *et al.*, 1997). Diss *et al.* (2008) identified the upregulation of Brn-3a in a large number of prostate cancer cases, finding that when Brn-3a is overexpressed within the cell, it has a direct effect on tumourigenesis, as it aids in cell growth (Ndisdang *et al.*, 1998). After analysing other research studies into the role of Brn-3a, they found that Brn-3a levels had only really been compared between noncancerous cells and cancerous cells, showing an increase in cancerous cells. They also found that the overexpression of Brn-3a increased the rate of cancer growth (Ndisdang *et al.*, 1998). Theil *et al.* (1993) found that the overexpression of Brn-3a in rate embryonic fibroblasts was oncogenic. Also, the POU homeodomain and an N-terminal activation domain, both present in the Brn-3a molecule, regulate the activity of some promoters within the cell, including Bcl-2, and both are present in breast cancer cells (Smith *et al.*, 1998). Diss *et al.* (2008) showed that the members of the POU homeobox containing Brn-3a play a major role in the tumourigenesis of prostate cancer. They believe that this role shows that Brn-3a also controls various other mechanisms in the progression of prostate cancer (Diss *et al.*, 2008).

Generally, when an aspect of a cell's genetic material changes, there are repercussions on the cellular activity, whether good or bad. Polyglutamine is prevalent in many if not all of the cells of the body. In order to adapt to the particular cellular needs, it undergoes changes in its protein sequences. One example is AR repeat sequences: the AR has a glutamine repeat within an ND, along with a central DNA-binding domain, a hinge region and a C-terminal ligand-binding domain (LBD). The ND polyglutamine repeat length mediates AR function: changes in repeat length may aid in the development of prostate cancer, as a change from the normal 20-length repeat unit to a much smaller one, such as a 9-length repeat unit, has been shown to cause greater ligand-binding activity and ND-LBD, leading to the formation of androgen-dependent prostate cancer (Saitoh *et al.*, 2002). Alternatively, when the polyglutamine repeat unit increases in size, androgen-dependent transcriptional activity has been known to decrease (Chi *et al.*, 2002). However, expansion of the CAG polyglutamine repeat in ataxin-3 from 22 to 64 units of length has shown a threefold decrease in the ability of ataxin-3 to bind to tetraubiquitin (Link *et al.*, 2008).

The AR controls division of the cells within the prostate gland by facilitating the interactins of testosterone and dihydrotestosterone (Coffey, 1979). The CAG polyglutamine repeat in the AR is present on exon 1 and codes for the area of the AR involved in DNA transcription (Chamberlain *et al.*, 1994). Androgen insensitivity becomes apparent when the CAG repeat is expanded (Arbizu *et al.*, 1983), which might be a result of transcriptional activity within the AR being decreased. This could also be the cause of prostate cancer development (Coffey, 1979). A study carried out by Irvine *et al.* (1995) led to the hypothesis that a patient with a short CAG repeat is more susceptible to developing prostate cancer. From this, Giovannucci *et al.* (1997) found that the repeat length variability of the CAG repeat has been linked to the development of aggressive forms of prostate cancer, which they think could be due to the activation of transcriptional activities within the AR.

A study carried out by Wellington *et al.* (1998) found that proteins which contained polyglutamine – atrophin-1, ataxin-3 and the AR – could be cleaved by caspases, leading to the truncation of these proteins and resulting in cell death. But it was Ikeda *et al.* (1996) who performed investigations on mice and showed that ataxin-3 can induce cell death. Other investigations have found that ARs with a number of CAG-repeat alterations can eventually lead to the progression of prostate cancer (Sørensen *et al.*, 1999).

Ebp1, a member of the PA2G4 family, has been found in a large number of different studies of prostate cancer cells. Its expression was monitored in C81 androgen-independent cells: at high expression, the ability of histidine-rich glycoprotein (HRG) cells to undergo expression was found to be inhibited, meaning that the C81 wells could not carry out their growth mechanisms; when the Ebp1 expression was decreased, the HRG cells were able to induce growth mechanisms within prostate cancer cells. Therefore, it has become apparent that if signals produced by HRG cells are in the presence of hormone-dependent prostate cancer cells, this may increase cellular growth (Wen *et al.*, 2000).

Zhang *et al.* (2008) investigated the AKT pathway and found that it had a mediator-like role in of HRG-induced cell survival. Phosphatidylinositol 3-kinase activates AKT and Ebp1 to undergo binding with the ErbB3 binding domain within the cytoplasm. Zhang *et al.* (2008) found that Ebp1 was able to suppress AKT signalling, inhibiting downstream Ebp1-induced growth. They concluded that even though the results showed Ebp1 might have some activity within the mitogen-activated signalling processes, the importance of the AKT pathway is still not fully understood: they could not be sure whether the effects of Ebp1 signalling within the cell were due to AR signalling, modification of the AR or modulation of Erb1 within the cell unrelated to AR signalling. Finally, Zhang *et al.* (2008) supported the role of Ebp1 as a regulator of the AR signalling pathway within the prostate.

Ebp1 can act both as a signalling hormone in prostate cancer (Scher and Sway, 2005) and as an AR co-repressor, which supports the theory that ErbB is a mediator in the activation of ARs. Zhang *et al.* (2008) found that Ebp1 restoration was able to increase the ability of C81 cells in prostate cancer and that the elimination of Ebp1 led the hormone-dependent prostate cancer cells to revert to their normal state.

16.5 Conclusions

Six genes – DLGAP2, TGFβR2, POU2F1, atatxin-3, PA2G4P2 and FLJ38379 – were highlighted in the analysis of our data set, which was run through various ANNs. These genes were all researched, and five of them showed some sort of direct link to a role within prostate cancer. One – atatxin-3 – was seen in a small number of prostate cancer cases. Its role is primarily in the development of neurodegenerative diseases, which suggests that metastasis could have resulted or that the CNS has a role to play in the progression of prostate cancer. Also, a single gene – FLJ38379 – was detected which is not elucidated in the literature and is said to be unclassified. This gene requires further investigation in order to determine what it is and what its roles within prostate cancer might be.

This investigation highlighted four genes – DLGAP2, TGFβR2, POU2F1, PA2G4P2 – that have been previously identified within prostate cancer cases and can be seen as good biomarkers for the diagnosis of prostate cancer, or of other cancers. However, a large amount of research must still be carried out to determine the actual mechanisms by which they are involved in the development of prostate cancer. FLJ38379 definitely needs to be further investigated, so that its identity can be found.

Overall, this investigation has achieved its aims in the sense that pluripotent stem cells have been identified within the data set and have been shown to have a strong role in the progression of prostate cancer. However, it has not achieved its aims in the sense that a specific pluripotent stem cell has not been identified. There is a large amount of literature available that highlights the role of stem cells within prostate cancer. Although all the theories point towards stem cells being the initiation factor in many cancers, including prostate cancer, there is still no hard evidence to support them. The final aim of determining whether or not Nanog could be used as a pluripotent stem cell marker for prostate cancer was unmet, as, even though it was present within the original data set, it was not one of the genes that were highlighted. Although a great amount of literature has been found to show that Nanog has a large role to play in the progression of prostate cancer, Nanog generally helps activate other factors, such as Oct4 or Sox2, which aid in the progression of cancer: Nanog's presence alone does not suggest the presence of prostate cancer.

References

Adams, I.R., McLaren, A. 2004. Identification and characterisation of mRif1: a mouse telomere-associated protein highly expressed in germ cells and embryo-derived pluripotent stem cells. Dev. Dyn. 229, 733–744.

Ambady, S., Malcuit, C., Kashpur, O., Kole, D., Holmes, W.F., Hedblom, E., *et al.*, 2010. Expression of NANOG and NANOGP8 in a variety of undifferentiated and differentiated human cells. Int. J. Dev. Biol. 54(12), 1743–1754.

Arbizu, T., Santamaría, J., Gomez, J.M., Quílez, A., Serra, J.P. 1983. A family with adult spinal and bulbar muscular atrophy, X-linked inheritance and associated testicular failure. J. Neurol. Sci. 59(3), 371–382.

Azhar, M., Schultz Jel, J., Grupp, I., Dorn, G.W. 2nd,, Meneton, P., Molin, D.G., *et al.*, 2003. Transforming growth factor beta in cardiovascular development and function. Cytokine Growth Factor Rev. 14, 391–407.

Berke, S.J., Schmied, F.A., Brunt, E.R., Ellerby, L.M., Paulson, H.L. 2004. Caspase-mediated proteolysis of the polyglutamine disease protein ataxin-3. J. Neurochem. 89, 908–918.

Boiani, M., Schöler, H.R. 2005. Regulatory networks in embryo-derived pluripotent stem cells. Nat. Rev. Mol. Cell Biol. 6, 872–884.

Boyer, L.A., Lee, T.I., Cole, M.F., Johnstone, S.E., Levine, S.S., Zucker, J.P., *et al.*, 2005. Core transcriptional regulatory circuitry in human embryonic stem cells. Cell 122(6), 947–956.

Burnett, B., Li, F., Pittman, R.N. 2003. The polyglutamine neurodegenerative protein ataxin-3 binds polyubiquitylated proteins and has ubiquitin protease activity, Hum. Mol. Genet. 12, 3195–3205.

Cartwright, P., Mclean, C., Sheppard, A., Rivett, D., Jones, K., Dalton, S. 2005. LIF/STAT3 controls ES cell self-renewal and pluripotency by a Myc-dependent mechanism. Development 132(5), 885–896.

Cavaleri, F., Schöler, H.R. 2003. Nanog: a new recruit to the embryonic stem cell orchestra. Cell 113, 551–552.

Chai, Y., Wu, L., Griffin, J.D., Paulson, H.L. 2001. The role of protein composition in specifying nuclear inclusion formation in polyglutamine disease. J. Biol. Chem. 276(48), 44 889–44 897.

Chai, Y., Shao, J., Miller, V.M., Williams, A., Paulson, H.L. 2002. Live-cell imaging reveals divergent intracellular dynamics of polyglutamine disease proteins and supports a sequestration model of pathogenesis. Proc. Nat. Acad. Sci. USA 99(14), 9310–9310.

Chai, Y., Berke, S.S., Cohen, R.E., Paulson, H.L. 2004. Poly-ubiquitin binding by the polyglutamine disease protein ataxin-3 links its normal function to protein surveillance pathways. J. Biol. Chem. 279(5), 3605–3611.

Chamberlain, N.L., Driver, E.D., Miesfeld, R.L. 1994. The length and location of CAG trinucleotide repeats in the androgen receptor N-terminal domain affect transactivation function. Nucleic Acids Res. 22(15), 3181–3186.

Chambers, I., Colby, D., Robertson, M., Nichols, J., Lee, S., Tweedie, S., Smith, A. 2003. Functional expression cloning of Nanog, a pluripotency sustaining factor in embryonic stem cells. Cell 113(5), 643–655.

Chambers, I., Silva, J., Colby, D., Nichols, J., Nijmeijer, B., Robertson, M., *et al.*, 2007. Nanog safeguards pluripotency and mediates germline development. Nature 450, 1230–1234.

Chen, L., Daley, G.Q. 2008. Molecular basis of pluripotency. Hum. Mol. Genet. 17(1), R23–R27.

Chen, Y.C., Hsu, H.S., Chen, Y.W., Tsai, T.H., How, C.K., Wang, C.Y., *et al.*, 2008. Oct-4 expression maintained cancer stem-like properties in lung cancer-derived CD133-positive cells. PLoS One 3(7), e2637.

Cher, M.L., Macgrogan, D., Bookstein, R., Brown, J.A., Jenkins, R.B., Jensen, R.H. 1994. Comparative genomic hybridization, allelic imbalance, and fluorescence *in situ* hybridization on chromosome 8 in prostate cancer. Genes Chromosom. Cancer 11, 153–162.

Chi, T.H., Wan, M., Zhao, K., Taniuchi, I., Chen, L., Littman, D.R., Crabtree, G.R. 2002. Reciprocal regulation of CD4/CD8 expression by SWI/SNF-like BAF complexes. Nature 418(6894), 195–199.

Chiou, S.H., Yu, C.C., Huang, C.Y., Lin, S.C., Liu, C.J., Tsai, T.H., *et al.*, 2008. Positive correlations of Oct-4 and Nanog in oral cancer stem-like cells and high-grade oral squamous cell carcinoma. Clin Cancer Res. 14(13), 4085–4095.

Çinar, M., Engin, M., Engin, E.Z., Ates, Y.Z. 2009. Early prostate cancer diagnosis by using artificial neural networks. Exp. Sys. Appl. 36(3), 6357–6361.

Coffey, D.S. 1979. Prostate Cancer, UICC Technical Report Series, Vol. 48. Geneva: International Union Against Cancer.

Cole, M.F., Johnstone, S.E., Newman, J.J., Kagey, M.H., Young, R.A. 2008. Tcf3 is an integral component of the core regulatory circuitry of embryonic stem cells. Genes Dev. 22, 746–755.

Cozzio, A., Passegué, E., Ayton, P.M., Karsunky, H., Cleary, M.L., Weissman, I.L. 2003. Similar MLL-associated leukemias arising from self-renewing stem cells and short-lived myeloid progenitors. Genes Dev. 17(24), 3029–3035.

De Wever, O., Nguyen, Q.D., Van Hoorde, L., Bracke, M., Bruyneel, E., Gespach, C., Mareel, M. 2004. Tenascin-C and SF/HGF produced by myofibroblasts in vitro provide convergent pro-invasive signals to human colon cancer cells through RhoA and Rac. FASEB J. 18, 1016–1018.

Dean, M., Fojo, T., Bates, S. 2005. Tumour stem cells and drug resistance. Nat. Rev. Cancer 5, 275–284.

Diss, J.K.J., Fraser, S.P., Walker, M.M., Patel, A., Latchman, D.S., Djamgo, M.B.A. 2008. β-subunits of voltage-gated sodium channels in human prostate cancer: quantitative in vitro and in vivo analyses of mRNA expression. Prostate Cancer Prostatic Dis. 11, 325–333.

Dodge, J.E., Kang, Y.K., Beppu, H., Lei, H., Li, E. 2004. Histone H3–K9 methyltransferase ESET is essential for early development. Mol. Cell Biol. 24, 2478–2486.

Dontu, G., Al-Hajj, M., Abdallah, W.M., Clarke, M.F., Wicha, M.S. 2003. Stem cells in normal breast development and breast cancer. Cell Prolif. 23, 59–72.

Duval, A., Hamelin, R. 2002. Mutations at coding repeat sequences in mismatch repair-deficient human cancers: toward a new concept of target genes for instability. Cancer Res. 62, 2447–2454.

Dyer, R.B., McMurray, C.T. 2001. Mutant protein in Huntington disease is resistant to proteolysis in affected brain. Nat Genet. 29(3), 270–278.

Fei, E., Jia, N., Zhang, T., Ma, X., Wang, H., Liu, C., *et al.*, 2007. Phosphorylation of ataxin-3 by glycogen synthase kinase 3beta at serine 256 regulates the aggregation of ataxin-3. Biochem. Biophys. Res. Commun. 357, 487–492.

Gannon, P.O., Koumakpayi, I.H., Le Page, C., Karakiewicz, P.I., Mes-Masson, A.-M., Saad, F. 2008. Ebp1 expression in benign and malignant prostate. Cancer Cell Int. 8, 18.

Gelsi-Boyer, V., Orsetti, B., Cervera, N., Finetti, P., Sircoulomb, F., Rougé, C., *et al.*, 2005. Comprehensive profiling of 8p11-12 amplification in breast cancer. Mol. Cancer Res. 3(12), 655–667.

Giovannucci, E., Stampfer, M.J., Krithivas, K., Brown, M., Brufsky, A., Talcott, J., *et al.*, 1997. The CAG repeat within the androgen receptor gene and its relationship to prostate cancer. PNAS 94(7), 3320–3323.

Glinsky, G.V. 2008. 'Stemness' genomics law governs clinical behavior of human cancer: implications for decision making in disease management. J. Clin. Oncol. 26, 2846–2853.

Gordon, S. 2003. Alternative activation of macrophages. Nat. Rev. Immunol. 3(1), 23–35.

Grady, W.M., Rajput, A., Myeroff, L., Liu, D.F., Kwon, K., Willis, J. 1998. Mutation of the type II transforming growth factor-receptor is coincident with the transformation of human colon adenomas to malignant carcinomas. Cancer Res. 58, 3101–3104.

Gu, G., Yuan, J., Wills, M., Kasper, S. 2007. Prostate cancer cells with stem cell characteristics reconstitute the original human tumor *in vivo*. Cancer Res. 67, 10.

Gusterson, R.J., Jazrawi, E., Adcock, I.M., Latchman, D.S. 2003. Molecular basis of cell and developmental biology: the transcriptional co-activators Creb-binding protein (CBP) and p300 play a critical role in cardiac hypertrophy that is dependent on their histone acetyltransferase activity. J. Biol. Chem. 278, 6838–6847.

Guo, Y., Kyprianou, N. 1999. Restoration of transforming growth factor β signaling pathway in human prostate cancer cells suppresses tumorigenicity via induction of caspase-1-mediated apoptosis. Cancer Res. 59, 1366–1371.

Han, J., Zhang, F., Yub, M., Zhao, P., Jia, W., Zhang, H., *et al.*, 2012. RNA interference-mediated silencing of NANOG reduces cell proliferation and induces G0/G1 cell cycle arrest in breast cancer cells. Cancer Lett. 321(1), 80–88.

Hanna, L.A., Foreman, R.K., Tarasenko, I.A., Kessler, D.S., Labosky, P.A. 2002. Requirement for Foxd3 in maintaining pluripotent cells of the early mouse embryo. Genes Dev. 16, 2650–2661.

Hart, A.H., Hartley, L., Ibrahim, M., Robb, L. 2004. Identification, cloning and expression analysis of the pluripotency promoting Nanog genes in mouse and human. Dev. Dyn. 230, 187–198.

Hata, A., Davis, B.N. 2009. Control of microRNA biogenesis by TGFbeta signaling pathway – a novel role of Smads in the nucleus. Cytokine Growth Factor Rev. 20, 517–551.

Hershko, A., Ciechanover, A. 1998. The ubiquitin system. Ann. Rev. Biochem. 67, 425–479.

Hornstein, M., Hoffmann, M.J., Alexa, A., Yamanaka, M. 2008. Protein phosphatase and TRAIL receptor genes as new candidate tumor genes on chromosome 8p in prostate cancer. Cancer Gen. Proteo. 5, 123–136.

Hu, P.P., Datto, M.B., Wang, X.F. 1998. Molecular mechanisms of transforming growth factor-beta signaling. Endocr. Rev. 19, 349–363.

Hughes, C., Murphy, A., Martin, C., Sheils, O., O'Leary, J. 2005. Molecular pathology of prostate cancer. J. Clin. Pathol. 58, 673–684.

Iacopetta, B.J., Welch, J., Soong, R., House, A.K., Zhou, X.P., Hamelin, R. 1998. Mutation of the transforming growth factor-type II receptor gene in right-sided colorectal cancer: relationship to clinicopathological features and genetic alterations. J. Pathol. 184, 390–395.

Ikeda, H., Yamaguchi, M., Sugai, S., Aze, Y., Narumiya, S., Kakizuka, A. 1996. Expanded polyglutamine in the Machado-Joseph disease protein induces cell death *in vitro* and *in vivo*. Nat. Genet. 13, 196–202.

Irvine, R.A., Yu, M.C., Ross, R.K., Coetzee, G.A. 1995. The CAG and GGC microsatellites of the androgen receptor gene are in linkage disequilibrium in men with prostate cancer. Cancer Res. 55(9), 1937–1940.

Jeffers, J.R., Parganas, E., Lee, Y., Yang, C., Wang, J., Brennan, J., *et al.*, 2003. Puma is an essential mediator of p53-dependent and -independent apoptotic pathways. Cancer Cell 4(4), 321–328.

Jeter, C.R., Badeaux, M., Choy, G., Chandra, D., Patrawala, L., Liu, C., *et al.*, 2009. Functional evidence that the self-renewal gene NANOG regulates human tumor development. Stem Cells 27(5), 993–1005.

Kang, J., Gemberling, M., Nakamura, M., Whitby, F.G., Handa, H., Fairbrother, W.G., Tantin, D. 2009. A general mechanism for transcription regulation by POU2F1 and POU5F1 in response to genotoxic and oxidative stress. Genes Dev. 23, 208–222.

Karanam, S., Moreno, C.S. 2004. CONFAC: automated application of comparative genomic promoter analysis to DNA microarray datasets. Nucleic Acids Res. 32, 475–484.

Karolchik, D., Baertsch, R., Diekhans, M., Furey, T.S., Hinrichs, A., Lu, Y.T., *et al.*, 2003. The UCSC Genome Browser Database. Nucleic Acids Res. 31, 51–54.

Kawaguchi, Y., Okamoto, T., Taniwaki, M., Aizawa, M., Miho Inoue, M., Katayama, S., *et al.*, 1994. CAG expansions in a novel gene for Machado-Joseph disease at chromosome 14q32.1. Nat. Gen. 8, 221–228.

Kim, J.B., Sebastiano, V., Wu, G., Araúzo-Bravo, M.J., Sasse, P., Gentile, L., *et al.*, 2009. Oct4-induced pluripotency in adult neural stem cell. Cell 135(3), 411–419.

Koeneman, K.S. 2006. Prostate cancer stem cells, telomerase biology, epigenetic modifiers, and molecular systemic therapy for the androgen-independent lethal phenotype. Urol. Oncol. – Semi. Ori. 24(2), 119–121.

Komander, D., Clague, M.J., Urbé, S. 2009. Breaking the chains: structure and function of the deubiquitinases. Nat. Rev. Mol. Cell Biol. 10(8), 550–563.

Koumakpayi, I.H., Diallo, J.S., Le Page, C., Lessard, L., Filali-Mouhim, A., Bégin, L.R., *et al.*, 2007. Low nuclear erbb3 predicts biochemical recurrence in patients with prostate cancer. BJU Int. 100(2), 303–309.

Lancashire, L.J., Mian, S., Ellis, I.O., Rees, R.C., Ball, G.R. 2005. Current developments in the analysis of proteomic data: artificial neural network data mining techniques for the identification of proteomic biomarkers related to breast cancer. Curr. Proteom. 2(15), 15–29.

Lancashire, L.J., Rees, R.C., Ball, G.R. 2008. Identification of gene transcript signatures predictive for estrogen receptor and lymph node status using a stepwise forward selection artificial neural network modelling approach. Art. Int. Med. 43, 99–111.

Lancashire, L.J., Powe, D.G., Reis-Filho, J.S., Rakha, E., Lemetre, C., Weigelt, B., *et al.*, 2009. A validated gene expression profile for detecting clinical outcome in breast cancer using artificial neural networks. Breast Cancer Res. Treat. 120(1), 83–93.

Latchman, D.S. 1999. POU family transcription factors in the nervous system. J. Cell Physiol. 179, 126–133.

Latil, A., Cussenot, O., Fournier, G., Driouch, K., Lidereau R. 1997. Loss of heterozygosity at chromosome 16q in prostate adenocarcinoma: identification of three independent regions. Cancer Res. 57, 1058–1062.

Lavial, F., Acloque, H., Bertocchini, F., Macleod, D.J., Boast, S., Bachelard, E., *et al.*, 2007. The Oct4 homologue PouV and Nanog regulate pluripotency in chicken embryonic stem cells. Development 134(19), 3549–3563.

Leblond-Francillard, M., Picon, A., Bertagna, X., De Keyzer, Y. 1997. High expression of the POU factor Brn3a in aggressive neuroendocrine tumors. J. Clin. Endocrinol. Metab. 82, 89–94.

Lemetre, C., Lancashire, L.J., Rees, R.C., Ball, G.R. 2009. Artificial neural network based algorithm for biomarker interactions modeling. In: Cabestany, J., Sandoval, F., Preito, A., Carcahdo, J.M. (eds) Bio-Inspired Systems (Lecture Notes in Computer Science, Theoretical Computer Science & General Issues, Vol. 5517), pp. 877–885.

Lessor, T.J., Yoo, J.Y., Xia, X., Woodford, N., Hamburger, A.W. 2000. Ectopic expression of the erbb-3 binding protein ebp1 inhibits growth and induces differentiation of human breast cancer cell lines. J. Cell Physiol. 183(3), 321–329.

Li, F., Macfarlan, T., Pittman, R.N., Chakravarti, D. 2002. Genes: structure and regulation: Ataxin-3 is a histone-binding protein with two independent transcriptional corepressor activities. J. Biol. Chem. 277, 45 004–45 012.

Liang, J., Wan, M., Zhang, Y., Gu, P., Xin, H., Jung, S.Y., *et al.*, 2008. Nanog and Oct4 associate with unique transcriptional repression complexes in embryonic stem cells. Nat. Cell Biol. 10(6), 731–739.

Link, K.A., Balasubramaniam, S., Sharma, A., Comstock, C.E., Godoy-Tundidor, S., Powers, N., *et al.*, 2008. Targeting the BAF57 SWI/SNF subunit in prostate cancer: a novel platform to control androgen receptor activity. Cancer Res. 68(12), 4551–4558.

Liu, B.Y., McDermott, S.P., Khwaja, S.S., Alexander, C.M. 2004. The transforming activity of Wnt effectors correlates with their ability to induce the accumulation of mammary progenitor cells. Proc. Nat. Acad. Sci. USA 101, 4158–4163.

Liu, Z., Ahn, J.Y., Liu, X., Ye, K. 2006. Ebp1 isoforms distinctively regulate cell survival and differentiation. Proc Nat. Acad. Sci. USA 103(29), 10 917–10 922.

Liu, W., Laitinen, S., Khan, S., Vihinen, M., Kowalski, J., Yu, G., *et al.*, 2009. Copy number analysis indicates monoclonal origin of lethal metastatic prostate cancer. Nat. Med. 15, 559–565.

Loh, Y.H., Wu, Q., Chew, J.L., Vega, V.B., Zhang, W., Chen, X., *et al.*, 2006. The Oct4 and Nanog transcription network regulates pluripotency in mouse embryonic stem cells. Nat. Genet. 38(4), 431–440.

Lopes, C.T., Franz, M., Kazi, F., Donaldson, S.L., Morris, Q., Bader, G.D. 2010. Cytoscape Web: an interactive web-based network browser. Bioinformatics 26(18), 2347–2348.

Lothe, R.A., Peltomäki, P., Meling, G.I., Aaltonen, L.A., Nyström-Lahti, M. 1993. Genomic instability in colorectal cancer: relationship to clinicopathological variables and family history. Cancer Res. 53, 5849–5852.

Lu, S.L., Kawabata, M., Imamura, T., Akiyama, Y., Nomizu, T., Miyazono, K., Yuasa, Y. 1998. HNPCC associated with germline mutation in the TGF-beta type II receptor gene. Nat. Genet. 19(1), 17–18.

Luedi, P.P., Dietrich, F.S., Weidman, J.R., Bosko, J.M., Jirtle, R.L., Hartemink, A.J. 2007. Computational and experimental identification of novel human imprinted genes. Genome Res. 17(12), 1723–1730.

Ma, Y., Liang, D., Liu, J., Axcrona, K., Kvalheim, G., Stokke, T., *et al.*, 2011. Prostate cancer cell lines under hypoxia exhibit greater stem-like properties. PLoS One 6(12), e29170.

Macheda, M.L., Rogers, S., Best, J.D. 2005. Molecular and cellular regulation of glucose transporter (GLUT) proteins in cancer. J. Cell. Physiol. 202, 654–662.

Mao, Y., Senic-Matuglia, F., Di Fiore, P.P., Hodsdon, M.E., De Camilli, P. 2005. Deubiquitinating function of ataxin-3: Insights from the solution structure of the Josephin domain. PNAS 102(36), 12 700–12 705.

Markert, E.K., Mizuno, H., Vazquez, A., Levine, A.J. 2011. Molecular classification of prostate cancer using curated expression signatures. Proc. Nat. Acad. Sci. USA 108(52), 21 276–21 281.

Massagué, J. 1998. TGF signal transduction. Ann. Rev. Biochem. 67, 753–791.

Masui, S., Nakatake, Y., Toyooka, Y., Shimosato, D., Yagi, R., Takahashi, K., *et al.*, 2007. Pluripotency governed by Sox2 via regulation of Oct3/4 expression in mouse embryonic stem cells. Nat. Cell Biol. 9(6), 625–635.

Matsumura, N., Huang, Z., Mori, S., Baba, T., Fujii, S., Konishi, I., *et al.*, 2011. Epigenetic suppression of the TGF-beta pathwayrevealed by transcriptome profiling in ovarian cancer. Genome Res. 21, 74–82.

McEvilly, R.J., Rosenfeld, M.G. 1999. The role of POU domain proteins in the regulation of mammalian pituitary and nervous system development. Prog. Nucleic Acid Res. Mol. Biol. 63, 223–255.

McNicol, A., Isral, S. 2008. Beyond hemostasis: the role of platelets in inflammation, malignancy and infection. Cardiovasc. Hematol. Disord. Drug Targets 8(2), 99–117.

Meng, H.M., Zheng, P., Wang, X.Y., Liu, C., Sui, H.M., Wu, S.J., *et al.*, 2010. Overexpression of Nanog predicts tumor progression and poor prognosis in colorectal cancer. Cancer Biol. Ther. 9(4), 295–302.

Miki, J., Furusato, B., Li, H., Gu, Y., Takahashi, H., Egawa, S., *et al.*, 2007. Identification of putative stem cell markers, CD133 and CXCR4, in hTERT-immortalized primary nonmalignant and malignant tumor-derived human prostate epithelial cell lines and in prostate cancer specimens. Cancer Res. 1(67), 3153–3161.

Milenkovic, T., Memiševic, V., Ganesan, A.K. 2010. Systems-level cancer gene identification from protein interaction network topology applied to melanogenesis-related functional genomics data. J. R. Soc. Interface 7, 423–437.

Mitsui, K., Tokuzawa, Y., Itoh, H., Segawa, K., Murakami, M., Takahashi, K., *et al.*, 2003. The homeoprotein Nanog is required for maintenance of pluripotency in mouse epiblast and ES cells. Cell 113(5), 631–642.

Mitsunaga, K., Araki, K., Mizusaki, H., Morohashi, K., Haruna, K., Nakagata, N., *et al.*, 2004. Loss of PGC-specific expression of the orphan nuclear receptor ERR-beta results in reduction of germ cell number in mouse embryos. Mech. Dev. Mar. 121(3), 237–246.

Mori, R., Dorff, T.B., Xiong, S., Tarabolous, C.J., Ye, W., Groshen, S., *et al.*, 2010. The relationship between proangiogenic gene expression levels in prostate cancer and their prognostic value for clinical outcomes. Prostate 70(15), 1692–1700.

Mu, Y., Gudey, S.K., Landström, M. 2012. Non-Smad signaling pathways. Cell Tiss. Res. 347(1), 11–20.

Muraoka-Cook, R.S., Dumont, N., Arteaga, C.L. 2005. Dual role of transforming growth factor beta in mammary tumorigenesis and metastatic progression. Clin. Cancer Res. 11, 937–943.

Muscheck, M., Sukosd, F., Pesti, T., Kovacs, G. 2003. High density deletion mapping of bladder cancer localizes the putative tumor suppressor gene between loci D8S504 and D8S264 at chromosome 8p23.3. Lab Invest. 80, 1089–1093.

Nagaraj, N.S., Datta, P.K. 2010. Targeting the transforming growth factor-β signaling pathway in human cancer. Expert Opin. Investig. Drugs. 19(1), 77–91.

Ndisdang, D., Morris, P.J., Chapman, C., Ho, L., Singer, A., Latchman, D.S. 1998. The HPV-activating cellular transcription factor Brn-3a is overexpressed in CIN3 cervical lesions. J. Clin. Invest. 101, 1887–1892.

Neville, P.J., Conti, D.V., Paris, P.L., Levin, H., Catalona, W.J., Suarez, B.K., *et al.*, 2002. Prostate cancer aggressiveness locus on chromosome 7q32-q33 identified by linkage and allelic imbalance studies. Neoplasia 4(5), 424–431.

Niwa, H., Miyazaki, J., Smith, A.G. 2000. Quantitative expression of Oct-3/4 defines differentiation, dedifferentiation or self-renewal of ES cells. Nat. Genet. 24, 372–376.

Okita, K., Ichisaka, T., Yamanaka, S. 2007. Generation of germline-competent induced pluripotent stem cells. Nature 448, 313–317.

Ozaki, K., Leonard, W.J. 2002. Cytokine and cytokine receptor pleiotropy and redundancy. J. Biol. Chem. 277(33), 29 355–29 358.

Parsons, R., Myeroff, L.L., Liu, B., Willson, J.K.W., Markowitz, S.D. 1995. Microsatellite instability and mutations of the transforming growth factor type II receptor gene in colorectal cancer. Cancer Res. 55, 5548–5550.

Pasche, B., Kolachana, P., Nafa, K., Satagopan, J., Chen, Y.G., Lo, R.S., *et al.*, 1999. TbetaR-I(6A) is a candidate tumor susceptibility allele. Cancer Res. 59(22), 5678–5682.

Paulson, H.L., Perez, M.K., Trottier, Y., Trojanowski, J.Q., Subramony, S.H., Das, S.S., *et al.*, 1997a. Intranuclear inclusions of expanded polyglutamine protein in spinocerebellar ataxia type 3. Neuron. 19(2), 333–344.

Paulson, H.L., Perez, M.K., Trottier, Y., Trojanowski, J.Q., Subramony, S.H., Das, S.S., *et al.*, 1997b. Intranuclear inclusions of expanded polyglutamine protein in spinocerebellar ataxia type 3. Neuron. 19(2), 333–344.

Pavlopoulos, G.A., Wegener, A.-L., Schneider, R. 2008. A survey of visualization tools for biological network analysis. BioData Min. 1, 12.

Pechlivanis, M., Kuhlmann, J. 2006. Hydrophobic modifications of Ras proteins by isoprenoid groups and fatty acids – more than just membrane anchoring. Biochim. Biophys. Acta 1764, 1914–1931.

Pecorino, L. 2012. Molecular Biology of Cancer: Mechanisms, Targets, and Theraputics, 3rd edn. Oxford: Oxford University Press.

Peri, S., Navarro, J.D., Amanchy, R., Kristiansen, T.Z., Jonnalagadda, C.K., Surendranath, V., *et al.*, 2003. Development of human protein reference database as an initial platform for approaching systems biology in humans. Genome Res. 13(10), 2363–2371.

Pesce, M., Wang, X., Wolgemuth, D.J., Schöler, H. 1998a. Differential expression of the Oct-4 transcription factor during mouse germ cell differentiation. Mech. Dev. 71(1), 89–98.

Pesce, M., Gross, M.K., Schöler, H.R. 1998b. In line with our ancestors: Oct-4 and the mammalian germ. Bioessays 20, 722–732.

Riccioni, R., Calzolari, A., Biffoni, M., Senese, M., Riti, V., Petrucci, E., *et al.*, 2006. Podocalyxin is expressed in normal and leukemic monocytes. Blood Cells Mol. Dis. 37, 218–225.

Rahimi-Ajdadi, F., Abbaspour-Gilandeh, Y. 2011. Artificial Neural Network and stepwise multiple range regression methods for prediction of tractor fuel consumption. Measurement 44(10), 2104–2111.

Rampino, N., Yamamoto, H., Ionov, Y., Li, Y., Sawai, H., Reed, J.C., Perucho, M. 1997. Somatic frameshift mutations in the BAX gene in colon cancers of the microsatellite mutator phenotype. Science 275, 967–969.

Ranta, S., Zhang, Y., Ross, B., Takkunen, E., Hirvasniemi, A., Chapelle, A., *et al.*, 2000. Positional cloning and characterisation of the human DLGAP2 gene and its exclusion in progressive epilepsy with mental retardation. Eur. J. Hum. Genet. 8, 381–384.

Reya, T., Morrison, S.J., Clarke, M.F., Weissman, I.L. 2001. Stem cells, cancer, and cancer stem cells. Nature 414, 105–111.

Reya, T., Duncan, A.W., Ailles, L., Domen, J., Scherer, D.C., Willert, K., *et al.*, 2003. A role for Wnt signalling in self-renewal of haematopoietic stem cells. Nature 423(6938), 409–414.

Rhodes, D.R., Chinnaiyan, A.M. 2005. Integrative analysis of the cancer transcriptome. Nat. Gen. 37, 531–537.

Richardson, G.D., Robson, C.N., Lang, S.H., Neal, D.E., Maitland, N.J., Collins, A.T. 2004. CD133, a novel marker for human prostatic epithelial stem cells. J. Cell Sci. 117, 3539–3545.

Ridley, A.J., Schwartz, M.A., Burridge, K., Firtel, R.A., Ginsberg, M.H., Borisy, G. 2003. Cell migration: integrating signals from front to back. Science 302, 1704–1709.

Riegman, P.H.J., Vietstra, R.J., van der Korpt, J.A.G.M., Brinkman, A.O., Trapman, J. 1991. The promoter of the prostate specific antigen gene contains a functional androgen responsive element. Mol. Endocrinol. 5, 1921–1930.

Robert, A.B., Wakefield, L. 2003. The two faces of transforming growth factor beta in carcinogenesis. Proc. Nat. Acad. Sci. USA 100, 8621–8623.

Roberts, A.B., Sporn, M.B. 1990. The transforming growth factor-betas. In: Sporn, M.B., Roberts, A.B. (eds) Peptide Growth Factors and Their Receptors. New York: Springer-Verlag, 419–472.

Rodda, D.J., Chew, J.L., Lim, L.H., Loh, Y.H., Wang, B., Ng, H.H., Robson, P. 2005. Transcriptional regulation of NANOG by OCT4 and SOX2. J. Biol. Chem. 280, 24 731–24 737.

Røe, K., Kakar, M., Seierstad, T., Ree, A.H., Olsen, D.R. 2011. Early prediction of response to radiotherapy and androgen-deprivation therapy in prostate cancer by repeated functional MRI: a preclinical study. Rad. Oncol. 6, 65.

Ronco, A.L., Fernandez, R. 1999. Improving ultrasonographic diagnosis of prostate cancer with neural networks. Ultrasound Med. Biol. 25(5), 729–733.

Ross, C.A., Wood, J.D., Schilling, G., Peters, M.F., Nucifora, F.C. Jr,, Cooper, J.K., *et al.*, 1999. Polyglutamine pathogenesis. Philos. Trans. R. Soc. Lond. B. Biol. Sci. 354(1386), 1005–1011.

Rumelhart, D.E., Hinton, G.E., Williams, R.J. 1986. Learning representations by back-propagating errors. Nature 323, 533–536.

Saitoh, M., Takayanagi, R., Goto, K., Fukamizu, A., Tomura, A., Yanase, T., Nawata, H. 2002. The presence of both the amino- and carboxyl-terminal domains in the AR is essential for the completion of a transcriptionally active form with coactivators and intranuclear compartmentalization common to the steroid hormone receptors: a three-dimensional imaging study. Mol. Endocrinol. 16(4), 694–706.

Salcido, C.D., Larochelle, A., Taylor, B.J., Dunbar, C.E., Varticovski, L. 2010. Molecular characterisation of side population cells with cancer stem cell-like characteristics in small-cell lung cancer. Br. J. Cancer 102(11), 1636–1644.

Scher, H.I., Sway, S.C. 2005. Biology of progressive, castration-resistant prostate cancer: directed therapies targeting the androgen-receptor signaling axis. J. Clin. Oncol. 23, 8253–8261.

Schilham, M.W., Oosterwegel, M.A., Moerer, P., Ya, J., De Boer, P.A., van de Wetering, M., *et al.*, 1996. Defects in cardiac outflow tract formation and pro-B-lymphocyte expansion in mice lacking Sox-4. Nature 380(6576), 711–714.

Schmid, H.P., Mcneal, J.E., Stamey, T.A. 1993. Observations on the doubling time of prostate cancer. The use of serial prostatespecific antigen in patients with untreated disease as a measure of increasing cancer volume. Cancer 71, 2031–2040.

Schopperle, W.M., DeWolf, W.C. 2007. The TRA-1-60 and TRA-1-81 human pluripotent stem cell markers are expressed on podocalyxin in embryonal carcinoma. Stem Cell 25, 723–730.

Schulz, T.J., Zarse, K., Voigt, A., Urban, N., Birringer, M., Ristow, M. 2007. Glucose restriction extends *Caenorhabditis elegans* life span by inducing mitochondrial respiration and increasing oxidative stress. Cell Metab. 6(4), 280–293.

Schutte, M., Hruban, R.H., Hedrick, L., Cho, K.R., Nadasdy, G.M., Weinstein, C.L., *et al.*, 1996. DPC4 gene in various tumor types. Cancer Res. 56(11), 2527–2530.

Sell, S. 2004. Stem cell origin of cancer and differentiation therapy. Crit. Rev. Oncol. Hematol. 51(1), 1–28.

Selwood, L., Johnson, M.H. 2006. Trophoblast and hypoblast in the monotreme, marsupial and eutherian mammal: evolution and origins. Bioessays 28, 128–145.

Shannon, P., Markiel, A., Ozier, O., Baliga, N.S., Wang, J.T., Ramage, D., Amin, N., *et al.*, 2003. Cytoscape: a software environment for integrated models of biomolecular interaction networks. Genome Res. 13(11), 2498–2504.

Shariat, S.F., Kattan, M.W., Traxel, E., Andrews, B., Zhu, K., Wheeler, T.M., Slawin, K.M. 2004. Association of pre- and postoperative plasma levels of transforming growth factor β(1) and interleukin 6 and its soluble receptor with prostate cancer progression. Clin. Cancer Res. 10, 1992–1999.

Shertser, K., Chertkow-Deutsher, Y., Klein, E., Ben-Shachar, D. 2012. The role of Dlgap2 in post-synaptic density (PSD) zone organization: implication to PTSD. RMMJ 3(Suppl. 1), 51.

Shi, Y., Massagué, J. 2003. Mechanisms of TGF – signaling from cell membrane to the nucleus. Cell 113, 685–700.

Silva, J., Chambers, I., Pollard, S., Smith, A. 2006. Nanog promotes transfer of pluripotency after cell fusion. Nature 441, 997–1001.

Singh, S.K., Hawkins, C., Clarke, I.D., Squire, J.A., Bayani, J., Hide, T., *et al.*, 2004. Identification of human brain tumour initiating cells. Nature 432(7015), 396–401.

Smith, M.D., Ensor, E.A., Coffin, R.S., Boxer, L.M., Latchman, D.S. 1998. Bcl-2 transcription from the proximal P2 promoter is activated in neuronal cells by the Brn-3a POU family transcription factor. J. Biol. Chem. 273, 16 715–16 722.

Sørensen, S.A., Fenger, K., Olsen, J.H. 1999. Significantly lower incidence of cancer among patients with Huntington disease an apoptotic effect of an expanded polyglutamine tract? Cancer 86(7), 1342–1346.

Srikantan, V., Sesterhenn, I.A., Davis, L., Hankins, G.R., Avallone, F.A., Livezey, J.R., *et al.* 1999. Allelic loss on chromosome 6q in primary prostate cancer. Int. J. Cancer 84, 331–335.

Steffan, J.S., Kazantsev, A., Spasic-Boskovic, O., Greenwald, M., Zhu, Y.Z., Gohler, H., *et al.*, 2000. The Huntington's disease protein interacts with p53 and CREB-binding protein and represses transcription. Proc. Nat. Acad. Sci. USA 97(12), 6763–6768.

Stewart, C.L. 2000. Oct-4, scene 1: the drama of mouse development. Nat. Genet. 24, 328–330.

Strachan, T., Read, A.P. 1999. Gene therapy and other molecular genetic-based therapeutic approaches. In: Strachan, T., Read, A.P. (eds) Human Molecular Genetics, 2nd edn. New York: Wiley-Liss.

Stuart, R.O., Wachsman, W., Berry, W.C., Wang-Rodriguez, J., Wasserman, L., Klacansky, I., *et al.*, 2004. *In silico* dissection of cell-type-associated patterns of gene expression in prostate cancer. PNAS 101(2), 615–620.

Suh, J., Rabson, A.B. 2004. NF-nB activation in human prostate cancer: important mediator or epiphenomenon? J. Cell Biochem. 91, 100–117.

Sutkowski, D.M., Fong, C.J., Sensibar, J.A., Rademaker, A.W., Sherwood, E.R., Kozlowski, J.M., Lee, C. 1992. Interaction of epidermal growth factor and transforming growth factor β in human prostatic epithelial cells in culture. Prostate, 21, 133–143.

Szmitko, P.E., Wang, C.H., Weisel, R.D., De Almeida, J.R., Anderson, T.J., Verma, S. 2003. New markers of inflammation and endothelial cell activation: part I. Circulation 108(16), 1917–1923.

Takahashi, K., Yamanaka, S. 2006. Induction of pluripotent stem cells from mouse embryonic and adult fibroblast cultures by defined factors. Cell 126, 663–676.

Tantin, D., Gemberling, M., Callister, C., Fairbrother, W. 2008. High-throughput biochemical analysis of in vivo location data reveals novel distinct classes of POU5F1(Oct4)/DNA complexes 1. Genome Res 18, 631–639.

Tao, R.S., Fei, E.K., Ying, Z., Wang, H.F., Wang, G.H. 2008. Casein kinase 2 interacts with and phosphorylates ataxin-3. Bull. Neurosci. 24, 271–277.

Theil, T., Mclean-Hunter, S., Zornig, M., Moroy, T. 1993. Mouse Brn-3 family of POU transcription factors: a new aminoterminal domain is crucial for the oncogenic activity of Brn-3a. Nucleic Acids Res. 21, 5921–5929.

Varshavsky, A. The ubiquitin system. Trends Biochem. Sci. 22(10), 383–387.

Vogelmann, R., Nguyen-Tat, M.D., Giehl, K., Adler, G., Wedlich, D., Menke, A. 2005. TGFbeta-induced downregulation of E-cadherin-based cell–cell adhesion depends on PI3-kinase and PTEN. J. Cell Sci. 118, 4901–4912.

Wang, Q., Li, L., Ye, Y. 2008. Inhibition of p97-dependent protein degradation by Eeyarestatin I. J. Biol. Chem. 283(12), 7445–7454.

Wang, Y., Xia, X.-Q., Jia, Z., Sawyers, A., Yao, H., Wang-Rodriquez, J., *et al.*, 2010. *In silico* estimates of tissue components in surgical samples based on expression profiling data. Cancer Res. 70(16), 6448–6455.

Wegner, B., Al-Momany, A., Kulak, S., Kozlowski, K., Obeidat, M., Jahroudi, N., *et al.*, 2010. CLIC5A, a component of the ezrin–podocalyxin complex in glomeruli, is a determinant of podocyte integrity, Am. J. Physiol. Renal Physiol. 298, 1492–1503.

Wellington, C.L., Ellerby, L.M., Hackam, A.S., Margolis, R.L., Trifiro, M.A., Singaraja, R., *et al.*, 1998. Caspase cleavage of gene products associated with triplet expansion disorders generates truncated fragments containing the polyglutamine tract. J. Biol. Chem. 273(15), 9158–9167.

Wen, Y., Hu, M.C., Makino, K., Spohn, B., Bartholomeusz, G., Yan, D.H., Hung, M.C. 2000. HER-2/neu promotes androgen-independent survival and growth of prostate cancer cells through the Akt pathway. Cancer Res. 60(24), 6841–6845.

Wilkinson, K.D., Lee, K.M., Deshpande, S., Duerksen-Hughes, P., Boss, J.M., Pohl, J. 1989. The neuron-specific protein PGP 9.5 is a ubiquitin carboxyl-terminal hydrolase. Science 246, 670–673.

Wood, J.D., Nucifora, F.C. Jr,, Duan, K., Zhang, C., Wang, J., Kim, Y., *et al.*, 2000. Atrophin-1, the dentato-rubral and pallido-luysian atrophy gene product, interacts with ETO/MTG8 in the nuclear matrix and represses transcription. J. Cell Biol. 150(5), 939–948.

Wyckoff, J.B., Jones, J.G., Condeelis, J.S., Segall, J.E. 2000. A critical step in metastasis: *in vivo* analysis of intravasation at the primary tumor. Cancer Res. 60, 2504–2511.

Xia, X., Lessor, T.J., Zhang, Y., Woodford, N., Hamburger, A.W. 2001. Analysis of the expression pattern of Ebp1, an erbb-3-binding protein. Biochem. Biophys. Res. Commun. 289(1), 240–244.

Xu, L., Blackburn, E.H. 2004. Rif1 protein binds aberrant telomeres and aligns along anaphase midzone microtubules. J. Cell Biol. 167, 819–830.

Yoo, J.Y., Wang, X.W., Rishi, A.K., Lessor, T., Xia, X.M., Gustafson, T.A. 2000. Interaction of the PA2G4 (EBP1) protein with ErbB-3 and regulation of this binding by heregulin. Br. J. Cancer 82, 683–690.

Yu, J., Vodyanik, M.A., Smuga-Otto, K., Antosiewicz-Bourget, J., Frane, J.L., Tian, S., *et al.*, 2007. Induced pluripotent stem cell lines derived from human somatic cells. Science 318(5858), 1917–1920.

Zbinden, M., Duquet, A., Lorente-Trigos, A., Ngwabyt, S.-N., Borges, I., Altaba, A.R. 2010. NANOG regulates glioma stem cells and is essential in vivo acting in a cross-functional network with GLI1 and p53. EMBO J. 29(15), 2659–2674.

Zeng, X., Miura, T., Luo, Y., Bhattacharya, B., Condie, B., Chen, J., *et al.*, 2004. Properties of pluripotent human embryonic stem cells BGO1 and BGO2. Stem Cells 22, 292–312.

Zhang, Y., Fondell, J.D., Wang, Q., Xia, X., Cheng, A., Lu, M.L., Hamburger, A.W. 2002. Repression of androgen receptor mediated transcription by the erbb-3 binding protein, Ebp1. Oncogene 21(36), 5609–5618.

Zhang, Y.X., Woodford, N., Xia, X.M., Hamburger, A.W. 2003. Repression of E2F1-mediated transcription by the ErbB3 binding protein Ebp1 involves histone deacetylases. Nucleic Acids Res. 31, 2168–2177.

Zhang, Y., Wang, X.W., Jelovac, D., Nakanishi, T., Yu, M.H., Akinmade, D. 2005. The erbb3-binding protein Ebp1 suppresses androgen receptor-mediated gene transcription and tumorigenesis of prostate cancer cells. Proc. Nat. Acad. Sci. USA 102(28), 9890–9895.

Zhang, Y., Linn, D., Liu, Z., Melamed, J., Tavora, F., Young, C.Y., *et al.*, 2008. EBP1, an ErbB3-binding protein, is decreased in prostate cancer and implicated in hormone resistance. Mol. Cancer Ther. 7(10), 3176–3186.

Zhang, Q., Helfand, B.T., Jang, T.L., Zhu, L.J., Chen, L., Yang, X.J., *et al.*, 2009. Nuclear factor-kappaB-mediated transforming growth factor-beta-induced expression of vimentin is an independent predictor of biochemical recurrence after radical prostatectomy. Clin. Cancer Res. 15(10), 3557–3567.

Zhang, Q., Chen, L., Helfand, B.T., Jang, T.L., Sharma, V., Kozlowski, J., *et al.*, 2011. TGF-β regulates DNA methyltransferase expression in prostate cancer, correlates with aggressive capabilities, and predicts disease recurrence. PLoS One 30, e25168.

Zoghbi, H.Y., Orr, H.T. 2000. Glutamine repeats and neurodegeneration. Ann. Rev. Neurosci. 23, 217–247.

Index

Principles of Stem Cell Biology: Index by Judith Reading

Locators in *italic* refer to figures and tables

Abbreviation EMT = epithelial–mesenchymal transition

Principles of Stem Cell Biology and Cancer: Future Applications and Therapeutics, First Edition.
Edited by Tarik Regad, Thomas J. Sayers and Robert C. Rees.